T0142290

Smart Innovation, Systems and Technologies

Volume 84

Series editors

Robert James Howlett, Bournemouth University and KES International, Shoreham-by-sea, UK
e-mail: rjhowlett@kesinternational.org

Lakhmi C. Jain, University of Canberra, Canberra, Australia;
Bournemouth University, UK;
KES International, UK
e-mails: jainlc2002@yahoo.co.uk; Lakhmi.Jain@canberra.edu.au

About this Series

The Smart Innovation, Systems and Technologies book series encompasses the topics of knowledge, intelligence, innovation and sustainability. The aim of the series is to make available a platform for the publication of books on all aspects of single and multi-disciplinary research on these themes in order to make the latest results available in a readily-accessible form. Volumes on interdisciplinary research combining two or more of these areas is particularly sought.

The series covers systems and paradigms that employ knowledge and intelligence in a broad sense. Its scope is systems having embedded knowledge and intelligence, which may be applied to the solution of world problems in industry, the environment and the community. It also focusses on the knowledge-transfer methodologies and innovation strategies employed to make this happen effectively. The combination of intelligent systems tools and a broad range of applications introduces a need for a synergy of disciplines from science, technology, business and the humanities. The series will include conference proceedings, edited collections, monographs, handbooks, reference books, and other relevant types of book in areas of science and technology where smart systems and technologies can offer innovative solutions.

High quality content is an essential feature for all book proposals accepted for the series. It is expected that editors of all accepted volumes will ensure that contributions are subjected to an appropriate level of reviewing process and adhere to KES quality principles.

More information about this series at http://www.springer.com/series/8767

Suresh Chandra Satapathy
Amit Joshi
Editors

Information and Communication Technology for Intelligent Systems (ICTIS 2017) - Volume 2

 Springer

Editors
Suresh Chandra Satapathy
Department of CSE
PVP Siddhartha Institute of Technology
Vijayawada, Andhra Pradesh
India

Amit Joshi
Sabar Institute of Technology for Girls
Ahmedabad, Gujarat
India

ISSN 2190-3018   ISSN 2190-3026   (electronic)
Smart Innovation, Systems and Technologies
ISBN 978-3-319-87603-0   ISBN 978-3-319-63645-0   (eBook)
DOI 10.1007/978-3-319-63645-0

© Springer International Publishing AG 2018
Softcover reprint of the hardcover 1st edition 2017
This work is subject to copyright. All rights are reserved by the Publisher, whether the whole or part of the material is concerned, specifically the rights of translation, reprinting, reuse of illustrations, recitation, broadcasting, reproduction on microfilms or in any other physical way, and transmission or information storage and retrieval, electronic adaptation, computer software, or by similar or dissimilar methodology now known or hereafter developed.
The use of general descriptive names, registered names, trademarks, service marks, etc. in this publication does not imply, even in the absence of a specific statement, that such names are exempt from the relevant protective laws and regulations and therefore free for general use.
The publisher, the authors and the editors are safe to assume that the advice and information in this book are believed to be true and accurate at the date of publication. Neither the publisher nor the authors or the editors give a warranty, express or implied, with respect to the material contained herein or for any errors or omissions that may have been made. The publisher remains neutral with regard to jurisdictional claims in published maps and institutional affiliations.

Printed on acid-free paper

This Springer imprint is published by Springer Nature
The registered company is Springer International Publishing AG
The registered company address is: Gewerbestrasse 11, 6330 Cham, Switzerland

Preface

This SIST volume contains the papers presented at the ICTIS 2017: Second International Conference on Information and Communication Technology for Intelligent Systems. The conference was held during March 25 and 26, 2017, Ahmedabad, India, and organized combinedly by ASSOCHAM Gujarat Chapter, G R Foundation, Association of Computer Machinery, Ahmedabad Chapter and supported by Computer Society of India, Division IV, Communication, and Division V, Education and Research. It targeted state-of-the-art as well as emerging topics pertaining to ICT and effective strategies for its implementation for Engineering and Intelligent Applications. The conference had a large number of high-quality submissions from many academic pioneering researchers, scientists, industrial engineers, students from all around the world, and it provided a common provide a forum to researcher to interact and exchange ideas. After a rigorous peer-reviewed process with the help of program committee members and external reviewers, 147 (Vol-I: 74, Vol-II: 73) papers were accepted with an acceptance ratio of 0.39. The conference featured many distinguished personalities such as Dr. Pankaj L. Jani, Hon'ble Vice Chancellor, Dr. Babasaheb Ambedkar University; Shri Sunil Shah, President, Gujarat Innovation Society; Dr. S.C. Satapathy, Chairman, Division V, Computer Society of India; Shri Vivek Ogra, President, GESIA IT Association; and Dr. Nilesh Modi, Chairman, ACM Ahmedabad Chapter. Separate invited talks were organized in industrial and academic tracks in both days. The conference also hosted few tutorials and workshops for the benefit of participants. We are indebted to ACM Ahmedabad Professional Chapter, CSI Division IV, V for their immense support to make this conference possible in such a grand scale. A total of 15 sessions were organized as a part of *ICTIS 2017* including 12 technical, 1 plenary, 1 inaugural session, and 1 valedictory session. A total of 113 papers were presented in 12 technical sessions with high discussion insights. Our sincere thanks to all sponsors, press, print, and electronic media for their excellent coverage of this conference.

April 2017
Suresh Chandra Satapathy
Amit Joshi

Organization

Organizing Committee Chairs

Bharat Patel	COO, Yudiz Solutions
Durgesh Kumar Mishra	Division IV, CSI
Nilesh Modi	ACM Ahmedabad Chapter

Organizing Secretary

Mihir Chauhan	ACM Professional Member

Program Committee Chairs

Malaya Kumar Nayak	IT and Research Group, London, UK
Shyam Akashe	ITM University, Gwalior, India
H.R. Vishwakarma	VIT, Vellore, India

Advisory Committee

Z.A. Abbasi	Department of Electronics Engineering, AMU, Aligarh, India
Manjunath Aradhya	Department of MCA, SJCE, Mysore
Min Xie	Fellow of IEEE
Mustafizur Rahman	Endeavour Research Fellow, Institute of High Performance Computing, Agency for Science, Technology and Research
Chandana Unnithan	Deakin University, Melbourne, Australia
Pawan Lingras	Saint Mary's University, Canada

Mohd Atique	Amravati, Maharashtra, India
Hoang Pham	Department of Industrial and Systems Engineering, Rutgers University, Piscataway, NJ
Suresh Chandra Satapathy	Division V, CSI
Naeem Hannoon	Universiti Teknologi Mara, Malaysia
Hipollyte Muyingi	Namibia University of Science and Technology, Namibia
Nobert Jere	Namibia University of Science and Technology, Namibia
Shalini Batra	Computer Science & Engineering Dept, Thapar University, Patiala, Punjab, India
Ernest Chulantha Kulasekere	University of Moratuwa, Sri Lanka
James E. Fowler	Mississippi State University, Mississippi, USA
Majid Ebnali-Heidari	ShahreKord University, Shahrekord, Iran
Rajendra Kumar Bharti	Kumaon Engg College, Dwarahat, Uttarakhand, India
Murali Bhaskaran	Dhirajlal Gandhi College of Technology, Salem, Tamil Nadu
Pramod Parajuli	Nepal College of Information Technology, Nepal
Komal Bhatia	YMCA University, Faridabad, Haryana, India
Lili Liu	Automation College, Harbin Engineering University, Harbin, China
S.R. Biradar	Department of Information Science and Engineering, SDM College of Engineering and Technology Dharwad, Karnataka
A.K. Chaturvedi	Department of Electrical Engineering, IIT Kanpur, India
Margaret Lloyd	Faculty of Education School of Curriculum, Queensland University of Technology, Queensland
Jayanti Dansana	KIIT University, Bhubaneswar, Odisha
Desmond Lobo	Computer Engineering Department, Faculty of Engineering at KamphaengSaen, Kasetsart University, Thailand
Sergio Lopes	Industrial Electronics Department, University of Minho, Braga, Portugal
Soura Dasgupta	Department of TCE, SRM University, Chennai, India
Apurva A. Desai	Veer Narmad South Gujarat University, Surat, India
Abrar A. Qureshi	University of Virginia's College at Wise, One College Avenue

V. Susheela Devi   Department of Computer Science
                                        and Automation Indian Institute of Science,
                                        Bangalore
Subhadip Basu   The University of Iowa, Iowa City, USA
Bikash Kumar Dey   Department of Electrical Engineering, IIT
                                        Bombay, Powai, Maharashtra
Vijay Pal Dhaka   Jaipur National University, Jaipur, Rajasthan
Mignesh Parekh   Kamma Incorporation, Gujarat, India
Kok-Lim Low   National University of Singapore, Singapore
Chandana Unnithan   Victoria University, Australia

Technical Program Committee Chairs

Suresh Chandra Satapathy   Division V, Computer Society of India
  (Chair)
Vikrant Bhateja (Co-chair)   SRMGPC, Lucknow, India

Members

Dan Boneh   Computer Science Dept, Stanford University,
                                        California, USA
Alexander Christea   University of Warwick, London, UK
Aynur Unal   Stanford University, USA
Ahmad Al-Khasawneh   The Hashemite University, Jordan
Bharat Singh Deora   JRNRV University, India
Jean Michel Bruel   Departement Informatique IUT de Blagnac,
                                        Blagnac, France
Ngai-Man Cheung   University of Technology and Design, Singapore
Yun-Bae Kim   SungKyunKwan University, South Korea
Ting-Peng Liang   National Chengchi University, Taipei, Taiwan
Sami Mnasri   IRIT Laboratory Toulouse, France
Lorne Olfman   Claremont, California, USA
Anand Paul   The School of Computer Science
                                        and Engineering, South Korea
Krishnamachar Prasad   Department of Electrical and Electronic
                                        Engineering, Auckland, New Zealand
Brent Waters   University of Texas, Austin, Texas, USA
Kalpana Jain   CTAE, Udaipur, India
Avdesh Sharma   Jodhpur, India
Nilay Mathur   NIIT Udaipur, India
Philip Yang   Price water house Coopers, Beijing, China
Jeril Kuriakose   Manipal University, Jaipur, India

R.K. Bayal   Rajasthan Technical University, Kota, Rajasthan,
                                      India
Martin Everett   University of Manchester, England
Feng Jiang   Harbin Institute of Technology, China
Prasun Sinha   Ohio State University Columbus, Columbus, OH,
                                      USA
Savita Gandhi   Gujarat University, Ahmedabad, India
Xiaoyi Yu   National Laboratory of Pattern Recognition,
                                      Institute of Automation, Chinese Academy
                                      of Sciences, Beijing, China
Gengshen Zhong   Jinan, Shandong, China
Abdul Rajak A.R.   Department of Electronics and Communication
                                      Engineering, Birla Institute of Dr. Nitika Vats
                                      Doohan, Indore, India
Harshal Arolkar   CSI Ahmedabad Chapter, India
Bhavesh Joshi   Advent College, Udaipur, India
K.C. Roy   Kautilya, Jaipur, India
Mukesh Shrimali   Pacific University, Udaipur, India
Meenakshi Tripathi   MNIT, Jaipur, India
S.N. Tazi   Govt. Engineering College, Ajmer, Rajasthan,
                                      India
Shuhong Gao   Mathematical Sciences, Clemson University,
                                      Clemson, South Carolina
Sanjam Garg   University of California, Los Angeles, California
Garani Georgia   University of North London, UK
Hazhir Ghasemnezhad   Electronics and Communication Engineering
                                      Department, Shiraz University of Technology,
                                      Shiraz, Iran
Andrea Goldsmith   Stanford University, California
Cheng Guang   Southeast University, Nanjing, China
Venkat N. Gudivada   Weisberg Division of Engineering and Computer
                                      Science, Marshall University Huntington,
                                      Huntington, West Virginia
Rachid Guerraoui   I&C, EPFL, Lausanne, Switzerland
Wang Guojun   School of Information Science and Engineering
                                      of Zhong Nan University, China
Nguyen Ha   Department of Electrical and Computer
                                      Engineering, University of Saskatchewan,
                                      Saskatchewan, Canada
Z.J. Haas   School of Electrical Engineering, Cornell
                                      University, Ithaca, New York
Hyehyun Hong   Department of Advertising and Public Relations,
                                      Chung-Ang University, South Korea
Qinghua Hu   Harbin Institute of Technology, China

Honggang Hu	School of Information Science and Technology, University of Science and Technology of China, P.R. China
Fengjun Hu	Zhejiang shuren university, Zhejiang, China
Qinghua Huang	School of Electronic and Information Engineering, South China University of Technology, China
Chiang Hung-Lung	China Medical University, Taichung, Taiwan
Kyeong Hur	Dept. of Computer Education, Gyeongin National University of Education, Incheon, Korea
Sudath Indrasinghe	School of Computing and Mathematical Sciences, Liverpool John Moores University, Liverpool, England
Ushio Inoue	Dept. of Information and Communication Engineering, Engineering Tokyo Denki University, Tokyo, Japan
Stephen Intille	Northeastern University, Boston, Massachusetts
M.T. Islam	Institute of Space Science, Universiti Kebangsaan Malaysia, Selangor, Malaysia
Lillykutty Jacob	Electronics and Communication Engineering, NIT, Calicut, Kerala, India
Anil K. Jain	Department of Computer Science and Engineering, Michigan State University, East Lansing, Michigan
Dagmar Janacova	Tomas Bata University in Zlín, Faculty of Applied Informatics nám. T.G, Czech Republic, Europe
Kairat Jaroenrat	Faculty of Engineering at KamphaengSaen, Kasetsart University, Bangkok, Thailand
S. Karthikeyan	Department of Information Technology, College of Applied Science, Sohar, Oman, Middle East
Michael Kasper	Fraunhofer Institute for Secure Information Technology, Germany
L. Kasprzyczak	Institute of Innovative Technologies EMAG, Katowice, Poland
Zahid Khan	School of Engineering and Electronics, The University of Edinburgh, Mayfield Road, Scotland
Jin-Woo Kim	Department of Electronics and Electrical Engineering, Korea University, Seoul, Korea
Muzafar Khan	Computer Sciences Department, COMSATS University, Pakistan

Jamal Akhtar Khan   Department of Computer Science College
                             of Computer Engineering and Sciences,
                             Salman bin Abdulaziz University Kingdom
                             of Saudi Arabia
Kholaddi Kheir Eddine   University of constantine, Algeria
Ajay Kshemkalyani   Department of Computer Science, University
                             of Illinois, Chicago, IL
Madhu Kumar   Computer Engineering Department, Nanyang
                             Technological University, Singapore
Rajendra Kumar Bharti   Kumaon Engg College, Dwarahat, Uttarakhand,
                             India
Murali Bhaskaran   Dhirajlal Gandhi College of Technology, Salem,
                             Tamil Nadu, India
Komal Bhatia   YMCA University, Faridabad, Haryana, India
S.R. Biradar   Department of Information Science and
                             Engineering, SDM College of Engineering
                             and Technology, Dharwad, Karnataka
A.K. Chaturvedi   Department of Electrical Engineering, IIT
                             Kanpur, India
Jitender Kumar Chhabra   NIT, Kurukshetra, Haryana, India
Pradeep Chouksey   TIT college, Bhopal, MP, India
Chhaya Dalela   JSSATE, Noida, Uttar Pradesh
Jayanti Dansana   KIIT University, Bhubaneswar, Odisha
Soura Dasgupta   Department of TCE, SRM University, Chennai,
                             India
Apurva A. Desai   Veer Narmad South Gujarat University, Surat,
                             India
Sushil Kumar   School of Computer and Systems Sciences,
                             Jawaharlal Nehru University, New Delhi,
                             India
Amioy Kumar   Biometrics Research Lab, Department
                             of Electrical Engineering, IIT Delhi, India
Qin Bo   Universitat Rovira i Virgili, Tarragona, Spain,
                             Europe
Dan Boneh   Computer Science Dept, Stanford, California
Fatima Boumahdi   Ouled Yaich Blida, Algeria, North Africa
Nikolaos G. Bourbakis   Department of Computer Science and
                             Engineering, Dayton, Ohio, Montgomery
narimene boustia   Boufarik Algeria
Jonathan Clark   STRIDe Laboratory Mechanical Engineering,
                             Tallahassee, Florida
Thomas Cormen   Department of Computer Science Dartmouth
                             College, Hanover, Germany
Dennis D. Cox   Rice University, Texas, USA

Marcos Roberto da Silva Borges	Federal University of Rio de Janeiro, Brazil
Soura Dasgupta	Iowa City, Iowa, USA
Gholamhossein Dastghaibyfard	College of Electrical & Computer Engineering, Shiraz University, Shiraz, Iran
Doreen De Leon	California State University, USA
Bartel Van de Walle	University Tilburg, Tilburg, Netherlands
David Delahaye	Saint-Martin, Cedex, France
Andrew G. Dempster	The University of New South Wales, Australia
Alan Dennis	Kelley School of Business Indiana University, Bloomington, Indiana, USA
Jitender Singh Deogun	Department of Computer Science and Engineering, University of Nebraska – Lincoln, Nebraska, USA
S.A.D. Dias	Department of Electronics and Telecommunication Engineering, University of Moratuwa, Sri Lanka
David Diez	Leganés, Spain, Europe
Zhang Dinghai	Gansu Agricultural University, Lanzhou, China
Ali Djebbari	Sidi Bel Abbes, Algeria
P.D.D. Dominic	Department of Computer and Information Science, Universiti Teknologi Petronas, Tronoh, Perak, Malaysia

Contents

Design of High-Speed LVDS Data Communication Link Using FPGA

Shraddha Shukla[✉], Jitendra P. Chaudhari, Rikin J. Nayak, and Hiren K. Mewada

Charotar University of Science and Technology, Changa, Gujarat, India
sshraddha50@gmail.com

Abstract. This paper proposed the design and implementation of high speed communication link between two FPGAs using LVDS driver using 100 MHz clock. Initially Asynchronous and Synchronous communication are discussed and then synchronous communication is used for LVDS data communication. The speed of the communication is sensitive to the noise and sampling of the data at the receiver end. Therefore, the differential signal voltage level and sampling at the falling edge of clock are proposed in the design to maintain the high speed.

Keywords: High speed communication · Asynchronous communication · Synchronous communication and setup and hold time · FPGA

1 Introduction

This paper mainly aims the discussion on the issues in the high speed communication using FPGA. FPGA has advantages of high speed processing, high performance capability and its reconfigurable properties over the processors [5, 6], so in this system FPGA is used as transmitting and receiving devices. There are two types of communication-Asynchronous communication and Synchronous communication, in this paper the advantages and disadvantages of this both type of communication is discussed and concluded the benefits of the synchronous communication for high speed communication. In high speed communication, there is more chance of corruption of the information so; the factors affecting the data should be taken into consideration and should be removed. Also in high speed synchronous communication parameters of clock signal, the issues coming into the clock signal transmission is discussed in the paper. The information data should be sampled correctly in communication so the method to sample data correctly is discussed in this paper.

In this paper, the design of the system consists of two FPGAs- one transmitter FPGA and the other is receiver FPGA. The data from the transmitter to the receiver is transmitted on the 100 MHz clock signal. The software design of the system is carried out in the VHDL language and is simulated using Xilinx ISE design suite.

© Springer International Publishing AG 2018
S.C. Satapathy and A. Joshi (eds.), *Information and Communication Technology for Intelligent Systems (ICTIS 2017) - Volume 2,* Smart Innovation, Systems and Technologies 84, DOI 10.1007/978-3-319-63645-0_1

2 Communication Methods and Its Significance Related to High Speed Communication

In high speed communication the control on the communication should be taken to achieve successful communication between two devices. There are two types of communication methods: Asynchronous communication and Synchronous communication. They have their advantages and disadvantages and are useful according to type of application of the communication.

Asynchronous communication is the data transfer method in which the data is transferred without use of external clock. While in Synchronous communication, transmitter sends the clock signal to the receiver and with the reference of that clock signal receiver will receive the data. In asynchronous communication, the data is not transferred into constant time intervals while in synchronous communication the data are sent for the constant time period of the clock signal. Asynchronous communication is also known as start and stop communication because one start bit is sent before the data bits and one stop bit is sent after the data bits. In asynchronous communication, there is no coordination of timing of individual bits of the signal between transmitter and receiver while in synchronous communication has. So in asynchronous communication the receiver has to obtain how the incoming data is organized without referring to the transmitter. Operation of asynchronous and synchronous communication is shown in the Figs. 1 and 2.

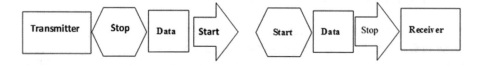

Fig. 1. Asynchronous communication

Asynchronous communication has several advantages: simplicity at the receiver side, lower power because of absence of the global clock, no delays regarding clock signal paths, no global time issues related to clock signals [9]. Though asynchronous communication provides these advantages it has disadvantages too. Asynchronous communication is preferred when small amount of data transfer, in larger amount of data, the traffic in the network will increase because larger data becomes much larger because of extra bits and hence extra bandwidth will be used [9]. Asynchronous communication is difficult in the ad hoc fashion of the communication network. In asynchronous communication there must be consideration for the dynamic state of the circuit to remove hazards.

These all the bugs in asynchronous communication will remove into the synchronous communication by sending the data on the time reference from the transmitter. In high speed communication, the chance of data corruption is at its peak. So, asynchronous communication is not preferred because of its disadvantages discussed above. Synchronous communication is suitable at high speed of communication because it minimizes the chance of hazard occurrence, minimizes the use of bandwidth, and increases accuracy in data sampling. In Fig. 2 transmitter send the data on the falling edge of the clock signal while receiver receives the data on the rising edge of the same clock signal.

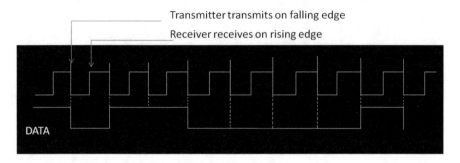

Fig. 2. Synchronous communication

2.1 Types of Synchronous Communication

There are mainly two synchronization methods: time synchronization and interpolation algorithm [1, 2]. Time synchronization requires a common clock source for reception of data information from different transmitters devices while interpolation algorithm doesn't need clock to receive the data and it mainly processes the digital signal based on synchronization algorithm after sampling.

Time synchronization includes three basic time synchronizing techniques used for communication between ICs—system-synchronous and source-synchronous and self-synchronous [8]. These three techniques are shown in the Figs. 3(a), 3(b) and 3(c).

These three techniques differ by their clock feeding methods. In system synchronous communication, the common clock signal is feed by the common clock provider to both the transmitter and receiver systems. This method looks inefficient when the transmitter and the receiver are located at the different distance from the common clock provider. This cause the skew delay in the path of the clock signal and hence wrong interpretation of the data occurs. In source synchronous communication, the clock signal is provided by the transmitter with the data. The transmitter will send the data on the reference of the same clock signal which is forwarded to the receiver. This will minimize the problem of skew delay and makes easier for receiver to sample the data correctly [3]. In self-synchronous communication, the clock signal is provided by the receiver to the transmitter and on the reference of that clock signal transmitter will send the data to the receiver.

For, high speed communication, source synchronous and self synchronous communication is more preferable than system synchronous communication because in synchronous communication, delays from the path of clock and data signal introduced in the system is more and for successful communication these delays should be taken into consideration and so the overall frequency of operation will decrease because maximum frequency is defined by the minimum cycle time and minimum cycle time is depends upon the minimum delays in the system.

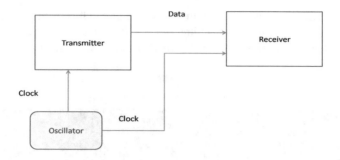

Fig. 3a. System synchronous communication

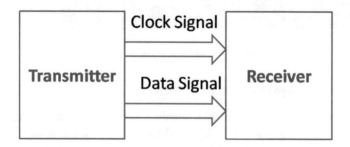

Fig. 3b. Source synchronous communication

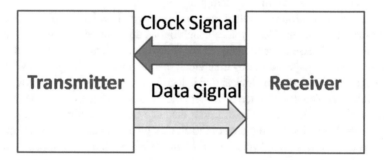

Fig. 3c. Self synchronous communication

3   Factors Affecting the High Speed Communication

3.1   Noise

In high speed communication, preliminary requirement is to handle high speed data which are coming from transmitter. By the experiment it has been observed that digital binary signal get distorted due to noise and other signal attributes as the transmission speed gets increase. To receive the data (to identify 0 s and 1 s) it is required to sample the signals such as high speed reference clock is required. Taking into consideration high speed of data, transmission and reception of the data should be in LVDS form. So, for driving the LVDS input, LVDS receiver is used receiver.

Single ended signal voltage level and differential signal voltage level is shown in the Fig. 4. As, we can see in the figure LVDS signal offers lower voltage swing, it can maintain higher speed without excessive slew rate [4, 7].

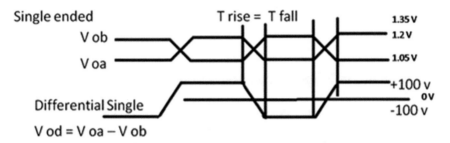

Fig. 4. LVDS signal level

Advantages of LVDS signaling for high speed communication:

- LVDS signal is immune to noise created by common node.
- Higher speed is achievable with low power consumption.
- It reduces the electromagnetic interference, because balanced differential lines have equal but opposite currents.

3.2 Sampling of the Data

If transmitter is transmitting the data on rising edge of the clock signal then the data at receiver side is sampled on the falling edge of the incoming clock, reason for sampling the data at falling edge at receiver side can be explained as below:

- Data is transmitted on the rising edge of the clock and the same clock is forwarded to the receiver.
- Now, as the rising edge comes, data is transmitted but the data will not be presented at the output side of the transmitter same time. There will some delay in time for arriving the data in accordance with the rising edge of clock. Reasons for this delay are set up and hold time. Now, as the clock arrived at the input side, valid data must be present for a minimum amount of time before to the input clock edge to guarantee successful data capture. This time is called set up time [10]. Setup time is not enough for successful capture the data. Data must be valid for a minimum amount of time after input clock edge to guarantee for successful capture of data. This time is called, Hold time [10]. Timing diagram is shown in the Fig. 5.
- So, if data is not presented perfectly on the rising edge of the clock at transmitter side, then if it will sample it at the receiver on rising edge, the incorrect data will be sampled. Due to this reason at receiver, data is sampled on the falling edge of the incoming clock.

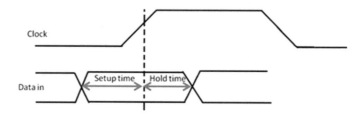

Fig. 5. Setup and hold time

4   Design of High Speed Communication Link

The design of the high speed communication link is given in the Fig. 6. The design includes one transmitter FPGA and one receiver FPGA. The data from the transmitter is transmitted through the LVDS driver and at the receiver side the data is received through the LVDS receiver to remove common node noise, electromagnetic interference. In the communication of data at high speed the design uses source synchronous communication. The clock signal is provided by the transmitter on which reference it is transmitting the data. Here, the transmitter transmits the data on 100 MHz clock frequency and forward the clock signal to the receiver. Here, the transmitter transmits the data on the rising edge of the clock signal while at the receiver side; the data is received on the falling edge of the clock signal provided by the transmitter.

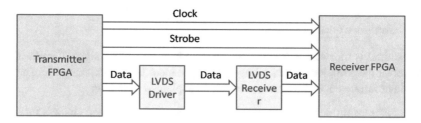

Fig. 6. Design of the system

In the transmitter side one signal is generated for indication of valid time period for transmission and reception. From the receiver side the data is only transmitted when the strobe signal is high and at the receiver side the data is only received when the strobe signal coming from the transmitter is high. LVDS driver is attached at the transmitter side and LVDS receiver is attached with the receiver FPGA. The data from the LVDS driver is received in the LVDS receiver connected to the receiver FPGA. At transmitter side, Spartan 6 FPGA [11] is used and at the receiver side Spartan 3E FPGA [12] is used. The software design is carried out in the VHDL language. TEXAS LVDS driver and receiver are used here in this design. SN55LVDS31 LVDS [14] driver and SN55LVDS32 [15] receiver are used to achieve high speed data communication.

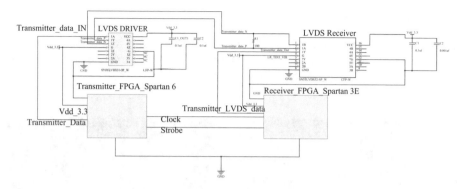

Fig. 7. Schematics of system design

4.1 Simulation Results

Simulation results for the transmitter and receiver are shown in the Fig. 8. As the simulation results for the transmitter shows, the data is transmitted on the reference of the 100 MHz clock frequency when the strobe signal goes high. The clock signal is generated using PLL IP core given in the directory of the clocking wizard for the Spartan 6 FPGA [13] device.

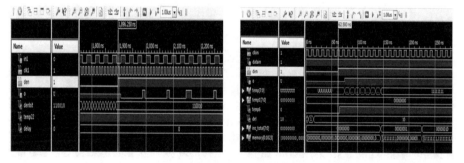

Fig. 8. Simulation waveform for transmitter FPGA

Fig. 9. Simulation waveform for receiver FPGA

Simulation waveform for the receiver is shown in the Fig. 9. At receiver side, as the Clock signal arrives, after reception of the strobe signal's logic high; the receiver receives the data and stores the data into internal RAM of the FPGA. In Fig. 7 the schematic design for the system design is shown.

The data signal form the transmitter goes at the input side of LVDS driver. From LVDS driver the differential signal is generated and this differential signal is fed into the LVDS receiver input. LVDS receiver will again convert differential signal into the single ended signal. Output from the LVDS receiver is given to the receiver FPGA.

4.2 Experimental Results

Figures 10 and 11 shows the output result of the implementation for communication on 100 MHz clock frequency. As discussed, at the receiver side the data is stored into the internal memory and one data from the memory (here data = "11010") is displayed on to the led.

Fig. 10. Displayed the received data on 100 MHz clock frequency stored that stored into the memory

Fig. 11. 100 MHz signal display on oscilloscope

5 Conclusion

The design achieves high speed communication using source synchronous communication method with its advantages over other communication methods. The design uses LVDS signaling for communication at high speed between two FPGA devices. The transmitter transmits the data using LVDS driver and the receiver uses LVDS receiver to receive the data. LVDS signaling between two FPGA devices reduces the probability of noise merging into the data at high speed, which is necessary for high speed communication. The paper also shows the significance of set up and hold time of the data and suggested solution to overcome this.

The system is designed into VHDL language. The design uses FPGA with its advantages of modular programming; high speed processing, its reconfigure capability, more logic resources compared to conventional processors. The design is reconfigurable and easy to implement. The design achieves data transmission and reception on 100 MHz between two FPGA devices with all necessary elements to be taken in care.

References

1. Chen, B., Wang, X., Yan, Z.: Design and implementation of data synchronous system for merging unit based on FPGA. In: 2011 4th International Conference on Electric Utility Deregulation and Restructuring and Power Technologies (DRPT), pp. 991–995, 6–9 July 2011

2. Tao, R., Jiang, B., Wang, C.: Sampling rate conversion and data synchronization in big merging unit. In: 2011 4th International Conference on Electric Utility Deregulation and Restructuring and Power Technologies (DRPT), pp. 531–534, 6–9 July 2011. doi:10.1109/DRPT.2011.5993949

3. Sivaram, A.T., Shimanouchi, M., Maassen, H., Jackson, R.: Tester architecture for the source synchronous bus. In: Proceedings of International Test Conference, ITC 2004, pp. 738–747, 26–28 October 2004

4. Tajalli, A., Leblebici, Y.: A slew controlled LVDS output driver circuit in 0.18 µm CMOS technology. IEEE J. Solid-State Circuits **44**(2), 538–548 (2009). doi:10.1109/JSSC.2008.2010788

5. Gangani, J., Samant, A., Rao, Y.S.: Reconfigurable blocks for digital power electronics applications in FPGA. In: 2011 Twenty-Sixth Annual IEEE Applied Power Electronics Conference and Exposition (APEC), pp. 2059–2064, 6–11 March 2011

6. Sugahara, K., Oida, S., Yokoyama, T.: High performance FPGA controller for digital control of power electronics applications. In: IEEE 6th International Power Electronics and Motion Control Conference, IPEMC 2009, pp. 1425–1429, 17–20 May 2009

7. Ng, E., Oo, K.: Low Power Gbit/sec Low Voltage Differential Signaling I/O System. Electrical Engineering and Computer Science, University of California, Berkeley

8. Xilinx.Com Abhijit Athavale "serialio-book". http://arrc.ou.edu/~rockee/RIO/

9. Synchronous vs. Asynchronous design. http://goo.gl/nNVFni

10. Intel.com: "Signal_Parameters". http://goo.gl/P7wQ5N

11. Xilinx.com: "Spartan 6 FPGA Family". http://goo.gl/1Id8xR

12. Xilinx.com: "Spartan 3E FPGA Family". http://goo.gl/vs5oVB

13. Xilinx.com: "The Clocking Wizard". http://goo.gl/HVJ1X4

14. ti.com: "SN55LVDS32-SP". http://goo.gl/DvqLJF

15. ti.com: "SN55LVDS31-SP". http://goo.gl/zYwwno

Face Super Resolution by Tangential and Exponential Kernel Weighted Regression Model

B. Deshmukh Amar$^{(\boxtimes)}$ and N. Usha Rani

Vignan University, Guntur 522213, Andhra Pradesh, India
amarbdeshmukh@gmail.com, Usharani.nsai@gmail.com

Abstract. The need of recognizing individual from the low resolution non-frontal picture is hard hassle in video surveillance. In an effort to alleviate the hassle of popularity in low decision photograph, literature presents unique strategies for face recognition after converting the low decision photograph to excessive resolution. For this reason, this paper provides a method for multi-view face video notable decision using the tangential and exponential kernel weighted regression model. In this paper, a brand new hybrid kernel is proposed to carry out non-parametric kernel regression version for estimation of neighbor pixel within the first-rate decision after the face detection is done the usage of Viola-Jones algorithms. The experimentation is finished with the U.S. Face video databases and the quantitative results are analyzed the usage of the SDME with the prevailing strategies. From the result final results, we prove that the most SDME of 77.3 db is obtained for the proposed approach compared with the existing techniques like, nearest interpolation, bicubic interpolation and bilinear interpolation.

Keywords: Super resolution · Face video · Face detection · Kernel · Viola-Jones algorithm · Second-Derivative-like Measure of Enhancement (SDME)

1 Introduction

In latest years, wise surveillance machine is broadly carried out in several fields which includes security and safety camera in which, the decision of the required face inside the photograph could be very low and so it can not provide the desired statistics. Furthermore, there are a few constrains within the imaging situations in certain situations and so acquiring face pics with high resolution isn't always viable. Therefore, the face pics captured by means of the camera may leave out many facial feature details which the human desires to be identified. So, photo resolution enhancement techniques, in particular, human face decision called face wonderful-decision is getting more interest [1]. Face great-decision (SR), which is likewise referred to as face hallucination approach hallucinating the excessive-decision (HR) face photo from its low-resolution (LR) photo. Baker and Kanade proposed the term face hallucination [2]. In general, face super-resolution techniques [3, 4] are mainly classified into two general categories such as learning-based techniques [5, 6] and reconstruction-based totally techniques [7, 8].

© Springer International Publishing AG 2018
S.C. Satapathy and A. Joshi (eds.), *Information and Communication Technology for Intelligent Systems (ICTIS 2017) - Volume 2*, Smart Innovation, Systems and Technologies 84, DOI 10.1007/978-3-319-63645-0_2

In gaining knowledge of-primarily based techniques, the prior statistics approximately the deviation of the excessive decision photos from the low resolution photographs are used to assemble the model and the constructed version is utilized in appearing face excellent resolution. Here, the main problem in learning based totally exceptional-decision method is the usage of particular previous information for the reconstructing high resolution image. To overcome this trouble, reconstruction-based totally techniques are used. There are techniques in reconstruction based totally techniques including worldwide method and nearby patch primarily based technique. Here, the snap shots are divided into overlapped patches and for each patch the closest buddies are used because the prior to create the required high decision patches. In international method approach, the whole face photos are taken into consideration as a worldwide version and in nearby technique, the face for organized patches are taken into consideration as a local version. In simulation, those methods gain top consequences while implemented to very low decision faces [9].

Nearby technique has appropriate subjective pleasant than international approach because of the clean face photograph [10, 11]. As the local picture patches are similar in nearby method, one photograph patch can be represented the use of a few neighbor patches which leads to nearby illustration of image patches. Moreover, there may be no noise in locality based illustration because the noisy photograph patch is replaced with comparable clear picture patch without synthesizing noisy photo patch as in LSR and SR [11]. Traditional parametric exquisite decision techniques are based totally on a certain version of the signal of interest and attempt to calculate the version parameters inside the presence of noise [12]. Then, a generative model primarily based at the expected parameters is generated because the high-quality estimate of the unique picture. On the contrary, nonparametric techniques are based on the facts itself to give an explanation for the shape of the version, and this model is known as a regression characteristic [13]. Because of the arrival of new system studying techniques, kernel strategies are becoming extra popular and most generally used for pattern detection and discrimination issues [14].

On this paper, we've provided a brand new hybrid kernel for non-parametric estimation of pixels the use of neighborhood patching system. The proposed technique performs multi-view face video first rate decision the use of tangential and exponential weighted regression version. At first, the input face video is study out and the frames are extracted from the face video. Once the frames are extracted, the face detection is accomplished the use of Viola-Jones set of rules which comprise the vital technique of rectangular characteristic extraction, training and trying out of AdaBoost classifier. Once the face place is detected, the low resolution face part is given to the final step in which, the multi-kernel regression version is implemented to attain exquisite resolution of the face component. The extraordinary decision picture is then analyzed with the assessment called SDME. The paper is prepared as follows: segment 2 offers the evaluate of literature and phase 3 gives the incentive in the back of the approach. Section 4 discusses the multi-kernel based totally regression model face multi-view face exceptional decision. Section 5 offers the unique experimentation and comparative results. Ultimately, the realization is given in Literature review.

Table 1 review the recent literature related to face super resolution. The most of the techniques are utilized the local neighbourhood-based estimation [1, 11, 15] to perform

Table 1. Literature review

Author	Method	Advantages	Disadvantages
Zeng and Huang [18]	Regression based method	Has better subjective quality due to more smooth face image	Require accurate prior for high-resolution image reconstruction
Tao et al. [10]	Shape clustering and subspace learning	Robust to noise	Maintain the shape in the reconstruction process is a problem worth to explore in future
Qu et al. [1]	Position patch neighborhood preserving	Considering spatial distance helps to obtain improved quality	Finding the distinct patch requires more time
Wang et al. [16]	Non-negative matrix factorization	The relation between high-resolution residue and low-resolution residue to better preserve high frequency details	Selecting NMF basis images corresponding to specific face parts is more challenging
Ma et al. [19]	Redundant transformation with diagonal loading	Offers robustness when dealing with the inputs that have different expressions, head poses, and illuminations	This approach is computationally intensive and sensitive to training examples
Jiang et al. [11]	Locality-constrained representation	It is very robust against noise in real surveillance scenarios	Fails to discover the intrinsic geometrical structure of the data set
Lu et al. [17]	Nonnegative matrix factorization	Better performance on local facial detailed features due to nonnegative part-based features	Factorization requires much computational effort
Hu et al. [15]	Local pixel structure-based method	Exhibits an impressive ability to infer the fine facial details and to generate plausible HR facial images from very small LR input	Require reference samples of HR estimation

the super resolution. Some techniques [16, 17] utilize the nonnegative matrix factorization to estimate the pixel even though it is computation overhead process. Also, some methods [10] utilize the learning algorithm for estimation of super resolution image even though the image prior information is required. From the literature, we identify that the estimation technique without the need of image prior and the computational efficient algorithms can be a better choice to proceed further in face video super resolution.

Zeng and Huang [18] have proposed regression based totally method for face outstanding resolution. It has higher subjective first-class because of more clean face photo however it require accurate earlier for high-resolution photo reconstruction. Tao et al. [10] have proposed form clustering and subspace getting to know-based totally version for face fantastic resolution. It's miles sturdy to noise however it continues the form inside the reconstruction technique which is a trouble worth to explore in future. Qu et al. [1] have proposed a way for face exceptional resolution. The position-primarily based patch neighborhood considers the spatial distance to gain stepped forward great however finding the distinct patch requires extra time. Wang et al. [16] have proposed a non-negative matrix factorization for face notable decision. This method continues the relation between high-decision residue and occasional-resolution residue to higher hold excessive frequency info however selecting NMF basis images similar to unique face components is greater tough.

Ma et al. [19] have advanced a face fantastic resolution by way of thinking about the redundant transformation with diagonal loading. It gives robustness whilst handling the inputs that have distinct expressions, head poses, and illuminations however this method is computationally in depth and sensitive to training examples. Jiang et al. [11] have proposed a Locality-constrained illustration for face excellent decision and it's miles very sturdy against noise in real surveillance eventualities but fails to discover the intrinsic geometrical shape of the information set. Lu et al. [17] have proposed a Nonnegative matrix factorization for face image awesome resolution. It has a higher overall performance on local facial particular functions because of nonnegative element-based capabilities but factorization calls for a good deal computational effort. Hu et al. [15] have proposed a neighborhood pixel structure-based approach for face wonderful decision and it exhibits an outstanding ability to infer the great facial info and to generate viable HR facial pix from very small LR input but it require reference samples for HR estimation.

2 Motivation Behind the Approach

2.1 Problem Definition

The main goal of this paper is to carry out the excellent decision at the face place that is obtained from the face video. So, the enter for the proposed gadget is the face video which contain multiple of frames. Each body is given to the face detection procedure which detects the face area. As soon as the face place is extracted, the fantastic decision system is accomplished at the face region to boom the range of pixels based on the up scaling factor without compromising the visual quality, As like, the input video V_i is represented as N frames as like,

$$V = \{V_i, \ 1 \le i \le N\} \qquad (1)$$

From every frame of video, the face region should be identified and it should be extracted.

$$I = VI(V_i) \qquad (2)$$

Where, $VI(\cdot)$ is the function to extract the face region. The extracted face regions can be represented as,

$$I_{face} = \{I_{ij}; \ 1 \le i \le ; \ 1 \le j \le n\} \qquad (3)$$

Where, $m \times n$ is the size of the face region. The problem considered here is to perform the super resolution of the face region by increasing the size of the region $(m \times n)$ to $(m * r \times n * r)$ where, r is the up scaling factor. The image obtained after performing super resolution is given as follows:

$$I_{\text{super}} = \left\{ I_{ij}^s;\ 1 \le i \le m \times r;\ 1 \le j \le n \times r \right\} \tag{4}$$

2.2 Challenges

Face amazing decision is an lively area of research due to the huge applicability of the technique in video surveillance device, particularly for security. While analysing the current works available within the literature for face great decision [15–20], the subsequent demanding situations are recognized.

Most of the super resolution frameworks require accurate image prior for high-resolution image reconstruction. This pose a challenge of identifying the accurate training face samples for performing effective learning task.

Utilizing the shape metrics for reconstruction can have good performance but the storing and maintaining of shape metrics is very difficult to handle.

Selecting the suitable size of neighbourhoods and shape of the neighbourhoods is challenging problem in super resolution because it is directly related to the estimating behaviour of the pixels.

Even though the most of the methods are face image super resolution, the important challenge in the surveillance camera is constructing super resolved faces from the low resolution videos because the face should be detected correctly from the multiple view and the estimated super resolved face should be useful for improving performance of face recognition task.

3 Proposed Methodology

Tangential and Exponential Kernel Weighted Regression Model for Multi-view Face Video Super Resolution

This section provides the multi-view face video outstanding decision the use of tangential and exponential kernel weighted regression model. Here, neighborhood patch estimation and multi-kernel weighted regression model is included for producing great resolved pixels. Basic, the proposed technique consists of three important steps. Within the first step, video is immediately read out and it's miles converted into a hard and fast of frames. In the 2nd step, each frame is serially taken and face element is detected the usage of Viola–Jones item detection [20] that is one of the famous strategies carried out for face detection. Once the face part is detected for the input body, the decision of the face detected component is improved using the proposed face awesome resolution approach in the 1/3 step. The proposed top notch decision approach integrates the manner of the multi-kernel weighted regression model and local patching procedure.

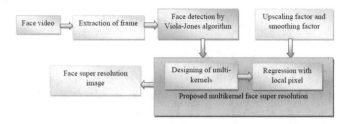

Fig. 1. Block diagram of the proposed multi-view face video super resolution

Right here, based totally at the scaling factor, the local patch length is determined and the estimation of pixels is completed the use of the exponential weighted regression version. This step is repeated for every frame and the awesome resolved faces are one by one saved for evaluation element. The block diagram of the proposed multi-view face video incredible decision is given in Fig. 1.

3.1 Video Reading and Frame Extraction

The input for the proposed multi-view face video super resolution is the video which is represented as the multiple frames. The video, is directly read out and every frame is taken out to find the face region. The input video may be in any of the file format like, AVI, MPEG, 3GPP and so on. The reading of video is performed by constructing VideoReader object and extracts one frame at a time associated with. Here, denote the frame index and represents the number of frames in the input video.

3.2 Face Detection by Viola-Jones Algorithm

The second step of the proposed multi-kernel weighted regression model is to detect the faces from every frames using Viola-Jones algorithm [20] which is the popular method for face detection. Even though many different methods are presented in the literature [21–23], Viola-Jones algorithm is taken here for the face detection because it can detect the multi-view faces effectively. This algorithm perform the face detection process using three important steps, like, (i) Feature extraction, (ii) Training of Adaboost classifier, (iii) Detection of face region.

(i) Feature Extraction: At first, the input frame $I(a,b)$ is given to the feature extraction step which finds the rectangular features using integral frame. The integral frame is computed by finding the summation of the pixel intensity from the above and left part of the location of the pixel. The process of finding the integral frame is given as follows:

$$II(a,b) = \sum_{a' \leq a, b' \leq b} I\left(a', b^i\right) \tag{5}$$

Where $II(a,b)$ is the integral frame and $I(a,b)$ is the original frame. Using the following pair of recurrences:

$$s(a,b) = s(a,b-1) + I(a,b) \tag{6}$$

$$II(a,b) = II(a-1,b) + s(a,b) \tag{7}$$

Where, $s(a,b)$ is the cumulative row sum, $s(a,-1) = 0$, and $II(-1,b) = 0$ the integral frame can be computed in one pass over the original frame. This integral frame is computed for every rectangular part of the images and it is stored as rectangular features. The dimension of the rectangular feature is usually high so the feature selection is important to avoid the computational overhead. Here, AdaBoost classifier [24] is utilized to select the important features and further to perform classification task.

(ii) Training of Adaboost Classifier: Once the features are extracted, the learning algorithm called, AdaBoost classifier [24] is trained based on the positive and negative samples. The training of classifier is performed by updating the weights of every iterations toward reaching the minimum error value. The error values are computed between the original class information with the ground truth label. The step behind the training of classifier is as shown below:

(a) Initialization: Initialize weights

$$w_{1,k} = \frac{1}{2p}, \frac{1}{2q} \text{ for } z_i = 0, 1 \text{ respectively.}$$

Where p and q are the number of negative and positives respectively
(b) Normalization: Normalize the weights,

$$w_{t,k} \leftarrow \frac{w_{t,k}}{\sum\limits_{l=1}^{m} w_{t,l}} \tag{8}$$

(c) Selection of weights: Select the best weak classifier with respect to the weighted error

$$\epsilon_t = \min_{g,r,\theta} \sum_{k} w_k |J(z_k, g, r, \theta) - z_k|. \tag{9}$$

where g_t, r_t and θ_t are the minimizers of ϵ_t.
(d) Update the weights:

$$w_{t+1,k} = w_{t,k} \alpha_t^{1-f_k} \tag{10}$$

where $f_k = 0$ if example z_k is classified correctly, $f_k = 1$ otherwise, and

$$\alpha_t = \frac{\epsilon_t}{1 - \epsilon_t}. \tag{11}$$

The above step is repeated for the required number of iterations and the final weight is selected to perform the classification task. The classification of the input rectangular feature is performed using the following equation.

$$C(z) = \begin{cases} 1 & \sum_{t=1}^{T} \beta_t J_t(z) \geq \frac{1}{2}\sum_{t=1}^{T} \beta_t \\ 0 & otherwise \end{cases} \tag{12}$$

where, $\alpha_t = \log\frac{1}{\beta_t}$. After performing the above process, face regions are extracted from every frame. The detected face region can be represented as,

$$I_{face} = \left\{ I_{ij}; 1 \leq i \leq; 1 \leq j \leq n \right\} \tag{13}$$

Where, $m \times n$ is the size of the face region.

3.3   Face Super Resolution by Tangential and Exponential Kernel Weighted Regression Model

This phase affords the tangential and exponential kernel weighted regression model for face incredible decision. The procedure of excellent decision is first executed via deriving the multi-kernel matrix primarily based on the weights to be assigned for the neighbour pixel values. Then, the derived kernel matrix is given for performing the face tremendous decision. The cause in the back of the choice of kernel methods is that it is well-known and used regularly for pattern detection and discrimination issues [10]. Even though the kernel regression techniques are familiar in information mining works, photo and video processing literature aren't a whole lot utilized these techniques. Also, kernel regression is the nonparametric estimation which lets in for tailoring the estimation trouble to the nearby traits of the information, whereas the same old parametric model is generally meant as a greater global in shape. 2nd, within the estimation of the neighborhood shape, better weight is given to the nearby records as compared to samples which are farther faraway from the centre of the analysis window. Additionally, this approach does not especially require that the statistics to observe a normal or equally spaced sampling shape [12].

Generation of Multi Kernel Matrix
The step one inside the regression technique is to build up the kernel matrix which does no longer depend upon no longer simplest the sample vicinity and density. Shape of the regression kernel is square in size and the dimensions of the kernel fixed based totally on the user exact parameters. Here, the scale of the kernel is constant upon based at the up scaling parameter given by way of the person. Then, the matrix is constructed by means of filling up the gap integer from the centre pixel. Each element on this matrix includes integers primarily based on the gap from the centre pixel. The matrix may be represented as follows;

$$F = \{f_{ij}; 0 \leq i, j \leq v\} \tag{14}$$

The distance-based integer matrix is then given for the kernel model such exponential and tangent function to convert the distance values into a kernel space. The exponential kernel function is represented as follows.

$$K_1 = \exp\left(-\left(\frac{1}{2 \times h^2}\right) * F\right) \tag{15}$$

Where, h is smoothing factor. The tangent kernel function is represented as follows:

$$K_2 = \tanh\left(-\left(\frac{1}{2 \times h^2}\right) \cdot F\right) \tag{16}$$

These two kernel matrix are effectively combined with the weighted formulae using the parameter called, alpha. The hybridization of two kernels are the new contribution proposed in this paper for kernel design. The advantage of the hybridized form of kernel ensures the advantage of exponential and tangent function in estimating the local neighbour values. The hybridized for of kernel function is given as follows;

$$K = \alpha K_1 + (1 - \alpha) K_2 \tag{17}$$

where, α is weighted constants.

Generation of Super Resolution Image

Once multi-kernel matrix is designed, the super resolution is performed by doing the interpolation with the extracted face image. At first, every pixel belonging to the $I^s(i,j)$ is generated by finding the sub image of $I^{ne}(a, b)$ corresponding to (i, j) the pixel. The size of this sub image $I^{ne}(a, b)$ is dependent on the size of the up scaling factor and the pixels without having the intensities is filled out with the neighbour values. Once the sub image is found out, the kernel matrix is multiplied with this sub image and the summation is taken as the representative pixel of the super resolution image, $I^s(i,j)$. The equation used to find the pixel of the super resolution image is given as follows,

$$I^s(i,j) = \frac{1}{k_1 * k_2} \sum_{a=1}^{k_1} \sum_{b=1}^{k_2} I^{ne}(a, b) * K(a, b) \tag{18}$$

where, $I^{ne}(a, b) \in (i, j)$; $k(a, b) \in (i, j)$, k_1 and k_2 are the size of the rows and columns of the sub image.

Algorithmic Description:

    i. Get the input video V and the input parameters, α, h
   ii. Extract the frames from the input video
  iii. Detect the face using Viola-Jones algorithm, I_{face} which is extracted face region

iv. Design multi-kernel matrix K using exponential and tangent kernel, α, h

v. Generate super resolution face image I^s

vi. Repeat step 3 to 5 for every frames.

Acknowledgments. The heading should be handled as a third level heading and should not be assigned quite a number.

4 Results and Discussion

This segment provides the experimental effects and the certain evaluation of the proposed multi-kernel-based totally regression model of face incredible decision.

4.1 Experimental Set Up

The proposed multi-view face video extremely good resolution is carried out the usage of MATLAB and the overall performance of the proposed approach and the existing technique is verified using SDME [25]. Here, the overall performance contrast is executed with the existing techniques like, Nearest-neighbor interpolation, Bicubic interpolation and Bilinear interpolation. For nearest neighbor interpolation, the block uses the fee of nearby translated pixel values for the output pixel values within the outstanding resolution photo. For bilinear interpolation, the block makes use of the weighted common of two translated pixel values for each output pixel cost. Bicubic interpolation is an extension of cubic interpolation for interpolating facts factors on a two dimensional everyday grid. The interpolated surface is smoother than corresponding surfaces received by bilinear interpolation or nearest-neighbor interpolation. Bicubic interpolation can be done using Lagrange polynomials, cubic splines, or cubic convolution algorithm.

Dataset Description: The experimentation is performed with the face video database available in [26] which is UCSD face video database. From the database, we have taken two different videos which have the multi view of a person.

Evaluation Metrics: The definition of SDME [2] is given as follows:

$$SDME = \frac{-1}{b_1 * b_2} \sum_{i=1}^{b_1} \sum_{j=1}^{b_2} 20 \ln \left| \frac{I_s^{max,i,j} - 2 * I_s^{centre,i,j} + I_s^{min,i,j}}{I_s^{max,i,j} + 2 * I_s^{centre,i,j} + I_s^{min,i,j}} \right| \qquad (19)$$

in which, the top notch decision photo is divided into b1 × b2 blocks, and are the maximum and minimal values of the pixels in every block one by one, and is the depth of the centre pixel in every block. Hence, the scale of the blocks have to be composed of an abnormal range of pixels consisting of 5 × 5, 7 × 7 or 9 × 9.

Parameters to Be Fixed: The proposed face super resolution contain three important parameters such as, smoothing factor h, weighted constant α and up scaling factor r. These parameters are extensively analyzed in the following section to find the best parametric value.

4.2   Experimental Results

This section presents the experimental results of the proposed face super resolution technique. Figure 2 shows the intermediate results of video 1 for side view position. Here, the frame extracted from video 1 in the is given in Fig. 2a. Figure 2b shows the face detected image through Viola-Jones algorithm and Fig. 2c represents the extracted face region from the frame. Finally, the output image of the face super resolution using the proposed multi-kernel-based regression model is given in Fig. 2d.

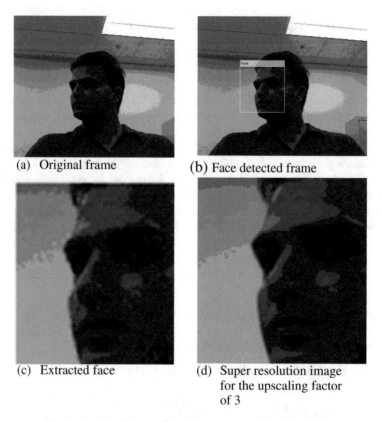

(a) Original frame   (b) Face detected frame

(c) Extracted face   (d) Super resolution image
                                  for the upscaling factor
                                  of 3

Fig. 2. Intermediate results of video 1 for side view position

4.3   Performance Evaluation

This section provides the performance assessment of the proposed multi-kernel regression model. Right here, parameters associated with the proposed approach are analyzed to pick out the quality fee for comparative evaluation. The primary parameter called, alpha is taken and the values are changed from 0.2 to one to locate the most fulfilling price. Discern four.a, the proposed method acquired the most SDME while the upscaling component is fixed to five. Until the alpha is equivalent to 0.8, the SDME cost is reduced while the alpha is multiplied. After that, the performance is accelerated. For the alpha fee of one, the proposed approach obtains the SDME fee of 70.01 db for the upscaling issue of 5 (Fig. 3).

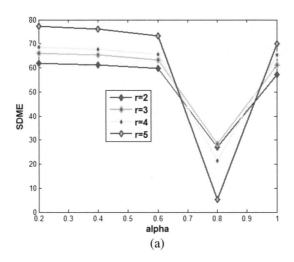

Fig. 3. Performance analysis based on alpha, (a) video 1

From Fig. 4a, we understand that the performance behaviour is almost similar for both the video. The maximum performance is achieved when the upscaling factor is equal to five and the smoothing factor is equal to 0.5.

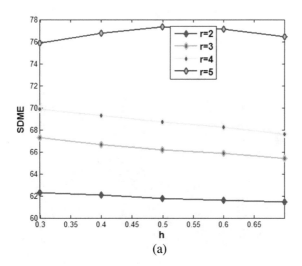

Fig. 4. Performance analysis based on global smoothing factor, (a) video 1

4.4   Comparative Analysis

This section discusses the comparative analysis of the proposed regression version for extraordinary decision with the present methods like, nearest interpolation, bicubic interpolation and bilinear interpolation. Discern 6.a indicates the overall performance of SDME for those techniques for the upscaling component of six is 46.96 db, 59.3 db, 61.1 db and 71.57 db. The overall performance of SDME in video 1 is plotted in discern 6.b. The same form of behaviour may be determined in video 1 additionally. The higher performance of 74.2 db is accomplished for the proposed technique when the upscaling issue is fixed to 5. The parent suggests that the performance of the strategies is increased while the upscaling factor is multiplied till it reaches a selected price. After that the overall performance is reduced even though the upscaling factor is increased (Fig. 5).

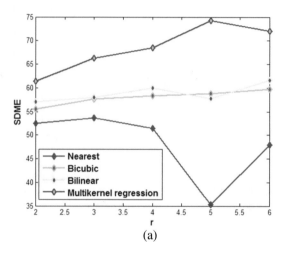

(a)

Fig. 5. Comparative analysis based on upscaling factor, (a) video 1

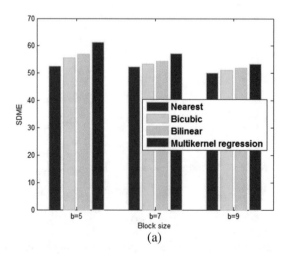

(a)

Fig. 6. Comparative analysis based on Block size, (a) video 1

Figure 6 shows for video 1, the performance is analyzed for all the techniques using SDME. From the figure, the nearest interpolation, bicubic interpolation, bilinear interpolation and proposed technique obtained the SDME of 52.49 db, 55.5 db, 57.08 db and 61.39 db when the block size is equal to five. Overall, the proposed technique outperformed the existing techniques for all the different size of blocks.

5 Conclusion

This paper offered as tangential and exponential kernel weighted regression model for multi-view face video great resolution. The closing contribution we made on this work is to layout a hybrid kernel for designing of regression model. The hybrid kernel makes use of the exponential and tangent feature with the weighted parameter. The designed hybrid kernel is then utilized to carry out the exquisite resolution of the face areas which is extracted from the input video the usage of Viola-Jones algorithm. The proposed face extremely good resolution is experimented with USA face video dataset and the performance is analyzed with the help of SDME. Additionally, the targeted parametric analysis is performed to find the better price to perform comparative analysis with the existing techniques like, nearest interpolation, bicubic interpolation and bilinear interpolation. From the effects, we proved that the proposed techniques obtained the maximum SDME of 77.3 db compared with the existing techniques. In future, the optimization algorithms can be blanketed to influence the regression version for better performance.

References

1. Qu, S., Hu, R., Chen, S., Chen, L., Zhang M.: Robust face super-resolution via position-patch neighborhood preserving. In: Proceedings of IEEE International Conference on Multimedia and Expo Workshops (ICMEW), pp. 1–5 (2014)
2. Baker, S., Kanade, T.: Hallucinating faces. In: IEEE International Conference on Automatic Face and Gesture Recognition, pp. 83–88 (2000)
3. Li, X., Hu, Y., Gao, X., Tao, D., Ning, B.: A multi-frame image super resolution method. Signal Process. **90**, 405–414 (2010)
4. Kim, K.I., Kown, Y.: Single-image super-resolution using sparse regression and natural image prior. IEEE Trans. Pattern Anal. Mach. Intell. **32**(6), 127–1133 (2010)
5. Zhuang, Y., Zhang, J., Wu, F.: Hallucinatingfaces: LPH super-resolution and neighbor reconstruction for residue compensation. Pattern Recognit. **40**, 3178–3194 (2007)
6. Chang, H., Yeung, D., Xiong, Y.: Super-resolution through neighbor embedding. In: IEEE Computer Society Conference on Computer Vision and Pattern Recognition, pp. 275–282 (2004)
7. Park, J., Lee, S.: An example-based face hallucination method for single-frame, low-resolution facial images. IEEE Trans. Image Process. **17**, 1806–1816 (2008)
8. Farsiu, S., Robinson, M., Elad, M., Milanfar, P.: Fast and robust multiframe super-resolution. IEEE Trans. Image Process. **13**, 1327–1344 (2004)

9. Lu, T., Hu, R., Jiang, J., Zhang, Y., He, W.: Super-resolution for surveillance facial images via shape prior and residue compensation. Int. J. Multimedia Ubiquitous Eng. **8**(6), 47–58 (2013)
10. Tao, L., Ruimin, H., Zhen, H., Yang, X., Shang, G.: Surveillance face super-resolution via shape clustering and subspace learning. Int. J. Signal Process. Image Process. Pattern Recogn. **5**(4), 107–116 (2012)
11. Jiang, J., Ruimin, H., Wang, Z., Han, Z.: Noise robust face hallucination via locality-constrained representation. IEEE Trans. Multimedia **16**(5), 1268–1281 (2014)
12. Takeda, H., Farsiu, S., Milanfar, P.: Kernel regression for image processing and reconstruction. IEEE Trans. Image Process. **16**(2), 349–366 (2007)
13. Wand, M.P., Jones, M.C.: Kernel Smoothing, ser. Monographs on Statistics and Applied Probability. Chapman & Hall, New York (1995)
14. Yee, P., Haykin, S.: Pattern classification as an ill-posed, inverse problem: a regularization approach. In: Proceedings of the IEEE International Conference on Acoustics, Speech, and Signal Processing, vol. 1, pp. 597–600 (1993)
15. Hu, Y., Lam, K.-M., Qiu, G., Shen, T.: From local pixel structure to global image super-resolution: a new face hallucination framework. IEEE Trans. Image Process. **20**(2), 433–445 (2010)
16. Wang, X., Lin, H., Xu, X.: Parts-based face super-resolution via non-negative matrix factorization. Comput. Electr. Eng. **40**(8), 130–141 (2014)
17. Lu, T., Hu, R., Han, Z., Jiang, J., Zhang, Y.: From local representation to global face hallucination: a novel super-resolution method by nonnegative feature transformation. In: Proceedings of Visual Communications and Image Processing (VCIP), pp. 1–6 (2013)
18. Zeng, X., Huang, H.: Super-resolution method for multiview face recognition from a single image per person using nonlinear mappings on coherent features. IEEE Signal Process. Lett. **19**(4), 195–198 (2012)
19. Ma, X., Song, H., Qian, X.: Robust framework of single-frame face superresolution across head pose, facial expression, and illumination variations. IEEE Trans. Hum. Mach. Syst. **45**(2), 238–250 (2015)
20. Viola, P., Jones, M.J.: Robust real-time face detection. Int. J. Comput. Vis. **57**(2), 137–154 (2004)
21. Fleuret, F., Geman, D.: Coarse-to-fine face detection. Int. J. Comput. Vis. **41**, 85–107 (2001)
22. Osuna, E., Freund, R., Girosi, F.: Training support vector machines: an application to face detection. In: Proceedings of the IEEE Conference on Computer Vision and Pattern Recognition (1997)
23. Roth, D., Yang, M., Ahuja, N.: A snow based face detector. In: Neural Information Processing, vol. 12 (2000)
24. Freund, Y., Schapire, R.E.: A decision-theoretic generalization of on-line learning and an application to boosting. In: Computational Learning Theory: Eurocolt 1995, pp. 23–37. Springer, New York (1995)
25. Panetta, K., Zhou, Y., Agaian, S., Jia, H.: Nonlinear unsharp masking for mammogram enhancement. IEEE Trans. Inf. Technol. Biomed. **15**(6), 918–928 (2011)
26. The UCSD face video database. http://vision.ucsd.edu/datasets/leekc/disk2/VideoDatabase/testing/

Monitoring of Distributed Resources Based on Client-Server (single-hop) Mobile Agents in Homogeneous Network of Interconnected Nodes

Rahul Singh Chowhan[(✉)] and Rajesh Purohit

M.B.M. Engineering College, J.N.V. University, Jodhpur, India
word2rahul@gmail.com, rajeshpurohit@jnvu.edu.in

Abstract. The internet in its broader sense has covered all dimensions of computer science and still continuing to grow dynamically with more subtle research possibilities and naïve implementation approaches. One such terminology in terms of balancing the load over a network of connected nodes is usage of mobile agents. Distributed applications, mobile devices and intermittent connections have fostered and provoked the need of the technology that support move of code and not of data. In this paper, the naive technique of load balancing that involves mobile agent paradigm is used. It promises to make full use of resources available within the network nodes based on their load handling capabilities.

Keywords: Distributed system applications · Intelligent computing · Machine availability and utilization

1 Introduction

In distributed and concurrent computing, networked machines coordinate and interact their working by using various message passing mechanism like RPC, message queues etc. The most essential attributes in distributed systems involves the concurrency in interaction of various components, failure of self-sustained modules, global synchronization in shared clock, and keeping the location transparency at times of upgrades and failures [1]. The communication by message passing was another phase to evolve concurrently into much better terminology that believes in migration of code and not data during interaction of various entities in distributed environment.

In load balancing, several computing resources like storage drives, clusters, CPUs etc. works collaboratively to improvise and optimize the usage of resources. Often the reliability is achieved through redundancy i.e. using multiple resources but this leads to frequent to-and-fro movement of tasks and processes for completion of their execution and resulting into a valid output/solution within a specified amount of time using the

© Springer International Publishing AG 2018
S.C. Satapathy and A. Joshi (eds.), *Information and Communication Technology for Intelligent Systems (ICTIS 2017) - Volume 2*, Smart Innovation, Systems and Technologies 84, DOI 10.1007/978-3-319-63645-0_3

appropriate memory and processing units. Multifarious load management algorithms works together in distributed environment to achieve a proper delegation of the tasks and processes to the required machines. This way it is cross checked that no machine has long waiting queues bombarded with huge processing loads while others are starving for execution of jobs sitting vacant [2].

In parallel and distributed system, uniform scheduling of load is a challenging aspect to be dealt with. Load scheduling is used to improve stability and performance over the entire realm of system loads [3]. In heterogeneous distributed system, efficient utilization of all the available resource worsens as the size of network increases. Evenly distributing the network load helps in increasing scalability and availability of systems to handle more incoming requests [4]. Load balancing is achieved by load monitoring and load distribution. With the load monitoring, various nodes coordinate with the server to submit their threshold. Later on, server uses load distribution algorithm over the results submitted after monitoring of resources. Load distribution algorithm picks up the over-loaded node from which the jobs are submitted to an under-loaded node in the network. This way the full consumption of resource available within the network is accomplished.

Mobile agents help to collaborate, communicate and cooperate in loosely coupled manner even for scalable distributed environments. Mobility of mobile agent allows it to migrate on to any remote machine permitted with its arrival [5]. Though these properties of mobile agents may be found in worms as well but FIPA and MASIF standardization allows the secure communication of mobile agents with underlying context [6, 7].

2 Methodology and Working Principle

In this experiment, we have used the client-server (single-hop) mobile agents that moves from server/network admin to next hop in the network to compute various parameters. The methodology introduces different time delays of 1 ms, 2 ms, 3 ms and 4 ms for both virtual and real machine scenarios separately. We tried to depict the relation between our matrices from the formula as:

$$A = f(C, M, P) \qquad (1)$$

where A denotes availability, C denotes CPU utilization, M denotes memory usage and P is delay assigned for a machine at that point of time.

In this proposed work, the mobile agents moves in single-hop fashion, it picks up the IP address of the respective hops from the server to find the computing capabilities and availability of that machine.

3 Proposed Algorithm for Load Monitoring and Event Flow Diagram

In this work, a naïve algorithm to monitor load has been proposed that make use of mobile agents to accomplish the phase1, i.e. load monitoring, to achieve load balancing in homogeneous distributed systems. The threshold value can be considered as an output of our algorithm as an input to phase2 i.e. in load distribution phase.

3.1 Load Monitoring Algorithm Using Client-Server (single-hop) Mobile Agent

(a) Start: Mobile agents are created on network admin or server node to determine Cn_i, Mn_i and Tn_i for each machine/node connected in the network.

$$Load\,Threshold(T) = .7 * (cpuLoad = getSystemCpuLoad())$$
$$+ .3 * (memLoad = getSystemMemoryLoad());$$

(b) When MA arrives on new machine:

Switch(Tn_i)
Case HIGH: if ($Tn_i = T_H$) then,
　　　save its load information & mark current node as "Heavily loaded";
　　　pick next URL;
　　　break;
Case LOW: if ($Tn_i = T_M$) then,
　　　save its load information & mark current node as "Lightly loaded";
　　　pick next URL;
　　　break;
　Case MED if ($Tn_i < T_L$) then,
　　　save its load information & mark current node as "Under-loaded";
　　　pick next URL;
　　　break;
Default: if (node failure || no node) then return to server;

(c) Multiple MA returns one by one as they moves in client server (single-hop) fashion, repeating step (b);
(d) Lastly, server sort the results using insertion sort function;
(e) Now, server may choose to use these refined results for distribution of load;
(f) End;

3.2 Event Flow Diagram

The event flow diagram for client-server (single-hop) mobile agent depicts that the mobile agent starts executing from network admin. The path to be travelled by each mobile agent is defined on the network admin node mobile agent picks the address of the destination node and returns back with appropriate results within the specified time. On arrival to a new node in its itinerary it finds the current percentage of load status usage and dispatches itself back to the server. After retrieving the results from all the mobile agents' server sorts them using insertion sort to know that which machine is under-loaded and which is overloaded so overall distribute the load and balance the whole system (Fig. 1).

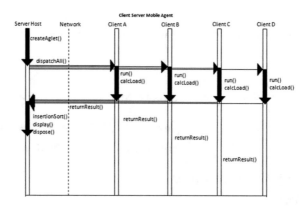

Fig. 1. Event flow diagram for client-server (single-hop) mobile agent

4   Experimental Setup and Execution Steps

In this experiment, combination of virtual machine and real machine is taken under consideration to analyze the final result. At the beginning, mobile agents move in an itinerary of two real machines and two virtual machines separately. After knowing the utilizations and threshold from first scenario i.e. from two machine combinations, it is again repeated for the three and four machines combinations to know the regression pattern and relation between various parameters by regulating the heterogeneity of virtual loads on the machines.

4.1   For Virtual Machine Setup

The server machine i.e. network admin has configuration of Intel(R) Core(TM) i3-4005U CPU @1.70 GHz processer and 4 GB DDR-2L RAM, and installed with Windows 7 Ultimate 64-bit Operating System. The client machines are configured with Windows XP 32-bit Operating System, 512 MB RAM. To enable the usage of mobile agents for purpose of utilization, we have extended the mobile agent system by integrating it with the NetBeans, as the core framework of IBM Aglets only provides with the interface and overriding of methods.

4.2   For Real Machine Setup

Both the server host and the client hosts are configured with installation of Windows 7 Ultimate, 64-bit Operating System and Intel(R) Core(TM) i3-4005U CPU @1.70 GHz processer and 4 GB DDR-2L RAM. The IDE used for programming the mobile agents was NetBeans8.1 and various inbuilt interfaces of IBM Aglets are used.

4.3   Execution Steps for Client-Server (single-hop) MA

1. Agent server called as agency is run over all the client hosts to allow visits of mobile agents.
2. Network Admin creates multiple mobile agents (MA) to be send to client hosts.
3. All the MA are associated with an IP they need to visit
4. Eachone of MA visits their respective nodes and computes the required paramters.
5. MA saves the computed state and dispatches from current context.
6. After recieving all the MAs on network admin machine, the monitored results are sorted (Fig. 2).
7. Based on results overloaded and idle node can be find.

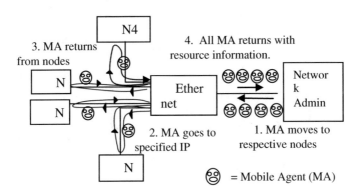

Fig. 2. Experimental setup showing client-server (single-hop) mobile agent

5   Results Analysis and Observations

The load balancer provides the information about the most available nodes in network. Availability is defined as the degree of an overall system that will keep it in stable state. To know the availability two different scenarios were drawn as graphs for real machines and virtual machines separately and they are as under:

5.1   Scenario-1: Various Combinations of Real Machines Using Client-Server (single-hop) Mobile Agent

The graph (a) and (b) shows the CPU percentage utilization and memory usage between two real machines and their availability. In the same way graphs (c), (d), (e) and (f) are showing CPU percentage utilization, Memory usage and machine availability for three and four real machines respectively. To observe the nature of homogeneous machine, the heterogeneity is introduced varying delays differently on different machines.

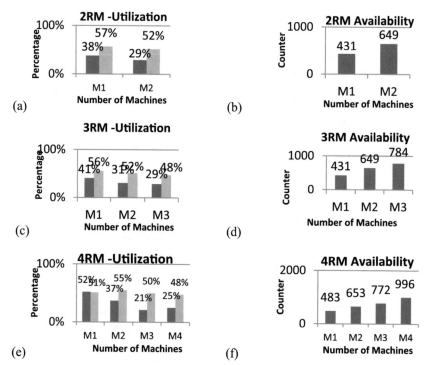

5.2   Scenario-2: Various Combinations of Virtual Machines Using Client Server (single-hop) Mobile Agent

In this scenario two, three and four virtual machines were arranged one by one in succession using VM-Ware. CPU and memory utilization are dependent on machine's running services and processes. Though creation of virtual environment let us monitor the parameters chosen for this experimental setup. This way the experiments based on parameters determines the actual behavior of the application.

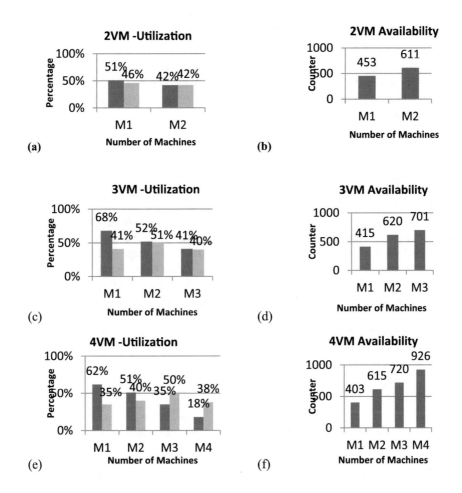

(a) 2VM -Utilization

(b) 2VM Availability

(c) 3VM -Utilization

(d) 3VM Availability

(e) 4VM -Utilization

(f) 4VM Availability

6 Conclusion and Future Scope

We have proposed a naive implementation of mobile agent to gather status information about the machines. The multiple mobile agents move in an itinerary on to all participating nodes in single-hop fashion. To conduct these experiments various combinations of real as well as virtual machines were considered. And for each scenario the number of machines varied from one to four. The conclusions drawn are as follows:

1. Client-server MA (single-hop) is taking more time because of congestion caused by multiple agents. The frequent context-switching happening is adding up more delay increasing the communication time of each agent. But in other case if MAs are send one by one then normal results as shown for real machines are received.

2. These results states that sending multiple agents for virtual machines installed on one real machine at a single point of time can increase the round trip time of all the MAs in network.

On the basis of availability counter, the relative availability is also calculated by taking the percentile of maximum available machine with less available machines in the network. Communication overhead for client server (single hop) mobile agent is calculated as:

$$.\text{Communication Overhead CS MA} = f(x) = \sum_{i=1}^{n} (2P_i)\text{pi}$$

Where P_i is propagation delay of mobile agent from server to the ith node, n denotes the total number of nodes in network.

An efficient monitoring technique is required to monitor the load of all connected nodes yet keeping the overall setup congestion free. The job of monitoring is not limited local system but to on-look and keep watch of incoming jobs and their assignment as it has also its part in distribution. The future scope of this novel research work may introduce a recovery mechanism for the data that mobile agents carry if they are lost in network. The static agents can also play an important role in this kind of scenario by sending the polls after every periodic interval of time. But this could be used for small network as larger the network grows more the congestion will be in network. Security is one of the major challenges in implementation of mobile agents so it holds more future research work and scope in this field.

References

1. Srinivasan, T., Vijaykumar, V., Chandrasekar, R.: A self-organized agent-based architecture for power-aware intrusion detection in wireless ad-hoc networks. In: 2006 International Conference on Computing & Informatics, pp. 1–6. IEEE (2006). doi:10.1109/ICOCI.2006.5276609
2. Gupta, D., Bepari, P.: Load sharing in distributed systems. In: Proceedings of the National Workshop on Distributed Computing (1999)
3. Kumar, R., Niranjanr, S., Singh, Y.: A review on mobile agent technology and its perspectives. J. Comput. Sci. Appl. 3(6), 166–171 (2015). doi:10.12691/jcsa-3-6-11
4. Thant, H.A., San, K.M., Tun, K.M.L., Naing, T.T., Thein, N.: Mobile agents based load balancing method for parallel applications. In: 6th Asia-Pacific Symposium on Information and Telecommunication Technologies, pp. 77–82. IEEE (2005)
5. Tripathi, A., Ahmed, T., Pathak, S., Carney, M., Dokas, P.: Paradigms for mobile agent based active monitoring of network systems. In: 2002 IEEE/IFIP Network Operations and Management Symposium (NOMS 2002), pp. 65–78. IEEE (2002). doi:10.1109/NOMS.2002.1015546
6. Lingaraj, K., Biradar, R.V., Patil, V.C.: A survey on mobile-agent middleware design principals and itinerary planning algorithms. In: 2015 International Conference on Applied and Theoretical Computing and Communication Technology (iCATccT), pp. 749–753. IEEE (2015). doi:10.1109/ICATCCT.2015.7456983
7. McDonald, J.T., Yasinsac, A., Thompson, W.C. III.: Mobile agent data integrity using multi-agent architecture. In: Air Force Inst of Tech Wright-Patterson AFB OH (2004)

Energy Balanced Clustering Protocol Using Particle Swarm Optimization for Wireless Sensor Networks

Sonu Jha[✉] and Govind P. Gupta

Department of Information Technology, National Institute of Technology,
Raipur, India
jha.sonu999@gmail.com, gpgupta3@gmail.com

Abstract. In a large scale Wireless Sensor Networks (WSNs), designing an energy balanced clustering protocol has become a challenging research issues. This is due to fact that design of an energy-balanced clustering for maximizing the network lifetime of WSNs is a NP-hard problem. For solving this NP-hard problem, many meta-heuristic approach based clustering protocols are proposed in the recent years. However, these existing clustering protocols suffer from unbalanced energy consumption problem. In this problem, cluster heads are not uniformly distributed and overloaded cluster heads die out faster than under-loaded cluster heads. In order to solve this problem, we have proposed an energy balanced clustering protocol using particle swarm optimization called EBC-PSO. In the proposed protocol, we have used a novel multi-objective fitness function which contains three constraints such as average intra-cluster distance, residual energy and average cluster size. A detailed evaluation and performance comparison of the EBC-PSO with the three most popular protocols such as LEACH, PSO-ECHS, and E-OEERP are included.

Keywords: Wireless Sensor Network · Clustering · PSO

1 Introduction

A Wireless Sensor Network (WSN) is an emerging distributed network which consists of a set of sensor nodes and one or more Sink nodes [1, 2]. In WSNs, sensor nodes are generally small size and low-cost device, due to this reason, they have limited energy resource. Generally, sensor nodes are operated using 2 AA batteries. This networks have become an emerging technology that play a very important role in realizing smart environment such as Smart Cities, Smart Grid, Smart Home, online monitoring and tracking system [2, 3]. Since sensor nodes are provided with limited energy and replacement of battery is almost impossible. Due to which energy conservation of sensor node is an important research issue that needs research effort for better solution in order to enhance the network lifetime.

For saving of the energy resource of the sensor node and enhancing the network lifetime, various methods are proposed; clustering is one of the well considered scheme which is employed in WSN for saving the energy of the network. In clustering process, nodes are organized into various groups known as clusters. Each cluster is provided

© Springer International Publishing AG 2018
S.C. Satapathy and A. Joshi (eds.), *Information and Communication
Technology for Intelligent Systems (ICTIS 2017) - Volume 2*, Smart Innovation,
Systems and Technologies 84, DOI 10.1007/978-3-319-63645-0_4

with a Cluster Head (CH) whose main work is to receive the sensed data from its cluster members (CMs), aggregate it and then transport the aggregated data to the Sink node [3]. Sink is connected to internet for public notifications of the sensed data. In each cluster, CH removes redundant data by doing data aggregation. Thus, proper selection of CH and its spatial distribution are very important issues in the energy balanced clustering process.

In recent years, various metaheuristic optimization algorithm based clustering protocols are proposed for WSNs. However, these existing clustering protocols are suffering from unbalanced energy consumption problem. In this problem, cluster heads (CHs) are not uniformly distributed and overloaded CHs die out faster than under load CHs. In addition, spatial distribution of CHs is not uniform. This is due to fact that average distance between CH and Sink is used in the fitness function which causes selection of all CHs near to the Sink. In order to solve this problem, we have proposed an energy balanced clustering protocol using particle swarm optimization called EBC-PSO. The main aim of EBC-PSO is to select a set of CHs that distributed uniformly over sensing area such that load on each CH is balanced. Fitness function used in PSO for the selection of CHs considers is a multi objective function which includes three parameters such as intra-cluster distance, residual energy and average cluster size. Performance evaluation of the EBC-PSO is compared with three well known clustering protocols such as LEACH [3], PSO-ECHS [9], and E-OEERP [10] in terms of total energy consumption by varying number of nodes, CHs, and number of rounds.

Content of this paper is arranged as follows: a brief review of the related clustering schemes is summarized in Sect. 2. An overview of the PSO concept is described in Sect. 2. In Sect. 4, system model and terminologies are discussed. Section 5 describes proposed energy balanced clustering protocol. In Sect. 6, simulation and result analysis of the proposed protocol and its comparison with the existing protocols are given. In last, we conclude the paper in Sect. 7.

2 Related Work

In this section, we have presented a brief outline of the existing clustering protocols proposed for WSNs. Generally, existing clustering protocols are categorized into two groups: heuristic based clustering and nature inspired based clustering [9, 10].

Several heuristic protocols have been developed for increasing lifetime of sensor network. Among these LEACH [3] is one of the most famous clustering protocols. LEACH selects CH based on some probability. Due to this, role of CH is transferred to other node after every round. The main limitation of LEACH is that low energy sensor node can be selected as CH. This causes CHs to die quickly. To improve the performance of LEACH number of protocol have been developed among these PEGASIS [4] and HEED [5] are popular. PEGASIS uses greedy approach to arrange sensor nodes into an ordered list such that each node can communicate with its adjacent nodes in the list. In HEED [5], residual energy of node is the main constrained used in the selection of CH. This protocol mainly highlighting on the energy efficient CH selection and reduction of the communication overhead. This causes to maximize the network lifetime. Many variants of LEACH have been also proposed to increase network life time.

TL-LEACH [6] organizes CH into two level- hierarchy, top level CH are called primary CH and second level CH are called secondary CH. Secondary CH instead of sending data to BS they forward data to primary

Different clustering algorithms were proposed based on nature inspired approaches. In LEACH-C (Centralized LEACH) [7] firstly sink node computes average node energy for each round and nodes having energy higher than average node energy is selected for becoming CHs. Cluster formation is done using simulated annealing. It performs better than LEACH as it selects CH based on energy, thus increases network lifetime. This protocol does not consider balancing size of cluster which leads to unbalance energy consumption problem. Latiff et al. [8] have discussed a clustering scheme, PSO-C. PSO-C derives a fitness function which contains two elements such as residual energy, CH-CMs distance. It does not consider cluster size which is important factor for reduction of energy consumption. In [9], a PSO based cluster head selection method, called PSO-ECHS, is proposed. The main limitation of PSO-ECHS is that it does not distribute CHs uniformly, thus unbalanced energy consumption was observed during its evaluation. Parvin et al. [10] have discussed a PSO based protocol for Clustered WSN, called E-OEERP. In E-OEERP, PSO based scheme was used for selection of CHs and focus on the problem of left out nodes during clustering process. This scheme is also suffered from unbalanced energy consumption problem.

3   Overview of Particle Swarm Optimization

Particle swarm optimization (PSO) is population based optimization technique which is inspired by social behavior of flock of birds [11]. In PSO, population or swarm both is synonym and representing the set of potential solutions. A particle in the population/ swarm stands for a solution position in the solution search space. In PSO, a particle can move in the search space to discover a new position with the best solution value. Initially, each particle is provided with position and velocity in the search space. Fitness function is designed based on the requirements of the problem. The main process of the PSO is to search the new position of the particle that gives best result of fitness function. In each iteration, a particle calculates its personal best and also global best. Each particle tries to reach global best solution by updating position and velocity by the use of personal best and global best. Following steps are used in the particle swarm optimization process.

- Initialize population with random position and initial velocity in problem space.
- Calculate fitness value of each particle.
- Compare current fitness value with particle's Pbest. If current fitness value is better than Pbest than update particle position to current value location position and Pbest with current fitness value.
- If current fitness value is better than Gbest than replace the value of Gbest with current fitness value and location.
- Velocity and position of particle is changed by Eqs. (1) and (2).

$$V_{id}(t) = w \times V_{id}(t-1) + C_1 \times N_1 \times (X_{Pbest_{id}} - X_{id}) + C_2 \times N_2$$
$$\times (X_{Gbest_{id}} - X_{id}) \qquad (1)$$

$$X_{id}(t) = X_{id}(t-1) + V_{id}(t) \qquad (2)$$

where w is known as inertia weight and its value is between 0 and 1. C_1 and C_2 are acceleration coefficient its value is between 0 and 2. N_1 and N_2 value range from 0 and 1. The value of Pbest and Gbest are updated by using Eqs. 3. and 4 as follows:

$$Pbest_i = \begin{cases} P_i, & if\,(\text{Fitness}(P_i) < Fitness(\text{Pbest}_i)) \\ Pbest_i, & otherwise \end{cases} \qquad (3)$$

$$Gbest_i = \begin{cases} P_i, & if\,(\text{Fitness}(P_i) < Fitness(\text{Gbest}_i)) \\ Gbest_i, & otherwise \end{cases} \qquad (4)$$

Particle P_i has initial position and velocity as $X_{i,d}$, $V_{i,d}$. Particle changes its position based on its memory. Position and velocity keeps on changing with every iteration until global position $X_{i,d}(S)$ is reached.

4   System Model

This research work considers a WSN system model where sensor nodes are deployed uniformly random in a square sensing area. We assume that there are N sensor nodes, one sink node and k cluster heads in the network. It is also assumed that all nodes are static stationary and their location coordinates are known. In the clustering process, a sensor node can joins to only one cluster. A sensor node can perform as cluster head (CH) or as a normal sensor node. In each round of data collection at the sink, CM of each cluster sends its sensed data to its CH. CH aggregates the received sensor data and forwards the aggregated data to the Sink using multi-hop routing from CH to the Sink. In this work, we have used an energy consumption model as proposed in [3].

5   Energy Balanced Clustering Algorithm

This section discusses an Energy-Balanced Clustering algorithm using Particle Swarm Optimization (EBC-PSO). First, we describe the derivation of a novel multi-objective fitness function which is used for the evaluation of the particle. Next, working of the EBC-PSO protocol is discussed.

5.1   Derivation of Fitness Function

In the proposed EBC-PSO protocol, we have used a novel multi-objective fitness function which contains mainly three components such as average intra-cluster distance, residual energy and average cluster size. The main goal of this multi objective fitness function is to optimize the combined effect of its components. Formulation of the fitness function is represented by Eq. 5.

$$fitness = a_1 * x_1 + a_2 * x_2 + a_3 * x_3 \qquad (5)$$

Where x_1, x_2 and x_3 represents average intra-cluster distance, residual energy and average cluster size respectively. a_1, a_2 and a_3 are constant and its value is between 0 and 1. Descriptions of the parameters of the fitness function are described as follows:

(a) **Average intra-cluster distance:** This is defined as the ratio of sum of distance of the entire sensor node that is in the transmission range of particular sensor node to the total node present in its transmission range. Our main aim is to minimize intra cluster distance by selecting CH which is closer to all sensor nodes.

$$x_1 = \frac{\sum_{i=1}^{m} dis(CM_i, CH)}{m} \qquad (6)$$

Where m is number of node that is in communication range of particular particle (i.e. CH). $dis(CM_i, CH)$ is distance between cluster member CM_i and CH (i.e. particle).

(b) **Residual Energy:** This is defined as the ratio of residual energy of CH to the total energy of cluster member. This function helps to select node as CH which has more energy than other sensor nodes.

$$x_2 = \frac{E_{residual}}{E_{total}} \qquad (7)$$

Where $E_{residual}$ is the residual energy of CH. E_{total} is the total energy of node that are in communication range of CH.

(c) **Average cluster size:** It is ratio of number of cluster member present in particular cluster to the total number of nodes. This function help to minimize number of un-clustered nodes and it also help to check the load balancing problem.

$$x_3 = \frac{C_n}{N} \qquad (8)$$

Where C_n is number of sensor nodes in a particular cluster. N is total number of nodes.

5.2 Energy-Balanced Cluster Formation Using PSO

The proposed energy balanced clustering scheme, EBC-PSO, is based on PSO meta-heuristic optimization algorithm. In EBC-PSO, we assume that all nodes send their location coordinate and value of residual energy to the sink during network setup phase. After getting the energy and location information, Sink node applied proposed

PSO based clustering algorithm for the selection of the optimal CHs. Following steps are follows for the implementation of the proposed clustering protocol (EBC-PSO).

Step 1. Initialization of parameters: In this step, initialize the value of different parameters used during PSO process such as size of particles, initial position of particles, initial value of inertia weight and initialization of some constant parameters.

Step 2. Evaluation of fitness value: Calculate fitness of each sensor nodes based on the Eq. 5. Sort the sensor nodes according to their fitness value and select m sensor nodes as CH candidates.

Step 3. Selection of Pbest and Gbest: Fitness value of these m CH candidates is their local best. It is called Pbaset.

Step 4. Selection of Gbest: Find out a particle (i.e. CH candidates) with maximum fitness value among the selected m CH candidates. This fitness value becomes global best (Gbest) for all particles.

Step 5. Updation of Velocity and Position of Particles: In this step, velocity and position of each CH is evaluated by using the Eqs. 1 and 2.

Step 6. Evaluation of fitness value for new position particles: In this step, fitness of new position particle is evaluated by using the Eq. 5.

Step 7. Selection of Pbest and Gbest: Using Eqs. 3 and 4 decides the value of Pbest and Gbest.

Step 8. Repeat: Go to the step 5 until the Maximum iteration criterion reached

After selection of optimal position CHs, process of cluster formation starts. In the cluster formation process, each non-cluster nodes joins the closest CH with the higher node energy. Table 1 shows a sample of the fitness value. In this example node 25 has highest fitness value so it is selected as CH. all the node that are in the communication range of node 25 joins cluster. Similarly next CH is node 89 which is selected as CH and nodes that are in communication ranges join it. Suppose next CH 76 is cluster member of previous CH 89 than node 76 does not become cluster head. In this way

Table 1. Sample fitness value

Node id	Fitness value
25	20
89	15
76	12
5	8
8	5
9	4

Table 2. Parameter list

Parameter	Value
Dimension of monitoring area	200×200
Location of the sink	(50–200, 50–200)
# of sensor nodes	100–300
Initial energy of a node	200 J
% of CHs	5–15
E_{elec}	50 nJ/bit
E_{fs}	10 pJ/bit/m^2
E_{mp}	0.0013 pJ/bit/m^4
d_0	87.00
d_{max}	25 m
Packet length	4000 bit

cluster is formed. If any CH dies then next node having fitness value higher next to CH is made CH. This help to increase network life time (Table 2).

6   Simulation Results and Discussion

This section present the simulation results of the proposed EBC-PSO protocol and performs comparative analysis with LEACH [3], PSO-ECHS [9], and E-OEERP [10] with respect to total energy consumption within the network. All simulations are performed using MATLAB R2014.

Figure 1 shows the performance of the EBC-PSO with LEACH [2], PSO-ECHS [9], and E-OEERP [10] with respect to consumption of node energy and by varying the number of rounds of data gathering from the deployed WSN. It is observed form the Fig. 1 that rate of energy consumption for the EBC-PSO is slower than other three. This is due to fact that EBC-PSO selects optimal position CHs that are uniformly distributed over sensing area and provide balanced energy consumption from the non-CH and CH nodes in the network.

Figure 2 shows the performance analysis of the EBC-PSO with the existing protocols with respect to energy consumption by varying the number of nodes from 100 to 300. This experiment evaluates the scalability of the proposed protocol. It is observed from the Fig. 2 that total energy consumption for the EBC-PSO is significantly lower than PSO-ECHS. This is due to fact that ECB-PSO uses a novel fitness function for selecting the CH which causes uniform and balanced energy consumption for all CH as well as cluster members.

Figure 3 demonstrates the performance comparison of the EBC-PSO with LEACH [2], PSO-ECHS [9], and E-OEERP [10] with respect to total energy consumption by varying the number of CHs. In this experiment, 100 nodes are deployed and number of CHs is varied from 5 to 25. This experiment observed the effect of varying the % of CHs in the performance of EBC-PSO. It can be viewed from the Fig. 3 that EBC-PSO outperforms the existing protocols and its rate of energy consumption is much lower than the other clustering protocols.

Figure 4 demonstrates the performance comparison of the EBC-PSO with LEACH [2], PSO-ECHS [9], and E-OEERP [10] with respect to total energy consumption by changing the location of the sink node. In this experiment, 100 nodes are deployed over a 200×200 sensing area. Location of the Sink is varied such as (100, 100), (150, 50) and (200, 200). It can be viewed from the Fig. 4 that total energy consumption is lower when location of the sink is at the center of the sensing area. This is due to fact that distance between CHs and the sink is almost equal in this network scenario. However, total energy consumption for the network scenario where sink node at the corner of the network, is greater that the first network scenario.

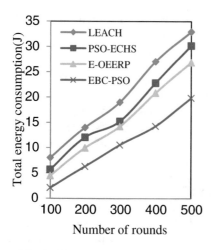

Fig. 1. Energy vs. number of rounds

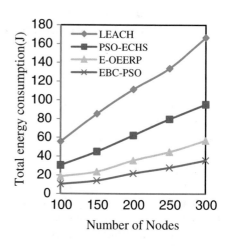

Fig. 2. Energy vs. number of nodes

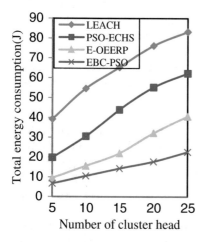

Fig. 3. Energy vs. number of CHs

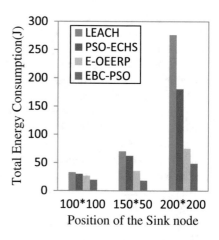

Fig. 4. Energy vs. various sink position

7   Conclusion

In this paper, energy balanced clustering using PSO (EBC-PSO) has been presented which consider the un-balanced energy consumption problem in WSNs. A novel multi-objective function is devised as a fitness function for selecting and distributing the CHs uniformly so that communication load for then are balanced. The performance comparisons of EBC-PSO with the three well known clustering protocols are descri-bed. EBC-PSO outperforms than LEACH [3], PSO-ECHS [9], and E-OEERP [10] in terms of total energy consumption. In future, proposed approach can be extended for mobile wireless sensor network and can be studied the effects of mobility over the performance of the clustering process.

References

1. Latiff, N.M.A., Tsimenidis, C.C., Sharif, B.S.: Energy-aware clustering for wireless sensor networks using particle swarm optimization. In: 2007 IEEE 18th International Symposium on Personal, Indoor and Mobile Radio Communications. IEEE (2007)
2. Gupta, G.P.: Efficient coverage and connectivity aware data gathering protocol for wireless sensor networks. In: 3rd IEEE International Conference on Recent Advances in Information Technology (RAIT-2016), pp. 50–55 (2016)
3. Heinzelman, W.R., Chandrakasan, A., Balakrishnan, H.: Energy-efficient communication protocol for wireless microsensor networks. In: Proceedings of the 33rd Annual Hawaii International Conference on System Sciences. IEEE (2000)
4. Lindsey, S., Raghavendra, C.S.: PEGASIS: power-efficient gathering in sensor information systems. In: IEEE Aerospace Conference Proceedings, vol. 3. IEEE (2002)
5. Younis, O., Fahmy, S.: HEED: a hybrid, energy-efficient, distributed clustering approach for ad hoc sensor networks. IEEE Trans. Mob. Comput. 3(4), 366–379 (2004)
6. Loscri, V., Morabito, G., Marano, S.: A two-levels hierarchy for low-energy adaptive clustering hierarchy (TL-LEACH). In: IEEE Vehicular Technology Conference, vol. 62. no. 3. IEEE (1999, 2005)
7. Heinzelman, W.B., Chandrakasan, A.P., Balakrishnan, H.: An application-specific protocol architecture for wireless microsensor networks. IEEE Trans. Wirel. Commun. 1(4), 660–670 (2002)
8. Rao, P.C.S., Jana, P.K., Banka, H.: A particle swarm optimization based energy efficient cluster head selection algorithm for wireless sensor networks. Wirel. Netw. 1–16
9. Rejina Parvin, J., Vasanthanayaki, C.: Particle swarm optimization-based clustering by preventing residual nodes in wireless sensor networks. IEEE Sens. J. 15(8), 4264–4274 (2015)
10. Kennedy, J.: Particle Swarm Optimization. Encyclopedia of Machine Learning, pp. 760–766. Springer, New York (2011)

Routing Protocol for Device-to-Device Communication in SoftNet Towards 5G

K. Rakshith[(✉)] and Mahesh Rao

Department of Electronics and Communication, Maharaja Institute of Technology, Mysore, India
{rakshithk_ece,maheshkrao_ece}@mitmysore.in

Abstract. Device to Device Communication (D2D) refers to a communication where devices communicate with each other directly without the need of cellular infrastructure. D2D is a new concept introduced under 3GPP; several researchers are working on providing solutions for D2D communication for the future network evolution such as IoT (Internet of Things) and IoE (Internet of Everything). Challenges associated with the D2D communication are security, interference management and resource allocation. Besides technical challenges, there are practical problems of incentivizing users to lend their devices to serve as relays for the traffic of others. To overcome these problems we have proposed a routing protocol for data packets of device communication called tree-based intelligent routing protocol based on SOFNET platform which results in support of IoT and IoE for the heterogeneous network. In this paper, we propose the architecture for routing in D2D communication. Simulation results on the proposed algorithm based on throughput are presented.

Keywords: IoT · IoE · Heterogeneous network · Device-to-device communication · SoftNet

1 Introduction

Fifth generation (5G) technology is the next biggest growth trend expected to rule the world in innovation and researchers are working to provide standardization. Figure 1 represents the different steps of the evolution which has happened to get to 5G. Cellular communication was originated initially with the advances in analog circuit –switched 1G network, which later brought development towards digitization towards 2G networks, followed by digital packet-switched network termed as 3G for voice and data network. 4G was a major revolution in wireless network with major developments in all IP based network architecture for handheld devices. Further the 5G technology will have focuses its development towards billions of devices connected together which has to have access to internet always.

Figure 1 shows wireless solutions like 802.11 and other following technology developments to support the cellular communication evolution. The other direction is Wi-Fi 802.11 have evolved in several different standards and techniques that supported the growth of cellular communication. Today's Wi-Fi 802.11 ac/ad/af is capable of handling

© Springer International Publishing AG 2018
S.C. Satapathy and A. Joshi (eds.), *Information and Communication
Technology for Intelligent Systems (ICTIS 2017) - Volume 2*, Smart Innovation,
Systems and Technologies 84, DOI 10.1007/978-3-319-63645-0_5

data rates upto 1 Gbps and will continue in expanding with higher data rates and longer ranges [1].

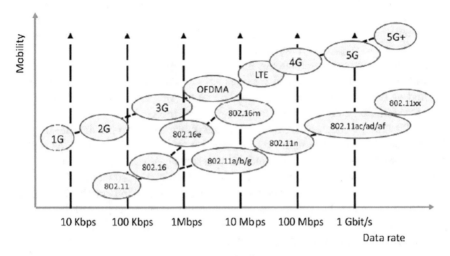

Fig. 1. Wireless networks and data rates

Routing is defined as selecting a particular path from the source node to destination node for efficient data transmission. A perfectly modeled routing algorithm is supposed to deliver the data packets in time without any delay. Several traditional routing algorithms were available for selecting a route, but they were not effective in this context. To overcome the problems in previous protocols and algorithms we designed a novel routing protocol for our framework. Due to the active participation of many types of users, it is very important to select the best route and so the data packets will be transmitted in time without any delay as per the requirements of different types of users. The data packets will be routed only through the selected best route [2].

The Routing algorithm should also be performed frequently whenever a user needs to transmit their data and capable of meeting user needs efficiently. In our proposed work we use a novel routing protocol named as Tree based Intelligence Routing protocol. In this routing protocol, a tree is constructed initially based upon the number of hops that are present to transmit the data. Further, the neighbors are identified if neighbors present then we use greedy multicast forwarding. By this routing, the best route is selected for data transmission through that route.

2   Related Works on Routing in M2M

Routing plays a major role in Machine to Machine (M2M) communication and hence this is of interest for many researchers. Xiaoying Zhang, et al. [3], focused on Energy-Aware Load-Balanced (EALB) routing protocol for M2M communications which deals with the major issue of energy consumption. Here the route discovery process is initiated by source node which broadcasts Route Request (RREQ) messages. Then with the Route

Reply (RREP) all the intermediate nodes are selected as relay node for packet transmission. In EALB protocol the RREQ packets are added with three fields such as *Route*$_{type}$, ϕ_{min} and ϕ_{total}. The term 'ϕ' is defined as the residual energy present in a node. This EALB protocol deals with two different mechanisms such as critical device selection mechanism and optimal route selection mechanism. Route is selected based in the parameters such as ϕ_{total}, delay, hop count and weight. This EALB routing protocol was designed for the purpose of reducing energy consumption and thereby improve the network lifetime. But this protocol was not able to show better performances when compared to the previous existing protocols (i.e.) their performance in throughput, energy consumption and delay are similar to previous protocols.

Jihua Lu et al. [4], dealt with routing and data transmission for efficient routing and ubiquitous connectivity. The entire routing involves with two phases such as routing phase (i.e.) performed by percolation and transmission phase (i.e.) based on the fountain codes. Here a small-world network is built with percolations based on the six degree of separation by distance rule. Here each node maintains a routing information table about all the close neighbors which includes their names, addresses, bandwidth of direct links, bandwidth of direct link between close neighbors, etc.,. After selecting a route, data transmission is performed based on fountain codes. Using fountain codes the data are encoded and then transmitted to intermediate nodes. Destination node on receiving the data packets, sends an acknowledgement to source. In this constructed network nodes consumes higher storage and the time taken for data transmission is also high. It involves with initiation, network discovery, network selection and MIH completion process for an effective handover framework.

Data aggregation and multiplexing scheme is focused by Safdar Nawaz Khan Marwat et al. [5]. Small data packets are collected in PDCP layer of relay node, for the purpose of maximizing multiplexing gain. PDDP layer exists in relay node, user equipment and also in DeNB, this layer can be considered as an interface in both control and user plane protocols. The relay node includes DPRS tunneling protocol, User datagram protocol and layer 2. This scheme was designed to improve the utilization of radio resources, when there is no expiry time. But this scheme did not consider Quality of Service and so all the user equipment are provided with equal priority. This was the major problem in this scheme, since in future the demand for QoS will be increased in M2M communication. One more thing is that, this entire scheme is applicable for only uplink of data transmission.

Shin-Yeh Tsai et al. [6], authors considered the energy consumption as a major constraint during data aggregation. This concept completely goes on through the buffering time in periodic per-hop timing control method. Here energy consumption is reduced by increasing the aggregated volume when the buffered packets increase. In periodic per hop, the collected data are sent to the application server through M2M gateway. The data collected are first put into its aggregation buffer, in case if the buffer is empty then buffering timer is started. Then if the buffered packet is equal to the maximum buffered packet, the buffering timer is stopped. Hence this periodic per hop not only consider the buffered packets, it also verifies the buffering tome of packets. Verification is performed for reducing the overlong delay. Finally simulation is performed in line-based and grid-based deployments, in which energy consumption is reduced.

Authors JianlinGuo et al. [7], focused on Resource-Aware adaptive mode Routing Protocol for Low-Power and Lossy Networks (RAM-RPL). This RAM-RPL involves with route discovery and data packet transmission. In this routing protocol Adaptive Mode Of operation (AMOP) is built for achieving the resource aware routing. When the devices are operated in AMOP, then the node configures its Mode of Operation with the resources and also the node can adaptively change the Mode of operation with respect to the resources usage in the network. This RAM-RPL make use of additional resources that are provided by different powerful nodes by which it moves on the workload from low powered nodes to high powered nodes.

Collection of data in Machine-to-Machine communication is illustrated by Andre Riker et al. [8], for controlling the network traffic by means of exploiting the capability of data aggregation. They proposed a Data Aggregation for energy harvesting Networks (DAV-NET). To reduce the energy consumption during data aggregation, the data are sent over paths that uses energy efficiently. DAV-NET works with two execution steps such as follows (i) determination of an aggregation tree and (ii) identification of aggregation level. Here it seems that each node maintains the energy level and hop gradient of its neighbors. This information is shared among nodes by sending Beacon messages in periodic intervals. With this information, the node starts with the process of selecting a parent node. The node with highest energy level and low hop gradient is selected as parent node. After construction of tree, each parent node will maintain a buffer for storing the messages that are received from the child nodes. Aggregation procedure is followed by payload extraction, payload merger and Application Layer Message (ALM) creation with Payload Concatenation. To regulate the network traffic the execution of payload merger is controlled. Further the aggregation level is determined by the energy level of the battery. Hereby even if the network traffic increases, the energy spent is not increased. But this was not able to control the unnecessary energy dissipated, if it was also controlled then this will be a more efficient way in reducing energy consumption.

Chih-Hua Chang et al. [9], were focused on effective data aggregation with minimized data distortion. Cross entropy method for identifying the optimal solution. It starts with the construction of Bernoulli probability vector and also a selection vector. Here the devices are randomly deployed and a device situated at the center is connected with other devices via direct link. A pre-defined set for power and rate allocation is estimated and so the feasible solution can be analyzed with the satisfaction of the constraints. By this the devices consuming higher power are discarded. In case if no device is satisfied by this constraint then they should be some other alternate to aggregate data. Hence this type of data aggregation is not applicable for all network scenarios.

Authors Jian Wang et al. [10], designed a scheme for improving the efficiency of data broadcasting between machine to machine communications. Authors designed Finite set Network coding Automatic Repeat request (FNARQ) scheme which gives efficient result. The packets are broadcasted based on the Time Division Multiple Access protocol which allocates time slots for broadcasting. Error occurs in packets due to the fading characteristics of channels. Here certain set of network coding strategies are preset in both based station and the user entities. The base station broadcasts all the data packets to the user entities, then the user entities on receiving the packets they try to decode those packets. According to the applied strategy, the entities decode the packets. But if the entity is idle then the data packets are neither received nor feedback is sent. If the packets are received then they compute decoding gain and choose the best one and send it as feedback message which is comprised of the chosen strategy's sequence number. Base station on receiving the feedback messages, determines the coverage rate, in case if the computed coverage rate is more than 95% then the base station intimates with broadcasting 'stop transmission' message. But if the coverage rate is low then the base station chooses the best network coding strategy with respect to the feedback received. But the feedback messages cannot be trusted all time, due to the increase of harmful attacker participation in the network.

M.G. Khoshkholgh et al. [11], concentrated on designing an energy-efficient protocol. This protocol was designed to be capable for large scale machine to machine communications. The designed transmission energy model includes with the functions as transmission power, packet size and link capacity. Data rate is defined as the function of transmission power between the sender and receiver devices along with noise power and inter-cell interference. Then Signal to Interference plus Noise Ratio is also computed for each device. Here all the devices transmit the data packets at fixed power level. Link capacity is estimated by using Shannon formula and hence the overall energy consumption is estimated by the link capacity, transmission energy and transmission power. The random variable considered here is characterized by either the Cumulative Distribution Function or Complementary Cumulative Distribution Function. Further interference model involves with defining the random variable by Laplace Transform and Poisson Point Process. This system dealt with highly complex computations.

Similarly authors Friedrich Pauls et al. [12], also modeled a system for reducing the power consumption during machine to machine communication. Two different machine to machine communication modes were designed for GPRS operation. The model is comprised into three parts such as setup, alive and data transmission. Initially the setup part is involved with selection of cell, frequency, burst, acquisition, and channel and frame synchronization. The two modes of operation are always on and On/Off mode. During always On mode the connection is always available in the network, but in this mode even if there is no transmission takes place the energy will be consumed. In on/off mode the devices are on only during their data transmission and in the remaining time they are in sleep state. In this on/off mode the power consumption is reduced during the keep alive phase that takes place in the devices. By this process the power is consumed but the requests from other devices will not reach if the device is in sleep condition. Hence based on the battery availability the operating modes should be used. This entire system is focused only for transmission of GPRS.

SiavashBayat et al. [13], developed an efficient data aggregation system corresponding to the urgency levels. This entire mechanism is based on coalitional game. Coalitional game algorithm is developed by two different rules such as merge and split which supports cooperative data transmission in Machine to machine communication. Here all the cellular users and eNBs will be equipped with single omni-directional antenna. In this algorithm each device will automatically decide whether to join or leave a coalition. Here the devices are capable to transmit urgent data along with the guaranteed data rate, but it will have certain cooperation cost. In case if the Machine-Machine Communication increases, then the number of coalitions has to be checked by the centralized algorithm, but this is not feasible. Since this developed algorithm involves with several computations and makes the entire system complex. This showed better energy efficiency when only compared with non-cooperative network.

3 Proposed Work

Based on the literature and other related work we propose a Tree Based Intelligent Routing Protocol (TIR). This TIR protocol is used for selecting the best path in the network and successfully transmitted the data the destination node without any delay. The algorithm for finding the best route in the network is defined in controller mechanism. The rules and the algorithm for finding the shortest and efficient route is present in the controller. The controller is responsible for determining the path and forwarding the packets based on the rules like traffic direction, changes in traffic path and Qos of the traffic. For routing process, the initial tree is constructed based on Steiner tree [12]. The tree is constructed based on identifying the total number of nodes present in the network. Once the nodes are identified the hops required to reach each node is calculated according to source and destination nodes. The hops calculated for reaching the destination node has to be minimum so that the shortest path is achieved. With the nodes available in hops calculate the tress is constructed for each node to transmit data to the destination. Once the tree is constructed the source nodes identifies the neighbors based on the tress constructed, of more neighbors are present to each node then the packets at the source node is split. The packets are split for delivering the data in through several nodes by reducing the load on each node. The forwarding of the split data packets are forwarded and received based on greedy multicasting forwarding technique. The set of next neighboring hop is identified by,

$$f(w) = \frac{\lambda |W|}{|N|} + \frac{(1 - \lambda) \sum_{d \in D} \min_{m \in w} l(m, d)}{\sum_{d \in D} l(s, d)}$$

Here, s is the forwarding node.

N is the set of all neighboring nodes.
W is the set of all subset of N.
D is the set of all destination nodes.
d is the destination d_1, d_2 or d_n.
$l(x, y)$ is the function measure distance between nodes.

The above equation consists of two parts, the first part number of neighbors that are present in the network to transmit the packet and the second part calculates the remaining distance to all the neighboring nodes from the forwarding node. The best route for the forwarding node is identified by calculating the delivery probability; the probability factor is estimated for selecting the best route to the destination. Kalman filter is used to identify the nodes that are in mobility, because the nodes in mobility carrying the data have the highest probability to reach maximum destination nodes. With the help of Kalman filter and delivery.

Figure 2 describes the pseudo code for routing process and Fig. 3 describes the flowchart of the routing process.

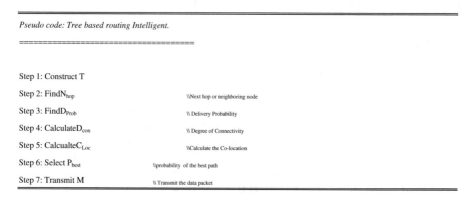

Pseudo code: Tree based routing Intelligent.

=====================================

Step 1: Construct T

Step 2: FindN$_{hop}$ \\Next hop or neighboring node

Step 3: FindD$_{Prob}$ \\ Delivery Probability

Step 4: CalculateD$_{con}$ \\ Degree of Connectivity

Step 5: CalcualteC$_{Loc}$ \\Calculate the Co-location

Step 6: Select P$_{best}$ \\probability of the best path

Step 7: Transmit M   \\ Transmit the data packet

Fig. 2. Pseudo code for Tree Based Intelligent Routing.

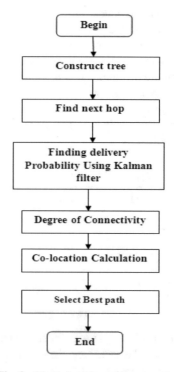

Fig. 3. Flow chart for routing process.

The Fig. 2 illustrates the Tree Based Intelligent Routing Process flow, initially the algorithm constructs the Steiner Tree (T) for the nodes available later it checks for the next hop for transmitting the data from forwarding node to the destination node Nhop. Delivery probability DProb is calculated based on Kalman filter followed by the degree of connectivity DCon and the Co-location CLoc of the neighboring nodes are identified for the route. Finally the best path Pbest for the packet is determined and the packet M is ready for transmission. Figure 3 show the flow diagram for the TIR protocol.

4 Handover

The role of handover plays an important role in wireless network. The user nodes will either be fixed or in mobility, there is need to provide connectivity for these mobility nodes. Handover is defined as a process of transferring the data session from one cell site to another without dropping the session. There are several approaches for handling the handover process, in this work we have considered two types of handover mechanism:

- Horizontal Handover
- Vertical Handover

Horizontal handover refers to a type where the UE's (User Equipment) moves in the same cell i.e. between the Base Station and GSM to GSM handover. Vertical handover refers to a network node (UE) that changes the type of connectivity from one network to another network, i.e., UMTS and WLAN.

The vertical handover process in initiated when the following events occur:

- Request from new service.
- Mobile Node identifies a new wireless link.
- Severe degradation in the current wireless link.
- When any requirement of resource blocks.
- Insufficient resources in some network.

The proposed handover mechanism is based on Fuzzy logic algorithm where the mobile nodes initiates the handover. The mobile nodes are always connected to either one of these networks LTE, WLAN or Wi-Max, The handover process is initiated when the node discovers a new link or may suffer degradation in the present link. When situation like these occur the decision on handing over the nodes to any preferred network is handled based on the Fuzzy rules. Figure 4, illustrated the Fuzzy Logic Controller architecture, and based on this logic controller the handover decision of choosing LTE, Wi-Fi or Wi-MAX network is determined. Fuzzy logic is used to deal with situation when there are similar conditions for handover which are not always recommended for handover, it deals with uncertainty cases.

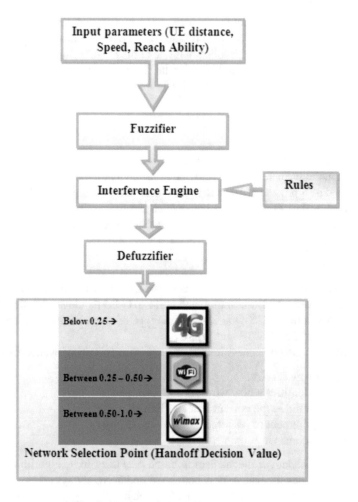

Fig. 4. Fuzzy logic based handover decision

5   Simulation Setup

We implement our proposed work using OMNET++ simulation framework, the simulator is supported by GUI (Graphical User Interface). This simulation tool helps us to support all the newly proposed protocols, mechanism and algorithms. Simulation parameters of the proposed work are as described in Table 1.

Table 1 provides the values of conditions set initially for simulation. For simulation, we have considered one eNodeB, one Server and 40 mobile nodes in the network. Each mobile node is always in mobility from one cell region to another. The simulation areas considered is about 2000 * 2000 m where the date rate for transmission is set to 100 Mbps. The node mobility speed is set to 10–100 Mbps, since it is based on device to device communication we have considered the ISM band for all communication hence

2.5 Ghz band is used. The normal operation speed of connected is in the range of 2–60 Mbps.

Table 1. Simulation setup parameters

Condition	Standards
Simulation area	$2000 * 2000 \text{ m}^2$
Mode	Duplex
Node mobility	10–100 Mbps
Connected address	Wireless core network
Bit rate	2–60 Mbps
Carrier frequency	2.4 GHz
Transmission rate	100 Mbps
Queue type	Drop tail queue
Management SSID	HOME

6 Performance Metric

Several performance parameters are considered for experimenting on the simulation and comparing our proposed system with the existing system. The parameters considers are:

- **Signaling Cost:**
  It is defined as the summation of packet delivery cost and cost of location update. The location update cost is calculated by

$$C_{LU} = \frac{E[M]CUH + CUh}{E[M]Tf}$$

  And packet delivery cost is obtained by

$$C_{PD} = \eta \lambda_a + \left(l_{h-lf} + l\right)\delta_D$$

  So,

$$C_{tot} = C_{LU} + C_{PD}$$

- **Delay:**
  It is defined as the amount of time taken for a bit of data to travel across the network from one node to another. This is calculated by

$$\text{Delay} = \frac{\sum (\text{Arrival time} - \text{Send time})}{\sum (\text{Number of connections})}$$

- **Throughput:**
  It is defined as the amount of data moved successfully from one place to another in a given period of time. Throughput can be measured for both Uplink and Downlink Traffic.

- **Bandwidth:**

  Bandwidth refers to the data rate supported by the network interface. Bandwidth represents the capacity of a network connection for supporting data transfer. The bandwidth is calculated with size of the data packet and the channel bit rate.

$$\text{Time of sending a file} = \frac{\text{Size of file to be sent (bits)}}{\text{Bit Rate}}$$

Figure 5 shows the ranges of network bandwidth during handover from one network to another network. When a mobile node moves in their network area that results in variation of bandwidth. Here the handover takes place from LTE to Wi-Fi, bandwidth varies initially and it is maintained at the constant level. While in handover from Wi-Fi to Wi-Max, there is a slight change on bandwidth with respect to time.

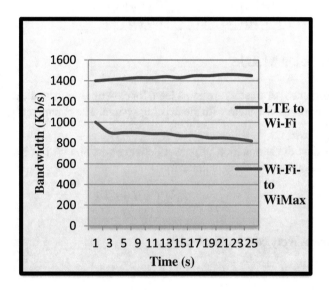

Fig. 5. Performance of bandwidth with respect to time during handover

Figure 6, shows the last packet transmission during handover of user. Delay is maintained at constant range i.e. no increase in delay with respect to time, during handover. In our TM2M5G mechanism, the handovers such as LTE to Wi-Fi and Wi-Fi to Wi-Max delay varies but stays constantly.

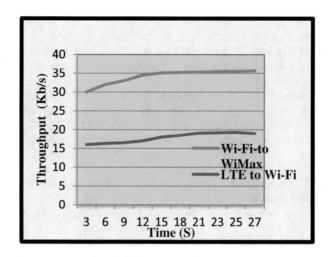

Fig. 6. Performance of delay with respect to time during handover

7   Conclusion

Our Tree Based Intelligent Routing (TIR) protocol is the best for selecting the best path for transmitting the data to reach the destination node without any delay. The protocol implemented on the Open flow controller was efficient in transmitting the data packets to the destination nodes. Further we would like to develop other algorithms for scheduling and security protocols for D2D communication and implement them and analyze the performance of delay, throughput, signaling overhead and bandwidth. Finally we would want to develop a complete new architecture for D2D communication and compare with the existing SoftNet based architecture. For further enhancement we have planned to test the multimedia transmission and evaluate the performance.

References

1. Adrian, L., Anisha, K., Byrav, R.: Network innovation using openflow: a survey. IEEE Commun. Surv. Tutor. doi:10.1109/SURV.2013.081313.00105
2. Bruno Astuto, A.N., Mendonça, M., Nguyen, X., Obraczka, K.: A survey of software-defined networking: past, present, and future of programmable networks. IEEE Commun. Surv. Tutor. (2014). doi:10.1109/SURV.2014.012214.00180
3. Zhang, X., Anpalagan, A., Guo, L., Shaharyar Khwaja, A.: An energy-aware load-balanced routing protocol for adhoc M2M communications. Trans. Emerg. Telecommun. Technol. (2015). doi:10.1002/ett.2963
4. Lu, J., An, J., Li, X., Yang, J., Yang, L.: A percolation based M2M networking architecture for data transmission and routing. KSII Trans. Internet Inf. Syst. **6**(2) (2012)
5. Khan, S.N., Marwat, M.Y., Görg, C., Timm-Giel, A.: Data aggregation of mobile M2M traffic in relay enhanced LTE-A networks. J. Wirel. Commun. Netw. EURASIP (2016). doi:10.1186/s13638-016-0598-0

6. Tsai, S., Sou, S., Tsai, M.: Reducing energy consumption by data aggregation in M2M networks. Wirel. Pers. Commun. **74**(4), 1231–1244 (2014). Springer

7. Guo, J., Orlik, P., Parsons, K., Ishibashi, K., Takita D.: Resource aware routing protocol in heterogeneous wireless machine-to-machine networks. In: Global Communications Conference (GLOBECOM). IEEE (2015)

8. Riker, A., Cerqueira, E., Curado, M., Monteiro, E.: Data aggregation for machine-to-machine communication with energy harvesting. In: 2015 IEEE International Workshop Measurements & Networking (M&N). doi:10.1109/IWMN.2015.7322977

9. Chang, C., Chang, R., Hsieh, H.: High-fidelity energy-efficient machine-to-machine communication. http://ieeexplore.ieee.org/xpl/mostRecentIssue.jsp?punumber=7116213, doi:10.1109/PIMRC.2014.7136139 (2014)

10. Wang, J., Xu, Y., Xu, K., Xie, W.: A finite set network coding automatic repeat request scheme for machine-to-machine wireless broadcasting, pp. 76–81 (2015). http://ieeexplore.ieee.org/xpl/mostRecentIssue.jsp?punumber=7422215s, doi:10.1109/BWCCA.2015.18

11. Khoshkholgh, M.G., Zhang, Y., Shin, K.G., Leung, V.C.M., Gjessing, S.: Modeling and characterization of transmission energy consumption in machine-to-machine networks. In: 2015 IEEE Wireless Communications and Networking Conference (WCNC). doi:10.1109/WCNC.2015.7127787

12. Pauls, F., Krone, S., Nitzold, W., Fettweis, G., Christopher Flores: evaluation of efficient modes of operation of GSM/GPRS modules for M2M communications. In: Vehicular Technology Conference (VTC Fall) (2013)

13. Bayat, S., Li, Y., Han, Z., Dohler, M., Vucetic, B.: Distributed data aggregation in machine-to-machine communication networks based on coalitional game, pp. 2026–2031 (2014). http://hdl.handle.net/10765/108909

14. Mauve, M., Füßler, H., Widmer, J., Lang, T.: Poster: position-based multicast routing for mobile ad-hoc networks. ACM SIGMOBILE Mob. Comput. Commun. **7**(3), 53–55 (2003)

15. Musolesi, M., Mascolo, C.: CAR: context-aware adaptive routing for delay tolerant mobile networks. IEEE Trans. Mob. Comput. **8**(2) (2009). ISSN 1536-1233, doi:10.1109/TMC.2008.107

A Survey of Computer Vision Based Corrosion Detection Approaches

Sanjay Kumar Ahuja[1(✉)] and Manoj Kumar Shukla[2]

[1] AIIT, Amity University, Noida, Uttar Pradesh, India
SanjayAhuja@India.com
[2] ASET, Amity University, Noida, Uttar Pradesh, India
MKShukla@Amity.Edu

Abstract. There are various destructive as well as non-destructive techniques available to detect corrosion in metallic surfaces. Digital Image Processing is widely being used for the corrosion detection in metallic surface. This non-destructive approach provides cost effective, fast and reasonably accurate results. Several algorithms have been developed by different researchers and research groups for detecting corrosion using digital image processing techniques. Several algorithms related to color, texture, noise, clustering, segmentation, image enhancement, wavelet transformation etc. have been used in different combinations for corrosion detection and analysis. This paper reviews the different image processing techniques and the algorithms developed and used by researchers in various industrial applications.

Keywords: Computer vision · Image processing · Corrosion detection

1 Introduction

We are using metal articles in our daily use, when these materials come in contact with humidity, the chances of corrosion get increased in Iron/steel objects. Corrosion is a chemical process which results in the destruction of metal surface. This results due to chemical and electrochemical reactions because of environmental conditions. As a result, there are loss of metallic components which may further lead to reduced efficiencies in the end use applications of the metal parts. This results in reduced life of metallic parts and hence increases in the maintenance cost. Study of corrosion growth helps in taking preventive measures to avoid such losses.

The introduction of digital image processing opens up expanded real life opportunities to sense in a variety of environments. While physical and chemical tests are highly effective for corrosion detection, however difficult to perform on large surface area. Digital Image processing based approaches are non-destructive approaches and are economical to perform on large areas like ships, metal bridges, electric pole etc.

Digital Image Processing techniques provides fast, reasonably accurate and objectives results. Various image processing techniques like image enhancement, noise removal, line detection, edge detection, registration, wavelet transformation, texture

© Springer International Publishing AG 2018
S.C. Satapathy and A. Joshi (eds.), *Information and Communication
Technology for Intelligent Systems (ICTIS 2017) - Volume 2*, Smart Innovation,
Systems and Technologies 84, DOI 10.1007/978-3-319-63645-0_6

detection, morphological functions, color analysis, segmentation, clustering and pattern recognition etc. are base functions that could be combined together to detect corrosion on metal articles.

This paper compares the several image enhancement segmentation, detection and classification techniques that researchers have adopted for corrosion detection. A typical image processing based process for corrosion detection is shown in Fig. 1.

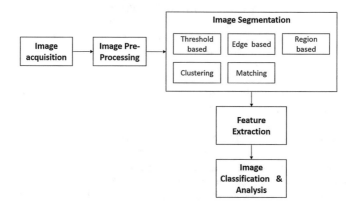

Fig. 1. Corrosion detection process

The typical steps involved in the process are:

1. **Image acquisition:** The first step of image processing is essentially capturing of the images with the help of suitable camera like digital, thermal etc. Images may be captured either manually or automatically. The images thus obtained are processed to study the corrosion. Automated Corrosion Detection System where the images are captured automatically are very fast and efficient as these require minimal manual intervention.

2. **Image Pre-Processing:** The images obtained generally have high background noise and unwanted reflections. To use these images these are required to be pre-processed by different filtering techniques like noise reduction, contrast improvement, image distortion reduction etc. The various Image processing filters can be Median Filter, Morphologic operations, Histogram equalisation, Image stitching, Shadow removal etc.

3. **Image Segmentation:** Once the image is pre-processed, the next step is to divide the image into meaningful structure for image analysis. Image Segmentation techniques uses general image processing methods such as Edge Detection, Region Growing, Thresholding and Clustering. The various categories used for pre-processing are:

   (a) **Threshold based:** In this technique segmentation is done by Histogram thresholding and slicing. It may require a combination of pre and post processing techniques also or may be applied directly.

   (b) **Edge detection:** In this technique, segmentation is done by detecting edges in the image. Once edges are detected, objects in the image are identified based on these detected edges. In this, first the boundaries are located to locate the object.

(c) **Region based:** In this technique, segmentation is done by starting from the center of the object and then growing the boundaries outward till it reaches the boundaries of the object.

(d) **Clustering:** These techniques are used to explore data analysis of various image patterns. Clustering is basically done by grouping together the patterns which have some similarity.

(e) **Pattern Matching:** In this technique, segmentation is done with the prior knowledge about an object's appearance. These patterns are matched in the image to locate the object.

4. **Feature Extraction:** After segmentation, the next step is the feature extraction. The features are related to color, texture, shape, motion. Few of the feature extraction filters are Haar wavelets, Hough transform, Laplacian of Gaussian (LoG), Histogram of Oriented Gradients (HoG), Gabor filtering, Canny edge detection, Background subtraction and Wavelet transform etc. The images are then processed to identify the corrosion area with series of steps.

5. **Image Classification and Analysis:** In this step, classification is based on certain known features such as size, position, contour measurement, texture measurements etc. Few image processing methods are: Percolation-based models, Graph-based search, Nearest neighbor, Principle Component Analysis, Generalized Hough transform, Line-tracing algorithms, Multi-temporal methods, Support vector machines etc. The image classifiers can be categorized based on Feature, Model and Pattern of the corroded area in the image. This step helps in detecting the corrosion in the images. Once corrosion is detected, the total corrosion area is calculated, to estimate whether the image is partially corroded or fully corroded to take appropriate action.

2 Related Work

Aijazi et al. [1], has proposed a method to form a 3D point cloud based on different positions and viewing angles of several images. The R, G, B values obtained from each image are then converted into HSV zones. This separates out the illumination color component and the image intensity. These parameters help to detect corroded area of different shapes and sizes, within a selected zone. The two methods used for detecting corrosion are based on histogram based distribution and adaptive thresholds respectively. The selection of these two methods is basically dependent on the level of corrosion in the image zone.

Petricca et al. [2], has used Deep Learning approach for automatic metal corrosion detection. He has used a classification based on the number of pixels containing specific red components of the image to implement one version and another deep learning model using Caffe to perform a comparison test.

Ortiz et al. [3] has proposed a solution for detecting corrosion/coating breakdown from the ship/vessel images to support surveyors in ship/vessel inspection. The solution is based on neural network to detects suspected pixels corresponding to defective/

corroded areas. He has used an aerial platform for capturing enhanced images. In this method he has extensively used behavior based high-level controls.

Igoe and Parisi [4] has proposed using smart phone based application to evaluate surface corrosion. In this study corrosion areas were characterized based in red color of iron using a smartphone sensor and java program. R, G, B models corresponding to corrosion were quantified for corrosion analysis. His study finds the 1:1 inverse relationship of red color in the corroded area to the green and blue responses having a quantifiable steeper regression. Errors in the color responses were within 5% range of errors found with the Perlin noise models.

Idris et al. [5] has proposed a vision based corrosion detection method for a pipeline inspection system. By using digital image processing based approach, the analogue signal loss due to the communication interference could be eliminated.

Many researchers have modified the Sobel and Canny edge detector for corrosion detection for pipe line inspection industry. The modified algorithms detected less false edges and also had improved intensity level of edges.

Son et al. [6], has proposed a method consisting of three steps: (1) color space conversion, (2) classification of corrosion area based on J48 decision tree algorithm and (3) determination of blasting area. They have used color space transformation from the Red–Green–Blue (RGB) to the Hue–Saturation–Intensity (HSI) color space and applied pixel level classification to detect the corroded areas. For pixel level classification they have used multiple approaches as Data set, Support-Vector-Machine (SVM), Back-Propagation-Neural-Network (BPNN), Decision tree (J48), Naive-Bayes (NB), Logistic-Regression (LR), K-Nearest-Neighbors (KNN) approaches.

Alkanhal [7], has proposed discrete wavelet packet transform and fractals for extracting Image feature parameters and analysis of pitting corrosion. Using these image processing techniques, he has analysed various characteristics as energy loss, Shannon entropy, fractal dimension and fractal intercept increase parameters with exposure time.

Idris and Jafar [8], has proposed using multiple image filters as Homomorphic, Bayer, Wavelet Denoising, Gaussian, Linear, Anisotropic Diffusion and using neural network to optimize the results. He has used Peak-Signal-to-Noise-Ratio (PSNR) and Mean-Square-Error (MSE) as two error metrics to compare image compression quality.

Corrosion on metal surfaces usually represent as a rough texture and reddish colors, Bonnin et al. [9], has proposed Bayesian framework based classifier and added roughness and color information to it to detect corrosion. He has used Weak-Classifier-color-based-Corrosion-Detector (WCCD) and AdaBoost-based-Corrosion-Detector (ABCD). WCCD is a supervised classifier built around a cascade scheme. This approach uses a chain of different fast classifiers with poor performance to obtain better global performance. He has used classifiers based on Gray-Level-Co-occurrence-Matrix (GLCM), Hue–Saturation–Value (HSV) and bi-dimensional histogram. ABCD method uses Adaptive Boosting paradigm (AdaBoost) for detection as well as classification of corroded areas. The decision trees in this work were based on Classification-and-Regression-Trees (CART) learning technique as weak classifiers.

Ranjan and Gulati [10], has proposed to use basic edge detection filters such as Sobel, Perwitt, Robert and Canny to detect corrosion areas. He then applied dilation and smoothing by eroding the image.

Acosta et al. [11], has proposed texture based classifier to identify corrosion area. He has simulated the corrosion area using Perlin Noise and used Bayesian classifier to identify corroded regions under different texture variations. He has proposed using Re-sampling, Holdout, Leave-one-out, K-fold cross-validation and Bootstrap methods for estimating corrosion area.

Fernndez et al. [12], has proposed image reconstruction approach based on wavelet reconstruction using normalized Shannon entropy calculation for automatic selection of the bands. He has proposed using Shannon Entropy for determination of texture decom-position level on different detail sub-images.

Jahanshahi et al. [13, 26], has evaluated image processing based approaches for corrosion detection for civil infrastructures like bridges, pillars, buildings etc. He eval-uated the effects of several image parameters like block sizes, color channels and color space on the performance of texture analysis based on wavelet texture algorithms. He has observed that a combination of color analysis and texture approached should be used to get better results. He has also concluded multi-resolution wavelet analysis as a more powerful tool for characterization of appropriate features for texture classification.

Daira et al. [14], has proposed using a combination of electrochemical and optical methods for analysing the corrosion processes and their dynamics. He has also done a comparative analysis to study the relation between obtained interferogram and the different polarization curves.

Sreeja et al. [15], has proposed using Gray-Level-Co-Occurrence-Matrix (GLCM) attributes and color attributes based approach to characterize the corroded regions in the image. He has used a Learning Vector Quantization (LVQ) based supervised clustering algorithm to detect corrosion.

Shen et al. [16], has proposed a segmentation approach based on color attribute and texture regions by combining the Fourier transform and image processing. In his research he proposed to adapt various background colors and overcome particular influ-ences for uniformly and non-uniformly illumination regions in corroded regions in the image.

Chen et al. [22], in one of his research has also proposed a combination of color attributes and texture features of images and corrosion detection using Fourier trans-formations. In another research paper he has proposed another approach [17] based on Support-Vector-Machine-based-Rust-Assessment-Approach (SVMRA). In this approach he proposed combination of Fourier Transform and Support Vector Machine (SVM) to provide a method for non-uniformly illumination regions in corroded regions in the image.

Liu et al. [18], has proposed thermal images for corrosion detection and using Prin-cipal-Component-Thermography (PCT). He proposed using Principal-Component-Analysis (PCA) to calculate signal amplitude, phase, principal components and their first and second derivatives for corrosion detection. He has used second principle component to determine depth of corrosion.

Motamedi et al. [19], proposed using CCTV and line laser method to capture images and applying morphological structuring elements. He has proposed using various shapes attributes such as solid line, periodic line, square, rectangle, octagon, diamond, arbitrary, pair and disk etc. to identify corrosion areas.

Ji et al. [20], has proposed using watershed transform applied over the gradient of gray level images.

Ghanta et al. [21], has proposed a corrosion detection method based on wavelet transformation applied to Red–Green–Blue (RGB) color plane of the image and then calculated the energy and entropy values in each sub-band B, (HH, LH, HL, LL) for the texture properties.

Zaidan et al. [23], has proposed a *stdfilt* filter based texture analysis. He also proposed application of structure element, edge detection and image dilation based approaches beside texture analysis for corrosion detection.

Medeiros et al. [24], has proposed color attributes and texture features based integrated approach to describe roughness on the metal surface and color changes because of corrosion. To detect the corrosion regions, he has proposed using Gray-Level-Co-occurrence-Matrix (GLCM) probabilities and Hue–saturation–intensity (HSI) color space statistics from images.

Xie [29], has proposed texture analysis techniques for detecting defects/corrosion on metallic surface. He proposed classification of approaches into four categories as statistical, structural, filter, and model based.

3 Summary and Conclusion

The existing corrosion detection and analysis techniques can be categorized into different categories as:

1. **Wavelet Domain:** This non-iterative approach uses wavelet transformations for calculating corrosion area. The concept is based on component analysis which results in entropy minimization for correction of illumination component of the image.
   The algorithm for detecting the corrosion has three steps as *(1) Feature Vectors Extraction, (2) Training, (3) Detection.* For energy detection and entropy classification feature vectors are used. The Energy and Entropy values are calculated after applying wavelet transform to RGB color planes. Corrosion area in the image can be detected with these extracted feature vectors.
2. **Classification using Support Vector Machine (SVM):** In this technique, classification is based on the color of the corroded area. The degree of the corrosion and the color of corroded area enables to detect the corrosion area.
3. **Damage Analysis based on NDE and SOM:** This technique is based on texture changes using Nondestructive Evaluation (NDE) method. The classification is done based on Self Organizing Mapping (SOM).
4. **Texture Analysis:** With the increase in corrosion the roughness of the metal surface increases. The gray value of the pixels at the edge is different than the gray value outside the boundary. Based on the segmentation detection for texture analysis, corroded and non-corroded parts are differentiated.

Researchers have applied different techniques on different types of images under different environmental conditions. Quality of images may not be good for large surface areas, so noise reduction filters need to be applied for better results. A combination of techniques can be applied on complex shapes to reduce the false positives and better detection rate.

Detection method	References
Color space based detection	[1, 2, 4, 6, 8, 15, 16, 21, 22, 24]
Wavelet transformation based feature vectors extraction, training and detection	[6–8, 12, 13, 21, 26]
Classification using support vector machine (SVM)	[6, 22]
NDE and SOM based analysis	[9, 15, 24, 25]
Texture analysis based detection and analysis	[9, 11–13, 16, 21–24, 29]

4 Future Work

Future work may be pursued to include:

– **Automated collection of Images** - In the current scenario, for the vision based corrosion detection systems the images are not captured automatically, which results in the poor quality of the images due to high back ground noise, image stabilization, poor image illumination etc.
– **Improved algorithms to support complex metal structures** - Most of the algorithms have been developed with the images from simple flat and curved metal surfaces. A lot is required to be researched taking into account complex metal geometries and structures. Future work can also be extended to working on the algorithms which take into account the complex metal geometries which are actually used in many practical scenarios like transportation, bridges, automotive, heavy machines etc.
– **Corrosion Growth Trend Analysis** - Computer vision based corrosion detection methods may also be used for the trend analysis and extrapolating the trends. This will help in taking timely preventive actions to stop the further deterioration of the metallic surfaces.
– **Machine Learning** - When corrosion analysis need to be done on large area, applying the image processing techniques with Machine learning can result in reduction in false positive based on prior history. Machine learning approach can also be applied for performing analysis on corrosion growth rate.

References

1. Aijazi, A.K., Malaterre, L., Tazir, M.L., Trassoudaine, L., Checchin, P.: Detecting and analysing corrosion spots on the hull of large marine vessels using colored 3D lidar point clouds. ISPRS Ann. Photogrammetry Remote Sens. Spat. Inf. Sci., 153–160 (2016)
2. Petricca, L., Moss, T., Figueroa, G., Broen, S.: Corrosion detection using AI: a comparison of standard computer vision techniques and deep learning model. In: CCSEIT, AIAP, DMDB, MoWiN, CoSIT, CRIS, SIGL, ICBB, CNSA-2016, pp. 91–99 (2016)

3. Ortiz, A., Bonnin-Pascual, F., Garcia-Fidalgo, E.: Visual inspection of vessels by means of a micro-aerial vehicle: an artificial neural network approach for corrosion detection. In: Second Iberian Robotics Conference, Robot 2015. Springer, Heidelberg (2016)
4. Igoe, D., Parisi, A.V.: Characterization of the corrosion of iron using a smartphone camera. Instrum. Sci. Technol. **44**(2), 139–147 (2016)
5. Idris, S.A., Jafar, F.A., Jamaludin, Z., Blar, N.: Improvement of corrosion detection using vision system for pipeline inspection. In: Applied Mechanics and Materials, vol. 761 (2015)
6. Son, H., Hwang, N., Kim, C., Kim, C.: Rapid and automated determination of rusted surface areas of a steel bridge for robotic maintenance systems. Autom. Constr. **42**, 13–24 (2014)
7. Alkanhal, T.A.: Image processing techniques applied for pitting corrosion analysis. Entropy Int. J. Res. Eng. Technol. **3**(1), (2014)
8. Idris, S.A., Jafar, F.A.: Image enhancement based on software filter optimization for corrosion inspection. In: 2014 5th International Conference on Intelligent Systems, Modelling and Simulation. IEEE (2014)
9. Bonnin-Pascual, F., Ortiz, A., Aliofkhazraei, D.M.: Corrosion detection for automated visual inspection. In: Developments in Corrosion Protection, pp. 619–632 (2014)
10. Ranjan, R.K., Gulati, T.: Condition assessment of metallic objects using edge detection. Int. J. Adv. Res. Comput. Sci. Softw. Eng. **4**(5), (2014)
11. Acosta, M.R.G., Daz, J.C.V., Castro, N.S.: An innovative image-processing model for rust detection using Perlin noise to simulate oxide textures. Corros. Sci. **88**, 141–151 (2014)
12. Fernndez-Isla, C., Navarro, P.J., Alcover, P.M.: Automated visual inspection of ship hull surfaces using the wavelet transform. Math. Probl. Eng. (2013)
13. Jahanshahi, M., Masri, S.: Effect of color space, color channels, and sub-image block size on the performance of wavelet-based texture analysis algorithms: an application to corrosion detection on steel structures. In: ASCE International Workshop on Computing in Civil Engineering (2013)
14. Daira, R., Chalvedin, V., Boulhout, M.: Detection of corrosion processes in metallic samples of copper by CND control. Mater. Sci. Appl. **4**(04), 238 (2013)
15. Sreeja, S.S, Jijina, K.P, Devi, J.: Corrosion detection using image processing. Int. Res. J. Comput. Sci. Eng. Appl. **2**(4), (2013)
16. Shen, H.K., Chen, P.H., Chang, L.M.: Automated steel bridge coating rust defect recognition method based on color and texture feature. Autom. Constr. **31**, 338–356 (2013)
17. Chen, P.H., Shen, H.K., Lei, C.Y., Chang, L.M.: Support-vector-machine-based method for automated steel bridge rust assessment. Autom. Constr. **23**, 9–19 (2012)
18. Liu, Z., Genest, M., Krys, D.: Processing thermography images for pitting corrosion quantification on small diameter ductile iron pipe. NDT & E Int. **47**, 105–115 (2012)
19. Motamedi, M., Faramarzi, F., Duran, O.: New concept for corrosion inspection of urban pipeline networks by digital image processing. In: 38th Annual Conference on IEEE Industrial Electronics Society, IECON 2012. IEEE (2012)
20. Ji, G., Zhu, Y., Zhang, Y.: The corroded defect rating system of coating material based on computer vision. In: Transactions on Edutainment VIII. LNCS, vol. 7220, pp. 210–220. Springer, Heidelberg (2012)
21. Ghanta, S., Karp, T., Lee, S.: Wavelet domain detection of rust in steel bridge images. In: 2011 IEEE International Conference on Acoustics, Speech and Signal Processing (ICASSP). IEEE (2011)
22. Chen, P.H., Shen, H.K., Lei, C.Y., Chang, L.M.: Fourier-transform-based method for automated steel bridge coating defect recognition. Procedia Eng. **14**, 470–476 (2011)
23. Zaidan, B.B., Zaidan, A.A., Alanazi, H.O., Alnaqeib, R.: Towards corrosion detection system. Int. J. Comput. Sci. **7**(3), 33–36 (2010)

24. Medeiros, F.N., Ramalho, G.L., Bento, M.P., Medeiros, L.C.: On the evaluation of texture and color features for nondestructive corrosion detection. EURASIP J. Adv. Sig. Process. **1**, 817473 (2010)
25. Bento, M.P., de Medeiros, F.N., de Paula Jr., I.C., Ramalho, G.L.: Image processing techniques applied for corrosion damage analysis. In: Proceedings of the XXII Brazilian Symposium on Computer Graphics and Image Processing, Rio de Janeiro, RJ (2009)
26. Jahanshahi, M.R., Kelly, J.S., Masri, S.F., Sukhatme, G.S.: A survey and evaluation of promising approaches for automatic image-based defect detection of bridge structures. Struct. Infrastruct. Eng. **5**(6), 455–486 (2009)
27. Planini, P., Petek, A.: Characterization of corrosion processes by current noise wavelet-based fractal and correlation analysis. Electrochim. Acta **53**(16), 5206–5214 (2008)
28. Mrillou, S., Ghazanfarpour, D.: A survey of aging and weathering phenomena in computer graphics. Comput. Graph. **32**(2), 159–174 (2008)
29. Xie, X.: A review of recent advances in surface defect detection using texture analysis techniques. ELCVIA Electron. Lett. Comput. Vis. Image Anal. **7**(3), 1–22 (2008)

Word Sense Ambiguity in Question Sentence Translation: A Review

Sanjay Kumar Dwivedi and Shweta Vikram[✉]

Babasaheb Bhimrao Ambedkar University, Lucknow, India
skd200@yahoo.com, shwetavikram.2009@rediffmail.com

Abstract. Word Sense Disambiguation has been a major challenge for various linguistic researches. Enough research has been carried in the past four decades. In machine translation, WSD plays a vital role in improving the accuracy of the translation. The automated translation of question papers from English to Hindi is one such key area which requires suitable WSD techniques to resolve ambiguity in a question word. When machine translates question sentences, it faces ambiguity problem that results in ambiguous translation. Identification of question type is important for remove ambiguity in the question paper. In this paper besides discussing WSD its approaches, resources for translation. We have also discussed question classification word sense disambiguation.

Keywords: Machine translation · Types of question · Word sense disambiguation · English language · Hindi language

1 Introduction

Word Sense Disambiguation (WSD) is a very challenging task of machine translation when source and target languages are different in many aspects. WSD is a process which selects the correct sense of the word with respect to context. It is an open problem of Natural Language Processing and it comes under the NP-complete problem. WSD is very helpful in machine translation. The term WSD was first time introduced by Warren Weaver in his famous memorandum on translation in 1949. Machine Translation refers to the use of the computer to automate some part of the task or the entire task of translating between human languages. It is a subfield of computational linguistics that investigates the use of computer software to translate text from one source language (SL) to target language (TL). Many attempts are being made all over the world to develop MT systems for various languages using rule-based, example-based, dictionary based, corpus-based and statistical-based approaches. MT systems can be designed either specifically for two particular languages, called a bilingual system, or for more than a single pair or languages, called a multilingual system. A bilingual system may be either unidirectional, from one source language into one Target Language, or may be bidirectional. Multilingual systems are usually designed to be bidirectional, but most bilingual systems are unidirectional.

© Springer International Publishing AG 2018
S.C. Satapathy and A. Joshi (eds.), *Information and Communication*
Technology for Intelligent Systems (ICTIS 2017) - Volume 2, Smart Innovation,
Systems and Technologies 84, DOI 10.1007/978-3-319-63645-0_7

2   Literature Review

In India one of the first attempts made to disambiguate the nouns of Hindi language using Hindi WordNet [11]. Corpus is necessary for many purposes such as training purpose, pattern identification. Different types of resources of corpora such as speech, text, and image [12, 13]. Creation of dictionaries, thesauri and corpora started in 1980. Corpora provided a vast amount of knowledge and information on various hands-on language parameters. WSD research has been ongoing for the four decades. Word sense disambiguation is relevant when a word has multiple senses. The first experiments work done on word sense disambiguation with machine translation activities (Hutchins 1999) [5, 25]. In this machine lookup information and translate a word into a target or desired language. Many senseval workshops held on NLP and they gave many unique ideas for WSD approaches. Word Sense Induction (WSI) [7] is a method which is introduced by Navigli, WSI comes under unsupervised techniques which aimed at automatically identifying the set of senses denoted by a word. In this method word senses from the text by clustering word occurrences based on the idea that a given word used in a specific sense which co-occurs with the same neighboring words [2]. Anusaaraka: Machine Translation can be categorizing into rule-based [16], statistically based, example based, and hybrid [1]. Rule-based systems are still higher than statistical systems.

Many works are done and ongoing in machine translation for one natural language to other natural languages. Here we give the brief introduction of work on machine translation in India. Matra[1] English to Hindi Machine Translation systems (MTS) started at CDAC Pune in 2004. Shakti-English to Hindi, Marathi, and Telugu MTS (combines rule-based and statistical approach) IISc Bangalore and IIIT Hyderabad started in 2004. Anglabharti MTS (English-Hindi, Tamil MTS), Angla-Hindi (combines example based approach and AnglaBharti approach) started 1991 and Anubhart Hindi-English MTS; Combines syntax started on 2004 at IIT Kanpur. Anglabharti is a pattern directed rule-based system with a context free grammar like structure for English (source language) that generates a 'pseudo-target' applicable to a group of Indian languages (target languages). Siva-English–Hindi translation started on 2004 at IISC Bangalore. Anusaaraka[2] is a unique approach to machine translation based on the theory of information dynamics inspired by the Paninian grammar formalism. MANTRA[3] (MAchiNe assisted TRAnslation tool) translates English text into Hindi in a specified area of personal administration, specifically, gazette notifications, office orders, office memorandums and circulars. MANTRA uses Lexicalized Tree Adjoining Grammar (LTAG) to represent the English as well as the Hindi grammar. Anglabharti uses a pseudo-interlingua approach. Many free translators available on the Internet which is: Google, Babylon, Yahoo Babel Fish, PROMT translation, Microsoft. Every Machine translation based on one or more than one approach.

---

[1] http://cdacmumbai.in/matra/.

[2] http://anusaaraka.iiit.ac.in/.

[3] http://en.wikipedia.org/wiki/MANTRA-Rajbhasha.

3   Approaches to WSD

Many approaches are used to disambiguate the senses:

- *Supervised WSD*: In this approach needs of supervision. These approaches used trained data set of machine-learning techniques [4, 8, 18].
- *Semi-supervised WSD*: Semi-supervised or minimally supervised approaches do not need to a tagged corpus. Semi-supervised WSD used labeled and unlabeled data [10].
- *Unsupervised WSD*: It is based on unlabeled corpora, and does not courage any manually sense-tagged corpus to provide a sense choice for a word in context. Clustering comes under this approach [23, 24].
- Other WSD: Example based (or Instance based), Bootstrapping, AI based, Hybrid WSD (Combination of all WSD approaches or some approaches is known as hybrid WSD approaches).

Gathering of information needed for word sense disambiguation and every machine translation based on a different approach. Every WSD approaches do not do anything without knowledge [6]. Comparison of the Word Sense Disambiguation approaches with some specific criteria is shown in Table 1 [6, 9, 14, 15, 20].

Table 1. Comparison of supervised, semi-supervised and unsupervised WSD approach

Specifications	Supervised WSD	Semi-supervised WSD	Unsupervised WSD
Type of data used in WSD	Secondary	Primary and secondary	Primary
Main data	Labeled data	Small set of labeled data	Unlabeled data
Time	It is time-consuming approach	It takes less time to supervised WSD	It takes less time
Output	It gives relevant output	Sometimes gives relevant output	No guarantee for relevant output
Representation	Tagged data in text form	Small set of tagged data in text form	Untagged data in text form
Cost nature	Expansive	In between Supervised and Unsupervised WSD	Cheap
Algorithm	Naive Bayes (NB), K-nearest neighbor (K-NN), Support Vector Machine (SVM), Neural Network (NN)	Decision list	Agglomerative, Divisive, K-means, Bisecting K-means
Requirement	Collection of very large data set	Medium size of data set	Small data set

4 External Knowledge Resources

The basic component of WSD is Knowledge. Knowledge resources provide data which are essential to associate senses with words. They can vary from corpora of texts, either unlabeled or annotated with word senses, to machine-readable dictionaries, thesauri, glossaries, ontology's, etc. [4]. External knowledge resource can be two types one is structure and other is unstructured [8].

4.1 Structured Resources

Arrangement of data in some definite or fixed order in resources is known as structure data. For example, Dictionary follows an alphabetical order.

- *Machine-readable dictionaries (MRDs)*, which have become a popular source of knowledge for natural language processing since the 1980s, when the first dictionaries were made available in electronic format: among these, we cite the Collins English Dictionary, the Oxford Advanced Learner's Dictionary of Current English, the Oxford Dictionary of English [26], and the Longman Dictionary of Contemporary English (LDOCE) [27].
- *Thesauri* provided information about the relationships between words, like synonymy antonym and, possibly, further relations.
- *Ontology* is the specifications of conceptualizations of specific domains of interest [28], usually including a taxonomy and a set of semantic relations.
- *Glossary* is an alphabetical list of a particular domain; it appears at the end of the book.

4.2 Unstructured Resources

Lacking a definite structure of data in the unstructured resource. Wikipedia and corpus do not have any definite structure.

- *World Wide Web:* It is the huge collection of online data.
  - *Wikipedia:* It is a huge collection of articles and this article written by many different Indian and another language. Wikipedia, mainly differentiates in three articles which are Feature, Good, and normal articles [24].
  - *Search Engine Result:* It has collection of searched data.
- *Corpora:* Corpora can be sense-annotated or raw (i.e., unlabeled). Both kinds of resources are used in WSD and are most useful in supervised and unsupervised approaches, respectively [22]. WordNet: WordNet [29, 30] is a computational lexicon of English based on psycholinguistic principles, created and maintained at Princeton University. It encodes concepts in terms of sets of synonyms its latest version, WordNet 3.1, so the English WordNet is a collection of English synsets. WordNet as a graph whose nodes represented by synset and whose edges represented the semantic relation between synset [19, 21].

- *Raw Corpus:* It is a collection of data in text format. It can be anything, such as articles, stories, and poems.
- *Sense-Annotated Corpora:* It is a set of data, but data have sensed. This corpus is very helpful for supervised learning.
- *Collocation resources*, which register the tendency for words to occur regularly with others.

An external knowledge resource is a vital part of machine translation similarly, the structure identification of source and target language is also important.

5   Question Classification

Question classification [3] is a process to identify the types of question sentences (such as wh- question, key word specific etc.) and it helps in machine translation. Interrogative sentences always have question marks at the end of sentences, but all question sentences do not follow the same pattern. Classification of question sentences helpful to identify the structure of question sentences such as Wh- question, Keyword specific question, and anomalous verb related question. Question classification is the vital part of WSD.

A question which requires reasoning and thus long explanations are identified by keyword like WH-word, explain, discuss, justify etc.

- Question related to quantity
- Question related to time.
- Question-related to person or place.
- Questions regarding to different passages like kaun-kaun, kya-kya, vibhinn.
- Sentences that ask question are called interrogative sentences. Many types of question in exam question papers: Yes/no Interrogative question. Example: Did Ram go the game Friday night?
- Alternative question - Example: Did Ram go Patna or Lucknow on Friday night?
- Interrogative wh- type question- Example: What is Ram doing?
- Tag question: Tag question is questions attached or tagged onto the ending of a declarative statement. They transform a declarative sentence into an interrogative sentence. Declarative sentences become question: Example: Ram live in the city, don't Ram?

  The computer is not working?

- Subjective questions: fill in the blank- Example: Delhi is a capital of _____
- Objective questions: Many different types of question come under this category such as fill in the blank (Different types of notation such as —, ..., , ( ), ___), matching, passage, true false. Example: Who amongst the following is considered to be the Father of 'Local Self-Government' in India? (a) Lord Dalhousie, (b) Lord Canning, (c) Lord Curzon, (d) Lord Ripon
- Keyword specific questions: In this types of the question have many different keywords. We collected some keyword from different resources such as exam question papers, online available data, and exercises of the book. Some keywords

are- Explain, discuss, justify, solve, find out, perform. Example: Describe the potential method for amortized analysis of an algorithm with a suitable example.

- Note type Question: Example: Write short notes on following:
  Cyber law in Indian context,
  Miller–Rabin method.

In modern English, only the anomalous verbs are normally inverted with the subject to form the interrogative [17].

6 Discussion on Question Sentence Ambiguity

This paper analyzes many types of question sentences in the English language which is discussed in Question Classification Section.

1. *Question sentences should be in text format:* Text format provides the uniformity to machine translation.
2. *Subject or area of the question sentences should be known:* It is very helpful in disambiguation because the meaning of the word also varies from subject to subject or context. For example word ring has a simple meaning in general language is a finger ring (Anguthi) but in physics, it has circulated (chhalla). Part of speech also changes the meaning of the word. Shabdkosh English to Hindi Dictionary shows the meaning of word "ring".
3. *Declare of nontranslate item (formula, abbreviation):* Non translates item helpful for understanding the meaning of context for example formula, abbreviation etc.
4. *Declare multiword expression:* Multiword expression (ME) or idiom is also change the meaning of the context. A collection of two or more words is known as the multiword expression. It is a pain in the natural language processing (NLP) for example "kick the bucket" means "to die (mar jaanaa)" in ME but in the word to word translation is "Balti ko laat maarnaa".
5. *Identification of the question type (WH question or key word specific):* Classification of question sentences helpful to identify the structure of question sentences such as Wh- question, Keyword specific question, and anomalous verb related question. Question classification is the vital part of WSD.

7 Conclusion

Machine translation is easy to disambiguate the word in simple translation because it identifies the pattern of sentences, but exam, question sentences does not have a pattern so the translation is more difficult. This paper reviews many types of question sentences and try to identify their pattern and discuss how to disambiguate their meaning. Questionable translation is very helpful for the student which is not friendly in English language.

References

1. Chaudhury, S., et al.: Anusaaraka: an expert system based machine translation system. In: 2010 International Conference on Natural Language Processing and Knowledge Engineering (NLP-KE), Beijing, pp. 1–6. IEEE. Print ISBN 978-1-4244-6896-6. (2010). doi:10.1109/NLPKE.2010.5587789
2. Navigli, R.: Meaningful clustering of senses helps boost word sense disambiguation performance. In: Proceeding of the 21st International Conference on Computational Linguistics and 44th Annual Meeting of the ACL, pp. 105–112, Sydney. Association for Computational Linguistics (2006)
3. Kumar, P., et al.: A Hindi question answering system for e-learning documents. In: ICISIP, Third International Conference on Intelligent Sensing and Information Processing, 2005, Bangalore, pp. 80–85. IEEE. Print ISBN 0-7803-9588-3. (2005). doi:10.1109/ICISIP.2005.1619914
4. Navigli, R.: Word sense disambiguation: a survey. ACM Comput. Surv. **41**(2), Article 10 (2009)
5. VidhuBhaha, R.V., Abirami, S.: Trends in Word Sense Disambiguation. Springer, Berlin (2012)
6. Ponzetto, S.P., Navigli R.: Knowledge-rich word sense disambiguation rivaling supervised systems. In: Proceedings of the 48th Annual Meeting of the Association for Computational Linguistics, Uppsala, Sweden, 11–16 July 2010, pp. 1522–1531. Association for Computational Linguistics (2010)
7. Navigli, R., Crisafulli, G.: Inducing word senses to improve web search result clustering. In: Proceedings of the 2010 Conference on Empirical Methods in Natural Language Processing, pp. 116–126. Association for Computational Linguistics (2010)
8. Faralli, S., Navigli, R.: A new minimally-supervised framework. In: Proceedings of the 2012 Joint Conference on Empirical Methods in Natural Language Processing and Computational Natural Language Learning, Jeju Island, Korea, 12–14 July 2012, pp. 1411–1422. Association for Computational Linguistics (2012)
9. Yadav, R.K., Gupta, D.: Annotation guidelines for Hindi–English word alignment. In: 2010 International Conference on Asian Language Processing, 978-0-7695-4288-1/10. IEEE (2010). doi:10.1109/IALP.2010.58
10. Mishra, N., Mishra, A.: Part of speech tagging for Hindi corpus. In: 2011 International Conference on Communication Systems and Network Technologies, 978-0-7695-4437-3/11. IEEE (2011). doi:10.1109/CSNT.2011.118
11. Agarwal, M., Bajpai, J.: Correlation-based word sense disambiguation. In: 2014 Seventh International Conference on Contemporary Computing (IC3), pp. 382–386. IEEE Conference Publications. (2014). doi:10.1109/IC3.2014.6897204
12. Ahmed, P., Dev, A., Agrawal, S.S.: Hindi speech corpora: a review. In: Oriental COCOSDA held Jointly with 2013 Conference on Asian Spoken Language Research and Evaluation (O-COCOSDA/CASLRE), 2013 International Conference, pp. 1–6. IEEE (2013)
13. Bansal, S., et al.: Determination of linguistic differences and statistical analysis of large corpora of Indian languages. In: Oriental COCOSDA held jointly with 2013 Conference on Asian Spoken Language Research and Evaluation (O-COCOSDA/CASLRE), 2013 International Conference, pp. 1–5. IEEE (2013)
14. Jurafsky, D., Martin, J.H.: Word sense disambiguation and Information Retrieval. In: Speech and Language Processing an Introduction to Natural Language Processing Computational Linguistics, and Speech Recognition. Pearson Education, Inc. ISBN 678-81-317-1672-4 (2000)

15. Pool, D., Mackworth, A., Goebel, R.: Computational Intelligence—A Logical Approach. Oxford University Press Inc. (1998). Fourth Impression (2012). ISBN 13:978-0-19-568572-5, ISBN 10:0-19-568572-5
16. Bharti, A., Chaitanya, V., Sangal, R.: Hindi grammar. In: Natural Language Processing—A Panninial Perspective. India Prentice-Hall (2010)
17. Wren, P.C., Martin, H., Prasada Rao, N.D.V.: High School English Grammar & Composition. S. Chand & Company Ltd., New Delhi (2010). ISBN 81-219-2197-X
18. Fulmari, A., et al.: A survey on supervised learning for word sense disambiguation. Int. J. Adv. Res. Comput. Commun. Eng. 2(12), (2013). (ISSN (Print) 2319–5940, ISSN (Online) 2278–1021)
19. Navigli, R., Lapata: Graph Connectivity Measures for Unsupervised Word Sense Disambiguation. In: IJCAI 2007, pp. 1683–1688 (2007)
20. Mante, R., et al.: A review of literature on word sense disambiguation. Int. J. Comput. Sci. Inf. Technol. (IJCSIT) 5(2), 1475–1477 (2014). (ISSN 0975-9646)
21. Kolte, S.G., et al.: WordNet: a knowledge source for word sense disambiguation. Int. J. Recent Trends Eng. 2(4), 213–217 (2009)
22. Montoyo, A., et al.: Combining knowledge- and corpus-based word-sense-disambiguation methods. J. Artif. Intell. Res. 23, 299–330 (2005).
23. Navigli, R., Lapata, M.: An experimental study of graph connectivity for unsupervised word sense disambiguation. IEEE Trans. Pattern Anal. Mach. Intell. 32, 678–692 (2010)
24. Turdakov, D.Y.: Word sense disambiguation methods. In: Programming and Computer Software, vol. 36, no. 6, pp. 309–326 (2010). ISSN 0361_7688 (Pleiades Publishing, Ltd., Original Russian Text ©, published in Programmirovanie)
25. Hutchins, W.J.: The development and use of machine translation systems and computer-based translation tools. International Conference on Machine Translation & Computer Language Information Processing. Proceedings of the conference, 26–28 June 1999, Beijing, China, ed. Chen Zhaoxiong, pp. 1–16 (1999)
26. Soane, C., Stevenson, A.: Oxford English Dictionary (2003)
27. Procter, P.: Longman Dictionary of Contemporary English Dictionary of Contemporary English. Cartermills Publishing (1978)
28. Gruber, T.R.: A translation approach to portable ontology specifications. Knowl. Acquisition 5(2), 199–220 (1993)
29. Fellbaum, C.: WorldNet. John Wiley & Sons, Inc. (1998)
30. Miller, et al.: Introduction to WordNet: an on-line lexical database. Int. J. Lexicography 3(4), 235–244 (1990)

Implementing a Hybrid Crypto-coding
Algorithm for an Image on FPGA

B.V. Srividya[1(✉)] and S. Akhila[2]

[1] Department of Telecommunication Engineering,
DayanandaSagar College of Engineering, Bangalore, Karnataka, India
srividyabv@gmail.com
[2] Department of Electronics and Communication Engineering,
BMS College of Engineering, Bangalore, Karnataka, India
akhila.ece@bmsce.ac.in

Abstract. This paper proposes a hardware design, implemented on an FPGA, for a hybrid selective encryption and selective error correction coding scheme. FPGA's are used as implementation platforms in image processing, as its structure exploits the temporal and spatial parallelism. The algorithm aims at implementing security and reliability in which encryption and encoding are performed in a single step using Bezier curve and Galois field GF (2^m). The system aims at speeding up the encryption and encoding operations without compromising either on security or on error correcting capability by using selective encryption and selective encoding. The coding for hybrid crypto-coding algorithm is carried out using VHDL. The algorithm is simulated and synthesized using Xilinx ISE 10.1 software. The algorithm is implemented on Spartan 3 FPGA device 3s1000fg676-5. The proposed scheme reduces the hardware as modular arithmetic operations are involved.

Keywords: Bezier curve · Galois field · Image · Encryption · Error correction · FPGA

1 Introduction

In order to obtain a high throughput rate in Image processing, the algorithms are implemented in Field Programmable Gate Array (FPGA), which is a reconfigurable hardware. Implementing on FPGA provides low power cost effective solution and a high data throughput.

Traditionally HDL languages such as VHDL and Verilog are used for implementing on FPGA. In this paper, using VHDL an image is encrypted using the concept of selective encryption that is based on Quartic Bezier Curve over Galois Field GF (2^m). Further the encrypted image is recovered from transmission errors using Low Density Parity Check Codes (LDPC). The hybrid crypto-coding algorithm is implemented on FPGA.

The following sections give a brief introduction to Bezier curves and Galois Field based on which the encryption and the error recovery algorithm are constructed and

© Springer International Publishing AG 2018
S.C. Satapathy and A. Joshi (eds.), *Information and Communication Technology for Intelligent Systems (ICTIS 2017) - Volume 2*, Smart Innovation, Systems and Technologies 84, DOI 10.1007/978-3-319-63645-0_8

implemented on FPGA. The Low Density Parity Check codes is also been discussed in the following section.

A. FPGA Overview

An FPGA is made up of an array of programmable logic cells that are interconnected using a network of interconnecting lines with switches amidst them. The reconfigurable interconnects allows the logic cells to be interconnected, thereby configuring the logic cells to perform the desired logical operations. Around the boundary of the chip, Input Output Cells exist. These I/O cells provide an interface between the external pins of the chip and the interconnecting lines. Indicating the logic function for each cell and for the switches is termed as programming an FPGA.

B. Introduction to Bezier Curves

Bezier curves are a method of designing polynomial curve segments [1, 2], where the shape of curves can be controlled using the control points. The control points (from P0 to Pn) of the Bezier curve determine the order 'n' of the curve. Bezier Curves can be classified as linear Bezier curve, Quadratic Bezier curve, Cubic and Quartic Bezier curves on the basis of the order 'n' [3].

- A Linear Bezier curve has n = 1 and its curve equation is given by Eq. (1)

$$B(t) = (1 - t)P_0 + tP_1, t \in [0, 1] \tag{1}$$

where there are two control points P0 and P1. Linear Bezier curve represents an interpolation between two points.

- A Quadratic Bezier curve has n = 2 and the curve Eq. (2) is given by

$$B(t) = (1 - t)^2 P_0 + 2t(1 - t)P_1 + t^2 P_2, t \in [0, 1] \tag{2}$$

where there are three control points P0, P1, and P2. The Quadratic Bezier curve represents a linear interpolate of the control points from P0 to P1 and also P1 to P2.

- The cubic Bezier curve is given by Eq. (3)

$$B(t) = (1 - t)^3 P_0 + 3t(1 - t)^2 P_1 + 3(1 - t)t^2 P_2 + t^3 P_3; t \in [0, 1] \tag{3}$$

where P0 to P3 are its control points.

- The Quartic Bezier curve B(t) having 5 control points from P0 to P4 is given by Eq. (4)

$$B(t) = (1 - t)^4 P_0 + 4t(1 - t)^3 P_1 + 6t^2(1 - t)^2 P_2 + 4t^3(1 - t)P_3 + t^4 P_4; t \in (0, 1) \tag{4}$$

C. Introduction to Galois Field

Evariste Galois is the inventor of Galois field. The number of elements is finite in GF (p^m). Some of the popular Forward Error Correcting codes like BCH Codes and Reed Solomon codes use finite fields for the purpose of encoding and decoding [4]. In cryptographic algorithms, the value of p is taken to be 2, and is represented as GF (2^m).

Every GF (2^m) has a primitive polynomial of degree m, which α and its conjugates to be its roots. From the primitive polynomial the elements of GF (2^m) can be constructed. The elements are $\{0, 1, \alpha\,\alpha^2, \alpha^3 \ldots \alpha^{m-2}\}$. Each element in GF (2^m) can be represented using m-bits. In coding theory and cryptographic algorithms, certain modular arithmetic operations are performed on the elements of the Galois Field. The following section shows the construction of the elements of the field and the arithmetic operations performed on the field elements.

Table 1. Elements of GF (2^4)

Element	Polynomial representation	Binary representation
0	0	(0000)
α^0	1	(1000)
α^1	X	(0100)
α^2	X^2	(0010)
α^3	X^3	(0001)
α^4	X + 1	(1100)
α^5	$X^2 + X$	(0110)
α^6	$X^2 + X^3$	(0011)
α^7	$1 + X + X^3$	(1101)
α^8	$1 + X^2$	(1010)
α^9	$X + X^3$	(0101)
α^{10}	$1 + X + X^2$	(1110)
α^{11}	$X + X^2 + X^3$	(0111)
α^{12}	$1 + X + X^2 + X^3$	(1111)
α^{13}	$1 + X^2 + X^3$	(1011)
α^{14}	$1 + X^3$	(1001)

Table 2. Addition in GF (2^4)

+	0	1	2	3	4	5	6	7	8	9	10	11	12	13	14	15
0	0	1	2	3	4	5	6	7	8	9	10	11	12	13	14	15
1	1	0	3	2	5	4	7	6	9	8	11	10	13	12	15	14
2	2	3	0	1	6	7	4	5	10	11	8	9	14	15	12	13
3	3	2	1	0	7	6	5	4	11	10	9	8	15	14	13	12
4	4	5	6	7	0	1	2	3	12	13	14	15	8	9	10	11
5	5	4	7	6	1	0	3	2	13	12	15	14	9	8	11	10
6	6	7	4	5	2	3	0	1	14	15	12	13	10	11	8	9
7	7	6	5	4	3	2	1	0	15	14	13	12	11	10	9	8
8	8	9	10	11	12	13	14	15	0	1	2	3	4	5	6	7
9	9	8	11	10	13	12	15	14	1	0	3	2	5	4	7	6
10	10	11	8	9	14	15	12	13	2	3	0	1	6	7	4	5
11	11	10	9	8	15	14	13	12	3	2	1	0	7	6	5	4
12	12	13	14	15	8	9	10	11	4	5	6	7	0	1	2	3
13	13	12	15	14	9	8	11	10	5	4	7	6	1	0	3	2
14	14	15	12	13	10	11	8	9	6	7	4	5	2	3	0	1
15	15	14	13	12	11	10	9	8	7	6	5	4	3	2	1	0

(i) **Elements in GF (2^m)**

The elements of Galois field GF (2^4) is constructed using the primitive polynomial P $(x) = x^4 + x + 1$ and is shown in Table 1

(ii) **Addition in GF (2^m)**

Galois field addition is explained with an example: The primitive polynomial of Galois Field GF (2^4) is $P(x) = x^4 + x + 1$. This primitive polynomial has α and its conjugates as the roots.

According to Table 1, each element of GF (2^4) is represented using 4-binary bits. Addition is performed using Bitwise XORing operation. For example:

$$\alpha^5 + \alpha^5 = (0110) + (0110) = (0000) = 0 = \alpha^0$$
$$\alpha^2 + \alpha^5 = (0010) + (0110) = (0100) = 1 = \alpha^1$$

There is a significant increase in the speed of addition, as there is no carry generation and carry propagation delay.

The addition table for the same is as shown in Table 2.

(iii) *Multiplication in GF (2^m)*

Modular multiplication is performed, by multiplying the polynomials and then performing modular reduction on the product. Let a(x), b(x) be the polynomial representation of two elements in GF (2^m), whose product needs to be computed and g(x) be the irreducible field generator polynomial, then modular multiplication is as illustrated in the following example.

Example: If g(x) = 1 + X + X^4, a(x) = 1 + X^3, b(x) = 1 + X^2

Then a(x) * b(x) = (1 + X^3) * (1 + X) = (1 + X^2 + X^3 + X^5)

Modular reduction of the above result is $(1 + X^2 + X^3 + X^5)$ mod $(1 + X + X^4) = X^3 + X + 1$.

Table 3 is the modular multiplication performed on the elements of GF (2^4) using the polynomial g(x) = 1 + X + X^4

Table 3. Multiplication in GF (2^4)

X	1	2	3	4	5	6	7	8	9	10	11	12	13	14	15
1	1	2	3	4	5	6	7	8	9	10	11	12	13	14	15
2	2	4	6	8	10	12	14	3	1	7	5	11	9	15	13
3	3	6	5	12	15	10	9	11	8	13	14	7	4	1	2
4	4	8	12	3	7	11	15	6	2	14	10	5	1	13	9
5	5	10	15	7	2	13	8	14	11	4	1	9	12	3	6
6	6	12	10	11	13	7	1	5	3	9	15	14	8	2	4
7	7	14	9	15	8	1	6	13	10	3	4	2	5	12	11
8	8	3	11	6	14	5	13	12	4	15	7	10	2	9	1
9	9	1	8	2	11	3	10	4	13	5	12	6	15	7	14
10	10	7	13	14	4	9	3	15	5	8	2	1	11	6	12
11	11	5	14	10	1	15	4	7	12	2	9	13	6	8	3
12	12	11	7	5	9	14	2	10	6	1	13	15	3	4	8
13	13	9	4	1	12	8	5	2	15	11	6	3	14	10	7
14	14	15	1	13	3	2	12	9	7	6	8	4	10	11	5
15	15	13	2	9	6	4	11	1	14	12	3	8	7	5	10

Another approach for performing modular multiplication, when the elements of the field are represented in binary values are as explained below.

The Binary representation of g(x) = (X^4 + X + 1) = (1101).

The modular multiplication in binary can be performed as illustrated in Table 3.

For example: If A = 9 and B = 9, then

$$AXB = 9 \times 9 = (1001) \times (1001) = (1010001)$$
$$(1010001) \bmod (1101) = (1011) = 13$$

Further, exponential operation can be performed using GF (2^m) as shown below.

$$5^7 = (5 \times 5 \times 5 \times 5 \times 5 \times 5 \times 5)\mathrm{GF}(2^4)$$
$$= (2 \times 2 \times 2 \times 5)\mathrm{GF}(2^4)$$
$$= (4 \times 10)\mathrm{GF}(2^4)$$
$$= 14$$

D. Low Density Parity Check Codes (LDPC)

Low density parity check codes fall into the category of linear block codes and is one of the popular error correcting codes, when data is transmitted over a noisy channel. The density of one's is smaller compared to that of zeros in LDPC [14]. There can be a regular or an irregular Parity matrix defined for an LDPC code. If the Parity matrix has a uniform row and column weight, then it is a Regular parity matrix P [16]. Every row and every column of the Regular Parity matrix has exactly the same number of elements. These conditions ensure that the parity matrix P has uniform row and column weights forming a Regular LDPC code. The Parity matrix P that does not adhere to the property of having uniform row and column weight forms an Irregular Parity matrix [16].

2   Related Work

The work on "Joint AES algorithm and LDPC codes" [5] by CP Gupta et al. discusses on achieving Security and error correction in a single step as, AES is secured and also LDPC codes retains full error correction capability. But, in symmetric key cryptosystems the two parties who are communicating need to share the secret key prior to the start of the transaction.

The authors of "Joint Encryption and Error Correction Technical Research Applied an Efficient Turbo Code" [6] Jianbin Yao et al., is effective in terms of security and reliability. But the system has not been verified for many attacks. The image recovery is achieved after several iterations.

The authors of "Implementation of High Security Cryptographic System with Improved Error Correction and Detection Rate using FPGA" [7] have discussed on achieving 100% Error detection, encryption scheme is effective and bandwidth is improved. But, the encryption and decryption delays increase as the input data is increased from 4-bits to 8-bits.

The proposed algorithm is on combining selective encryption and selective encoding to obtain an secured error free data. The encryption algorithm is a public key cryptosystem, where in encryption [8] and decryption [8] operations are performed with a pair of mathematically related keys [9, 10] based on the Galois Field GF (p^m). In performing, Selective Encoding the complexity of the hardware is simplified, a better performance of the decoder is achieved even when output is zero. Further there is a reduction in the area as the Hardware used is XOR gates. The algorithm uses, K-P modulo-additions for decoding, where K being the length of the information digits and P is the number of non zero digits of the Parity matrix.

The following Sect. 3 discusses on the proposed system, Sect. 4 is on results and discussion. Section 5 discusses the conclusion arrived for the proposed system.

3 Proposed Work

The structure of the proposed encryption and encoding scheme is as shown in Fig. 1. Security is achieved using the encryption algorithm as discussed in [20].

For correcting the errors, the modified Low density parity check codes have been explained in [21].

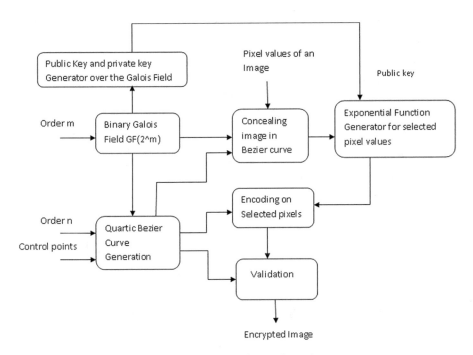

Fig. 1. Encryption and encoding scheme

A secured error free image can be obtained as explained according the following algorithm:

- The control points, of the n-order Bezier curve and the m-order finite field are shared between the sender and the receiver before the start of the transaction. Generate the Bezier image from the n-order Bezier curve. This image has the same size as that of the data image. The data image is concealed in the Bezier image as shown in the Fig. 2, and is denoted as I. This image I is further selectively encrypted as explained in the following section.

- **Selective Encryption:**
  The Selected Pixel values of the aggregated image I is exponentially raised to the power of m^e, over GF (2^m) as given by the relation (5)

  $$C1 = (I)^{m^e} \bmod GF(2^m). \tag{5}$$

  I is the selected pixel values of the concealed image. The order of the finite field 'm' is the secret key and the public key is 'e'. Only pixel values greater than the threshold value are encrypted resulting in selective encryption. The 2^m digits which are selectively encrypted are selectively encoded as explained in the following section.

- **Data Reliability using Selective Encoding:**
  The Data Reliability is achieved by the construction of LDPC codes based on n-order Bezier Curve over Galois Field GF (2^m) [13, 15] to obtain full error correcting capability. LDPC codes have a better performance when combined with Galois Field [17, 18]. Non-Binary LDPC codes is a better choice when more number of errors needs to be corrected [13, 19]. The data that needs to be selectively encoded is the selectively encrypted 2^m digits. If this 2^m digits of selectively encrypted data has two consecutive digits data (i) and data (i + 1) to be the same, then the first digit data (i) is replaced with a zero. This process continues till all the 2^m digits of data have been checked for repetition with its adjacent value. This selected data denoted as Cs has zeros when adjacent values are the same. This selected data Cs is encoded, by performing modular multiplication of Cs and the Generator matrix G.

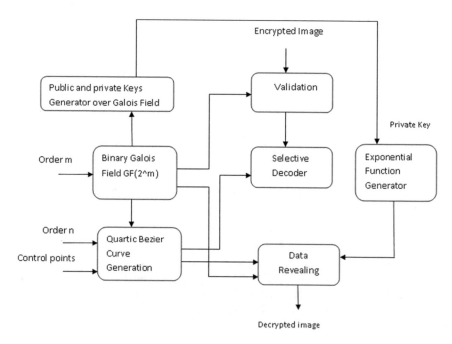

Fig. 2. Decryption and decoding scheme

- **Validation at transmitter:**
  The selectively encoded data CE is converted to CE1 after validation.
  Mean of P1, P2, P_{n-1} is computed, where P1, P2,
  P_{n-1} are the control points on the curve. P_0 and P_n are the starting point and the ending point of the n-order Bezier curve respectively.

$$\Delta = (xt, yt) = \frac{(P_n - mean)}{P_n - P0} \tag{6}$$

Defined over the order 'm' of the Galois field. This value of Δ, as shown in Eq. (6) is used for the purpose of validation.

$$CE1 = \frac{\Delta}{CE} \bmod GF(2^m) \tag{7}$$

Where CE1 as shown in Eq. (7), is the 2^{m+1} digits of cipher text and CE is the selectively encoded value.

Decryption with Decoding embedded:
Figure 2 shows the Decryption of the cipher text with a decoder embedded to retrieve the plain text.

The receiver upon receiving every 2^{m+1} digits of cipher text of an image, checks for the authentication of the data using the value of Δ, calculated using the control points. After checking for validation, the modified LDPC decoder makes the received data to be error free. The error free data is exponentially raised to $(m^{-1})^d$ over GF (2^m), with 'd' being the private key. These exponentially raised pixel values are algebraically combined with the Bezier image by the receiver to obtain the data image.

- **Validation at receiver:** The receiver calculates the selectively encoded value using Eq. (8).

$$CE = \frac{\Delta}{CE1} \bmod GF(2^m) \tag{8}$$

where CE1 is the 2^{m+1} digits of cipher text and CE is the selectively encoded value. The value of Δ is used for the purpose of validation. The value of Δ is derived from the control points of the chosen n-order Bezier curve. If the value of Δ is not known to the receiver, then the erroneous information cannot be corrected, thereby preserving the security of the information.
- **Selective decoding:**
  After validation is performed, the received vector R contains checksum which is 2^m digits and the erroneous pixel values which is 2^m digits. The decoder corrects the image from the received erroneous image and makes it error free. To achieve this, the FEC codes are applied. In LDPC, each row of the encoded image has 2^{m+1} digits of data that is given as input to the decoder. Syndrome is calculated by the decoder to determine the position of errors in the received information. The syndrome S is a modular multiplication of the Received data R and the parity check matrix H. If the

Syndrome S is Zero, then the received vector is error free else, the decoder determines the location of the error. The error location is determined by referring to the Parity Check matrix H. The erroneous data is corrected using the appropriate checksum. After correcting the errors, the consecutive zeros will be replaced by the right most non zero pixel value.

- **Selective Decryption:**
  The Error free image V is selectively decrypted using the following logarithmic equation, to obtain the concealed image I, as shown in Eq. (9).

$$I = (V)^{(m^{-1})^d} \bmod GF(2^m) \tag{9}$$

where 'm^{-1}' is the inverse of the secret key and the private key for decryption is 'd'. In public key cryptosystem, the pairs of keys used for encryption and decryption are related mathematically. The relation between the public key e, that is used for encryption and the private key d that is used for decryption is given by Eq. (10)

$$m^e(m^{-1})^d = 1 \bmod GF(2^m - 1) \tag{10}$$

- **Data Revealing:**
  The original image is embedded in the Bezier image. This concealed image is selectively encrypted, and after error recovery, the same is selectively decrypted. After selective decryption, the original image is retrieved from the concealed image.

4   Results and Discussion

The proposed hybrid crypto-coding algorithm is coded in MATLAB and also in VHDL.

The image considered for experimental purpose is lena.jpg which is of size 256×256 pixels.

The results are obtained are using Quartic Bezier curve and Galois field GF (2^8).

The results are discussed for an encrypted image that is affected with White Gaussian Noise having SNR = 0.5 dB

The following Fig. 3 is a snapshot of the data image, the encrypted image, the received erroneous image and the decrypted image from MATLAB.

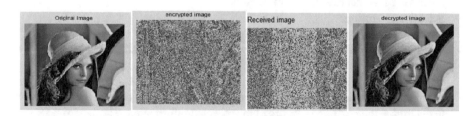

Fig. 3. (a) Original image (b) Encrypted image (c) Received image (d) Decrypted image

From Fig. 3, it can be observed that the received image which has been encrypted and transmitted has been modified due to the presence of Gaussian noise. This modified image has non-zero syndrome values from S1 to S256. Only the authenticated user can correct these modified pixel values. After the valid authentication check the errors have been eliminated using the decoding algorithm. It can be seen that the Decrypted image is same as the original image. 39,322 modified pixel values have been detected and corrected using the proposed algorithm.

The coding for hybrid crypto-coding algorithm is also carried out in VHDL, simulated and synthesized using the Xilinx ISE. The algorithm is implemented on FPGA Spartan 3 3s400ft256-5. The synthesis results obtained are shown in Fig. 4.

# ROMs	256x256-bit ROM	2
# Multipliers	8x8-bit multiplier	5
Adders/Subtractors	8-bit adder	1
# Registers	Flip-Flops	10
# Latches	3-bit latch	45
# Decoders	1-of-256 decoder	14

Device utilization summary: Selected Device: 3s400ft256-5

Number of Slices	48 out of 3584	1%
Number of Slice Flip Flops	70 out of 7168	0%
Number of 4 input LUTs	86 out of 7168	1%
Number of IOs	147	
Number of bonded IOBs	46 out of 173	84%
IOB Flip Flops	20	
Number of GCLKs	3 out of 8	37%

Timing Summary:

Minimum period: 3.003ns (Maximum Frequency: 332.967MHz)
Maximum output required time after clock: 6.141ns
Total 3.003ns (1.760ns logic, 1.243ns route) (58.6% logic, 41.4% route)

Fig. 4. Advanced HDL synthesis report

Figure 4 shows the hardware utilized when implemented on FPGA Spartan 3 3s400ft256-5. The timing summary indicates that 58% of the time is used for logic and 41% is used for routing.

The Fig. 5 shows the sample snapshot of the simulation results for combined selective encryption and selective encoding for an 8 × 8 pixel value of the original image using ModelSim.

Figure 5 shows that the crypto-coding algorithm is successful, as the encrypted data is made error free and then decrypted by the receiver.

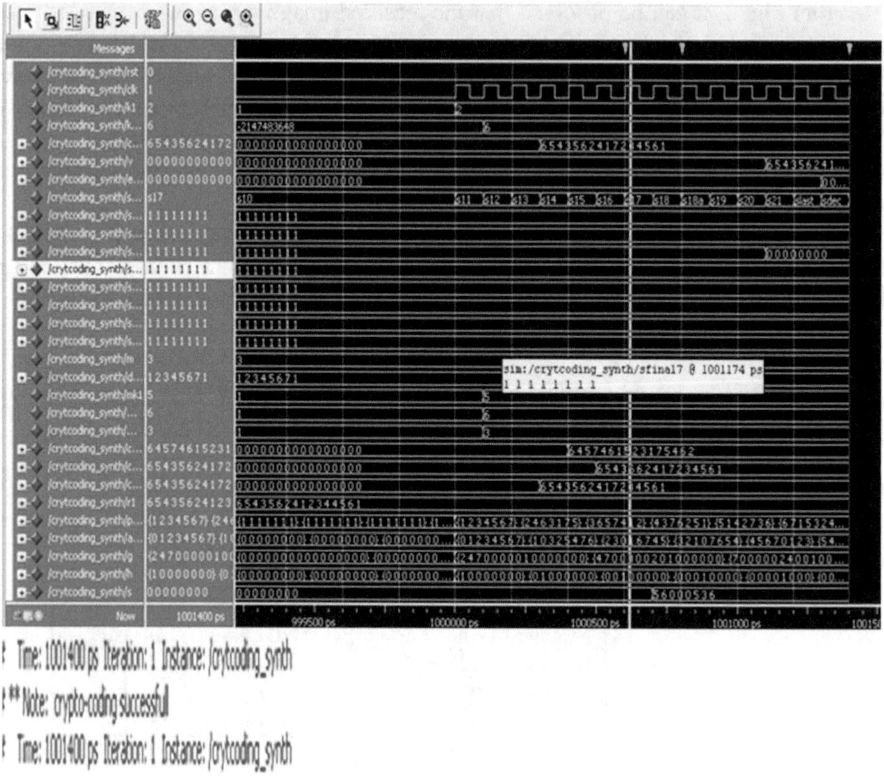

Fig. 5. Modelsim results of crypto-coding

5   Conclusion

This paper, establishes the working of combining Selective Encryption and selective Error Correction using modified LDPC with Bezier curve and Galois field GF (p^m). The Bezier curve points generated with the Galois field is the parity matrix P. This Parity matrix P is used for the construction of the generator matrix G and parity check matrix H. The original data image is concealed in the Bezier image by taking the aggregate of the pixel values. The pixel values of the concealed image above a certain threshold value are only encrypted. Thus Selective encryption is performed on the concealed image. Further, the proposed algorithm uses Selective Encoding, where in the repeating consecutive pixel values of the image are replaced by zeros. Using this approach, it is possible to encode only a few non-zero pixel values. The Encoding and Decoding involves modular arithmetic operations. The proposed decoder can handle BER = 1200/2048. The crypto-coding is done using one hot state encoding in VHDL and implemented on FPGA.

References

1. Caglar, H., Akansu, A.N.: A generalized parametric PR-QMF design technique based on Bernstein polynomial approximation. IEEE Trans. Sig. Process. **41**(7), 2314–2321 (1993)
2. [Online notes]: Bernstein; Visualization and Graphics Research Group. Department of Computer Science, University of California. www.idav.ucdavis.edu/education/CAGDNotes/Bernstein-Polynomials.pdf
3. Weisstein, E.W.: "Bézier Curve", From MathWorld—A Wolfram Web Resource. http://mathworld.wolfram.com/BezierCurve.html
4. Cameron, P.J.: The Encyclopedia of Design Theory: Galois Fields 30 May 2003
5. Gupta, C.P., Gautam, S.: Joint AES algorithm and LDPC codes. Int. J. Sci. Eng. Res. **4**(7), 603–606 (2013)
6. Yao, J., Liu, J., Yang, Y.: Joint encryption and error correction technical research applied an efficient turbo code. Int. J. Secur. Appl. **9**(10), 31–46 (2015). doi:10.14257/ijsia.2015.9.10.03
7. Babu, N., Noorbasha, F., Gunnam, L.C.: Implementation of high security cryptographic systems with improved error correction and detection rate using FPGA. IAES Int. J. Electr. Comput. Eng. **6**(2), 602–610 (2016). ISSN 2088-8708
8. Stallings, W.: Cryptography and Network Security: Principles and Practice, 4th edn. Prentice-Hall Press, Upper Saddle River (2006)
9. Abdul Elminaam, D.S., Abdul Kader, H.M., Hadhoud, M.M.: Performance evaluation of symmetric encryption algorithms. Commun. IBIMA **8**, 58–63 (2009). ISSN 1943-7765
10. Jakhar, S.: Comparative analysis between DES and RSA algorithms. IJARCSSE **2**(7), 386–390 (2012)
11. Amounas, F., El Kinani, E.H., Hajar, M.: A novel approach for enciphering data based ECC using Catalan numbers. Inst. Adv. Eng. Sc. Int. J. Inf. Netw. Secur. (IJINS) **2**(4), 339–347 (2013). ISSN 2089-3299
12. Kute, V.B., Paradhi, P.R.: A software comparison of RSA and ECC. Int. J. Comput. Appl. **2**(1), 61–65 (2009)
13. Ganepola, V.S., et al.: Performance study of non-binary LDPC codes over Galois field. In: CSNDSP08. IEEE (2008)
14. Wasule, P.U., Ugale, S.: Review paper on decoding of LDPC codes using Advanced Gallagers algorithm. IJAICT **1**(7), 622–625 (2014)
15. AlinSindhu, A.: Galois field based very fast and compact error correcting technique. Int. J. Eng. Res. Appl. **4**(1), 94–97 (2014). www.ijera.com. ISSN 2248-9622 (Version 4)
16. Fossorier, M.P.C., Mihaljevic, M., Imai, H.: Reduced complexity iterative decoding of low density parity check nodes based on belief propagation. IEEE Trans. Commun. **47**(5), 673–680 (1999)
17. Chen, J.P., Fossorier, M.P.C.: Density evolution for two improved BP-based decoding algorithm for LDPC codes. IEEE Commun. Lett. **6**(5), 208–210 (2002)
18. Xu, M., Wu, J., Zhang, M.: A modified offset Min-sum decoding algorithm for LDPC codes. In: 3rd IEEE International Conference on Computer Science and Information Technology, (ICCSIT), vol. 3 (2010)
19. Kim, J., Ramamoorthy, A.: The design of efficiently encodable rate-compatible LDPC codes. IEEE Trans. Commun. **57**(2), 365–375 (2009)

20. Srividya, B.V., Akhila, S.: A heuristic approach for secured transfer of image based on Bezier curve over Galois field GF(p^m). In: IEEE International Conference on Circuits, Communication, Control and Computing (2014)
21. Srividya, B.V., Akhila, S.: Bezier curves with low density parity check codes over Galois field for error recovery in an image. In: IEEE International Conference on Communication and Signal Processing (2016)

Realization of FPGA Based PID Controller for Speed Control of DC Motor Using Xilinx SysGen

Akanksha Somani[1(✉)] and Rajendra Kokate[2]

[1] Maharashtra Institute of Technology, Aurangabad, Maharashtra, India
akanksha_somani@rediffmail.com
[2] Government College of Engineering, Jalgaon, Maharashtra, India
rdkokate@gmail.com

Abstract. PID controllers can be designed using analog and digital methods; digital PID controllers are most significantly used in the industries. FPGA based PID Controllers are preferred because of their improved settling time and are small in size, consume power efficiently and provide high speed of operation compared to software based PID controllers or microprocessor/microcontroller based PID controllers. The effort has been taken to implement the digital PID controller using FPGA device based on multiplier principle, which is implemented using MATLAB Simulink and system generator. The controller is designed for speed control of dc motor and implementation is accomplished on Xilinx Spartan 3 FPGA chip to achieve the settling time of 10 s this shows 5 s early settlement compared to basic PID controller. The resources consumption of the scheme is also presented.

Keywords: Proportional Integral Derivative (PID) · Field Programmable Gate Array (FPGA) device · System generator (SysGen)

1 Introduction

An industrial control system is divided in two areas a plant and a unit required to control the behavior of plant i.e. plant is the utility which is controlled by the controller [1]. There are various controllers available to control the plant like proportional, derivative and integral controller separately or in the combination, amongst them Proportional Integral Derivative controller is one of the most utilizable type of feedback controller which is used for controlling the dynamic plants [2]. The digital PID controller can be implemented using microprocessors, microcontrollers or programmable logic controllers (PLC), but the drawback of these systems is the speed of operation, as the more time is required to fetch, decode, execute the code. Wherein FPGA based systems poised to be simpler and they provide high processing speed and high reconfiguration level [2]. A FPGA deals with high clock frequency, accuracy and programmable features, has noteworthy returns compared with various types of controllers. Moreover, an FPGA-based system can process signals from different kinds of sensors and FPGA development boards are often equipped with high number of input pins and they are easily available

© Springer International Publishing AG 2018
S.C. Satapathy and A. Joshi (eds.), *Information and Communication
Technology for Intelligent Systems (ICTIS 2017) - Volume 2*, Smart Innovation,
Systems and Technologies 84, DOI 10.1007/978-3-319-63645-0_9

at cheaper price. In the proposed FPGA-based system, Xilinx® XC3S400 FPGA works as the foundation of the system. The system is designed using Simulink and then hardware in realized in Xilinx by programming with Very-High-Speed Integrated Circuit Hardware Description Language (VHDL). Thus system achieve the better settling time by controlling the speed of dc motor wherein this is achieved by parallel, operations and architectural design of digital controllers [7].

The paper is focused on the design of FPGA based digital PID Controller with multipliers; the computations are performed using multipliers [1].

2   PID Controllers

The PID controller is combination of proportional, integral and derivative control. The design needs basic mathematical operations, based on the increase in complexity and this has been overcome to some extent by using the digital PID controller [2].

The general PID equation is given as:

$$u(t) = K\left(e(t) + \frac{1}{T_i} \int_0^t e(t)dt + T_d \frac{de(t)}{dt} \right) \tag{1}$$

Where,
  $e(t)$ = Error signal
  $u(t)$ = Command signal
  K = Gain or proportional gain
  T_i = Integration time or rise time
  T_d = Derivative time.

System wherein the PID controller is operating as feedback element is shown in Fig. 1, where Proportional, integral and derivative controller works with feedback loop. The PID controller is defined by Eq. 1, in which e is the error signal which is sent to PID controller. The signal passed to system is combination of proportional, integral and derivative terms related to error signal.

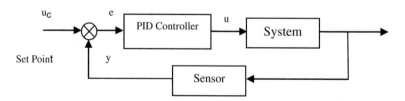

Fig. 1. PID based feedback control system

A proportional term helps to cut down the rise time, but it never get rid of the steady-state error, while an integral term eliminate the steady-state error, but it could show the worst transient response and a derivative term enhances the stability of the system, reduce the overshoot, and improves the transient response [3].

PID controller feedback control system implemented using Simulink is shown in Fig. 2.

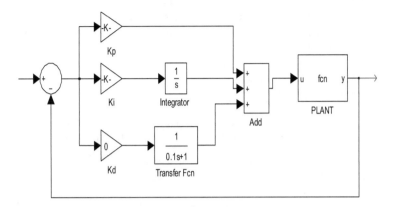

Fig. 2. PID controller with feedback using Simulink

Taking laplace transform of Eq. (1) we get

$$U(s) = K\left(E(s) + \frac{1}{sT_i}E(s) + sT_DE(s)\right)$$ (2)

$$U(s) = KE(s)\left(1 + \frac{1}{sT_i} + sT_D\right)$$ (3)

Where, if $K = K_P$, then $K_I = K_P/T_I$ and $K_D = K_P T_D$. Then the equation can be rewritten as [1]

$$U(s) = K_PE(S) + \frac{K_I}{S}E(S) + K_DSE(S)$$ (4)

To take in the system by FPGA, the system is needed to be converted to discrete form this is achieved by transformation technique, and thus the continuous-time system is converted to discrete-time system [5],

$$D(z) = \frac{U(Z)}{E(Z)} = K_P + K_I\frac{T_s}{1 - Z^{-1}} + K_D\frac{1 - Z^{-1}}{T_s}$$ (5)

Where, T_s is the sampling time.

3   Plant Implementation

DC motors are usually very high-speed, revolving at several thousand revolutions per minute (rpm); they are simple to operate and their initial torque is large, because of which they are preferred for several traction operations for velocity control and in high speed control applications.

The dc motor has special characteristic to operate either on a.c. or d.c. supply based on different applications. The moment of inertia, damping, electromotive force, electric resistance, and electric inductance are the parameters considered for the control of dc motor, works as a plant (Table 1).

Table 1. DC motor parameters

Moment of inertia in the rotor	$J = 0.02 \text{ kg·m}^2$
Damping (friction) of the mechanical system	$b = 0.15 \text{ Nms}$
Electromotive force constant	$K = 0.011 \text{ Nm/A}$
Electric resistance	$R = 1.2 \ \Omega$
Electric inductance	$L = 0.6 \text{ H}$

The input to the plant is armature voltage V in volts. Measured variables are the angular velocity of the shaft ω in radians per second and the shaft angle θ in radians [6]. Transfer function is for the dc motor is represented by Eq. 6.

$$Transfer\ Function = \frac{0.011}{0.012S^3 + 0.114S^2 + 0.1801S} \tag{6}$$

4   Design of Multiplier Based PID Controller

System generator is a tool for designing high performance DSP systems using FPGA. Figure 3 shows PID controller built using system generator toolbox. The PID controller is created according to the Eq. 7 [4] (Fig. 4).

$$U(s) = K_P e + K_I \left(I_{n-1} + edt \right) + K_D \frac{e - e_{n-1}}{dt} \tag{7}$$

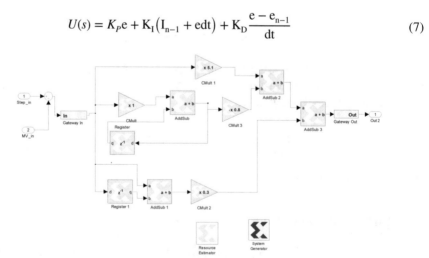

Fig. 3. Multiplier based PID controller

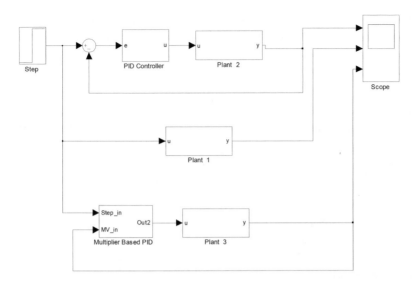

Fig. 4. Simulink implementation of various systems (basic PID controller, system without controller, multiplier based PID controller)

5   Results

The results of the model are presented in Fig. 5, where comparison between basic PID controller, system behaviour without PID Controller and multiplier based PID controller is presented. The graphical representation shows that Settling time of multiplier based PID controller is 5 s before the basic PID controller. Hence this design model will help to control the speed of dc motor efficiently.

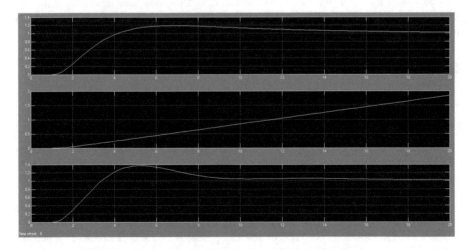

Fig. 5. Output comparison for various PID controllers

The designed Multiplier based PID controller is synthesized using Xilinx ISE 10.1. The target device used to implement the design is Spartan 3 XC3S400 FPGA. The summary of recourses utilized is presented in Table 2.

Design statistics: Minimum period: 5.031 ns (Maximum frequency: 198.768 MHz).

Table 2. Device utilization summary

	Used	Available	Utilization
Logic utilization			
Number of slice flip flops	40	7,168	1%
Number of 4 input LUTs	587	7,168	8%
Logic distribution			
Number of occupied Slices	328	3,584	9%
Number of slices containing only related logic	328	328	100%
Number of slices containing unrelated logic	0	328	0%
Total number of 4 input LUTs	624	7,168	8%
Number used as logic	587		
Number used as a route-thru	37		
Number of bonded IOBs	68	141	48%
Number of BUFGMUXs	1	8	12%
Number of RPM macros	4		

6   Conclusion

The current model achieves improved accuracy, compactness with time constraints by the effective use of FPGA and hardware utilization. The improved performance of multiplier based PID controller gives decrease in settling time as compared to the basic PID controller by 5 s. The future work implies to achieve effective resource utilization, less power consumption and increase in speed of operation with the use of DA based or multiplierless PID controller.

References

1. Gupta, V., Khare, K., Singh, R.P.: Efficient design and FPGA implementation of digital controller using Xilinx SysGen. Int. J. Electron. Eng. **2**, 99–102 (2010)
2. Khan, H., Akram, M.H., Khalid, I., Salahuddin, H., Aziz, A., Qureshi, A.: Comparative analysis and design of diverse realizations for FPGA based digital PID controller using Xilinx SysGen. Int. J. Comput. Electr. Eng. **4**(6), 908–911 (2012)
3. Soliman, W.G., Patel, V.P.: Modified PID implementation on FPGA using distributed arithmetic algorithm. Int. J. Adv. Eng. Res. Dev. **2**(11), 44–52 (2015)
4. Hock, O., Cuntala, J.: PID Regulator Using System Generator. University of Zilina, Faculty of Electrical Engineering, Department of Mechatronics and Electronics
5. Sravanthi, G., Rajkumar, N.: PID controller implementation for speed control applications using FPGA. Int. J. VLSI Syst. Des. Commun. Syst. **02**(06), 0364–0370 (2014)

6. Babuska, R., Stramigioli, S.: Matlab and Simulink for Modeling and Control. Delft University of Technology, Netherlands (1999)
7. Sultan, M., Siddiqui, M., Sajid, A.H., Chougule, D.G.: FPGA Based Efficient Implementation of PID Control Algorithm. In: International Conference on Control, Automation, Communication and Energy Conservation (2009)

E-Negotiation: Emerging Trends in ADR

Dinesh Kumar[(✉)]

Department of Laws, Panjab University, Chandigarh, India
dinesh@pu.ac.in

Abstract. Negotiation is an unconsciously has become a part of our life and we don't even realize in our life that when we start negotiating. It starts from our childhood to adolescents from a chocolate to bike for getting good marks in the examination. But besides this we never realize that sometimes this acumen can help us becoming one of the successful negotiators in our professional life. Earlier in case of disputes the corporate houses prefer to have arbitration clause to avoid court hassle but now the companies are moving step forward to seek out the differences through negotiation that they are not turned into disputes with time. The communication technology has also helped the same by providing an e-platform in the form of electronically mediated negotiation. The paper is an attempt to discuss the relevance and steps of negotiation in commercial disputes and relevance of e-negotiation in the same lines.

Keywords: Alternative dispute resolution · e-Negotiation · Traditional negotiation · Commercial contracts

1   Introduction

Negotiation means getting the best of your opponent.

Marvin Gaye

Alternative dispute resolution has emerged as one of the powerful tool for dispute resolution which doesn't have judicial character. This alternative dispute resolution has become a part of our personal and professional life. If someone is struck up in matrimonial dispute the court will first refers the parties for mediation before even starting the case as far as commercial contracts are concerned the corporate houses have a specific clause in almost all the contracts relating to arbitration in case of dispute and these days even for negotiation to sought out the difference. In some cases, with our consent and in other cases unconsciously we are adopting the alternative dispute resolution techniques as a part of our day to day life. The four common mode of dispute resolution are:

1. Mediation
2. Negotiation
3. Conciliation
4. Arbitration

© Springer International Publishing AG 2018
S.C. Satapathy and A. Joshi (eds.), *Information and Communication Technology for Intelligent Systems (ICTIS 2017) - Volume 2*, Smart Innovation, Systems and Technologies 84, DOI 10.1007/978-3-319-63645-0_10

Indian legal system has recognised Mediation as a method to resolve dispute between the parties. Mediation is a form of a third party intervention with the consent of the parties to a dispute. It is a process which is not as participatory as a negotiation or conciliation process; it is a process in which parties do not directly negotiate, but rely on the finding of the mediator. Mediator acts as a facilitator and helps the parties to resolve the dispute in a peaceful manner. A mediator's function is to provide solution to a dispute in the hope that the parties will accept it. The mediator is a neutral person with knowledge and expertise in the subject-matter of the dispute and helps the parties to reach a best possible solution. The decision of mediator is signed by the parties and enforceable in the court of law. On the other hand the negotiation is initiated by the parties on its own and the negotiator can be the parties itself or the negotiator can help the parties to break deadlock and begin negotiation on any issue of their common interest which can be business deal or personal matter or state issues if it is between two international parties.

The purpose of conciliation is not to change the nature of a dispute, that is, whether it is a legal or paralegal dispute, but to point out the strengths and weaknesses of the application of the conciliation process in settling a dispute or a difference. In a conciliation process the need for accommodation of ideas and concessions must be emphasised for mutual benefits. Quantification of losses both in financial terms and non-financial terms must be identified. A conciliator's function is therefore to clarify to the parties their respective position and to convince them of the legal consequences if a dispute is not settled amicably. In other words, it is the function of a conciliator to alert the parties of the adverse consequences that might take place if a dispute is not settled outside the court [1].

Arbitration is another method to resolve commercial disputes. It is procedure in which a commercial dispute is submitted, by agreement of the parties, to an arbitrator/s to decide. The decision of the arbitrator will be binding on the parties and it can be changed in the court of law only if it contrary to law and on very specific grounds. In arbitration, the party are free to choose jurisdiction i.e. seat of arbitration, rules of law by which parties intend to be governed or decide the dispute. The arbitrator is also bound to give the award in time bound manner. The arbitration is different from negotiation and mediation because in arbitration there is no win-win situation for both the parties and the arbitrator is duty bound to decide on merits. Whereas in negotiation both the parties feel satisfied with the outcome and the final outcome is not subject to any challenge in any court of law.

The same has been recognised by the Arbitration and Conciliation Act, 1996 which had been drafted on the UNCITRAL Principles. The disputes need to be resolved with ease, quickly and low cost. Though arbitration has become a clause in standard form of commercial contracts but negotiation and mediation is still a discretionary option which has been underestimated as dispute resolution mechanism. Only, when the litigation is pending under Section 89 of Civil Procedure Code, 1908 the Court observes that the dispute can be resolved through settlement between the parties the Court can pursue the same.[1] But in this alternative also negotiation is missing.

2 Negotiation as a Method of ADR

Negotiation is the one mode of ADR which is under estimated. In this case the parties to the disputes agree upon courses of action, bargain for collective advantage, or make an effort with an outcome which serves the mutual interests. Negotiation is a part of our life and we experience the same in all roles of life from matrimonial disputes to all aspects of life. The basic difference between negotiation and mediation is that whereas in the former process the parties settle their differences through active participation, the latter is more non-participatory and is reliant on a third party's intervention. Of the two processes of settling disputes, perhaps negotiation is more effective than mediation, in a negotiation process the parties, by consent agree how to settle their disputes, whereas in a mediation process the parties' consent is depends on the role of mediator [1].

The role of a negotiator is to determine the grounds of common interest as well as differences in the subject matter. A trained negotiator regularly works to trash out the alterations amongst the parties by providing most suitable solution for the parties. In this procedure, the negotiator endeavours to bargain the outcome of both the parties by adjusting to the demands. A positive negotiation in the advocacy approach is when the negotiator is able to attain a desired result without affecting the future business and this type of approach is fruitful to both parties. For example the Ambani brothers the property worth rupees 99,000 crore business empire was bifurcated with the help of family friend K.V. Kamath, Chief Executive Officer (CEO) and Managing Director of

---

[1] Section 89 of Civil Procedure Code, 1908 Settlement of disputes outside the Court.

1. Where it appears to the court that there exist elements of a settlement which may be acceptable to the parties, the court shall formulate the terms of settlement and give them to the parties for their observations and after receiving the observations of the parties, the court may reformulate the terms of a possible settlement and refer the same for—
   a. arbitration;
   b. conciliation;
   c. judicial settlement including settlement through Lok Adalat; or
   d. mediation
2. Where a dispute had been referred-
   a. for arbitration or conciliation, the provisions of the Arbitration and Conciliation Act, 1996 shall apply as if the proceedings for arbitration or conciliation were referred for settlement under the provisions of that Act
   b. to Lok Adalat, the court shall refer the same to the Lok Adalat in accordance with the provisions of sub-section (1) of section 20 of the Legal Services Authority Act, 1987 and all other provisions of that Act shall apply in respect of the dispute so referred to the Lok Adalat;
   c. for judicial settlement, the court shall refer the same to a suitable institution or person and such institution or person shall be deemed to be a Lok Adalat and all the provisions of the Legal Services Authority Act, 1987 shall apply as if the dispute were referred to a Lok Adalat under the provisions of that Act;
   d. for mediation, the court shall effect a compromise between the parties and shall follow such procedure as may be prescribed.

ICICI. This is classic and most successful example of property settlement without moving to the Court of Law. This type of settlement generally aims to results into win-win situation for both parties. Negotiation is emerging as one of the powerful tools even the law firms are recommending the same as a part of commercial contracts. The sample of the same is mentioned below for discussion purposes [2]:

> *All disputes arising out of or in connection with this Agreement shall to the extent possible be settled amicably by negotiation between the parties within [30] days from the date of written notice by either party of the existence of such a dispute. If the parties do not reach settlement within period of [30] days, they will attempt to settle it by mediation in accordance with the Centre for Effective Dispute Resolution (CEDR) Model Mediation Procedure. Unless otherwise agreed between the parties, the mediator will be nominated by CEDR. To initiate the mediation a party must give notice in writing (ADR Notice) to the other [party OR parties] to the dispute requesting a mediation. A copy of the ADR Notice should be sent to CEDR. The mediation will start no later than [x] days after the date of the ADR Notice.*

The clause definitely lays down the negotiation as the first option whenever the dispute arises. The negotiation doesn't have any set rules or steps it depends from person administrative skills to manage the things in pressure but generally there are certain steps which are taken into consideration by everyone who is negotiating.

1. The first golden trio of negotiation is PPT i.e. Person, Place and Time. Before starting the negotiation the negotiator is bound to verify that the person which whom he want to start negotiation is authorised to negotiate on behalf of other party. Place refers to where the negotiator starts the negotiation the issues is negotiable or not. Negotiator has to invest time if he wants to come out with fruit bearing results for both the parties. When such kind of process starts you can't specify how much sitting are required. It is also called the opening statement for the purposes of negotiation.
2. Before initiating the negotiation the parties should know the background of each other, areas of common interest, areas of differences and the impact of their differences in long run. If the parties are unable to sense the impact in long run the negotiator should be a person who could sensitize the parties on these issues.
3. The second step is the most sensitive one i.e. seeking clarification or information regarding the issues on which the parties has locked the horns and all the issues affecting the information. The role of negotiator is very crucial in this regard as it is the ice breaking session between the parties. In this parties are also made aware of each other interest, limitation and the justification for the bids negotiated between the parties. This step is also very significant to identify the difference between the parties which helps the negotiator to bargain more effectively.
4. The third step is *Moving Forward* which means the talks initiated between the parties don't lead to stagnation. The negotiator should have the capability to keep the bargaining moving.
5. Final step in this regard is fixing the deal means the terms or condition bargained and the parties have agreed to settle the dispute.

3 Emerging Trends of E-Negotiation

The Digital India project is the dream project and the first priority of Central government to transform India to digital and knowledge based society. The objective of the project is to provide solution in every sphere of life through E-means i.e. E-solutions through E-Governance. The object of Indian government is to ensure all work is done electronically and become a complete paperless nation by 2019. Though this dream seem to be distant but still as per the Goldman Sachs [3] study predicts India as the third world economy by 2050 after United States and China. The strong consumer base of 175 million in India will result into $60 billion in gross merchandise value (GMV) by 2020 [4]. Another study of Goldman Sachs proposed that India's overall e-commerce market was expected to crack the $100-billion mark by 2020. It had said the overall online market in the country including travel, payments and retail could reach $103 billion, of which the e-tail segment would be valued at $69 billion [5]. This economy boom has laid down another aim to Indian Government i.e. resolving of disputes. With the digitalization of market and increase in e-commerce the necessity of online dispute resolution has also turned to be in demand. There are many economies and commercial contracts which are moving towards e-negotiation especially in case of online shopping. It is a more effective measure than the traditional negotiation mechanism because the pricing mechanism in e-negotiation is automated and intelligent. In this world of cyber negotiation the e-mail negotiation is nascent and cogent solution for the parties at distance. The e-negotiation is different from traditional negotiation on the various aspects which are as follows:

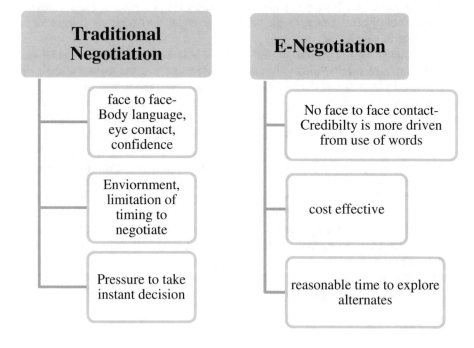

Though e-negotiation is not a settled concept but in its nascent stage still this method has some both positive and negative aspects. E-negotiation has certain clear advantages over the traditional negotiation. E-negotiation addresses two steps of negotiation i.e. time and distance in a positive manner. This type of negotiation does not have to wait till all the parties come together on a table. The negotiation can be initiated by just writing an e-mail to the authorised person on behalf of the other party. In this negotiation, the parties can negotiate even they are located into opposite directions or different countries. It is both cost effective and time effective as the time taken in travelling and in conducting the negotiation is reduce drastically. In negotiation the most important factor is communication but in e-negotiation the writing and use of words play crucial role. As body language and conduct of the parties are missing to support the negotiation. The language of communication should be simple and in synchronized manner. This type of negotiation can only be initiated if it had been properly planned. Before starting this negotiation the negotiator is aware of background and difference between the parties and also has a tentative aim to achieve through bargaining. If the bargaining aims are not clear, this type of negotiation may result in to a monotonous and useless exercise that will frustrate the entire settlement process. Without setting an aim in e-negotiation will be like firing without aim. This type of negotiation can also be done by making the parties the part of this process. It affords an opportunity to the parties to observe the negotiation and help them to obtain a target beneficial to both parties.

There have been several studies [6] conducted on traditional negotiation vis-a-vis e-negotiation to check the effectiveness of one over the other. It has been observed that electronic mediated negotiation does not yield any significant difference outcome of the negotiation. E-negotiation has been proved beneficial with respect to price and time duration.

4   Conclusion

The Arbitration and Conciliation Act, 1996 recognize arbitration and conciliation as methods of ADR. The mediation is also recognised tool of ADR and supported by the rules made by different High Courts and Supreme Court from time to time to provide a binding force behind them. Though there has been specific rules drafted for the purposes of enforcement of conciliation and mediation agreement but however, negotiation or e-negotiation has no such legal sanction or enforceability. Though with the changing time and cloud computing we are developing the techniques on automated negotiation systems and there is a continuous research is to find a dependable e-negotiation system.

On the other hand the corporate houses are working on their drafting skills to make a dispute resolving with multi-party settlements, saving the time and handling the difference before they are converted into disputes. Though negotiation and e-negotiation is an underrated instrument of ADR but it can be proved very effective in many aspects like jurisdiction, time, distance, cost, procedural technicalities of law and lifelong commercial relationship of the parties are not spoiled. But there is a need to make negotiation an effective tool to resolve the dispute at par with other methods of ADR.

References

1. Chatterjee, C., Lefcovitch, A.: Alternative Dispute Resolution: A Practical Guide, p. 19. Routledge, London (2008)
2. Ford, J.: Managing Commercial Disputes. It has been written in a blog of Cripps LLP which is a law firm having registered office in Kent and London listed in 2016 legal 500 firms and has been shortlisted in four categories in the South East Deal Makers Awards (2017). http://www.cripps.co.uk/managing-commercial-disputes/
3. http://www.goldmansachs.com/gsam/docs/instgeneral/general_materials/whitepaper/india_revisited.pdf
4. E-commerce to reach US$ 60-bn GMV by 2020. http://www.indianembassy.nl/docs/1465626631IEN%20June%202016.pdf
5. Business Standard, 01 June 2016
6. Galin, A., Gross, M., Gavriel, G.: E-negotiation versus face-to-face negotiation what has changed—if anything? Comput. Hum. Behav. **23**, 787–797 (2007)

Enhancement of Security in AODV Routing Protocol Using Node Trust Path Trust Secure AODV (NTPTSAODV) for Mobile Adhoc Network (MANET)

Suchita Patel[1($\boxtimes$)], Priti Srinivas Sajja[1], and Samrat Khanna[2]

[1] Department of Computer Science, ISTAR-(SICART), Sardar Patel University,
Vallabh Vidyanagar, Gujarat, India
suchita.mca@gmail.com, priti@pritisajja.info
[2] SEAS, Rai University, Ahmedabad, Gujarat, India
sonukhanna@yahoo.com

Abstract. Mobile ad hoc network (MANET) is a collection of wireless mobile devices that can communicate with each other and forming a temporary network without requiring any fixed infrastructure or centralized administration. The nodes of MANETs are always susceptible to compromise. In such scenario, designing a secure routing protocol has been a major challenge for last many years. In this paper, we propose Node Trust Path Trust Secure AODV routing protocol (NTPTSAODV) to enhance the security for MANETs through a new approach 2-tier trust based model for Intrusion Detections (IDs) techniques. To implement NTPTSAODV, Ad hoc On-demand Distance Vector (AODV) routing protocol has been modified for making it secure and thwart black hole attacks. In order to make result more accurate, performance of NTPTSAODV, AODV and black hole detection AODV (BDAODV) was tested in presence of multiple attacker cases and after observations of performance analysis, it can be concluded that NTPTSAODV is capable of delivering packets even in the presence of malicious nodes in the network. To evaluate the network performance, packet delivery ratio (PDR), Average End-to-End Delay (AED), Average Throughput (AT) and routing overhead are considered criteria.

Keywords: AODV · IDs · BDAODV · NTPTSAODV · PDR · AED · AT

1 Introduction

Mobile ad hoc network (MANET) is a group of mobile devices. There is no centralized infrastructure or access point such as a base station for communication. As in [1, 2] each mobile node acts as a host when requesting/responding information from/to other nodes in the network, and acts as router when discovering and maintaining routes for other nodes in the network. These nodes generally have a limited transmission range. Topology in MANET is very dynamic and ever-changing where nodes are free to join or leave arbitrarily. Based on the routing information update mechanism, the MANET routing protocols can be categorized as follows.

© Springer International Publishing AG 2018
S.C. Satapathy and A. Joshi (eds.), *Information and Communication
Technology for Intelligent Systems (ICTIS 2017) - Volume 2*, Smart Innovation,
Systems and Technologies 84, DOI 10.1007/978-3-319-63645-0_11

Proactive/Table-Driven Routing Protocol– (DSDV) Destination Sequenced Distance Vector Routing Protocol, (WRP) Wireless Routing Protocol, and (GSR) Global State Routing.

Reactive Routing Protocol– (AODV) Ad-Hoc On-Demand Distance Vector Routing, (CBRP) Cluster based Routing Protocols), (DSRP) Dynamic Source Routing Protocol, (TORA) Temporally Ordered Routing Algorithm.

Hybrid Routing Protocol– (ZRP) Zone routing protocol.

1.1 Working of AODV

AODV routing protocol works as a reactive so never stores paths to nodes information that are not collaborating. AODV [11, 13, 14] protocol mainly works for path generating and data transmission from source node to destination node via mediator nodes. Path finding messages are called routing messages includes Route Request (RREQ), Route Reply (RREP) and Route Error (RERR) which are broadcasted by source node in the network. These all messages also consider as a packets that are used to discover route from source to destination node. Whole process starts from source node which sends RREQ packets to neighboring node, it first checks its own routing table to discover if available route for the destination node, if route is available then it considers that route as a source to destination path and starts data transmission, else neighboring node again sends RREQ packets to neighboring node and this process continues until destination not found. If destination node found, then generates unicast RREP packets send to the source node from destination node via followed path stored in routing table information. Whenever RREQ packet received by any node it saves the path which has transmitted from source node in its routing table. Some times when any link broken RERR packets form of messages are propagated to the source node. Each node

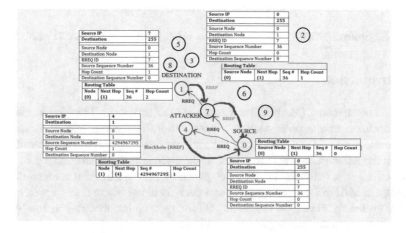

Fig. 1. Working of AODV routing protocol

maintains and update its routing table that contains data fields while receiving a routing messages. Whenever RERR packets propagate from reverse route and all intermediate nodes removes its entry in their routing table. AODV always periodically checks and maintain link between neighbor nodes by sending hello packets [1]. Figure 1 represents working of AODV routing process.

1.2   Security Issues with AODV-Black Hole Attack

MANET routing is very challenging task because of dramatic changes in mobility of network nodes resulting changes in topology and absence of centralized network controller. Mobile ad hoc network always suffers from many internal or external attackers. As name implies Black hole attacker node identifies own identity as having a shortest and fresh path to the destination node with highest sequence number [6–8, 15]. So source node misleads and assumes that node is having the best shortest route in the direction of destination and sends the data packets to it. This type of attacker node is known as Black Hole attacker node as it does not forward the data packets to its neighboring nodes, while on receiving data packets it simply drops the all.

2   Related Work

In [2], Balaji proposed a trust based approach by AODV protocol but they don't consider data packets as parameters to calculate trust. They only worked with control packets and network acknowledgement. Solution given by them fails because black hole attack can even drop data packets by perfectly transmitting control packets.

In [3], Chaubey et al. have proposed Trust Based Secure On Demand Routing Protocol(TSDRP) and studied the impact of Black hole attack with that of AODV routing protocol for making it secure. TSDRP protocol is capable of delivering packets to the destinations node even in the presence of malicious node while increasing network size.

In [4] Ph.D. Thesis – A trusted AODV routing protocol for mobile Adhoc networks covers TAODV protocol based on trust calculates on basis of TREQ and TREP. They also add three more fields to routing table for calculate trust level of nodes. TAODV uses digital signature which is an additional overhead.

In [5] Payal et al. given solution a DPRAODV which finds a threshold value and compares that with DSN of RREP packet and entry which is stored in routing table. If value is higher than a threshold, then reply given node consider as a black hole node and entry stored in black list. Negative point of this protocol has higher routing overhead due to addition of ALARAM packets.

3   Two-Tier Trust Model

From the study of existing work discussed in Sect. 2, helped us in identifying the following problems: Many solutions can detect single attack but can't work for cooperative black hole attacks; Some detection techniques only detect attacks but never prevent network from it if it is than with false alarm; Many techniques increase delay and overhead which degrades network performance; Mobility also important concern for MANET also some uses digital signature concepts which was again additional overhead. Some trust based solution detect and prevent network from black hole attacks but our model providing qualitative work with light weight statistical formulas which only considers most trustable nodes and most trustable path for secure data transmission.

Proposed model aims to provide the following purposes:

(i)   To propose an innovative approach for securing and improving the AODV Routing Protocol for MANET security.

(ii) To acquire more information about intrusion (black hole attack) detection techniques under AODV routing protocol.

(iii) To propose an effective solution "NTPTSAODV: 2 -Tier Trust based model" under modification of BDAODV.

(iv) To make comparative qualitative performance analysis for normal AODV, BDAODV (Black hole Detection AODV) and NTPTSAODV under Black Hole attacker cases.

3.1   Solution Proposed

This section describes details of our proposed Node Trust Path Trust Secure AODV routing protocol (NTPTSAODV) including its Structure, Architecture 2-Tier Trust based model, Node Trust Computation Parameters and Computation on Node Trust (NT). Node trust and path trust plays a very important role for MANET routing under AODV protocol. The main focus of our propose approach is to find trust factors on identifying the misbehaving nodes which are not suitable for reliable routing and also finding trustworthy path to carry data messages successfully from the source to destination. 2-Tier Trust based model of our proposed NTPTSAODV routing protocol is to provide more trusted and stable route with avoidance of route break for MANETs. Our proposed model implemented by using a very popular networking research tool NS 2.35 [16]. The Structure of 2-Tier Trust Model of our proposed protocol for all nodes trust and path trust shown in Fig. 2.

AODV routing protocols uses two types of packets Control Packets and Data Packets to establish the path from source to destination and send data packets. The proposed model name implies 2-Tier trust model; we compute path trust followed by node trust. The new approach of 2-Tier trust model which performs all derivations and

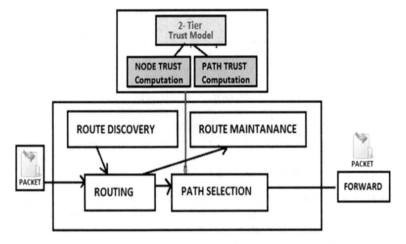

Fig. 2. The structure of 2-tier trust model

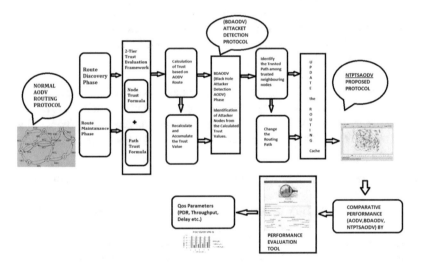

Fig. 3. Proposed architecture - NTPTSAODV: 2-tier trust based model under AODV for MANET security

calculations are shown in the Fig. 3. Each node will maintain one hop neighbor trust value to which packets have been sent for forwarding [9, 10, 15]. Several researchers work with control packets to calculate trust value, in our propose technique, control packets and data packets are considered. Table 1 shows all parameters taken into consideration to compute Node trust value.

Table 1. Node trust computation parameters

Count type	Received count	Forward count
Control packet		
RREQ	RREQ	RREQ_T
RREP	RREP	RREP_T
RERR	RERR	RERR_T
Data packet	DATA	DATA_T
Node stability	Velocity	Pause time

3.2   Computation on Node Trust (NT)

Node Trust computation are done mainly on three factors: (i) Trust Calculation from Control Packets (T(cp)) (ii) Trust Calculation from Data Packets (T(dp)) and (iii) Trust Calculation from Node Stability (NS)

Each node has to maintain count for control packets as well as the data packets to get final value of Trust, which is stored in each node routing cache. Formula for calculations of each node control packets and data packet are given in following equations. **Control Packet Count maintenance:** Each node maintains the received count of RREQ, RREP, RRERR and forward count of RREQ, RREP, RRERR. **Data Packet Count Maintenance:** Each node maintains the received count of DATA and forward count of DATA.

3.2.1   Final Value and Weight Factor Calculation of Control Packets

Here, F1, F2, F3, F4 are final values that are used to calculate nodes route request rate, route reply rate, route error rate and data packet transmission rate. The derived trust represented by T(cp) and T(dp). Also W1, W2, W3, W4 are weight factor of control as well as data packets

- **RREQ Packet -** The formula is final value of RREQ packet by each node

$$F1 = [1/Min(RREQ, RREQ\_T)] * [(RREQ - RREQ\_T)2/(RREQ\_T)] \qquad (1)$$

RREQ(RREQ_Fail) is difference between the Number of Route Request packets received by node and Number of Route Request packets forwarded by the node. RREQ_T(RREQ_count) is total Route Request packets received by node. F1 is Final value to calculate node Route Request rate. The formula is weight factor of RREQ packet by each node. W1 = Weight factor of RREQ packets

$$W1 = [(RREQ + RREQ\_T)/Min(RREQ, RREQ\_T)] \qquad (2)$$

- **RREP Packet** - The formula is final value of RREP packet by each node

$$F2 = [1/\text{Min}(RREP, RREP\_T)] * [(RREP - RREP\_T)2/(RREP\_T)] \quad (3)$$

RREP(RREP_Fail) is difference between the Number of Route Reply packets received by node and Number of Route Reply packets forwarded by the node. RREP_T (RREP_count) is total Route Reply packets received by node F2 is Final value to calculate node Route Reply rate. The formula is weight factor of RREQ packet by each node. W2 = Weight factor of RREP packets

$$W2 = [(RREP + RREP\_T)/\text{Min}(RREP, RREP\_T)] \quad (4)$$

- **RRERR Packet** - The formula is final value of RRERR packet by each node

$$F3 = [1/\text{Min}(RRERR, RRERR\_T)] * [(RRERR - RRERR\_T)2/(RRERR\_T)] \quad (5)$$

RRERR(RRERR_Fail) is difference between the Number of Route Error packets received by node and Number of Route Error packets forwarded by the node. RRERR_T(RRERR_count) is total Route Error packets received by node. F3 is Final value to calculate node Route Error rate. The formula is weight factor of RRERR packet by each node. W3 = Weight factor of RRERR packets

$$W3 = [(RRERR + RRERR\_T)/\text{Min}(RRERR, RRERR\_T)] \quad (6)$$

3.2.2 Final Value and Weight Factor Calculation of Data Packets

- **DATA packets** - The formula is final value of DATA packet by each node

$$F4 = [1/\text{Min}(DATA, DATA\_T)] * [(DATA - DATA\_T)2/(DATA\_T)] \quad (7)$$

DATA(DATA_Fail) is difference between the Number of Data packets received by node and Number of Data packets forwarded by the node. DATA_T (DATA_count) is total Data packets received by node. F4 is Final value to calculate node Data rate. The formula is weight factor of DATA packet by each node. W4 = Weight factor of DATA packets

$$W4 = [(DATA + DATA\_T)/\text{Min}(DATA, DATA\_T)] \quad (8)$$

- **NODE TRUST VALUE**

T(cp) - Trust calculation from Control packets of each node

$$T(cp) = W1 * F1 + W2 * F2 + W3 * F3 \tag{9}$$

T(dp) - Trust calculation from Data packets of each node

$$T(dp) = W4 * F4 \tag{10}$$

3.2.3   Final Value and Weight Factor Calculation of Node Stability

This factor represents the stability value of nodes. We can consider that if higher the value of this factor manner more trusty node. Link failure always due to major reason of high mobility of network nodes. That reason in proposed model we consider network stability factor under calculating trust value and achieve our goal. Node Stability is calculated as following:

$$NS = Tpause/Vnode \tag{11}$$

Where Tpause(Time_pause) is Pause Time; Vnode(speed) is Relative Velocity of node and NS is Node Stability of node - (0 = Vnode = Vmax).

After getting all three factor computations (Control packets trust value, Data Packets trust value and Node Stability trust value) our final node trust value as follow shown in equation

- **FINAL NODE TRUST VALUE**

$$Node\ Trust\ Value(Node\_Trust) = T(cp) * T(dp) * NS \tag{12}$$

However, each node in network calculate value of trust and updates value in routing cache. We may say that node's highest value of trust considers more reliable node for packet transmission.

4   Algorithm for Black Hole Detection and Alternate Path Selection During Route Setup

The source node Current_Node (7) starts a route discovery process by broadcasting a Route Request (RREQ) packet to all its neighboring nodes.

RREQ packet contains the Node_Trust of Current_Node (7).

If (Node_TrustCurrent_Node (7) > ß) then
Broadcast RREQ packet to neighbors with Node_TrustCurrent_Node (7).
Else
Not able to generate RREQ packet.

Neighboring node n receives RREQ packet. n may be an intermediate node or a

destination node.

If (Node_Trust[n] > ß) then //This is a new RREQ packet
{
        If (n is an intermediate node) then
                n calculates the Node_Trust form trust formula discuss above by three parameters Trust_Value(Current_Node (7), n).

                Update Node_Trust   field in routing table, update Node_Trust field in RREQ packet and forward it to allneighboring nodes of n.

                If (n receives other RREQ packets for same sequence number from some other sending Current_Node (7)) then
                        Discard these RREQ packets.

        Else
        //n is a destination node.

                If (n receives multiple RREQ packets within a time window, which starts from the first arrival of RREQ) then

                n picks the largest Node_Trust of RREQs. //Destination will select highest Node_Trust from RREQ packets from   among all received RREQ packets.
}
If (Node_Trust[n] > ω) then
{
  n sends back reply packet as a RREP to the source node Current_Node (7).
Else
n discards the RREP packet.
    n node considers as a black hole attacker node. Alternate path selection done.
Else
        n discards the RREQ packet
}

4.1 Methodology of Evaluation

Based on algorithm steps we consider data transmission between source node 7 and destination node 58 as well as detected black hole attacker node 47 due to less trust value compare to other route nodes. Data transmission between source and destination starts via the path 7-3-48-38-47-58. Black hole attacker node is node 47 indicated by red color which drops the udp packet. Each node maintains the received count of RREQ, RREP, RRERR and forward count of RREQ, RREP, RRERR, DATA, and Node stability parameters.

The trust is computed for control packets, data packets and node stability factors as described above. In this model, When the malicious next hop is found the alternate path is selected and to switch over for data transmission. The alternate path has highest trust value nodes when compared to attacker path. Table 2 represents Node_Trust value and how alternate path selection possible in place of attacker path (7-3-48-38-47-58).

Table 2. Black hole node detection and alternate path selection values.

Total Number of Nodes in Network: 70			
Source Node (7) ; Destination Node (58)			
Route1: 7-3-48-38-47-58		Alternate Route: 7-3-48-38-21-58	
Output.tr file shows all route node trust_value		Output.tr file shows all route node trust_value	
Node	**Node_Trust**	**Node**	**Node_Trust**
7	1951.674984	7	1951.674984
3	473.474331	3	473.474331
48	8423.716327	48	8423.716327
38	7191.826081	38	7191.826081
47(Blackhole_node)	31.886506	21(alternate_nexthop)	419.555648
58	Destination Node	58	Destination Node
Black Hole Attacker Node: 47			

This confirms that all the nodes in the active route with high trust nodes are involved in data transmission. Attacker nodes always drops the data packets.

4.1.1   Black Hole Attack Nodes Detection

If Node trust value of next hop is less than threshold(ß) means that it is malicious node otherwise it is normal node. Here at time 4.92444 s Node_Trust: Current_Node 47 is only 31.886506 which is very less value and compare to threshold also less so we are considered node 47 as a attacker node.

4.1.2   Alternate Path Data Transmission

(7-3-48-38-21-58) When the attacker next hop is found the alternate path is selected and to switch over for data transmission. The alternate path has highest trust value nodes when compared to attacker path.

5   Simulation Results

In our simulation, we compare the performances of NTPTSAODV with that of the existing AODV and BDAODV using Network Simulator – 2.35 (NS-2). Graphs are used to compare the results of the normal AODV, existing solution for detection black hole attack BDAODV and proposed NTPTSAODV protocol and comparative result clearly shows the improvement of the proposed model. Simulation results were obtained and compared via performance evaluation tool which is shown in Fig. 4 developed via pure python code.

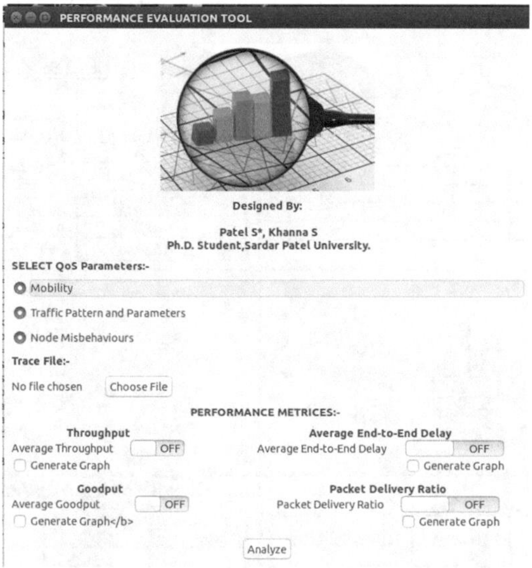

Fig. 4. Performance evaluation tool for result analysis

The proposed NTPTSAODV protocol has shown good performance over the QoS parameters like Packet delivery, Throughput & End-to-End Delay. Graphs are used to compare the results of the normal AODV, existing solution for detection black hole attack BDAODV and proposed NTPTSAODV protocol and comparative result clearly shows the improvement of our proposed protocol. Simulation results were obtained and compared via performance evaluation tool which is shown in Fig. 4 developed in python code. Three QoS Parameters of the network: Impact of speed; Impact of traffic load connections in the network and Impact of attacker nodes are considered. The proposed protocol has shown improved QoS parameters values where trust values are used to identify the attacker nodes in the route and immediately take action for an alternate path to successfully complete the routing. This approach of the proposed NTPTSAODV protocol has resulted in an increased packet delivery ratio and a decreased End-to-End Delay involved in routing. In terms of packet delivery ratio and throughput, the protocol NTPTSAODV provides the highest performance when compared to existing protocol BDAODV and AODV. Table 3 shows comparative result analysis of AODV, BDAODV and proposed NTPTSAODV. The Routing Overloads, End-to-End Delay and Packed Dropped Ratio for the NTPTSAODV protocol are reduced and increased packet delivery ratio. Data shown in Table 3 (3.1, 3.2, 3.3, 3.4, 3.5) as below.

5.1 Comparative Study by Graph: AODV, BDAODV and NTPTSAODV

Table 3 having table and figures represent the impact under Blackhole attack on Packet Delivery, Packet Drop, routing load, Throughput and End to End Delay of our proposed Node Trust Path Trust Secure AODV routing protocol (NTPTSAODV), black hole

Table 3. Comparative results analysis of AODV, BDAODV and NTPTSAODV

Impact of Speed

Table 3.1 Routng Overhead

No. of Connections	Routing_Overhead_AODV	Routing_Overhead_BDAODV	Routing_Overhead_NTPTSAODV
10	1226	920	907
15	1155	919	908
20	1159	921	907
25	1221	917	906
30	1143	907	893

Table 3.2 End-to-End Delay

No. of Connections	Average End-to-End_Delay_AODV	Average End-to-End_Delay_BDAODV	Average End-to-End_Delay_NTPTSAODV
10	0.096383	0.0519326	0.0333099
15	0.0574298	0.0598269	0.0327957
20	0.0279111	0.0488012	0.0264372
25	0.0184231	0.0244021	0.0144261
30	0.0139222	0.0222034	0.0111011

Routing Overhead

Average End-to-End Delay

Impact of Attackers

Table 3.3 Packet Delivery Ratio (%)

No. of Malicious Nodes	AODV_PDR	BDAODV_PDR	NTPTSAODV_PDR
1	2.81	60.56	92.95
2	19.64	75	91.07
3	27.63	79.6	88.15
4	34.53	81.43	89.67
5	44.45	86.78	91.34

Table 3.4 Packet Dropped Ratio

No. of Malicious Nodes	Packet_Dropped_Ratio_AODV	Packet_Dropped_Ratio_BDAODV	Packet_Dropped_Ratio_NTPTSAODV
1	97.18	39.43	7.04
2	80.35	25	8.92
3	72.36	20.39	11.84
4	67.32	18.45	9.34
5	62.23	9.02	4.22

PACKET DELIVERY RATIO (%)

PACKET DROPPED RATIO (%)

Impact of Connections

Table 3.5 Throughput

Times	Throughput_AODV	Throughput_BDAODV	Throughput_NTPTSAODV
5 Sec	0.096383	0.0319326	0.0333099
10 Sec	0.0574298	0.0398269	0.0327957
15 Sec	0.0279111	0.0288012	0.0264372
20 Sec	0.0472268	0.0367269	0.0372587
25 Sec	0.0293311	0.0289112	0.0264372

Throughput

detection AODV (BDAODV) and AODV routing protocol. Packet deliver ratio of our proposed NTPTSAODV is consistently maintained between 88% and 93% whereas that of AODV falls down to 2% and BDAODV 60% while the number of malicious nodes increase 1–5. Average Throughput of our proposed NTPTSAODV is always good and consistently maintained than that of AODV and BDAODV while increasing node speed 5–25 s in the network. Routing overhead of our proposed NTPTSAODV is always small and that of AODV is fluctuating 1143–1226 and BDAODV 907–921 while increasing traffic load, connections in the network.

6 Conclusion

In this paper, a new NTPTSAODV: 2-tier trust based model has been proposed and implemented against similar existing Intrusion detection technique approach. Our proposed technique not only detects intrusion (black hole attacker nodes), but also responds in finding trusted routes based on 2-tier statistical formula with different factors for MANET security. If node is attacker, then it finds to an alternate path selection based on highest trust value for further reliable routing. Limitation of this work is threshold value taken for comparing node trust. In future work, we will extend our proposed technique for wireless Mesh Networks and Sensor Networks.

References

1. Perkins, C.E., Das, S.R., Royer, E.: Ad-I-Ioe on Demand Distance Vector (AODV). RFC 3561
2. Marthi, S., Giuli, T.J., Lai, K., Baker, M.: Mitigating routing misbehavior in mobile adhoc network. In: Proceeding of 6th Annual International Conference on Mobile Computing and Networking (MOBICOM), vol. 6, no. 11, pp. 255–265 (2000)
3. A Trusted AODV Routing Protocol for Mobile Ad Hoc Networks. Ph.D. thesis, Department of Computer Science and Engineering, The Chinese University of Hong Kong (2003)
4. Pirzada, A.A., McDonald, C.: Establishing trust in pure ad-hoc networks. In: Proceedings 27th Australasian Computer Science Conference, ACSC 2004, Dunedin, New Zealand, vol. 26, no. 1, pp. 47–54 (2004)
5. Aggarwal, A., Gandhi, S., Chaubey, N.: A study of secure routing protocol in mobile ad hoc networks. In: Proceedings of National Conferences on Advancement in Wireless Technology and Applications, 18–19 December 2008, SVNIT, Surat, India
6. Raj, P.N., Swadas, P.B.: DPRAODV: a dynamic learning system against black hole attack in AODV based MANET. IJCSI Int. J. Comput. Sci. Issues **2**, 54–59 (2009)
7. Su, M.-Y., Chiang, K.-L., Liao, W.-C.: Mitigation of black-hole nodes in mobile ad hoc networks. In: 2010 International Symposium on Parallel and Distributed Processing with Applications (ISPA), pp. 162, 167, 6–9 September 2010
8. Alem, Y.F., Xuan, Z.C.: Preventing black hole attack in mobile adhoc networks using anomally detection. In: 2nd International Conference on Future Computer and Communication (2010)
9. Balaji, N., Shanmugam, A.: A trust based model to mitigate black hole attacks in DSR based MANET. Eur. J. Sci. Res. **50**(1), 6–15 (2011)
10. Chaubey, N., Aggarwal, A., Gandhi, S., Jani, K.A.: Performance analysis of TSDRP and AODV routing protocol under black hole attacks in MANETs by varying network size. In: 2015 Fifth International Conference on Advanced Computing & Communication Technologies, pp. 320–324, 21–22 February 2015
11. Tan, S., Kim, K.: Secure route discovery for preventing black hole attacks on AODV-based MANETs. In: 2013 IEEE International Conference on High Performance Computing and Communications & 2013 IEEE International Conference on Embedded and Ubiquitous Computing
12. Aggarwal, A., Gandhi, S., Chaubey, N., Jani, K.A.: Trust based Secure on Demand Routing Protocol (TSDRP) for MANETs. In: 2014 Fourth International Conference on Advanced Computing & Communication Technologies

13. Khin, E.E., Phyu, T.: Impact of black hole attack on AODV routing protocol. Int. J Inf. Technol. Model. Comput. IJITMC **2**(2), May 2014Marc Greis' Tutorial for the UCB/LBNL/VINT Network Simulator "ns". http://www.isi.edu/nsnam/ns/tutorial/
14. Gupta, A., Rana, K.: Assessment of various attacks on AODV in malicious environment. In: 2015 1st International Conference on Next Generation Computing Technologies, NGCT-2015, Dehradun, India, 4–5 September 2015
15. Ranjan, R., Singh, N.K., Singh, A.: Security issues of black hole attacks in MANET. In: International Conference on Computing, Communication and Automation, ICCCA 2015
16. Network Simulator - 2 (NS-2). http://mohit.ueuo.com/NS-2.html

Natural Language Interface
for Multilingual Database

Sharada Valiveti[✉], Khushali Tripathi, and Gaurang Raval

Institute of Technology, Nirma University, Ahmedabad 382481, Gujarat, India
{sharada.valiveti,15mcei29,gaurang.raval}@nirmauni.ac.in

Abstract. India, a country where unity in diversity is practiced through
several cultural, social, linguistic and religious adaptations in every facet
of life. Here, communication between different states requires common
platform or language of interpretation. Language happens to be a barrier
in handling many societal issues including security of a state. The inter-
state border activities reduce the efficiency of security deployments as
the common goal of handling the miscreant is not available after he
crosses the state border. So there is a need of natural language interface
which can support different Indian languages. Any application catering
to inter-state domain has a backend to process the information. Such
databases use regional language to store and retrieve information. These
regional languages are not user friendly for non-native users. Our goal is
to design Natural Language Database Interface for the conversion of one
of the Indian Languages i.e. Gujarati to English.

Keywords: Natural language interface · Database · Natural language
translation · Natural language interface for database

1 Introduction

Natural Language Processing not only concerns with translation of natural language to computer understandable language, but NLP can also be used in automated speech recognition, semantic search engines, language translation and
information retrieval systems. Our main goal is to concentrate on language translation applications.

India has many states. Each state has several linguistic and slang differences
within the same regional language. Communication between different states
demands a common language of interpretation for mutual understanding. Language happens to be the barrier for many societal issues including security of a
state. The inter-state border activities reduce the efficiency of security deployments as the common goal of handling the miscreant is time consuming if he
crosses the state border. The goal of the study is to design a language interface
capable of supporting various Indian languages.

Modern applications use databases for storing and processing data. Without
database, management of the data is quite complex. Search results may not be

© Springer International Publishing AG 2018
S.C. Satapathy and A. Joshi (eds.), *Information and Communication
Technology for Intelligent Systems (ICTIS 2017) - Volume 2*, Smart Innovation,
Systems and Technologies 84, DOI 10.1007/978-3-319-63645-0_12

accurate too. Database reduces search complexity, but increases user complexity. This increment is because of the compulsion of formal language. Thus, for such databases, knowledge of formal languages is mandatory. Input in the form of formal language query is less preferred then the language query interface as the formal query is dependent on training parameters and the knowledge bases. As a solution to this problem, we have designed database interface for natural language. In this database, user is allowed to insert query into natural language. This natural language query is then processed and translated into SQL language. Query result will be translated into natural language and presented to the user.

In this paper, we have surveyed research papers on processing natural language database. Section 2 is the literature review. In this section different papers are discussed in detail. Section 3 discusses the tools and technologies used by the said system. Useful tools and technologies for creating database for natural language is discussed. Section 4 concludes the paper.

2 Literature Review

Researchers across the globe have contributed to several applications that deal with the language conversion. Many papers on the natural language database interface and natural language translation are available. As per the need of the domain, it is important to understand the natural language to database query translation. So, study is navigated in the direction of identifying applications that deal with the database interface for language conversion. This section includes two different subsections. In the first subsection, we elaborate upon the study made in the domain of database interface. And second subsection contains study of language translation.

In 2005, Li et al. [1] have implemented NaLIX. NaLIX is a natural language query interface to an XML database. It is a stand-alone Java application which supports schema-free XQuery and term expansion. In this system, Timber [2] is used as XML database and MINIPAR [3] as a natural language parser. It allows an English sentence as a query. This query can include aggregation, nesting, value joins and such other grammatical requirements. The system provides meaningful feedback and helpful rephrasing suggestions to the user to generate the proper query which can be understood unambiguously by the system. Its architecture contains two parts, Query translation and Query formulation. Query translation includes parse tree classification, parse tree validation and parse tree translation. In parse tree classification, word in the parse tree can be mapped into components of XQuery. If match occurs then the word will classify as a token otherwise as a marker. Thenafter, the parse tree validator checks whether the parsed tree is valid or not. If not valid, the error message is sent to the user. In parse tree translation step, they have used core token for an effective semantic grouping. Meaningful Lowest Common Ancestor Structure is used to find relationship between given keywords. Feedback messages and warnings are generated to solve improper user query. This system also includes query repository to store user's queries. Users can access their old queries as well as create template of queries. Results can be displayed in text view, tree view or grid view. Thus, NaLIX is an interactive query interface. But limitation is, it is used for only XML queries, no other query languages are supported.

In 2008, Djahantighi et al. [4] implemented natural language interfaces to database. This system required syntactic knowledge as well as semantic knowledge. Syntactic knowledge is for restoring data in natural language processing. Some functions are used to score the text and then data restoring is done by shallow parsing. System needs preprocessor for changing of the words in input. And the preprocessor creates semantic database. For presenting equal query of different sentences, semantic database is used. They have used expert system instead of expert individuals. The expert system knowledge base is used for making decisions in expert system. This expert system includes knowledge base and inference engine. They have used knowledge bases, semantic and syntactic approaches to solve similar words ambiguity in transformation. Such expert system is implemented in prolog which supports synonymous words in any language. The process includes user input, tokenizing and syntax parsing, semantic analysis and then SQL query output. In future, they will implement image processing system to detect questions and sentences automatically.

In 2009, Cheng et al. [5] extracted the unstructured information and converted it into structured format. They have observed and modified Hobbs architecture [6] for information extraction and designed some more modules such as text zoner, filter, named entity recognition, fragment combiner, semantic interpreter, lexical disambiguation, discourse processing and template generator. Text zoner segments data into three parts - header, body and footnotes.

In 2009, Zhang et al. [7] have used syntax based approach instead of keyword based approach. They have compared their spatial query interface with the Google Maps. Two main advantages of the system are, (1) Query conditions can be expressed in natural language and (2) Large number of spatial predicates are supported. Query is processed in order of part of speech tagging, semantic parsing and schema matching. In part of speech tagging, they have used classic viterbi algorithm for sequence tagging. This algorithm finds out tag possibilities of each word and generates best sequence tagging. The algorithm is more efficient due to dynamic programming framework coupled with Markov assumptions. In semantic parsing step, they have determined three types of keywords such as target object, spatial predicate and reference object. Rule based semantic parsing is used. In schema matching step, target object and reference object are derived from the database. Distance function is used for matching the objects. The authors have converted results into Keywhole Markup Language [8]. In [7], their system includes PostgreSQL backend spatial database, logic layer is implemented using Perl scripts and query interface is implemented using PHP language. The result of their system is precise and nearly 60% whereas Google Map's result is precisely 3.3%, which shows that their result is more precise than Google Maps. Limitation of their system is, it is designed for small datasets and limited number of user queries are supported.

In 2012, Iqbal et al. [9] have designed search engine to support negation queries in natural language interfaces. This engine is tested on Mooney data set and it satisfies almost all the negation queries of the set. The system supports full natural language interface rather than restricted natural language interface. A sophisticated algorithm is used with the set of techniques and transformation rules to understand the intention of the user. Set of rules are governed according to the natural language query entered by the user. The negation query handling engine performs the processes such as identification of negation keywords, detection of co-ordinating conjuctions, identification of ontological resources and natural language query transformation.

In 2014, Reinaldha et al. [10] have implemented system which solved the limitations of Wibisono. Wibisono is a natural language interface for database. Wibisono is unable to process question-type queries while the unit conversions are solved. First step of the process is to identify whether user's input is question type or directive type. As per the type of input, corresponding process will be implemented.

In 2009, Chaware et al. [11] have proposed the techinique to process for multilingual information. They have used English database as a backend. Hindi, Marathi or Gujarati languages are supported for front-end users. Their system supports efficient and easy way of converting local language keyword into English language. They have proposed a tolerance algorithm for matching the string by different writing styles while maintaining the similar semantics. Their application domain is a Shopping Mall. They have explained different input mechanisms such as Multilingual Physical Keyboards, Multilingual On-screen Keyboards, IME (Input Method Editor). In their system, they have used IME for input keywords of local languages. This inputted keyword can be stored in different encoding form. They have explained encoding form such as ASCII Encoding, ISCII Encoding, Unicode Encoding, The Nchar data type. The system contains four modules, local language interface module, the query formation module, the database module and the display module. Local language interface module used ASP and JavaScript to implement user interfaces. In database module, different tables are maintained in MS-Access. The system has three steps of processing the data. In the first step, the input keyword will be interpreted and parsed according to vowels, consonants and special characters. In the second step, SQL query will be formed and applied on stored databases. In step three, each character will be mapped with its ASCII code and then converted from English to local language. They considered tolerance as phonetically matching words which can be maximum possible errors. Their focus is on vowels matching to achieve the phonetic mapping. Efficiency of the tolerance algorithm has increased exponentially because of IME approach of input keyword. Figure 1 shows the tolerance algorithm.

The tolerance algorithm takes local language string as an input. All vowels are extracted from the string. Later on, all possible vowels combinations are used to construct the strings. These combinations are searched from the back-end

```
Input: Local language string, S₁
Output: English language string, Eₛ
  1.  Extract the vowels from the string
  2.  Construct all possible strings
      using vowels combinations.
  3.  Search the database based on all
      combinations.
  4.  Return the result equivalent to all
      English strings.
  5.  Display the exact match from the
      database.
```

Fig. 1. Tolerance algorithm

database. This returns the equivalent English statement. Finally, exact match of the query statement is retrieved.

Table 1 shows the list of research papers which are explained above.

Table 1. List of papers

Title	Authors	Year	Organization
Information retrieval in multilingual environment	Chaware, SM and Rao, Srikantha	2009	IEEE
A natural language interface for crime-related spatial queries	Zhang, Chengyang and Huang, Yan and Mihalcea, Rada and Cuellar, Hector	2009	IEEE
NaLIX: an interactive natural language interface for querying XML	Li, Yunyao and Yang, Huahai and Jagadish, HV	2005	ACM
Negation query handling engine for natural language interfaces to ontologies	Cheng, Tin Tin and Cua, Jeffrey Leonard and Tan, Mark Davies and Yao, Kenneth Gerard and Roxas, Rachel Edita	2009	IEEE
Natural Language Interfaces to Database (NLIDB): question handling and unit conversion	Reinaldha, Filbert and Widagdo, Tricya E	2014	IEEE
Using natural language processing in order to create SQL queries	Norouzifard, M and Davarpanah, SH and Shenassa, MH and others	2008	IEEE

3   Tools and Technologies

Different tools are available for natural language translation and input parsing. Most of the researchers are supported by programming languages like Python, Java, Prolog and many more. The main problem with all these programming languages is that they are not useful under all circumstances. So it is important to choose a suitable tool. This happens to be an essential requirement before proceeding for an application that deals with natural language translation.

Table 2. Summary of tools used

Authors	Tools	Pros	Cons
Chaware et al. [11]	ASP, JavaScript, MS-Access	The system supports Hindi, Marathi or Gujarati to English translation and vice versa. Used tolerance algorithm	System is domain specific
Zhang et al. [7]	Part of Speech is classic viterbi algorithm used on Penn Treebank dataset; In backend spatial database, logic layer is designed using perl scripts	Query conditions can be expressed in natural language, Large number of spatial predicates are supported, result is more precise than Google Maps	This system is designed for only small datasets and limited number of user queries
Yunyao Li et al. [1]	Wordnet for term expansion, Timber as XML database Minipar as natural language parser	This system solved two problems such as automatically understanding natural language and translating parsed natural language query into a correct formal query	This system supports only XML queries. Queries in other languages are not allowed
Iqbal et al. [9]	Mooney data set is used for numbers of negation queries	The system reduces ambiguity between terms solved by RSS by interacting with the user	The query engine is tested only on Mooney dataset
Cheng et al. [5]	LingPipe was used as a part of speech tagger	Segmenting the extracted documents and processed the in structured format	Leads to the problem of over filtering in few cases thereby filtering the relevant text
Reinaldha et al. [10]	JScience library for unit conversion, Stanford dependency parser	Overcome limitations of Wibisono such as question type queries and unit conversion	The system is domain dependent
Djahantighi et al. [4]	This expert system is implemented in Prolog	It transforms parsed input sentence into SQL language	Unable to handle question type inputs
Xinye Li et al. [12]	Chinese Dependency parsing algorithm	It supports semantic search using XML language	Focus is only on dependency parsing algorithm and XML language
Sharef et al. [13]	SPARQL	This system supports Deeper understanding of semantic natural language through twelve linguistic patterns	No guarantee of correct SPARQL construction for composite questions

Table 2 shows tools and technologies used by different researchers for creating natural language interface for databases for language translation purpose. A review of the advantages and disadvantages of each tool is described in Table 2.

Viterbi algorithm [14] is used for part of speech tagging, which is a dynamic programming algorithm. This algorithm is used to find the sequence of hidden states (Viterbi Path). Part of speech tagging is a process of marking up words based on its definition as well as its context. Mooney dataset [15] contains data which can be useful for handling negation queries.

Penn Treebank dataset [16] contains list of part of speech tags. Wordnet [17] is a lexical database of English language. Lingpipe [18] is a set of Java libraries which can be used for linguistic analysis. JScience [19] library is a Java library for the scientific community. This library can be used to create synergy between all sciences. Minipar [3] is an online natural language parser, which is used to parse the processed data.

Other tools and technologies used by researchers are ASP, JavaScript, MS-Access, Perl Scripts, Mooney Dataset, Stanford Dependency Parser, Prolog, Chinese Dependency Parser, SPARQL etc.

4   Conclusions

Natural language processing is a very interesting topic for many researchers. Lots of researches have done research work in this field and still many more activities are going on in this domain. Applications of natural language processing are automated speech recognition, semantic search engines, language translation as well as information retrieval systems. Among them, natural language translation is a popular application. Many natural language interfaces have been implemented for translation purpose as well as search purpose. Some natural language interfaces for database are NaLIX, FREya, Orakel, AquaLog, semsearch, Wibisono etc. Sampark TDIL is Indian Language translation application.

Many challenges have been faced during natural language translation such as challenge of restricted natural language input, formulating input data into structured query format, ambiguities in the data, inputing text with negative effect and many more. Researchers have been focusing on such challenges. As per the review paper, we can say that some of the challenges are solved or partially solved and many challenges are still a big puzzle and remain unsolved.

References

1. Li, Y., Yang, H., Jagadish, H.: Nalix: an interactive natural language interface for querying xml. In: Proceedings of the 2005 ACM SIGMOD International Conference on Management of Data, pp. 900–902. ACM (2005)
2. Jagadish, H.V., Al-Khalifa, S., Chapman, A., Lakshmanan, L.V., Nierman, A., Paparizos, S., Patel, J.M., Srivastava, D., Wiwatwattana, N., Wu, Y., et al.: Timber: A native xml database. VLDB J. Int. J. Very Large Data Bases 11(4), 274–291 (2002)

3. Lin, D.: Latat: language and text analysis tools. In: Proceedings of the First International Conference on Human language Technology Research, pp. 1–6. Association for Computational Linguistics (2001)
4. Norouzifard, M., Davarpanah, S., Shenassa, M. et al.: Using natural language processing in order to create sql queries. In: International Conference on Computer and Communication Engineering, ICCCE 2008, pp. 600–604. IEEE (2008)
5. Cheng, T.T., Cua, J.L., Tan, M.D., Yao, K.G., Roxas, R.E.: Information extraction from legal documents. In: Eighth International Symposium on Natural Language Processing, SNLP 2009, pp. 157–162. IEEE (2009)
6. Hobbs, J.R., Appelt, D., Bear, J., Israel, D., Kameyama, M., Stickel, M., Tyson, M.: Fastus: a cascaded finite-state transducer for extracting information from natural-language text, arXiv preprint cmp-lg/9705013 (1997)
7. Zhang, C., Huang, Y., Mihalcea, R., Cuellar, H.: A natural language interface for crime-related spatial queries. In: IEEE International Conference on Intelligence and Security Informatics, ISI 2009, pp. 164–166. IEEE (2009)
8. Nolan, D., Lang, D.T.: Keyhole markup language. In: XML and Web Technologies for Data Sciences with R, pp. 581–618. Springer, New York (2014)
9. Iqbal, R., Murad, M.A.A., Selamat, M.H., Azman, A.: Negation query handling engine for natural language interfaces to ontologies. In: International Conference on Information Retrieval & Knowledge Management (CAMP), pp. 249–253. IEEE (2012)
10. Reinaldha, F., Widagdo, T.E.: Natural language interfaces to database (nlidb): question handling and unit conversion. In: International Conference on Data and Software Engineering (ICODSE), pp. 1–6. IEEE (2014)
11. Chaware, S., Rao, S.: Information retrieval in multilingual environment. In: 2009 Second International Conference on Emerging Trends in Engineering & Technology, pp. 648–652. IEEE (2009)
12. Li, X., Yu, X.: Xml semantic search with natural language interface. In: IEEE International Conference on Computer Science and Automation Engineering (CSAE), vol. 1, pp. 38–41. IEEE (2012)
13. Sharef, N.M., Noah, S.A.M.: Linguistic patterns-based translation for natural language interface. In: 2014 International Conference on Information Science & Applications (ICISA), pp. 1–5. IEEE (2014)
14. Viterbi algorithm. https://en.wikipedia.org/wiki/Viterbi_algorithm
15. Mooney dataset. https://sites.google.com/site/naturallanguageinterfaces/freya/data
16. Penn treebank dataset. https://www.ling.upenn.edu/courses/Fall_2003/ling001/penn_treebank_pos.html
17. Wordnet. https://wordnet.princeton.edu/wordnet/
18. Lingpipe. http://alias-i.com/
19. Jscience. http://jscience.org/

A Simple and Efficient Text-Based CAPTCHA Verification Scheme Using Virtual Keyboard

Kajol Patel[1] and Ankit Thakkar[2]([✉])

[1] Department of Computer Engineering, Institute of Technology,
Nirma University, Ahmedabad 382 481, Gujarat, India
15MCEI21@nirmauni.ac.in
[2] Department of Information Technology, Institute of Technology,
Nirma University, Ahmedabad 382 481, Gujarat, India
ankit.thakkar@nirmauni.ac.in

Abstract. Digital media becomes an effective way of communication which is available round the clock to everyone including humans and machines. This put the requirement for machines to differentiate between human and machine as far as access of the website or its relevant services is concerned. CAPTCHA (Completely Automated Public Turing test to tell Computer and Human Apart) is a test that helps machines (or programs) to differentiate between human and machine. CAPTCHA should be easy for users to solve and difficult for bots to attack. In this paper, a simple and efficient text-based CAPTCHA verification scheme is proposed which is easy for human and hard for bots. The proposed scheme uses virtual keyboard, eliminates input-box, and does verification on the basis of the positions of the characters.

Keywords: CAPTCHA · Virtual keyboard · Position based verification

1 Introduction

CAPTCHA is used in websites to prevent automated interactions by bots. For example, Gmail improves its service by blocking access to automated spammers, eBay blocks automated programs that flood the websites, and Facebook protects its site by limiting the creation of fraudulent profiles [1]. In November 1999, slashdot.com released a poll for voting to select the best college of CS in the US. In this poll, automated programs were created by students of the Carnegie Mellon University and the MIT that repeatedly voted for their colleges. This incident put the requirement of using CAPTCHA for online polls to ensure that only humans are allowed to participate in polls [2]. CAPTCHA are used in many web applications (or web services) like search engines, password systems, online polls, account registrations, prevention of spam, blogs, messaging and phishing attack detection etc. [3].

CAPTCHA can broadly be classified as text-based CAPTCHA, image-based CAPTCHA, audio-based CAPTCHA and video-based Captcha. This paper

© Springer International Publishing AG 2018
S.C. Satapathy and A. Joshi (eds.), *Information and Communication Technology for Intelligent Systems (ICTIS 2017) - Volume 2*, Smart Innovation, Systems and Technologies 84, DOI 10.1007/978-3-319-63645-0_13

focuses on text-based CAPTCHA only. Text-based CAPTCHAs are widely used as it is simple and user-friendly. Few examples of text-based captcha are Gimpy, EZ-Gimpy, MSN-CAPTCHA, and Baffle-Text etc. In Gimpy CAPTCHA, ten random words are selected from a dictionary and displayed to the user. These words are displayed to the user using distorted images. Noise can be added to the images so that it would be difficult for a machine to identify the CAPTCHA. To access web service, the user must correctly enter the characters of the given images. In EZ-Gimpy CAPTCHA, only one word is selected from a dictionary and displayed to the user after applying misshape/distortion. Mori *et al.* [4] used shape context matching method to identify the words in images of EZ-Gimpy and Gimpy CAPTCHAs. These CAPTCHAS were cracked by object recognition algorithms with success rate of 92% and 33% respectively.

The characters of CAPTCHA images are distorted so that it would be difficult for bots to crack CAPTCHAs. This also creates ambiguity to users to identify characters of the CAPTCHA. This may result in multiple attempts by the user which may lead to security issues [6]. In [12], various tests were conducted and examined that how different CAPTCHAs and their complexity affects the user experience.

Microsoft Captcha was vulnerable to low-cost segmentation attack. A number of text-based CAPTCHAs were cracked with an overall success rate of 60% and success rate of 90% is achieved with segmentation [7]. Yan *et al.* [5] carried out an attack on visual CAPTCHAs provided by Captchaservice.org and was successful in cracking them. They exploited design errors and using simple naive pattern recognition algorithms. In [8], authors have targeted the bank of China website that uses text-based CAPTCHA. They have explored vulnerabilities like fixed-character length of a text, use of only lower-case characters, etc. Pengpeng Lu *et al.* [10] proposed a new segmentation method for connected characters using BP (Back Propagation) neural network and drop-falling algorithms. This method can solve CAPTCHAs having connected characters but fails if its characters is seriously distorted or overlapped.

Starostenko *et al.* [9] discussed the novel approach to break text-based CAPTCHA with variable text and orientation. SVM classifier is used in recognizing straightened characters. Segmentation success rate of 82% for reCaptcha 2011 is achieved by this method. Financial institutions also use CAPTCHAs for protecting their services. Wang *et al.* [11] proposed an algorithm to defeat rotated text-CAPTCHA by transformation and segmentation using an adaptive system.

Most of the text-Captchas are cracked by OCR attacks and to prevent from OCR-attacks, CAPTCHA can be made more complex with noise and distortion that affects the usability of users. Simplicity makes text-based CAPTCHA as preferred choice of implementation, at the same time there is a need to protect the CAPTCHAs from boats. Hence, this paper proposes a new method for CAPTCHA verification that makes easy for users to read and input the characters of the CAPTCHA, and at the same time, it makes difficult for bots to input the CAPTCHA characters. This approach uses a virtual keyboard to take input from the user, eliminate the use of input box and compares the CAPTCHA character based on the position of the characters instead of contents of the CAPTCHA. The rest of the paper is organized as follows: Proposed approach

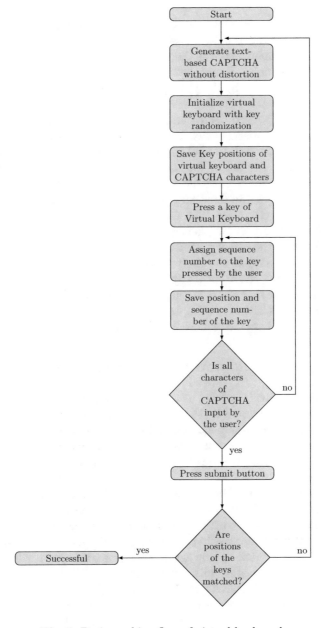

Fig. 1. Basic working flow of virtual keyboard

is discussed in Sect. 2, Simulation setup and result discussion is given in Sect. 3, and concluding remarks and future scope is given in Sect. 4.

2   Proposed Method

To reduce bot attacks, more complex CAPTCHAs are generated with distortions and noise that affects the usability of users. Users get frustrated because of refreshing the CAPTCHA many times as they face difficulty in reading characters of the CAPTCHAs due to noise. Hence, instead of making CAPTCHA more complex, security can be increased by developing a new CAPTCHA verification method which is difficult for the bots but easy for the humans to pass the verification process. This paper has proposed a new approach using virtual keyboard.

In the proposed approach, text-based CAPTCHA is created without noise that makes easy for the user to read and pass the test in a single attempt in most cases. The user uses the virtual keyboard to input CAPTCHA word. However, this word is stored in the form of the position of the characters. The keys pressed by the users are also highlighted and the sequence number is assigned to each character pressed by the user. The sequence number of a particular character can be viewed by the user by placing the mouse pointer over the specific key. This method avoids the use of textbox to take input from the user which makes it difficult for bots to input the CAPTCHA characters.

In addition to that, the proposed approach adds complexity by randomization of keys of virtual keyboard. It should be noted that the proposed approach compares the CAPTCHA text using key positions of the key pressed by the user rather actual value of keys. The flowchart of proposed method is shown in Fig. 1.

3   Simulation Setup and Result Discussions

The proposed approach is verified using JAVA language. A CAPTCHA image of random characters generated by the server is displayed to the user. It should be noted that the proposed approach does not generate fixed length CAPTCHA. In addition to that noise is removed to increase readability of the CAPTCHA. Both of this help to overcome the weaknesses of text-based CAPTCHA generation schemes discussed in Sect. 1.

When a page is loaded, the positions of the characters are saved. When the user clicks the submit button, the position and sequencing of the CAPTCHA-text is compared with the position and sequencing of the virtual keyboard keys pressed by the user. If the positions are matched, the user gets the access of the required services provided by the server otherwise, the page will be refreshed and a CAPTCHA test begins with a new text-CAPTCHA and keyboard.

An example of CAPTCHA test is shown in the Fig. 2. The characters of the keyboard get highlighted as the user clicks the character of the keyboard. This helps the user to identify the characters which have been input by the user. This can be evident through Fig. 3. Each character is assigned a unique sequence number as soon as the user clicks on it. This sequence number helps the user to order the characters which have been input by him. The sequence number of character 'q' is shown to the user when a mouse hovers on the character 'q'. This can be evident through Fig. 4. A user is required to click on submit button when all characters are input by the user.

Fig. 2. Snapshot of CAPTCHA and keyboard

Fig. 3. Snapshot of keyboard with keys highlighted

Fig. 4. Snapshot of sequence displayed on mouseover

4   Conclusion and Future Scope

This paper presents a simple and efficient CAPTCHA verification scheme that differentiate between human and machine. The proposed approach generates a simple text-based CAPTCHA which is easy to read by humans and hence, humans can pass the test in a single attempt as far as possible. At the same time, use of virtual keyboard along with randomized key positions makes it difficult for machines to pass the CAPTCHA test. The proposed approach uses virtual keyboard to take input for CAPTCHA verification, eliminates the inputbox that makes difficult for boats to decide where to input CAPTCHA text, and uses of position-based verification in place of comparing contents of the CAPTCHA text.

In future, the proposed approach can be extended by randomizing positions of the CAPTCHA and virtual keyboard, and both can take any position on the screen. Use of handwritten characters to initialize the virtual keyboard can also be considered as future scope. In addition to that, response time analysis can make the proposed approach much stronger and can be considered as a future scope.

References

1. Bursztein, E., Martin, M., Mitchell, J.: Text-based CAPTCHA strengths and weaknesses. In: Proceedings of the 18th ACM Conference on Computer and Communications Security, pp. 125–138 (2011)
2. Choudhary, S., Saroha, R., Dahiya, Y., Choudhary, S.: understanding CAPTCHA: text and audio based CAPTCHA with its applications. Int. J. Adv. Res. Comput. Sci. Softw. Eng. **3**(6) (2013)
3. Banday, M.T., Shah, N.A.: A study of CAPTCHAS for securing web services. arXiv preprint arXiv:1112.5605 (2011)
4. Mori, G., Malik, J.: Recognizing objects in adversarial clutter: breaking a visual CAPTCHA. In: 2003 IEEE Computer Society Conference on Computer Vision and Pattern Recognition, Proceedings, pp. 1–134. IEEE (2003)
5. Yan, J., Ahmad, E., Salah, A.: Breaking visual captchas with naive pattern recognition algorithms. In: Twenty-Third Annual Computer Security Applications Conference, ACSAC 2007, pp. 279–291. IEEE (2007)
6. Yan, J., Ahmad, E., Salah, A.: Usability of CAPTCHAS or usability issues in CAPTCHA design. In: Proceedings of the 4th Symposium on Usable Privacy and Security, pp. 44–52. ACM (2008)
7. Yan, J., Ahmad, E., Salah, A.: A low-cost attack on a Microsoft CAPTCHA. In: Proceedings of the 15th ACM Conference on Computer and Communications Security, pp. 543–554. ACM (2008)
8. Ling-Zi, X., Yi-Chun, Z.: A case study of text-based CAPTCHA attacks. In: 2012 International Conference on Cyber-Enabled Distributed Computing and Knowledge Discovery (CyberC). IEEE (2012)
9. Starostenko, O., Cruz-Perez, C., Uceda-Ponga, F., Alarcon-Aquino, V.: Breaking text-based CAPTCHAS with variable word and character orientation. Pattern Recogn. **48**, 1101–1112 (2015). Elsevier
10. Lu, P., Shan, L., Li, J., Liu, X.: A new segmentation method for connected characters in CAPTCHA. In: 2015 International Conference on Control, Automation and Information Sciences (ICCAIS). IEEE (2015)
11. Wang, Y., Lu, M.: A self-adaptive algorithm to defeat text-based CAPTCHA. In: 2016 IEEE International Conference on Industrial Technology (ICIT), pp. 720–725. IEEE (2016)
12. Gafni, R., Nagar, I.: CAPTCHA-security affecting user experience. In: Issues in Informing Science and Information Technology (2016)

Outcome Fusion-Based Approaches for User-Based and Item-Based Collaborative Filtering

Priyank Thakkar$^{(\boxtimes)}$, Krunal Varma, and Vijay Ukani

Institute of Technology, Nirma University, Ahmedabad, India
{priyank.thakkar,krunalvarma,vijay.ukani}@nirmauni.ac.in

Abstract. Collaborative Filtering (CF) is one of the most effective approaches to engineer recommendation systems. It recommends those items to user which other users with related preferences and tastes liked in the past. User-based and Item-based Collaborative Filtering (IbCF) are two flavours of collaborative filtering. Both of these methods are used to estimate target user's rating for the target item. In this paper, these methods are implemented and their performance is evaluated on the large dataset. The major attention of this paper is on exploring different ways in which predictions from UbCF and IbCF can be combined to minimize overall prediction error. Predictions from UbCF and IbCF are combined through simple and weighted averaging and performance of these fusion approaches is compared with the performance of UbCf & IbCF when implemented individually. Results are encouraging and demonstrate usefulness of fusion approaches.

Keywords: User-based collaborative filtering · Item-based collaborative filtering · Recommender systems

1 Introduction

Information overload has definitely become a serious problem as the volume of information is growing at much higher rate than our ability to process it. Collaborative filtering is one of the most promising technologies that helps us examine all the available information and figure out the most useful.

UbCF and IbCF are two popular variants of collaborative filtering. UbCF focuses on finding nearest neighbours to the target user and then uses ratings assigned to the target item by these nearest users to predict target user's rating for the target item. Similarly, focus of IbCF is on finding nearest neighbours to the target item and then use ratings given by the target user to these nearest items to estimate the target user's rating for the target item.

Success of Collaborative Filtering (CF) is evident in both research and practice. It is being used successfully in both information filtering and E-commerce applications. Users also need recommendation that they can trust and use to find

© Springer International Publishing AG 2018
S.C. Satapathy and A. Joshi (eds.), *Information and Communication Technology for Intelligent Systems (ICTIS 2017) - Volume 2*, Smart Innovation, Systems and Technologies 84, DOI 10.1007/978-3-319-63645-0_14

items that they will like. Therefore, research community is continuously striving for improving the quality of the recommendations. In this paper, an effort is made to deal with this challenge by fusing predictions from UbCF and IbCF. Initial results are encouraging and motivating.

The remainder of the paper is structured as follows. Section 2 presents a review of earlier work while collaborative filtering is discussed in Sect. 3. Section 4 discusses proposed fusion approaches and experimental evaluation is discussed in Sect. 5. Section 6 completes the paper with concluding remarks.

2   Related Work

Academia and the industry both have witnessed development of many collaborative filtering systems. Perhaps, the first recommender system was Grundy system [17] which used streotypes for building user models. These models used very less information about each user. Collaborative filtering algorithms were first used by GroupLens [10,15], Video Recommender [8] and Ringo [19] to automate predictions. Book recommendation system used by amazon.com was also based on collaborative filtering. An interesting example was a tag recommender discussed in Yagnik et al. which was based on concepts from model-based collaborative filtering [21]. Number of different recommender systems were presented in a special issue of Communications of the ACM [16]. Examples of user-based collaborative filtering includes [3,7,9,15] while [6,18] are significant examples of item-based collaborative filtering. Patel et al. explored both user-based and item-based collaborative filtering [13]. They explored different possibilities through which user and item profiles can be formulated and predictions can be made.

Advances in collaborative filtering were further discussed in [11]. They discussed about utilization of implicit feedback and temporal models to improve model accuracy. Rao et al. approached collaborative filtering with graph information for low rank matrix completion [14]. They formulated and derived a highly efficient conjugate gradient based minimizing scheme. Recently, Liu et al. proposed a kernalized matrix factorization for collaborative filtering which had distinct advantages over conventional matrix factorization method [12].

Content-based and collaborative filtering were combined in a hybrid approach in [5]. Their hybrid approach was based on Bayesian network. A hybrid content-based and item-based collaborative filtering approach for recommending TV programs was proposed in [2]. They used singular value decomposition for enhancement. Wang et al. attempted to unify user-based and item-based collaborative filtering approaches through similarity fusion [20]. Their fusion framework was probabilistic in nature. This paper also attempts to combine estimations from user-based and item-based collaborative filtering, but in a much natural way. The idea is to investigate fruitfulness of such fusion. Positive results can be encouraging and can also motivate research community to address the problem with more sophisticated methods.

3   Collaborative Filtering

This section concisely discusses user-based and item-based collaborative filtering. If we assume P users and Q items, dimensionality of user-item matrix X is $P \times Q$. Element $x_{i,j} = r$ specifies that i^{th} user has rated j^{th} item with rating r, where $r \in \{1, 2, \cdots, |r|\}$. Element $x_{i,j} = \phi$ shows that i^{th} user has not rated j^{th} item. Each row of X represents corresponding user's profile in terms of ratings assigned by this user to various items. Each column of X depicts profile of corresponding item in terms of ratings received by this item from different users.

3.1   User-Based Collaborative Filtering

User-based collaborative filtering estimates rating of the test user for the test item based on ratings given to the test item by other users with similar profile. Users with similar profile are typically known as nearest neighbours/neighbouring users. There are different ways in which this similarity can be calculated. Pearson correlation is one such method, and in this paper, similarity/correlation between users u_1 and u_2 is computed using it through the Eq. 1 [1].

$$sim(u_1, u_2) = \frac{\sum_{i \in I_{u_1 u_2}} (x_{u_1,i} - \overline{x}_{u_1})(x_{u_2,i} - \overline{x}_{u_2})}{\sqrt{\sum_{i \in I_{u_1 u_2}} (x_{u_1,i} - \overline{x}_{u_1})^2 \sum_{i \in I_{u_1 u_2}} (x_{u_2,i} - \overline{x}_{u_2})^2}} \tag{1}$$

Here, $I_{u_1 u_2}$ is a set of items corated by u_1 and u_2 and $\overline{x}_{u_1}$ designates the average rating of user u_1. Once, this similarity values are computed, user i's rating for the item j can be computed in different ways, and in this paper, Eq. 2 is used for this task [1].

$$x_{i,j} = \overline{x}_i + \frac{\sum_{u' \in \hat{U}} sim(i, u') \times (x_{u',j} - \overline{x}_{u'})}{\sum_{u' \in \hat{U}} |sim(i, u')|} \tag{2}$$

Here, $\hat{U}$ represents set of N users. These N users have rated item j and they are the most similar to user i.

3.2   Item-Based Collaborative Filtering

Item-based collaborative filtering estimates rating of the test user for the test item based on ratings given by test user to other items with similar profile. Items with similar profile are popularly known as nearest neighbours/neighbouring items. Nearest neighbouring items can be found in several ways. In this paper, to find similarity/correlation between items i_1 and i_2, Pearson correlation is used and it is computed using Eq. 3 [18].

$$sim(i_1, i_2) = \frac{\sum_{u \in U} (x_{u,i_1} - \overline{x}_{i_1})(x_{u,i_2} - \overline{x}_{i_2})}{\sqrt{\sum_{u \in U} (x_{u,i_1} - \overline{x}_{i_1})^2 \sum_{u \in U} (x_{u,i_2} - \overline{x}_{i_2})^2}} \tag{3}$$

Here, U designates a set of users. These users are the ones who have rated both i_1 and i_2. Average rating of item i_1 is represented by $\overline{x}_{i_1}$. Once, this similarity values are computed, user i's rating for the item j can be computed in different ways, and in this paper, Eq. 4 is used for this task.

$$x_{i,j} = \overline{x}_j + \frac{\sum_{i' \in \hat{I}} sim(j, i') \times (x_{i,i'} - \overline{x}_{i'})}{\sum_{i' \in \hat{I}} |sim(j, i')|} \qquad (4)$$

Here, $\hat{I}$ denotes set of N items. These items are the ones which have been rated by user i and are the most similar to item j.

4   Proposed Approach

This paper proposes to combine estimates from user-based and item-based collaborative filtering. The idea is depicted in Fig. 1. It is evident from Fig. 1 that predictions are first made using user-based and item-based collaborative filtering. This paper proposes to combine these predictions to get the final prediction as defined in Eq. 5.

$$x_{i,j} = w_1 \times x_{(i,j)(UbCF)} + w_2 \times x_{(i,j)(IbCF)} \qquad (5)$$

Here, $x_{i,j}$ is used to designate the final prediction of ith user for j^{th} item through fusion techniques. $x_{(i,j)(UbCF)}$ indicates prediction of ith user for j^{th} item using user-based collaborative filtering while $x_{(i,j)(IbCF)}$ represents the same prediction using item-based collaborative filtering. In this paper, this fusion is achieved through two methods. The first method relies on simple averaging and therefore uses $w_1 = w_2 = 0.5$. Second method uses weighted averaging and optimal weight

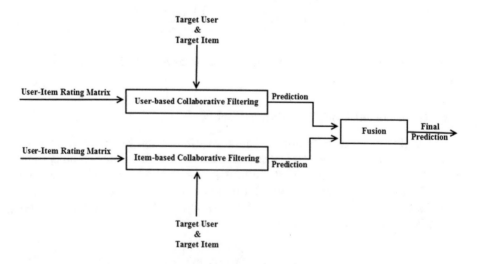

Fig. 1. Proposed approach

values are decided through performance of user-based and item-based collaborative filtering during five-fold cross validation of training set. Mean Absolute Percentage Error (MAPE) is employed as the performance measure during cross-validation. MAPE is first mapped on the scale of 0 to 1 and then w_1 and w_2 are decided using Eqs. 6 and 7 respectively.

$$w_1 = \frac{1 - MAPE_{UbCF}}{(1 - MAPE_{UbCF}) + (1 - MAPE_{IbCF})} \tag{6}$$

$$w_2 = \frac{1 - MAPE_{IbCF}}{(1 - MAPE_{UbCF}) + (1 - MAPE_{IbCF})} \tag{7}$$

where, $MAPE_{UbCF}$ and $MAPE_{IbCF}$ are MAPE of user-based and item-based collaborative filtering respectively during cross-validation.

5   Experimental Evaluation

Dataset and evaluation measures used are briefly described in this section. It also throws light on experimental methodology and various experiments carried out. Results of different experiments are also discussed towards the end of the section.

5.1   Dataset

Hetrec2011-movielens-2k dataset dated May 2011 [4], (http://www.imdb.com/, http://www.rottentomatoes.com/, http://www.grouplens.com/) has been used for experimentation. This dataset was published by research group named GroudpLens. Dataset origin are in actual MovieLens10M dataset. The data set contains 2,113 users and 10,197 movies. It contains 855,598 user ratings in the range of 0.5 to 5.0, in steps of 0.5. This results in a total of 10 discrete rating values. Average number of ratings per user and per movie are 405 and 85 respectively. The dataset was pre-processed to construct user-movie rating matrix.

5.2   Evaluation Measures

Mean Absolute Error (MAE), Mean Absolute Percentage Error (MAPE) and Mean Squared Error (MSE) were used to evaluate performance and their formulas are shown in Eqs. 8, 9 and 10.

$$MAE = \frac{1}{n} \sum_{t=1}^{n} |A_t - F_t| \tag{8}$$

$$MAPE = \frac{1}{n} \sum_{t=1}^{n} \frac{|A_t - F_t|}{|A_t|} \times 100 \tag{9}$$

$$MSE = \frac{1}{n} \sum_{t=1}^{n} (A_t - F_t)^2 \tag{10}$$

where, A_t is actual value, F_t is predicted value and n is total number of test examples.

5.3   Experimental Methodology

To carry out various experiments, first of all, we figured out users who had rated 100–120 movies. We could find 87 such users. We considered them as target users. For each of the target users, we randomly selected 25 movies/items as target items. This implied that we had a test set of 87 users and 25 movies for which predictions were made and performance was evaluated. From actual user-item matrix, these 87 × 25 ratings were masked and the remaining matrix was used as the training set. Experiments were carried out using UbCF, IbCF and two fusion approaches. Fusion approach 1 employed simple averaging while Fusion approach 2 used weighted averaging as discussed in Sect. 4.

5.4   Results and Discussion

Table 1 shows result of different techniques. It can be seen that for each of the techniques, experiments were performed for 12 different values of Nearest Neighbours (NN). MAPE and MSE had also been measured but due to space limitations, only MAE is reported.

Table 1. MAE in UbCF, IbCF and fusion approaches

Sr.	NN	UbCF	IbCF	Fusion approach 1	Fusion approach 2
1	1	0.911	0.813	0.719	0.717
2	2	0.779	0.725	0.656	0.654
3	5	0.662	0.647	0.599	0.598
4	10	0.622	0.618	0.582	0.581
5	20	0.603	0.611	**0.577**	**0.577**
6	30	0.598	**0.610**	0.579	0.579
7	50	**0.595**	0.612	0.580	0.580
8	60	0.596	0.615	0.583	0.583
9	70	0.596	0.619	0.585	0.585
10	80	0.597	0.622	0.586	0.586
11	90	0.597	0.624	0.587	0.587
12	100	0.597	0.624	0.588	0.588

Minimum MAE achieved in different techniques is summarized in Fig. 2. Minimum MAE in both the fusion approaches is 0.577. This is definitely better when compared with UbCF and IbCF, with minimum MAE of 0.595 and 0.610 respectively. This encourages to investigate fusion problem with more sophisticated techniques.

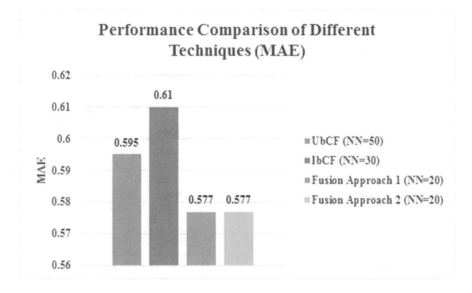

Fig. 2. Performance comparison of different techniques

6 Conclusion

This paper described fusion of user-based and item-based collaborative filtering to minimize error in prediction. Experiments were performed on a real-life dataset. Predictions were made through UbCF, IbCF and two fusion approaches. The first fusion approach was based on simple averaging while the second one exploited weighted averaging. It is evident from the results that fusion approaches perform better than independent UbCF and IbCF approaches. This improvement is encouraging and provides useful direction to research community to explore. One can address this fusion problem as an optimization problem and sophisticated soft computing techniques can be tried to solve the optimization problem and further minimize error. It is also interesting to notice that there is not much of the difference in performance between two fusion approaches. This may also be further investigated in future.

References

1. Adomavicius, G., Tuzhilin, A.: Toward the next generation of recommender systems: a survey of the state-of-the-art and possible extensions. IEEE Trans. Knowl. Data Eng. **17**(6), 734–749 (2005)
2. Barragáns-Martínez, A.B., Costa-Montenegro, E., Burguillo, J.C., Rey-López, M., Mikic-Fonte, F.A., Peleteiro, A.: A hybrid content-based and item-based collaborative filtering approach to recommend tv programs enhanced with singular value decomposition. Inform. Sci. **180**(22), 4290–4311 (2010)

3. Breese, J.S., Heckerman, D., Kadie, C.: Empirical analysis of predictive algorithms for collaborative filtering. In: Proceedings of the Fourteenth conference on Uncertainty in artificial intelligence, pp. 43–52. Morgan Kaufmann Publishers Inc. (1998)
4. Cantador, I., Brusilovsky, P., Kuflik, T.: Second workshop on information heterogeneity and fusion in recommender systems, HetRec2011. In: RecSys, pp. 387–388 (2011)
5. De Campos, L.M., Fernández-Luna, J.M., Huete, J.F., Rueda-Morales, M.A.: Combining content-based and collaborative recommendations: a hybrid approach based on Bayesian networks. Int. J. Approx. Reasoning **51**(7), 785–799 (2010)
6. Deshpande, M., Karypis, G.: Item-based top-n recommendation algorithms. ACM Trans. Inform. Syst. (TOIS) **22**(1), 143–177 (2004)
7. Herlocker, J.L., Konstan, J.A., Borchers, A., Riedl, J.: An algorithmic framework for performing collaborative filtering. In: Proceedings of the 22nd Annual International ACM SIGIR Conference on Research and Development in Information Retrieval, pp. 230–237. ACM (1999)
8. Hill, W., Stead, L., Rosenstein, M., Furnas, G.: Recommending and evaluating choices in a virtual community of use. In: Proceedings of the SIGCHI Conference on Human Factors in Computing Systems, pp. 194–201. ACM Press/Addison-Wesley Publishing Co. (1995)
9. Jin, R., Chai, J.Y., Si, L.: An automatic weighting scheme for collaborative filtering. In: Proceedings of the 27th Annual International ACM SIGIR Conference on Research and Development in Information Retrieval, pp. 337–344. ACM (2004)
10. Konstan, J.A., Miller, B.N., Maltz, D., Herlocker, J.L., Gordon, L.R., Riedl, J.: Grouplens: applying collaborative filtering to usenet news. Commun. ACM **40**(3), 77–87 (1997)
11. Koren, Y., Bell, R.: Advances in collaborative filtering. In: Recommender Systems Handbook, pp. 77–118. Springer, Boston (2015)
12. Liu, X., Aggarwal, C., Li, Y.F., Kong, X., Sun, X., Sathe, S.: Kernelized matrix factorization for collaborative filtering. In: SIAM Conference on Data Mining, pp. 399–416 (2016)
13. Patel, R., Thakkar, P., Kotecha, K.: Enhancing movie recommender system. Int. J. Adv. Res. Eng. Technol. (IJARET), ISSN 0976–6499 (2014)
14. Rao, N., Yu, H.F., Ravikumar, P.K., Dhillon, I.S.: Collaborative filtering with graph information: Consistency and scalable methods. In: Advances in Neural Information Processing Systems, pp. 2107–2115 (2015)
15. Resnick, P., Iacovou, N., Suchak, M., Bergstrom, P., Riedl, J.: Grouplens: an open architecture for collaborative filtering of netnews. In: Proceedings of the 1994 ACM conference on Computer supported cooperative work, pp. 175–186. ACM (1994)
16. Resnick, P., Varian, H.R.: Recommender systems. Commun. ACM **40**(3), 56–58 (1997)
17. Rich, E.: User modeling via stereotypes. Cogn. Sci. **3**(4), 329–354 (1979)
18. Sarwar, B., Karypis, G., Konstan, J., Riedl, J.: Item-based collaborative filtering recommendation algorithms. In: Proceedings of the 10th International Conference on World Wide Web, pp. 285–295. ACM (2001)
19. Shardanand, U., Maes, P.: Social information filtering: algorithms for automating "word of mouth". In: Proceedings of the SIGCHI Conference on Human Factors in Computing Systems, pp. 210–217. ACM Press/Addison-Wesley Publishing Co. (1995)

20. Wang, J., De Vries, A.P., Reinders, M.J.: Unifying user-based and item-based collaborative filtering approaches by similarity fusion. In: Proceedings of the 29th Annual International ACM SIGIR Conference on Research and Development in Information Retrieval, pp. 501–508. ACM (2006)
21. Yagnik, S., Thakkar, P., Kotecha, K.: Recommending tags for new resources in social bookmarking system. Int. J. Data Min. Knowl. Manag. Process 4(1), 19 (2014)

Ontology Merging: A Practical Perspective

Niladri Chatterjee$^{(\boxtimes)}$, Neha Kaushik, Deepali Gupta,
and Ramneek Bhatia

Department of Mathematics, Indian Institute of Technology Delhi,
Hauz Khas, New Delhi 110016, India
niladri.iitd@gmail.com, swami.neha@gmail.com,
deepaligupta0737@gmail.com, ramneek1995@gmail.com

Abstract. Digitization of data has now come in a big way in almost every possible aspects of modern life. Agriculture as a domain is no exception. But digitization alone does not suffice, efficient retrievability of the information has to be ensured for providing web services including question-answering. However, building an ontology for a vast domain as a whole is not straightforward. We view creation of an ontology as an incremental process, where small-scale ontologies for different sub-domains are expected to be developed independently, to be merged into a single ontology for the domain. The paper aims at designing a framework for ontology merging. The method is described with agriculture as the primary domain with several subdomains such as crop, fertilizer, as subdomains among others. The supremacy of the scheme over Protégé, a well-known ontology management software is demonstrated.

Keywords: Ontology · Ontology merging · Protégé

1 Introduction

With the advent of more data-centric applications, ontologies as a knowledge representation technique have gained much popularity in the last one decade or so. Ontologies allow creation of annotations in which information is organized as a machine readable and machine understandable content. An ontology [1], by definition, is explicit specification of

- Concepts (classes) in a domain, e.g. Crop, Soil
- Relationships that exist between concepts, e.g. grows_in (Crop, Soil) gives information on which Crop grows well in which type of Soil.
- Attributes (also called as roles, properties or slots) of the concepts, e.g. SowingTime (for Crop), Moisture_Content (for Soil)
- Instances, e.g., Broccoli (for Crop), Loamy (for Soil)

Ontology development is rapidly growing to facilitate reuse of knowledge. Many applications, such as information retrieval, question-answering, document retrieval, text summarization, are carried out efficiently using the domain ontologies. Domain-specific ontology development has taken up a fast speed and can result in a

© Springer International Publishing AG 2018
S.C. Satapathy and A. Joshi (eds.), *Information and Communication Technology for Intelligent Systems (ICTIS 2017) - Volume 2*, Smart Innovation, Systems and Technologies 84, DOI 10.1007/978-3-319-63645-0_15

gamut of ontologies in a particular domain. These ontologies are needed for various web services including query-answering among others.

Ontology development is sprouting in agricultural domain also. Agriculture is a vast domain consisting of many subdomains, with varying terms being used across regions and with time. It is difficult to build an ontology for such a domain at one go. Hence, a practical way is to build the ontologies incrementally in various subdomains and then merge them for deriving results. Figure 2 shows 12 core subdomains of Indian agriculture.

[2, 3] outline the need for designing an agricultural ontology. End user queries involve information retrieval from different subdomains. One example query is "*Which fertilizer is good for wheat crop?*"

Ontologies aid in efficient query processing and information retrieval. Merging of similar and cross-/overlapping- domain ontologies is required to effectively solve the purpose of ontologies in agriculture. Distributed and heterogeneous ontologies should be inter-related to make them interoperable. Various operations to inter-relate two ontologies O1 and O2 are: *merging, mapping, alignment, refinement, unification,* and *integration* [4].

Merging means "coming together (the act of joining together as one)". In merging, two original ontologies, O1 and O2, are joined together to create a single merged ontology. The original ontologies cover similar or overlapping (sub) domains. For example, with respect to agriculture, both O1 and O2 may belong to crops subdomain. However, ontologies of overlapping subdomains such as fertilizers and crops can also be merged.

Alignment in ontology stands for creating links between O1 and O2. Ontology alignment aims to achieve consistency between O1 and O2. It does not unite the two ontologies into one. Ontology alignment is carried out between ontologies of the complementary domains. For example, in agricultural domain, one may choose to have O1 and O2 from soil and crops domain respectively, keeping them separate, still serving to answer queries like which crop grows well in which soil.

The present paper focuses on merging of agricultural ontologies belonging to different subdomains such as crops, fertilizers, and soil. Merging can be performed only after accurately aligning the concepts of the source ontologies.

Section 2 outlines various tools and methods available for ontology merging. Section 3 presents the motivation behind this research work. Section 4 presents the proposed scheme. A review and analysis of the proposed scheme is presented in Sect. 5. Section 6 concludes the paper with some future directions.

2   Literature Survey

Many algorithms and tools for ontology merging have been worked upon. It creates a problem for naïve researchers in the field as they face exorbitant text relating to ontology merging tools and methods. [4–6] provide a comprehensive survey of ontology merging and alignment methods and tools.

One of the earliest tools in ontology merging is SMART [7]. It identifies linguistically similar classes and creates a list of initial linguistic similarity based on class-name similarity. Examples of linguistic similarity measures used are synonym, common suffix and prefix, and shared sub-string etc. It is a semi-automated tool and generates suggestions for matches, users need to validate those suggestions.

AnchorPROMPT [8] is based on graph structure of ontologies. It traverses the path between related term-pairs, called as anchors in [8], in the source ontologies and identifies the similar terms along this path. Using this information, AnchorPROMPT finds new anchors.

PROMPT [9] is an ontology management tool. It facilitates ontology merging, alignment, and versioning. PROMPT provides merging suggestions to the user. These merging suggestions are based on linguistic and structural knowledge. PROMPT also presents aftereffects of applying these merging suggestions to the ontology.

SMART, anchorPROMPT and PROMPT are developed as a plugin for Protégé-2000[1]. PROMPT is a popularly used tool for ontology merging.

A knowledge based translator, named OntoMorph, to facilitate ontology merging is presented in [10]. OntoMorph specifies mappings in the form of rule language. It uses both, syntactic and semantic rewriting. Syntactic rewriting uses pattern matching and works with sentence level transformations. Semantic rewriting uses logical inference on semantic models.

HICAL (Hierarchical Concept Alignment System) employs machine learning techniques for alignment of concept hierarchies [11]. It infers mappings from the overlap of data instances between two taxonomies.

CMS (Crosi Mapping System) uses semantics of the OWL constructs for structure matching in ontology alignment [12]. FCA-Merge [13] is based on bottom-up ontology merging. A merged concept lattice is obtained using formal concept analysis and application-specific ontology instances (belonging to the ontologies which are to be merged) which is converted to a merged ontology by human intervention. Chimaera [14] is an interactive tool for ontology merging. It assists the users for ontology editing, merging and testing.

Protégé, an ontology editing environment, also provides automatic ontology merging service among other options. It provides GUI based ontology merging.

Ontology Alignment Evaluation Initiative (OAEI)[2] is a standard platform for evaluating ontology merging/matching/mapping/alignment tools. OAEI aims at improving ontology matchers by assessing their weaknesses and strengths. It also provides comparison of various matchers. OAEI benchmark datasets available for 2016 campaign[3] are *benchmark, anatomy, conference, multiform, interactive machine evaluation, large biomedical ontologies, disease and phenotype, process model matching,* and *instance matching.* Agriculture domain is not present in these benchmark datasets yet, we look forward to test the system developed during current work in OAEI campaigns.

Availability of so many tools and methods for ontology merging proves to be a motivation for identifying an optimal solution for merging cross-domain agricultural ontologies.

---

[1] http://protege.stanford.edu.

[2] http://oaei.ontologymatching.org/.

[3] http://oaei.ontologymatching.org/2016/.

3   Motivation

There was a need to develop a new algorithm and tool for merging as none of the tools and research work highlighted in the previous section are currently functional. This is attributed primarily to the fact that most of the tools for ontology merging are developed as part of some research activity. On an average, the research goes on for 3–4 years on a specific target [15]. Thereafter, the tools/plugins get outdated and hence discarded. One such example is PROMPT[4] which is a pioneer work in ontology merging. The website shows PROMPT 3.0 as the last release of the plugin which is compatible with Protégé3.3.1. The plugin is not available for download now (last checked on 20 January 2017). It has been 10 years since the website was updated. The Refactor merging tool in Protégé does not perform merging correctly and suffers from various problems, which the proposed scheme aims to overcome, as discussed further in Sect. 5.

Another motivation is to meet the demand of web services. Although knowledge resources like AGROVOC[5], NAL thesaurus[6], Agropedia[7] are available for agricultural domain, these do not fulfil the requirements of answering queries by common users and suffer from some limitations. AGROVOC is a vast thesaurus and hence contains many terms which are not relevant from farmers' perspective of agricultural domain. Examples are-*curriculum*, *indigenous knowledge*, *computer software*, *vocabulary* etc. It also lacks some relevant agricultural terms, e.g., *coriander* (an Indian herb), *neutral fertilizer* (type of fertilizer), *straight fertilizer* (type of fertilizer), *complex fertilizer* (type of fertilizer) etc. Target users of NAL thesaurus are agricultural researchers as it contains too scientific terms. For example, NAL thesaurus provides two options for *search type- terms contain text* and *terms begin with text*. When we searched the term *coriander* with both options, NAL thesaurus displayed thousands of results with none matching the word "*coriander*". Agropedia provides knowledge models for few crops as image and pdf form. These knowledge models are created by teams of domain experts with great efforts. Ontology creation for agricultural domain is still in its infancy and requires attention. WordNet can also be used to aid the merging process.

4   The Proposed Scheme

The following algorithm (Fig. 1) illustrates the proposed scheme of ontology merging. The algorithm works with 'n' number of input ontologies (in the form of owl files) and gives a final merged ontology as output. The algorithm makes use of element-level as well as structure level matching techniques for merging the concepts, instances and relations (as explained in Sect. 1) of the input ontologies.

---

[4] http://protegewiki.stanford.edu/wiki/PROMPT).

[5] http://aims.fao.org/vest-registry/vocabularies/agrovoc-multilingual-agricultural-thesaurus.

[6] http://agclass.nal.usda.gov/.

[7] http://agropedia.iitk.ac.in/.

```
1. Consider owl ontologies Oₐ, 1≤a≤n, n is the number of
   ontologies to be merged. Oₘ is the final merged
   ontology.
2. Parse owl files to extract the set of concepts and
   relations in Oₐ , 1≤a≤n.
3. Use string based and informal resource based matching
   techniques to match the similar concepts in Oₐ and O_b,
   1≤a≤n, 1≤b≤n.
4. Resolve the matching concepts C_ai and C_bj into C_Mk in O_M,
   where 1≤i,j≤n_m, where n_m is the no. of concepts in m^th
   ontology
5. Find the matching relation pairs in Oa, 1≤a≤n, using
   structure-level matching techniques.
6. Resolve the matching relations of all the ontologies.
7. Find out the related concepts and their relations among
   the unresolved concepts and relations using relevant
   portions of the text. This is done using NLP techniques
   and statistical analysis.
8. Merge the related concepts into O_M.
9. Copy the remaining concepts and relations from O_a,
   1≤a≤n, into O_M.
10.Generate the owl file of the final merged ontology
   using an Ontology Editor.
```

Fig. 1. The proposed algorithm

The proposed scheme applies basic alignment schemes like prefix and suffix matching, edit distance, n-grams (3-grams in particular), tokenization and lemmatization and common knowledge thesauri (WordNet). After applying these techniques on the concepts of ontologies, the alignment scores between different pairs of concepts, normalized between 0 and 1 are obtained. A value closer to 1 suggests better alignment. These techniques are then combined after analysing the scores obtained using different matching techniques. It is observed that WordNet shows almost all words as similar because they belong to the same domain- agriculture. Hence WordNet similarity is used only as threshold maintained at 0.9. The degree of similarity is calculated by assigning equal weight to other techniques and checking with a threshold value of 1.2. Duplication of aligned classes is removed and symmetricity of alignment is ensured, i.e., the order in which the sources ontologies are loaded in the tool should not matter.

Element-level techniques do not suffice for merging ontologies. For example, *WeatherCondition* in *Ontology1* (Fig. 3) is very similar to *EnvironmentalFactor* of *Ontology2* (Fig. 4). However, this is not apparent by any element-level technique. *Type* as sub-class of *Soil* in *Ontology1* is merged with *Type* (sub-class of *fertilizer*) in *Ontology2*. This should not be happening as *Type* holds different meanings in the two domains. Structure-level techniques are used to deal with such issues by:

- Addition of Concepts- All non-aligned concepts are aligned only if they have a large co-topic similarity and a substantial number of common children concepts
- Removal of Concepts- Alignments obtained using element-level techniques are removed if the concepts have very less co-topic similarity and non-matching children concepts.

These techniques are first applied on concepts of the source ontologies and subsequently on the properties and relations of the concepts in the source ontologies.

Fig. 2. Subdomains of Indian agriculture, *Source*: [16, 17]

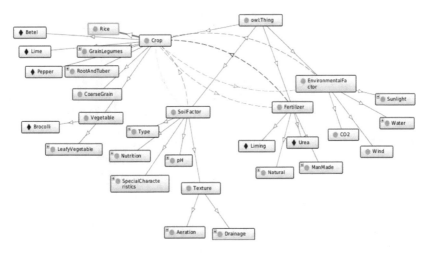

Fig. 3. Graphical representation of *Ontology1*: Source Ontology

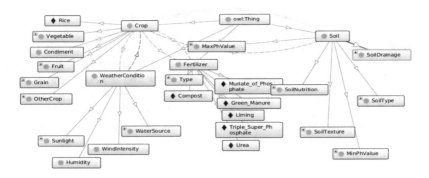

Fig. 4. Graphical representation of *Ontology2*: Source Ontology

5   Results and Analysis

In this section, we present the details of results obtained with ontology merging using Protégé and the proposed scheme. A lot of merging tools and methods have already been worked upon as outlined in Sect. 2. Protégé, a well-known system for ontology management is also bundled with an ontology merging feature.

We have examined the performance for merging two ontologies *Ontology1* (Fig. 3) and *Ontology2* (Fig. 4). These ontologies have been generated from agricultural text available over websites, such as, agricoop.nic.in, farmer.gov.in, agrifarming.in to name a few. The algorithm presented in [18] has been used to extract terms and relationships for construction of ontologies. The extracted terms and relationships are then fed to Protégé to generate owl files of the ontologies. The graphical view of ontologies presented here is generated using OntoGraf plugin of Protégé. Table 1 explains the representation of ontologies using OntoGraf with examples from *Ontology1* shown in Fig. 3. Same representation scheme has been followed for *Ontology2*, and also for *Ontology3*, and *Ontology4* which are discussed below.

Table 1. Representation of ontologies using OntoGraf

Meaning	Example from *ontology1* in Fig. 3	Type of element
Ontology concept	`Crop, SoilFactor, Fertilizer, Vegetable, Rice, Wind`	Rectangular node having yellow dot with the label
Instances in the ontology	`Pepper, Lime, Broccoli, Urea`	Rectangular node having purple diamond with the label
Hierarchical relations between concepts	`Leafy Vegetable` is a subclass of `Vegetable`, `Vegetable` is a subclass of `Crop`	Directed solid links
Domain-specific relations between the concepts	Purplle dotted line represents the relation `works_on(Fertilizer, Crop)`	Directed dotted links

These ontologies are created from the domains of Crop, Weather, Fertilizer and Soil. These are created in a way so as to enable checking of merging scheme for removing duplicates, alignment accuracy, detection of matching concepts with dissimilar names, etc. The proposed scheme has been implemented using Python, Owlready[8] library to extract the concepts, object properties and data properties from the source ontologies. Similarity techniques (as discussed in Sect. 4) are implemented in Python to find the alignments between the extracted elements of the ontologies. After applying the alignments obtained, owl file for the final merged ontology is also

---

[8] https://pypi.python.org/pypi/Owlready.

Table 2. Metrics of *Ontologies*

	Ontology1	*Ontology2*	*Ontology3*	*Ontology4*
No. of concepts	22	22	44	29
No. of object properties	6	6	12	6
No. of data properties	7	8	15	8
No. of instances	41	43	84	56

exported to *Ontology4.owl* using Owlready in Python. Table 2 shows some essential metrics of the two source ontologies and resultant ontologies.

It can be seen in Fig. 5 that Protégé is just inserting the concepts of one ontology into another. Thus, the resulting ontology's metrics are simply a sum of the source ontologies' metrics. Also, it results in duplication of concepts as can be seen for *crop, sunlight, fertilizer, etc.* In *Ontology1* and *Ontology2*, *crop* is a common concept. It is expected that it is considered as one concept in *merged Ontology*. However, Protégé shows *crop* (*Ontology1*) and all its subclasses as different from crop (*Ontology2*) in the *Ontology3* (Fig. 5). This tool does not link the concepts which are expected to play a similar role on a structural level. It is also to be noted that Protégé does not merge the relations and instances of concepts in the source ontologies. This is evidenced in Fig. 5.

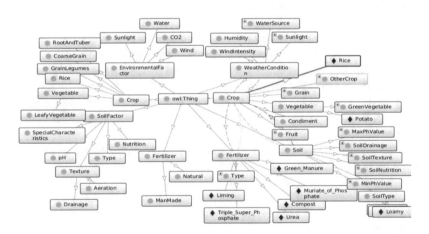

Fig. 5. Graphical representation of *Ontology3*: merged ontology using Protégé

Figure 6 shows *Ontology4*, merged ontology obtained using the proposed scheme. The metrics for this ontology are smaller than the sum of the original ones, hence, this tool is memory efficient. This scheme removes duplicate concepts in the merged ontology and also merges structurally similar concepts like *WeatherCondition* and *EnvironmentalFactor*. It also merges the instances and relations of concepts in the source ontologies. The scheme ensures that every aspect of source ontologies is present in the merged ontology. Thus, the proposed scheme obtained much better merged ontology in comparison with the merged ontology obtained using Protégé.

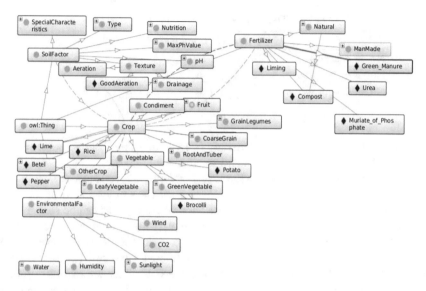

Fig. 6. Graphical representation of *Ontology4*: merged ontology using the proposed scheme

6 Conclusion and Future Work

The present work examines existing tools and proposes a new scheme for ontology merging. Protégé provides a good interface for ontology creation. It automatically generates OWL code from the information provided by the user regarding the classes, data properties, object properties, annotations etc. A lot of literature about ontology merging tools is available. However, the literature does not provide guidance for practical applications of the same. This paper equips the reader about a practical experience of ontology merging.

Protégé gives just a concatenation of source ontologies after merging. Moreover, it does not merge the instances and relations of the concepts of source ontologies. The scheme presented in this paper uses both element-level as well as structure-level techniques for alignment of concepts, instances and relations. It takes care of duplicity of concepts as well as incorrect alignments in the merged ontology. The merged ontology retains every aspect of the source ontologies.

The paper presents a good scheme for ontology merging taking practical example from agriculture domain. The scheme overcomes the shortcomings faced while using an existing tool for ontology merging. This work is being more rigorously tested using data-driven ontology evaluation and application-based ontology evaluation techniques for future work and improvement in the scheme.

Acknowledgements. The work has been supported by Department of Electronics and Information Technology, Ministry of Communication and Information Technology, Government of India in the form of a sponsored project entitled "Development of Tools for Automatic Term Extraction and RDFization of Agriculture Terms with focus on Crops sub-domain".

References

1. Gruber, T.R.: Toward principles for the design of ontologies used for knowledge sharing? Int. J. Hum. Comput. Stud. **43**(5), 907–928 (1995)
2. Lata, S., Sinha, B., Kumar, E., Chandra, S., Arora, R.: Semantic web query on e-governance data and designing ontology for agriculture domain. Int. J. Web Semant. Technol. **4**(3), 65 (2013)
3. Malik, N., Sharan, A., Hijam, D.: Ontology development for agriculture domain. In: 2nd International Conference Computing for Sustainable Global Development (INDIACom), pp. 738–742. IEEE (2015)
4. Choi, N., Song, I.-Y., Han, H.: A survey on ontology mapping. ACM Sigmod Rec. **35**(3), 34–41 (2006)
5. Predoiu, L., Feier, C., Scharffe, F., de Bruijn, J., Martín-Recuerda, F., Manov, D., Ehrig, M.: D4. 2.2 state-of-the-art survey on ontology merging and aligning V2. In: EU-IST Integrated Project IST-2003-506826 SEKT, p. 79 (2005)
6. Shvaiko, P., Euzenat, J.: Ontology matching: state of the art and future challenges. IEEE Trans. Knowl. Data Eng. **25**(1), 158–176 (2013)
7. Noy, N.F., Musen, M.A.: SMART: automated support for ontology merging and alignment. In: Proceedings of the 12th Workshop on Knowledge Acquisition, Modelling, and Management (KAW 1999), Banf, Canada (1999)
8. Noy, N.F., Musen, M.A.: Anchor-PROMPT: using non-local context for semantic matching. In: Proceedings of the Workshop on Ontologies and Information Sharing at the International Joint Conference on Artificial Intelligence (IJCAI), pp. 63–70 (2001)
9. Noy, N.F., Musen, M.A.: The PROMPT suite: interactive tools for ontology merging and mapping. Int. J. Hum. Comput. Stud. **59**(6), 983–1024 (2003)
10. Chalupsky, H.: Ontomorph: a translation system for symbolic knowledge. In: KR, pp. 471–482 (2000)
11. Ichise, R., Takeda, H., Honiden, S.: Rule induction for concept hierarchy alignment. In: Workshop on Ontology Learning (2001)
12. Kalfoglou, Y., Hu, B.: CROSI Mapping System (CMS) - result of the 2005 ontology alignment contest. In: Ashpole, B., Ehrig, M., Euzenat, J., Stuckenschmidt, H. (eds.) Proceedings of the K-CAP 2005 Workshop on Integrating Ontologies, pp. 77–85 (2005)
13. Stumme, G., Maedche, A.: FCA-merge: bottom-up merging of ontologies. IJCAI **1**, 225–230 (2001)
14. McGuinness, D.L., Fikes, R., Rice, J., Wilder, S.: The chimaera ontology environment. In: AAAI/IAAI 2000, pp. 1123–1124 (2000)
15. Otero-Cerdeira, L., Rodríguez-Martínez, F.J., Gómez-Rodríguez, A.: Ontology matching: a literature review. Expert Syst. Appl. **42**(2), 949–971 (2015)
16. Sinha, B., Chandra, S.: Semantic web query on e-governance data for crop ontology model of Indian agriculture domain. In: Dutta, B., Madalli, D.P. (eds.) International Conference on Knowledge Modelling and Knowledge Management (ICKM), Bangalore (Bengaluru), pp. 56–66 (2013a)
17. Sinha, B., Chandra, S.: Semantic web ontology model for Indian agriculture domain. In: Dutta, B., Madalli, D.P. (eds.) International Conference on Knowledge Modelling and Knowledge Management (ICKM), Bangalore (Bengaluru), pp. 101–111 (2013b)
18. Chatterjee, N., Kaushik, N.: A practical approach for term and relationship extraction for automatic ontology creation from agricultural text. In: International Conference on Information Technology (ICIT), Bhubaneshwar, pp. 241–247 (2016)

Financial Time Series Clustering

Kartikay Gupta$^{(\boxtimes)}$ and Niladri Chatterjee

Indian Institute of Technology Delhi, New Delhi, India
{maz158144,niladri}@iitd.ac.in

Abstract. Financial time series clustering finds application in forecasting, noise reduction and enhanced index tracking. The central theme in all the available clustering algorithms is the dissimilarity measure employed by the algorithm. The dissimilarity measures, applicable in financial domain, as used or suggested in past researches, are correlation based dissimilarity measure, temporal correlation based dissimilarity measure and dynamic time wrapping (DTW) based dissimilarity measure. One shortcoming of these dissimilarity measures is that they do not take into account the lead or lag existing between the returns of different stocks which changes with time. Mostly, such stocks with high value of correlation at some lead or lag belong to the same cluster (or sector). The present paper, proposes two new dissimilarity measures which show superior clustering results as compared to past measures when compared over 3 data sets comprising of 526 companies. *abstract* environment.

Keywords: Clustering · Finance · Time series · Cross-correlation · Forecasting

1 Literature Review

In the last two decades a significant amount of work has been done on time series clustering [7]. Financial Time series clustering is a subject that has also gained lots of attention in the last decade [14]. It is an important area of research that finds wide applications in noise reduction, forecasting and enhanced index tracking [1]. A good forecast of future prices is desired by financial companies especially the algorithmic trading firms. Index track funds are low cost funds that closely track the returns of one particular index. In a typical financial time series clustering procedure, each time series is considered as an individual object and inter-object dissimilarities are then calculated. Subsequent clustering is done using one of the many clustering algorithms. Rosario [2] used a well-known dissimilarity measure based on the Pearson's correlation coefficient. They then cluster the data set using single linkage hierarchical clustering. Saeed [4] used symbolic representation for dimensionality reduction of financial time series data, and then used longest common subsequence as similarity measurement. Guan [5] proposed a similarity measure for time series clustering which relied on the signs (positive or negative) of the logarithmic returns of the stock

© Springer International Publishing AG 2018
S.C. Satapathy and A. Joshi (eds.), *Information and Communication
Technology for Intelligent Systems (ICTIS 2017) - Volume 2*, Smart Innovation,
Systems and Technologies 84, DOI 10.1007/978-3-319-63645-0_16

prices. It did not take into account the size of movement. Marti et al. [6] tried to find answer to the question, what should be an appropriate length of a time series for the clustering procedure. John et al. [8] proposed a shape based dissimilarity for time series clustering which relied on the cross correlation coefficients. This work comes closest to our approach, but still there is a significant difference between the two approaches. While calculating their dissimilarity measure, they do not break the time series into smaller segments. In the present approach, time series are broken into smaller parts and then a different procedure is followed. This is because lead – lag relationship between two time series may change over time. Further, we propose another dissimilarity measure that gives similar or better results as compared to the first dissimilarity measure proposed in the present paper.

2   Preliminaries

In the present work, hierarchical clustering is used to form clusters from the inter object dissimilarity matrix computed using a dissimilarity measure. Linkage method is an important aspect of hierarchical clustering algorithm. In the present work, we choose 'single' linkage and 'ward' linkage (implemented as ward.D2 linkage in R 'stats' package) for our analysis. This is because 'single' linkage has been a preferred choice of researchers in financial time series clustering papers, e.g., [2,3]. Ward linkage [10] was used for financial time series clustering by Guan [5]. Details about pre-existing dissimilarity measures i.e., Correlation based dissimilarity measure (COR) and Temporal correlation based dissimilarity measure (CORT) are given in the Appendix. Additionally, taking inspiration from [16], lead/lag time between two time series X_T and Y_T is defined as the integer 'k' which maximizes cross correlation between the two time series.

3   Proposed Dissimilarity Measures

3.1   Cross-Correlation Type Dissimilarity Measure (CCT)

Given two time series X_T and Y_T (each of length T), here the interest is in finding a dissimilarity measure between them. Let 'm' be the maximum value of lead or lag being taken into consideration for calculation of dissimilarity between the two time series. We consider a segment of time series X_T starting from 'm+1' and ending at 'T-m-1', and divide this segment into 'n' equal parts each of length 'p'. Here, we conveniently choose 'm', 'n' and 'p' such that $2m+np+1 = T$. This is required to make sure that all the data points in time series are utilized for the calculation of dissimilarity measure. Though it would still suffice if '2m+np+1' is slightly less than T. Another couple of restrictions are that $m \leq p$ and $p \geq 15$. These restrictions have been imposed to avoid unwanted cross-correlations. Now, with this background we define our CCT similarity measure which is given by the following:

$$CCT = \frac{1}{n} \times \sum_{l=1}^{n} max\{CCT_k(m+1+p \times (l-1)) \quad | -m \leq k \leq m\}, \quad (1)$$

where $CCT_k(i)$ is defined as follows

$$CCT_k(i) = \frac{\sum_{j=i}^{i+p-1}(x_{j+1} - x_j)(y_{j+k+1} - y_{j+k})}{\sqrt{\sum_{j=i}^{i+p-1}(x_{j+1} - x_j)^2}\sqrt{\sum_{j=i}^{i+p-1}(y_{j+k+1} - y_{j+k})^2}} \qquad (2)$$

The value of CCT_k is same as the correlation between returns of a segment of time series Y_t at a lead k with respect to a segment of time series X_t. The motivation behind this similarity measure is that similar financial time series would have more sub-segments which are highly correlated with each other at some lead or lag. Stock prices of similar assets in different exchanges exhibit such pattern. This phenomena is perfectly captured by CCT similarity measure. Notice that value of each of $CCT_k(i)$ lies in the interval $[-1,1]$. Hence, the value of CCT similarity measure also lies in the interval $[-1,1]$. This similarity measure is then converted into a dissimilarity measure using a function, which is given by:

$$\phi(u) = \frac{2}{1 + e^{4 \times u}} \qquad (3)$$

We choose to do our analysis with the above function as opposed to the function $\sqrt{2(1-x)}$ for conversion of similarity measure into dissimilarity measure. Through our data experiments, we see that even this function gives similar results as the function in Eq. 3. Though we have not given details of those experiments in this paper. The dissimilarity measure thus obtained after conversion can be used for clustering of financial time series data.

As an example for CCT similarity measure consider the two time series data given in Table 1. Since the length of each time series is 37, one suitable choice for 'p', 'n' and 'm' may be 15, 2 and 3, respectively. In subsequent calculations it is found that for the first segment of series 1 (i.e., 4th to 18th data point in Series 1) the max $CCT_k(4)$ value is 0.57 which exists at k = 3. For the second segment (i.e., 19th to 33rd data point in Series 1) the max $CCT_k(19)$ value is 0.27 which exists at k = 0. Thus the value of CCT measure comes out to be 0.42.

Table 1. Hypothetical data set for two time series each of length 37.

S. no.	Series 1	Series 2	S. no.	Series 1	Series 2	S. no.	Series 1	Series 2	S.no.	Series 1	Series 2
1	0.81	0.15	11	0.96	0.96	21	0.58	0.68	31	0.13	0.47
2	0	0.76	12	0.69	0.36	22	0.2	0.83	32	0.23	0.14
3	0.92	0.01	13	0.05	0.7	23	0.69	0.01	33	0.75	0.98
4	0.94	0.1	14	0.85	0.66	24	0.76	0.01	34	0.63	0.85
5	0.97	0.56	15	0.55	0.37	25	0.77	0.44	35	0.29	0.36
6	0.77	0.88	16	0.65	0.07	26	0.13	0.01	36	0.33	0.55
7	0.26	0.91	17	0.39	0.63	27	0.89	0.05	37	0.65	0.39
8	0.68	0.05	18	0.32	0.35	28	0.78	0.61			
9	0.13	0.26	19	0.7	0.79	29	0.95	0.69			
10	0.82	0.5	20	0.39	0.13	30	0.56	0.13			

3.2   Cross-Correlation Type-II Dissimilarity Measure (CCT-II)

In another version of this dissimilarity measure, we ignore those intervals whose 'maximum $CCT_k(i)$' value is less than a given threshold (denoted by 'Thr'). We mean that instead of the 'maximum $CCT_k(i)$' value we put '0', when that is less than the threshold. Rest of the computations remains the same. The expression for this similarity measure is as follows:

$$\text{CCT-II} = \frac{1}{n} \times \sum_{l=1}^{n} (M(m+1+p \times (l-1)) \times I_{[Thr,1]}(M(m+1+p \times (l-1)))) \quad (4)$$

$$M(i) = max\{CCT_k(i) \quad | -m \leq k \leq m\} \quad (5)$$

where $CCT_k(i)$ is same as defined in (3)

and $I_{[Thr,1]}(x)$ denotes the indicator function of the set $[Thr, 1]$.

This similarity measure is then converted into dissimilarity measure using the function given in Eq. (3).

As an example, consider the time series data given in Table 1. We now compute CCT-II similarity measure with the same set of values (as used for computing CCT similarity measure) i.e. p = 15, n = 2 and m = 3. The threshold value is set as 0.50. Again it is noticed that max $CCT_k()$ values are 0.57 and 0.27. But since 0.27 lies below the threshold thus it is replaced by 0. Thus, the value of CCT-II similarity measure comes out to be 0.29.

The motivation behind this dissimilarity measure is that often the stock prices are weakly correlated to each other at some lead or lag. This may be due to a general trend followed by all stocks or it may be a spurious relation. This leads to error while clustering the data set using CCT dissimilarity measure. This noise can be removed by not considering the value of cross-correlations whose value is less than a given threshold.

In order to apply CCT-II measure of dissimilarity between two time series we need to fix the threshold value beforehand. The threshold value obviously lies in the interval (0, 1). Additionally, it can't be close to 0 as then there would remain very little difference between the values of CCT and CCT-II dissimilarity measures. It can't be close to 1 as then most of the information regarding the behavior of two time series would be ignored and hence will not be considered in the evaluation of the final expression of CCT-II dissimilarity measure. Thus, threshold value should be kept close to 0.5. For this work we have chosen the range for the threshold to be [0.35, 0.65]. If a test data set is available then it can be used to find the optimal value of threshold. If test data set is not available, then choose threshold value as any random value in the range [0.35, 0.65].

The time complexity of CCT and CCT-II dissimilarity measure is O(pmn). Calculation of $CCT_k(i)$ value requires O(p) computations, as we are calculating cross-correlation value at lead 'k' between two time series segments each of length 'p'. Since we need to evaluate this expression for all k such that $-m \leq k \leq m$, thus it takes O(pm) time to evaluate '$max\{CCT_k(i) \quad | -m \leq k \leq m\}$'. This process is repeated 'n' times thus, CCT dissimilarity measure is of O(pmn) time complexity. Similarly, it can be argued that CCT-II dissimilarity measure is of O(pmn) time complexity. This time complexity is better than time complexity of Dynamic Time Wraping dissimilarity measure, which takes $O(T^2)$ time [15] to find its final expression, where T is the length of time series. Since 'np+2m+1' is equal to T or slightly less than T, thus, this time complexity is equivalent to $O(n^2p^2)$. This time complexity is clearly greater than O(pmn) as $m \leq p$. Lower time complexity of proposed measures enable its computations to be carried out faster.

4   Experiments and Analysis

Experiments are conducted on 3 data sets one by one. Each data set consists of 'End of Day' (EOD) stock prices of some companies. Figure 1 depicts the format of data sets used for the experiments in the present paper. While clustering, each time series associated with a company is considered as an individual object. Inter object dissimilarities are then calculated with each of the 4 measures i.e., COR, CORT, CCT, CCT-II. A dissimilarity matrix is thus created, which is used for further analysis. Single linkage hierarchical clustering and Ward linkage hierarchical clustering are then employed to create the corresponding dendograms (see Fig. 2). The dendogram then can be cut at any level to form desired number of clusters of the data set.

Company Name	Date 1	Date 2	Date 3			Date T
Company 1	0.45	0.46	0.43			0.52
Company 2	1.56	1.59	1.58			1.42
Company 3	2.89	2.85	2.91			2.50
⋮						
Company N	1.07	1.08	1.09			1.05

Fig. 1. EOD stock prices of the companies.

A cluster evaluation measure is then used to compare the clustering results obtained through the different dissimilarity measures. The present paper uses the cluster evaluation measure defined in [9,16]. This cluster evaluation measure lies in the range [0,1]. Higher value of cluster evaluation measure corresponds to better clustering results.

Here, we discuss results corresponding to each data set in a different subsection. R version 3.1.1 was used for preparation of all the figures presented in this paper. Since the number of data points in all time series in all the experiments is 2014, hence the parameter values of 'n', 'p' and 'm' are taken to be 18,100 and 100 respectively in all the experiments mentioned below. In these

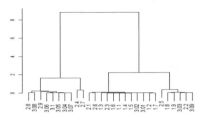

(a) Dissimilarity Measure: COR

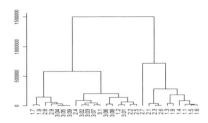

(b) Dissimilarity Measure: CORT

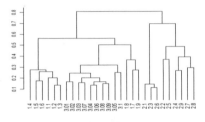

(c) Dissimilarity Measure: CCT

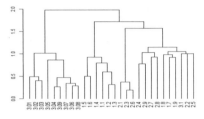

(d) Dissimilarity Measure: CCT-II

Fig. 2. The above figures give dendograms corresponding to different dissimilarity measures. Ward linkage hierarchical clustering has been used for forming the dendograms. Each object in the clusters is represented by a number whose unit's place indicate it's true cluster value.

experiments, the threshold value for the CCT-II measure concerning Indian companies (the first data set) is taken to be higher as compared with the companies traded in USA (the second and third data set). This is because prices of the Indian companies tend to show more noise or spurious cross-correlations as compared to prices of American companies. Thus, threshold value for the first data set is taken to be 0.65 and threshold value for the second & third data set is taken to be 0.35.

4.1 Indian Data Set

The EOD stock prices of the companies listed in Table 2, form the first data set of our experiments. The time-span of the prices is from 5th August 2008 till 2nd May 2016. This data set can be originally clustered into 3 clusters as indicated in Table 2. Dendograms obtained on this data set using Ward linkage hierarchical clustering are shown in Fig. 2. Table 3 gives cluster evaluation measure corresponding to different representations (number of clusters formed through dendogram) and different dissimilarity measures. Amongst all the dendograms shown in the Fig. 2, the best result is seen in Fig. 2 (d). Figure 2(d) shows dendogram formed using CCT-II dissimilarity measure. Cluster evaluation measure

Table 2. The name of the companies whose stock prices are part of the Indian data set. These companies can be divided into 3 broad categories.

Cluster 1		Cluster 2		Cluster 3	
Top Private Banks in India		Top Oil Companies in India		Top Public Banks in India	
Name	Symbol	Name	Symbol	Name	Symbol
HDFC	1.1	Indian Oil Corp. Ltd.	2.1	SBI	3.01
ICICI	1.2	ONGC	2.2	Bank of Baroda	3.02
AXIS	1.3	Bharat Petroleum	2.3	PNB	3.03
Kotak Mahindra Bank Ltd.	1.4	Essar oil Ltd.	2.4	IDBI Bank Ltd.	3.04
Indusind Bank Ltd.	1.5	Cairn India Ltd.	2.5	Central Bank of India	3.05
Yes Bank Ltd.	1.6	Hindustan Petroleum Corp. Ltd.	2.6	Canara Bank	3.06
The Federal Bank Ltd.	1.7	Aban offshore Ltd.	2.7	Union Bank of India	3.07
Karur Vysya Bank Ltd.	1.8	Hindustan Oil exploraion company Ltd.	2.8	Bank Of India	3.08
South Indian Bank Ltd.	1.9	Mangalore Refinery & Petrochemicals Ltd.	2.9	Syndicate Bank	3.09
				Indian Bank	3.10

Table 3. Cluster evaluation measure for the different dissimilarity measures. This table corresponds to the Indian data set. 'Number of Clusters' represent the number of clusters formed using the dendogram. Higher the measure the closer is the representation to the true clusters. 'COR', 'CORT', 'CCT' and 'CCT-II' stand for correlation based dissimilarity measure, temporal correlation based dissimilarity measure, first proposed measure and second proposed measure respectively.

Cluster evaluation measure				
Number of clusters	COR	CORT	CCT	CCT-II
2	0.56	0.59	0.78	0.74
3	0.56	0.58	**0.71**	**0.86**
4	0.57	0.56	0.82	0.79

verifies that the clusters obtained through the proposed measures are better than the pre-existing measures. Cluster evaluation measure is consistently higher for the proposed measures, irrespective of how many clusters we form out of the dendograms.

Clustering results obtained on first data set using Single linkage hierarchical clustering also verify the superiority of CCT and CCT-II measures. In this case also, Cluster evaluation measure values for CCT and CCT-II measures are higher than the COR and CORT dissimilarity measures.

4.2   S&P500 Data Set

The EOD stock prices of the companies listed in S&P500 index form the second data set. These companies are traded in USA. The time span of the prices for this data set is from 12th September 2008 till 23rd august 2016. The companies whose prices were not available for the complete time span under consideration, were removed from the data sets. This data set can be originally clustered into 10 clusters, where each cluster represents the sector of the company.

In the experiment associated with this data set, Cluster evaluation measure values are given in Tables 4 and 5. The threshold value is taken to be 0.35 for CCT-II measure in this experiment. In the case of ward linkage (as seen in Table 4) CCT and CCT-II measures clearly give better results as compared to COR and CORT dissimilarity measures. In the case of single linkage (as seen in Table 5), all measures give similar results.

Table 4. Cluster evaluation measure corresponding to different dissimilarity measures and number of clusters. Ward linkage hierarchical clustering has been used for clustering the datasets. This table corresponds to the S&P500 data set which can be originally divided into 10 clusters.

Cluster evaluation measure				
Number of clusters	COR	CORT	CCT	CCT-II
9	0.23	0.24	0.69	0.65
10	0.23	0.24	**0.69**	**0.70**
11	0.23	0.24	0.69	0.68

Table 5. Cluster evaluation measure corresponding to different dissimilarity measures and number of clusters. Single linkage hierarchical clustering has been used for clustering the datasets. This table corresponds to the S&P500 data set which can be originally divided into 10 clusters.

Cluster evaluation measure				
Number of clusters	COR	CORT	CCT	CCT-II
9	0.18	0.18	0.21	0.21
10	0.18	0.18	**0.21**	**0.21**
11	0.18	0.18	0.21	0.21

Table 6. Cluster evaluation measure for the different dissimilarity measures. 'Number of Clusters' represent the number of clusters formed using the dendogram. Higher value corresponds to better clustering results. This table corresponds to the DJIA data set which can be originally divided into 8 clusters.

Cluster evaluation measure				
Number of clusters	COR	CORT	CCT	CCT-II
7	0.39	0.39	0.67	0.54
8	0.40	0.43	**0.65**	**0.56**
9	0.42	0.43	0.66	0.58

4.3   DJIA Data Set

The EOD stock prices of the companies listed in Dow Jones Industrial Average index form the third data set of our experiments. The time span of the prices for this data set is from 12th September 2008 till 23rd august 2016. This data can be originally clustered into 8 clusters. Here, the threshold value is taken to be 0.35 for CCT-II measure.

As can be seen in Table 6, the proposed measures show better or similar results as compared to pre-existing measures when hierarchical clustering is done using Ward linkage. In the case, when clustering is done using Single linkage hierarchical clustering for this data set, the proposed measures give similar or slightly inferior results as compared to COR & CORT measures. Though in this case also, maximum value of cluster evaluation measure (amongst all possible number of clusters) is seen corresponding to CCT measure.

Now, we discuss some of the directions in which future work could be carried out related to this research paper. Determination of optimal values of threshold, n, p and m could be carried out in future. Also, optimal function for conversion of similarity measure into dissimilarity measure needs to be determined.

5   Conclusion

Financial time series clustering is an important area of research and finds wide applications in noise reduction, forecasting and index tracking. In this paper, two new dissimilarity measures have been proposed for financial time series clustering. These dissimilarity measures are used to cluster time series belonging to 3 data sets. One data set consists of EOD stock prices of 28 Indian companies. The second data set consists of EOD stock prices of 468 companies listed in S&P500 Index. The third data set consists of EOD stock prices of companies listed in Dow Jones Industrial Average Index. Overall, the data sets consist of 526 companies which is a fairly large number. Clustering is done and it is shown that our proposed dissimilarity measures outperform existing dissimilarity measures.

Acknowledgments. Kartikay Gupta was supported by Teaching Assistantship Grant by Ministry of Human Resource Development, India.

A   Appendix

A.1   Correlation Based Dissimilarity Measure (COR)

A simple dissimilarity measure for time series clustering is based on Pearson's correlation factor between time series X_T and Y_T given by

$$COR(X_T, Y_T) = \frac{\sum_{t=1}^{T}(X_t - m_X)(Y_t - m_Y)}{\sqrt{\sum_{t=1}^{T}(X_t - m_X)^2}\sqrt{\sum_{t=1}^{T}(Y_t - m_Y)^2}}$$

where m_X and m_Y are the average values of the time series X_T and Y_T respectively. The dissimilarity measure is then given by

$$Dissimilarity_{COR}(X_T, Y_T) = \sqrt{2(1 - COR(X_T, Y_T))}$$

For more information regarding this distance measure, one may refer to [9].

A.2   Temporal Correlation Based Dissimilarity Measure (CORT)

As introduced by Douzal [12], the similarity between two time series is evaluated using first order temporal correlation coefficient [13] given by,

$$CORT(X_T, Y_T) = \frac{\sum_{t=1}^{T-1}(X_{t+1} - X_t)(Y_{t+1} - Y_t)}{\sqrt{\sum_{t=1}^{T-1}(X_{t+1} - X_t)^2}\sqrt{\sum_{t=1}^{T-1}(Y_{t+1} - Y_t)^2}}$$

The dissimilarity proposed by Douzal [12] modulates the 'dissimilarity value' between X_T and Y_T using the coefficient $CORT(X_T, Y_T)$. Specifically, it is defined as follows.

$$Dissimilarity_{CORT}(X_T, Y_T) = \phi_k[CORT(X_T, Y_T)] \times Dissimilarity(X_T, Y_T)$$

where ϕ_k is an adaptive function given by,

$$\phi_k(u) = \frac{2}{1 + e^{ku}}, \ k \geq 0 \tag{6}$$

and $Dissimilarity(X_T, Y_T)$ refers to dissimilarity value computed using any of the available dissimilarity measures like Euclidean, DTW etc. In this paper, we choose DTW as the preferred dissimilarity measure. This is because DTW effectively takes into account slight shape distortions while calculating its dissimilarity measure value. For more details regarding DTW, readers are referred to [9,11].

References

1. Dose, C.: Clustering of financial time series with application to index and enhanced index tracking portfolio. Phys. A **355**, 145–151 (2005)
2. Mantegna, S.: An Introduction to Econophysics Correlations and Complexity in Finance. Cambridge University Press, Cambridge (1999)
3. Basalto, N., et al.: Hausdorff clustering of financial time series. Phys. A **379**, 635–644 (2007)
4. Saeed, T.: Effective clustering of time-series data using FCM. Int. J. Mach. Learn. Comput. **4**(2), 170–176 (2014)
5. Guan, J.: Cluster financial time series for portfolio. In: Proceedings of the 2007 ICWAPR, Beijing, China, 2–4 November 2007
6. Marti, A., Nielson, D.: Clustering financial time series: how long is enough? In: Proceedings of the Twenty-Fifth IJCAI (2016)

7. Aghabozorgi, S., Shirkhorshidi, A.S., Wah, T.Y.: Time-series clustering - a decade review. Inform. Syst. **53**, 16–38 (2015)
8. John, G.: k-shape: efficient and accurate clustering of time series. SIGMOD Rec. **45**(1), 69–76 (2016)
9. Montero, V.: TSclust: an r package for time series clustering. J. Stat. Softw. **62**(1), 1–43 (2014)
10. Murtagh, L.: Ward's hierarchical agglomerative clustering method: which algorithms implement Ward's criterion? J. Classif. **31**, 274–295 (2014)
11. Berndt, D.J., Clifford, J.: Using dynamic time warping to find patterns in time series. In: KDD, Workshop, pp. 359–370 (1994)
12. Chouakria, A.D., Nagabhushan, P.N.: Adaptive dissimilarity index for measuring time series proximity. Adv. Data Anal. Classif. **1**(1), 5–21 (2007)
13. Chouakria-Douzal, A.: Compression technique preserving correlations of a multivariate temporal sequence. In: Advances in Intelligent Data Analysis, pp 566–577. Springer, Heidelberg (2003)
14. Gavrilov, M., et al.: Mining the stock market: which measure is best ? In: Proceedings of the Sixth ACM SIGKDD International Conference on Knowledge Discovery and Data Mining, KDD, pp. 487–496 (2000)
15. Giorgino, T.: Computing and visualizing dynamic time warping alignments in R: the dtw package. J. Stat. Softw. **31**(7), 1–24 (2009)
16. Hoffmann, M., et al.: Estimation of the lead-lag parameter from non-synchronous data. ISI/BS Bernoulli **19**(2), 426–461 (2013)

Routing Protocols for Wireless Multimedia Sensor Networks: Challenges and Research Issues

Vijay Ukani$^{(\boxtimes)}$, Priyank Thakkar, and Vishal Parikh

Institute of Technology, Nirma University, Ahmedabad, India
{vijay.ukani,priyank.thakkar,vishalparikh}@nirmauni.ac.in

Abstract. Due to miniaturization of hardware and availability of low-cost, low-power sensors, Wireless Sensor Network and Multimedia Sensor Network applications are increasing day by day. Each application has a specific quality of service and experience requirements. The design of routing and MAC protocol which can fulfill the requirements of the application is challenging given the constrained nature of these devices. Considerable efforts are directed towards the design of energy efficient QoS-aware routing protocols. In this article, we present state of the art review of routing protocols for Wireless Multimedia Sensor Networks while addressing the challenges and providing insight into research issues.

Keywords: Wireless multimedia sensor network · Routing protocols · QoS · QoE · Geographic routing

1 Introduction

Wireless Sensor Network (WSN) has emerged as one of the most prominent technologies with applications in almost all walks of life like industrial process control, health monitoring, target tracking, vehicle traffic monitoring, surveillance etc. [5,22]. WSN is treated as one of the most important technologies in surveillance and monitoring applications [6,7].

Recent developments in low-cost CMOS cameras, microphones and small-scale array sensors which ubiquitously captures multimedia contents have promoted the development of a low-cost network of video sensors. These sensors can be integrated with traditional WSN [10,21,24]. Wireless Multimedia Sensor Network (WMSN) is capable of capturing, processing and disseminating visual information over a network of sensor nodes. Capturing of multimedia contents from specialized sensors incurs a higher cost than traditional scalar sensors. Video data transmission requires high bandwidth due to a large number of bits that are required to represent the video. It also results in higher energy consumption for transmission. Most applications of WMSN like mission critical surveillance are real-time in nature where timely delivery of the information is very critical for the success of the application. Routing and MAC protocols assume a critical role here.

© Springer International Publishing AG 2018
S.C. Satapathy and A. Joshi (eds.), *Information and Communication Technology for Intelligent Systems (ICTIS 2017) - Volume 2*, Smart Innovation, Systems and Technologies 84, DOI 10.1007/978-3-319-63645-0_17

A routing protocol is responsible for forwarding the packet over the most suitable path. Routing protocols in WSNs Differ depending on the application and network architecture. Protocols for traditional WSN like LEACH [20], PEGASIS [18], SPIN [16], Directed Diffusion [2] etc... were either based on energy efficient delivery or content based delivery of the sensed information. Similar approaches are not suitable for WMSN where Quality of Service (QoS) requirements like timely delivery and reliability of the application is of prime importance. Several efforts were made to design QoS-aware routing protocols where the routing decision was stationed on either energy efficiency, deadline of the packet, reliability, packet type or combination of these parameters. The objective of this article is to study working principle of these protocols, compare them, and provide insight into their selection and further research issues.

2   Routing Challenges for WMSN

A WMSN is a resource constrained heterogeneous collection of tiny sensor devices. A typical architecture of WMSN contains scalar sensor nodes, multimedia sensors, multimedia processing hubs for in-network processing [3]. All devices operate on limited power supply making it difficult to deploy complex protocols in operation on these devices. A routing protocol needs to be energy efficient in general over specific requirements of the application. Besides limited energy, there are other factors like limited memory and processing capabilities that hinder the goal of achieving application QoS.

WMSNs are characterized by dense deployment leading to redundancy in sensing and transmission. While redundant transmission can be considered as a type of multipath forwarding providing reliability, it consumes critical resources like bandwidth and energy. In network processing techniques like data fusion might help in reducing the redundancy. In typical WMSN, correlated video contents from different sensors can be fused together or visual contents can be represented in meaningful scalar quantity to reduce redundancy. However, these techniques increase the amount of processing in the network and introduce latency, which complicates the QoS fulfillment for delay sensitive applications.

Sensor nodes are prone to failure. Improper assignment of routing and sensing task might drain energy on few nodes quickly leading to change in network topology. Load balancing remains a desired characteristics of a routing protocol. Many applications of WMSNs are concentrated towards monitoring or surveillance of certain area. Due to high deployment density, the same event triggers many sensors. In case event of interest, affected sensors are required to report the event as quickly as possible to the base station which might lead to congestion in the network. Handling congestion with limited available tools is a challenge in the design of the protocol.

Multimedia content like video or audio needs to be encoded before transmission in the network. Compressed contents generally create chunks of diverse importance, eg. video compression typically creates key frames and difference frame where key frame carries more information than its counterpart. Packets carrying key frames expects better reliability in transmission. Differential handling of packets based on the high-level description needs special attention at routing and MAC layer. Mission critical applications like process monitoring are delay sensitive in nature. Every packet can have an associated deadline within which the packet should be delivered to the destination.

The routing protocol in WMSN has to deal with challenging task of providing variable QoS guarantee depending upon whether the packet carries control information, low-rate scalar data or various high rate video traffic. Each of the traffic class has its own requirements which must be taken care by the routing process. The task becomes challenging due to lack of global knowledge, limited energy, and computational ability of the nodes.

3 Classification of Routing Protocols

Traditionally routing protocols for WSN were classified into three categories based on underlying network structure [4]:

- Flat: All nodes assumes identical roles
- Hierarchical: Nodes will play different roles
- Location-based: Position of the nodes are exploited to aid packet routing

Based on the protocol operation, they can be classified as:

- Multipath-based: Protocols identifying multiple paths
- Query-based: Nodes propagates query for data
- Negotiation-based: Negotiation messages before actual packet forwarding to suppress redundant forwarding
- QoS-based: Satisfy certain QoS metrics when delivering data
- Coherent-based: In-network processing (aggregation) based forwarding

The same classification holds for WMSN but the requirements and thus, metric changes due to change in the type of applications. Protocols proposed specifically for QoS awareness includes Sequential Assignment Routing (SAR) [23], Energy-Aware QoS Routing (EAQoS) [1], Directional Geographic Routing (DGR) [8], RAP [19], RPAR [9], SPEED [14], Multipath Multi-SPEED (MMSPEED) [12], Power Aware SPEED (PASPEED) [26]. Typically, there is no IP-like addressing scheme in WSN. If location information is available, routing protocols can utilize it to reduce the latency and energy consumption of the network. Geographic routing protocols work under the assumption that each node is location aware. In sensor networks, such location-awareness is necessary to make the sensor data meaningful. It is, therefore, natural to utilize geographic position of the node in packet routing. Most of the routing protocols for WMSN like SPEED, PASPEED, RAP, DGR, and MMSPEED uses location information packet forwarding. These protocols are specifically designed for WMSN to fulfill either real-time delivery or reliability or power-awareness or combination of these parameters. Several surveys were conducted involving comparison of routing protocols for WMSNS [4,13,17].

4   QoS Aware Protocols

Most WMSN applications generate real-time traffic. The real-time transmission has specific QoS requirements like deadline-driven transmission, reliability, and Quality of Experience (QoE) on end user part. The routing protocols used should be capable enough to fulfill these demands. Many QoS based routing protocols were proposed in the literature to support QoS transmission of multimedia traffic. They mostly deal with delay and reliability requirements of the application. They assure delivery of packets in time by assigning priority based on deadline to reach destination [14,25,26]. Shorter the deadline to reach the destination, higher is the priority. Based on the importance, different packets may have different reliability requirements. Protocols which deals with providing reliability to the packets had two primary options. They either sent multiple copies of the packet over multiple disjoint paths or estimated the path quality of multiple paths and mapped packets [11] on either of that path based on reliability requirements. These protocols differ in the way multiple paths were computed and the way path quality was estimated.

EAQoS assumes that the transmission of image and video data through WSN requires both energy and QoS awareness. The proposed protocol provides required QoS to real-time traffic at the same time support best effort traffic. It looks for a delay-constrained path with the least cost. The cost is a composite metric involving many parameters like the distance between the nodes, residual energy, time until battery drainage, relay enabling cost, sensing-state cost, max connections per relay, error rate etc. A path is to be chosen from all available path which meets the end-to-end delay requirement of real-time traffic and maximizes the throughput for best effort traffic.

RAP [19] is geographic routing protocol using Velocity Monotonic Scheduling(VMS) policy for packet scheduling in a node. Every packet request some velocity with which it should be transmitted to meet the delay requirements. Each packet is expected to be delivered within deadline if it can travel at the requested velocity. The packets are prioritized based on the requested velocity. Higher the velocity requirement, higher will be the priority of the packet. Packets are forwarded using geographic forwarding. All greedy geographic routing protocols suffer from void/hole where no further progress is possible. RAP re-routes the packets around void by using perimeter routing mode as used in GPSR [15]. The packets which can not meet the deadline even if it transferred over fastest path are bound to miss the deadline. Such packets are dropped to avoid wasting bandwidth.

Transmission power affects the transmission delay of the packets. Experiments were performed by O. Chipara et al. in Real Time Power Aware Routing (RPAR)[9] to measure the effect of transmission power on communication delay. The observations was that increase in transmission power increased the delivery velocity of the packet. Power control is at the core of RAP. For each packet to be forwarded, the required velocity is computed based on remaining deadline and distance to destination. Possible forwarding nodes are evaluated at certain power level. Based on transmission power required, the energy requirements for

transmission is estimated. If none of the neighbors could provide required velocity, RPAR starts power adaptation, which dynamically increases the transmission power to increase velocity provided by that node. Nodes which are already
working at maximum transmission power are ineligible for power adaptation.
The transmission power of a node is decreased if it satisfies velocity requirement
of a packet. Transmission power is reduced to alleviate congestion. However,
power reduction does not solve the congestion issue. Packet redirection towards
non-congested area is required at the network layer to handle congestion.

Directional Geographical Routing (DGR) [8] is a multipath routing protocol
designed to support streaming of video over WSN. It assumed H.26L encoded
video to be transmitted. The network is assumed to be unreliable. The reliability is provided by protecting the transmission using FEC. To cater to high
bandwidth demands of video transmission, the video stream was separated into
multiple streams which can be transferred in parallel over multiple disjoint paths.
DGR constructs the application-specified number of multiple disjoint paths to
the destination and mapped these FEC protected streams on to it. This lead to
energy balancing across the network and better QoS and QoE achieved due to
transmission over multiple paths.

SPEED [14] is a static priority deadline-driven routing protocol. It uses Stateless Non-deterministic Geographic Forwarding (SNGF) as the primary routing
mechanism. The real-time communication is achieved by maintaining desired
delivery speed. Unlike few other QoS-aware routing protocol, SPEED does not
require any specific support from at MAC layer and can work with any MAC
layer protocol. It diverts traffic at routing layer and locally regulates packets
sent to the MAC layer. Thus, it maintains the desired delivery speed across sensor networks. The SNGF chooses the next node that supports desired delivery
speed. Neighbor tables are maintained by exchanging beacons carrying location
information. Routing tables are not required to maintained as the next hop is
selected from 1-hop neighborhood. The memory required is thus proportional to
Neighbor Set (NS). Delay of a link between two nodes is estimated by measuring
the time between data packet sent and ACK received. Unlike other traditional
approaches, congestion in the network is estimated by the delay of the link
instead of the queue size. SPEED ignores the energy available on next hop node
while making a routing decision. PASPEED [26] is a power aware version of
SPEED which maintains the energy available on each neighbor and exploits this
information while choosing the next node.

Felemban and Lee in [11,12], proposed Multi-path and Multi-Speed Routing
Protocol (MMSPEED), which provide QoS differentiation in timeliness and reliability domain. Service differentiation in timeliness domain is provided by multiple
network-wide speed options as compared to single speed provided by SPEED.
Variable reliability is offered by probabilistic multipath forwarding depending
on packet's reliability requirement.

Almost all geographic routing protocols are reactive protocols as they maintain 1-hop neighborhood information. This very local information helps them to
reduce the memory requirements but decreases the accuracy of path estimation,

Table 1. Comparison of routing protocols

Routing protocol	Performance metrics	Packet prioritisation	Time management	Reliability support	Hole bypassing/void avoidance	Location awareness	Congestion control support	MAC prioritisation	Load balancing
RAP	End-to-End deadline miss ratio	Velocity Monotonic Scheduling	Packet carries slack deadline in header and updated every hop	No	Uses perimeter mode of GPSR	Yes	No support	IEEE 802.11e	No
RPAR	Energy Consumption, Dead line miss ratio	Yes (based on required velocity)	Packet carries slack deadline in header and updated every hop	No	Increase power to cover void else perimeter routing	Yes	Stop increasing power to control congestion	No	No
EAQoS	Throghput, Energy	RT and NRT	No	No	No	No	No	No	Between NRT and RT traffic
SPEED	End-to-End delay, Dead line miss ratio, Energy consumption, overhead	Yes (based on dealine and distance to sink)	Delay estimated as RTT - Rec overhead	No	Backpressure re-routing	Yes	1.locally drop packet 2. packet rerouting	Not needed	Yes by dispersing packet to large relay area
PASPEED	End-to-End delay, Dead line miss ratio, Energy consumption, overhead	Yes (based on dealine and distance to sink)	Delay estimated as RTT - Rec overhead	No	Backpressure re-routing	Yes	1.locally drop packet 2. packet rerouting	Not needed	Yes by dispersing packet to large relay area
MMSPEED	Average end to end delay, Overhead, Reliability	Yes (based on speed value)	Packet carries slack deadline. Delay estimation by marking packets	Yes (multi-path forwarding)	Backpressure re-routing	Yes	Drop packets and backpressure	IEEE 802.11e	Multipath routing
DGR	Average delay, Reliability, PSNR	No	No	Yes (multi-path routing)	Uses right hand thumb rule	Yes	No due to load balancing	No	Multipath routing

thus, increasing the chances of encountering void in the network. They fail in predicting the presence of void on selected path due to lack of global knowledge. A QoS-aware geographic routing protocol was proposed in [25] which was based on SPEED but maintaining 2-hop neighborhood. The 2-hop neighborhood information helped in estimating the presence of void early in the network. It led to the reduction in end-to-end delay and more number of packets meeting deadline.

Many routing protocols for WMSN exist in literature. A comparative analysis of routing protocols is presented in Table 1.

5 Conclusions

WSN and WMSN applications has specialized hardwares and diverse requirements. Generic solutions may not necessarily be optimal. Highly specific solutions are advocated for efficient working of an application. Routing protocols remains a critical design decision for any application. In this article, we provided an overview of WSN and WMSN and factors that affect the design of routing protocols. Several protocols are explained in brief highlighting its working principle and shortcomings. A summary of these protocols is also provided to compare them on a variety of aspects.

References

1. Akkaya, K., Younis, M.: An energy-aware qos routing protocol for wireless sensor networks. In: Proceedings of the IEEE Workshop on Mobile and Wireless Networks, MWN 2003, pp. 710–715 (2003)
2. Akyildiz, I.F., Su, W., Sankarasubramaniam, Y., Cayirci, E.: A survey on sensor networks. IEEE Commun. Mag. **40**(8), 102–114 (2002)
3. Akyildiz, I., Melodia, T., Chowdhury, K.: Wireless multimedia sensor networks: a survey. IEEE Wirel. Commun., 32–39 (2007)
4. Al-Karaki, J.N., Kamal, A.E.: Routing techniques in wireless sensor networks: a survey. IEEE Wirel. Commun. **11**(6), 6–28 (2004)
5. Baggio, A.: Wireless sensor networks in precision agriculture. In: Aquino-Santos, R., Rangel-Licea, V. (eds.) Precision Agriculture, pp. 1–22 (2005)
6. Benzerbadj, A., Kechar, B., Bounceur, A., Pottier, B.: Energy efficient approach for surveillance applications based on self organized wireless sensor networks. Procedia Comput. Sci. **63**, 165–170 (2015)
7. Bokareva, T., Hu, W., Kanhere, S., Ristic, B.: Wireless sensor networks for battlefield surveillance. In: Proceedings of the Land Welfare Conference (2006)
8. Chen, M., Leung, V.C.M., Mao, S., Yuan, Y.: Directional geographical routing for real-time video communications in wireless sensor networks. Comput. Commun. **30**(17), 3368–3383 (2007)
9. Chipara, O., He, Z., Xing, G., Chen, Q., Wang, X., Lu, C., Stankovic, J.A., Abdelzaher, T.F.: Real-time power-aware routing in sensor networks. In: IWQoS, pp. 83–92. IEEE (2006)
10. CMUcam: Cmucam: open source programmable embedded color vision sensors. http://www.cmucam.org/projects/cmucam4/. Accessed 14 Oct 2013

11. Felemban, E., Lee, C.G., Ekici, E., Boder, R., Vural, S.: Probabilistic qos guarantee in reliability and timeliness domains in wireless sensor networks. In: 24th Annual Joint Conference of the IEEE Computer and Communications Societies, INFOCOM 2005, Proceedings IEEE, vol. 4, pp. 2646–2657, March 2005
12. Felemban, E., Lee, C.G., Ekici, E.: Mmspeed: multipath multi-speed protocol for QoS guarantee of reliability and timeliness in wireless sensor networks. IEEE Trans. Mob. Comput. **5**(6), 738–754 (2006)
13. Gürses, E., Akan, Ö.B.: Multimedia communication in wireless sensor networks. Annales des Télécommunications **60**(7–8), 872–900 (2005)
14. He, T., Stankovic, J.A., Lu, C., Abdelzaher, T.F.: Speed: a stateless protocol for real-time communication in sensor networks. In: International Conference on Distributed Computing Systems, ICDCS 2003 (2003)
15. Karp, B., Kung, H.: GPSR: greedy perimeter stateless routing for wireless networks. In: Proceedings of the Sixth Annual ACM/IEEE International Conference on Mobile Computing and Networking, MobiCom 2000, Boston, Massachusetts, pp. 243–254, August 2000
16. Kulik, J., Heinzelman, W., Balakrishnan, H.: Negotiation-based protocols for disseminating information in wireless sensor networks. Wirel. Netw. **8**(2/3), 169–185 (2002)
17. Kumhar, M., Ukani, V.: Survey on qos aware routing protocols for wireless multimedia sensor networks. Int. J. Comput. Sci. Commun. **6**(1), 121–128 (2015)
18. Lindsey, S., Raghavendra, C.S.: Pegasis: Power-efficient gathering in sensor information systems. In: Proceedings, IEEE Aerospace Conference, vol. 3, pp. 3-1125–3-1130 (2002)
19. Lu, C., Blum, B.M., Abdelzaher, T.F., Stankovic, J.A., He, T.: Rap: a real-time communication architecture for large-scale wireless sensor networks. In: IEEE Real Time Technology and Applications Symposium, pp. 55–66. IEEE Computer Society (2002)
20. Patel, R., Pariyani, S., Ukani, V.: Energy and throughput analysis of hierarchical routing protocol (leach) for wireless sensor network. Int. J. Comput. Appl. **20**(4), 32–36 (2011)
21. Rahimi, M., Baer, R., Iroezi, O.I., Garcia, J.C., Warrior, J., Estrin, D., Srivastava, M.: Cyclops: in situ image sensing and interpretation in wireless sensor networks. In: SenSys, pp. 192–204. ACM Press (2005)
22. Rothenpieler, P., Krüger, D., Pfisterer, D., Fischer, S., Dudek, D., Haas, C., Zitterbart, M.: Flegsens - secure area monitoring using wireless sensor networks. In: Proceedings of the 4th Safety and Security Systems in Europe (2009)
23. Sohrabi, K., Gao, J., Ailawadhi, V., Pottie, G.J.: Protocols for self-organization of a wireless sensor network. IEEE Pers. Commun. **7**(5), 16–27 (2000)
24. Tavli, B., Bicakci, K., Zilan, R., Barcelo-Ordinas, J.: A survey of visual sensor network platforms. Multimedia Tools Appl. **60**(3), 689–726 (2012)
25. Ukani, V., Thacker, D.: Qos aware geographic routing protocol for multimedia transmission in wireless sensor network. In: 2015 5th Nirma University International Conference on Engineering (NUiCONE), pp. 1–6, November 2015
26. Ukani, V., Kothari, A., Zaveri, T.: An energy efficient routing protocol for wireless multimedia sensor network. In: International Conference on Devices, Circuits and Communication, September 2014

Simplified Process of Obstructive Sleep Apnea Detection Using ECG Signal Based Analysis with Data Flow Programming

Jyoti Bali[✉] and Anilkumar V. Nandi

B.V.B College of Engineering and Technology, KLE Technological University,
Hubballi, Karnataka, India
jyothipatil2@gmail.com, anilnandy@gmail.com

Abstract. The work is focused on detection of Obstructive Sleep apnea (OSA), a condition of cessation of breathing during night sleep caused by blockage of upper respiratory tract in an individual. ElectroCardioGram (ECG) signal is one of the clinically established procedures that can be relied on for deciding on the presence or absence of sleep apnea along with its severity in the subject at an earlier stage, so that the expert can advise for the relevant treatment. Earlier detection of OSA, can avoid the severe consequences leading to hypertension, Atrial-Fibrillation and day-time sleepiness that can affect the patient. ECG signal recordings from Apnea database from Physiobank, MIT website have been used for the purpose. The ECG signal based methods like QRS complex detection, RR interval variability, Respiratory Variability, Heart rate variability parameters used to detect OSA are compared and evaluated in order to select the most accurate method. Here we present the stepwise procedures, results and analysis of implementation methods used for detection of sleep apnea based on ECG signal using robust dataflow programming feature available in LabVIEW2014. Results indicate that accuracy, specificity and sensitivity of Heart Rate based detection method of OSA are 83%, 75% and 88% respectively and thus rated as one of the simple and reliable ways of detecting OSA.

Keywords: Sleep apnea · Obstructive Sleep Apnea · Apnea database · Physiobank · QRS complex detection · RR interval variability · Respiratory variability · Heart rate variability · Accuracy · Specificity and Sensitivity

1 Introduction

Sleep is an important activity essentially required for the human being to overcome the fatigue and rejuvenate the different physiological processes related to physical, emotional and psychological health. Sleep Apnea is defined as the undesired interrupted sleep behavior. In Sleep apnea conditions the affected person suffers from interruptions in sleep due to disordered breathing. As the breathing is stopped for a certain amount of time, oxygen level in the blood reduces. This is sensed by the Autonomic Nervous system (ANS) which commands the arteries supplying oxygen along with blood to constrict and increase the rate of flow of blood causing in turn higher blood pressure to suffice the demand. If this

© Springer International Publishing AG 2018
S.C. Satapathy and A. Joshi (eds.), *Information and Communication*
Technology for Intelligent Systems (ICTIS 2017) - Volume 2, Smart Innovation,
Systems and Technologies 84, DOI 10.1007/978-3-319-63645-0_18

continues, the person ends up developing a high blood pressure even during day time. Atrial Fibrillation is also one of the commonly occurring problems for persons suffering from sleep apnea. Sleep-apnea is classified into three types Obstructive Sleep Apnea (SA), Central Sleep Apnea (CSA) and Mixed Sleep Apnea (MSA) as can be seen in Fig. 1. Obstructive Sleep Apnea is caused by the breathing disorder resulting in cessation of breathing completely or partially due to blockage of upper airway tract in spite of satisfactory respiratory effort. Obstructive sleep apnea indicates the complete blockage of upper airway, whereas Obstructive Sleep Hypo-apnea stands for partial blockage of upper airway. OSA affects adults normally in the middle age group. Central apnea caused by breathing disorder resulted out of lack or absence of respiratory effort from ANS. Among the different types of sleep apnea, Obstructive Sleep Apnea/Hypo-apnea are considered to be prevalent. We focus here on the study of OSA detection methods. The studies indicate that the patients suffering from OSA tend to have Atrial Fibrillation with a higher probability. If OSA is unattended, it contributes to hypertension causing heart related disorder. General symptoms of sleep apnea are excessive sleepiness, depression, impaired concentration, seen during day time where as the symptoms like nocturnal choking, heavy snoring sound, sweating and restless sleep behavior can be observed during night time. Early detection of OSA can help the affected person to get treated for his abnormal sleep behavior, which in turn can improve his health with the corrected sleep behavior. There is a lot of work done on Sleep apnea detection techniques proposed by researchers across the globe to find out a reliable and a cost effective way of detecting the OSA [1–3].

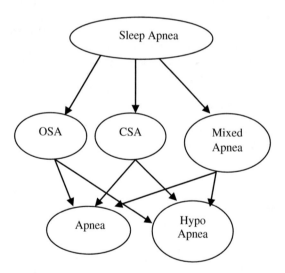

Fig. 1. Classification of types of Sleep Apnea

2   Survey of Sleep Apnea Detection Methods

There is a gold standard test called as Polysomnography (PSG) used for sleep studies which requires sophisticated and expensive laboratory setup. PSG test involves the procedures involving recording of sleep behavior of the subject for an overnight duration. The technique proves to be very cumbersome and inconvenient as the subject needs to undergo test in highly restricted conditions. Hence the need is to evolve an accurate, non-invasive and reliable alternative technique for sleep studies. There are several methods evolved over years and ECG signal is said to have emerged as one such option. As ECG signal has proven to be a standard, reliable test conducted through sophisticated equipments and commonly available facility in hospital setups, it is taken as an important diagnostic aid for many of the disease diagnostics. The ECG signal is modulated in its amplitude and frequency by breathing activity and thus carries relevant information about the respiration signal. This is due to the movement of ECG electrodes placed on chest caused by respiration action. ECG signal tapped from human being can be used to study the respiration signal hidden in it, using which we can infer on interrupted sleep effect due to respiratory disorder. Hence there is a need to analyze the variations in the parameters of the ECG signal caused by respiratory action, as well extract only the respiration information from the ECG signal. There are various signal processing and analysis methods employed to detect sleep apnea.

As per the literature survey, we could review the non-invasive methods based on ECG signal practiced for detection of Sleep Apnea. The popularly used signal processing methods are investigated and analyzed for respiration monitoring [1, 2]. The methods evolved to acquire respiration signal from single lead and multi-lead ECG signals are experimented. Clinical validation of the methods evolved to derive the respiration rate from the ECG derived Respiration signal and its estimation are discussed [3–5]. The popularly used signal processing techniques to extract ECG parameters like QRS complex are PanTompkins algorithm, Wavelet transform and Hilbert Transform techniques etc. [6, 7]. The initial activity of pre-processing the ECG signal can be done based on the different kinds of noise drastically affecting the quality of the acquired ECG signal [8–11]. One of the standard sources used by researchers for experimentation on various bio-signals across the world are from physiobank.org, having databases on ECG Arrhythmia, ECG-Apnea, EEG signals, EMG signals etc. In our proposed work, we have used databases on ECG Arrhythmia and ECG-Apnea, [12]. Further the data flow programming techniques used in LabVIEW2014 provide a robust set of built-in functions and tool kits that enable the researcher to test the algorithm in a much easier fashion and interpret the results in a very interactive way [13]. The standards are used by researchers to know the ideal ranges of ECG parameters in terms of amplitudes and time intervals, normal variants from the scientific statements released by Associations like American Heart Association Electrocardiography and Arrhythmias Committee, Council on Clinical Cardiology; the American College of Cardiology Foundation; and the Heart Rhythm Society [14].

3 Plan and Implementation of Methodologies

The plan of implementation of the detection of Sleep Apnea can be realized using the various well established methods as presented in Fig. 2. ECG being a noninvasive technique has a large number of hidden parameters that help in detection of the respiratory behavior, in turn the presence and absence of Sleep Apnea. The first and the foremost step is to get the ECG signal either using data acquisition hardware like analog front end from the electrodes attached to human being or download bio-signal recordings from healthy subjects as well those with ailments. Next step is the preprocessing of ECG signal to eliminate the undesired noise from it, so that the pure ECG signal can be easily analyzed further for the accurate detection of sleep apnea. Preprocessing involves cascaded stages of filtering process involving a low pass filter (LPF) with cut off 150 Hz used to remove baseline wander noise, a high pass filter with cutoff 0.05 Hz to remove motion artifacts and finally a notch filter with cutoff 50 Hz to remove power line interference as shown in Fig. 3. Based on the strategies that can be employed for ECG signal analysis, the important are the QRS complex detection, ECG feature Extraction, Heart Rate Variability and ECG Derived Respiration (EDR) Signal extraction. For each of the strategies followed, the important signal processing methods used are the Pan Tompkins algorithm, under time domain approach and Fourier Transform, Wavelet Transform and Hilbert Transform based techniques under frequency domain approach. The first step common to all the methods is the QRS complex and ventricular beat detection followed by morphological and rhythm analysis [6–11]. The parameters analyzed under each method differ from each other.

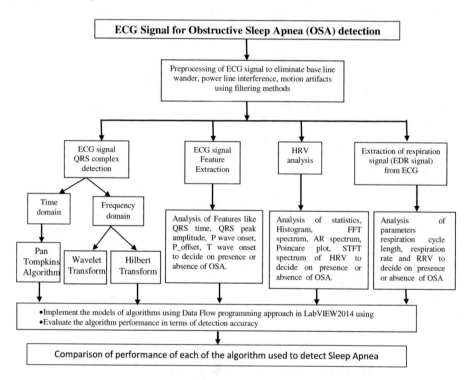

Fig. 2. Survey of ECG based OSA detection methods

Further the generic algorithm meant for sleep apnea detection involves the first step as preprocessing of ECG signal followed by the detection of QRS complex, its amplitude and location. Apart from that, even the ECG beat morphology features help the user get the ECG parameters accurately. Next is to detect the heart rate from the array of peaks

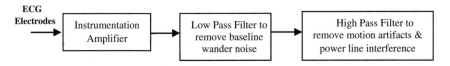

Fig. 3. Stepwise processes involved in data acquisition of ECG signal

and also the range of RR intervals. Mean heart rate provides the information to distinguish the normal case with that of apnea condition. But if there are repeated episodes of bradycardia and tachycardia condition, then it is the presence of sleep apnea. Based on the number of episodes of apnea it can be decided as mild OSA, moderate OSA, severe OSA or simply normal if the heart rate is well within the ideal range of heart beat between 60 to 70 bpm. The implementation of QRS complex detection, which can be implemented either using Pan-Tompkins algorithm as shown in Fig. 4 or using wavelet transformation technique as shown in Fig. 5. Amplitude and location of QRS complex is important to detect other parameters of ECG helpful in enhancing the accuracy of QRS detection. The collection of a set of parameters required to gather characteristics of ECG signal with respect to its respiration behavior.

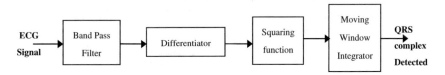

Fig. 4. Pan Tompkins algorithm for QRS complex detection

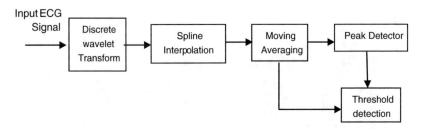

Fig. 5. Stepwise process of discrete wavelet transform method

The above discussed procedures are implemented using LabVIEW2014 from TI providing the robust environment for data flow programming providing user friendly built-in functions [13].

4   Implementation and Discussion of Results

The algorithms selected for ECG signal analysis as shown in Fig. 6 to detect sleep apnea are implemented using the Advanced Signal processing toolkit and Biomedical toolkit using LabVIEW2014. The below Fig. 7a shows the data acquision of ECG signal done in LabVIEW2014 using MIT ECG Apnea database and further of treatment of noise caused by artifacts, baseline wander and power line interference. Thus the purified ECG signal is now fed to the peak detector implemented using either using Wavelet transform or the Pan Tompkins algorithm. The stepwise procedure as shown in Fig. 8(a, b, c and d) is used to detect the QRS complex present in the signal and detect the set of peak amplitudes, its location and the time of occurrence can be thus used to compute the heart rate for every cycle. QRS peak detection process performed using wavelet transform technique is as shown in Fig. 9.

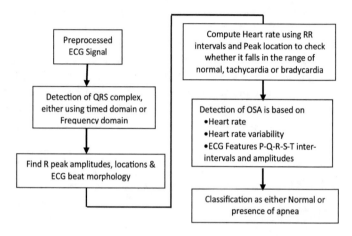

Fig. 6. Algorithm for detection of Sleep apnea based on ECG signal

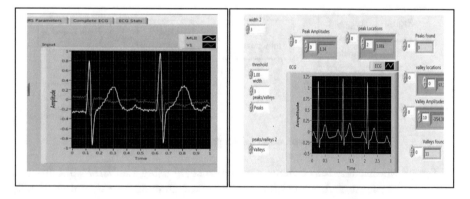

Fig. 7. (a) Reading the ECG signal with apnea (b) Identifying the peaks and valleys in the ECG signal input

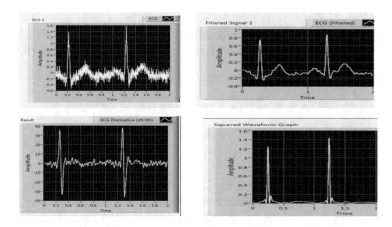

Fig. 8. Step wise procedure of identifying QRS complexes (a) Reading the ECG signal with apnea (b) Filtering to remove noise (c) Differentiation of the filtered output (d) Squaring the differentiated output to identify the peaks distinctly

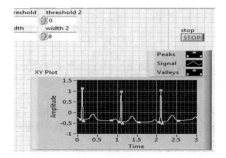

No. of records	12
TP	7
TN	3
FP	1
FN	1
Accuracy	83%
Specificity	75%
Sensitivity	88%

Fig. 9. Peaks and valleys detected by Wavelet transforms technique

Fig. 10. Performance measures of detection algorithm

Further the mean heart rate can be used to identify the slow heart rate, fast or normal heart rate, which is an important indicator of sleep apnea. One more observation is that the continuous episode of bradycardia followed by tachycardia is a case of sleep apnea. Thus the mean heart rate computed from the waveform measurements is done using the Feature extraction program. The algorithm for detection of apnea is based on the well defined ranges of ECG parameters for accurate classification. Thus the heart rate and HRV characteristics are used to determine the variation in RR intervals for the 12 sample records from ECG Apnea database and the detection process is summarized as shown in Fig. 9. The measures of performance of the algorithm are computed as 83% accuracy, 75% of specificity and 88% of sensitivity. We have used abbreviations for output status variables such as TP, TN, FP and FN in place of True Positive, True Negative, False Positive, and False Negative respectively. Here TP stands for the status, when the input ECG recording is judged by algorithm has significant number of apnea episodes, whereas TN stands for the input recording being judged as a normal one without any apnea episode, both matching with the expert opinion. FP and FN represents the algorithm output being contradicting the expert opinion in judging the status of the input ECG recordings wrongly. The performance measures of sleep apnea detection algorithm are computed using the formulae given below and are tabulated as in Fig. 10.

$$\text{Accuracy} = (TN + TP)/(\text{Total no. of recordings})$$

$$\text{Specificity} = TN/(TN + TP) \ \& \ \text{Sensitivity} = TP/(TP + FN)$$

5 Conclusion

Testing of algorithms can be effectively done through dataflow programming approach of LabVIEW2014, as it helps researchers to have interactive hands-on with signals, signal processing and analysis techniques. We could compare the advantages of using the best possible combination of QRS complex detection techniques, followed by signal analysis technique for basic heart rate based decision making algorithm. Further the results show that the algorithm can be improved by utilizing the ranges of values used for accurate classification using Fuzzy logic decision making.

References

1. Zhao, Y., Zhao, J., Li, Q.: Derivation of respiratory signals from single-lead ECG. In: International Seminar on Future Biomedical Information Engineering, pp. 15–18. IEEE Computer Society (2008)
2. Moody, G.B., Mark, R.G., Zoccola, A. Mantero, S.: Derivation of respiratory signals from multi-lead ECGs. In: Computers in Cardiology 1986, 8–11 September. IEEE Computer Society Press (1985)
3. Moody, G.B., Mark, R.G., Bump, M.A., Weinstein, J.S., Berman, A.D., Mietus, J.E., Goldberger, A.L.: Clinical validation of the ECG-derived respiration (EDR) technique. In: Computers in Cardiology 1986, vol. 13, pp. 507–510. IEEE Computer Society Press, Washington, DC (1986)
4. Brown, L.F., Arunachalam, S.P.: Real-time estimation of the ECG-derived respiration (EDR) signal. In: Rocky Mountain Bioengineering Symposium & International ISA Biomedical Sciences, Instrumentation Symposium, 17–19 April 2009, Milwaukee, Wisconsin (2009)
5. Rangayyan, R.M.: Biomedical Signal Analysis. Wiley, New York (2002)
6. Tompkins, W.J.: Biomedical Digital Signal Processing. Prentice-Hall, Upper Saddle River (1995)
7. Broesch, J.D.: Digital Signal Processing Demystified. LLH Technology Publishing, Eagle Rock (1997)
8. Lin, Y.D., Hu, Y.H.: Power-line interference detection and suppression in ECG signal processing. IEEE Trans. Biomed. Eng. 55(1), 354–357 (2008)
9. Zhang, D.: Wavelet approach for ECG baseline wander correction and noise reduction. In: Proceedings of 27th Annual International Conference of the IEEE Engineering in Medicine and Biology Society, September 2005, pp. 1212–1215 (2005)
10. Faezipour, M., Tiwari, T.M., Saeed, A., Nourani, M., Tamil, L.S.: Wavelet-based denoising and beat detection of ECG signal. In: Proceedings of IEEE-NIH Life Science Systems and Applications Workshop, April 2009, pp. 100–103 (2009)

11. Pan, J., Tompkins, W.J.: A real-time QRS detection algorithm. IEEE Trans. Biomed. Eng. **32**, 230–236 (1985)
12. Physionet website: MIT-BIH Arrhythmia Database Directory (2010). http://www.physionet.org/physiobank/database/mitdb
13. Correia, S., Miranda, J., Silva, L., Barreto, A.: Labview and Matlab for ECG acquisition, filtering and processing. In: 3rd International Conference on Integrity, Reliability and Failure, Porto/Portugal, pp. 20–24 (2009)
14. Kligfield, P., Gettes, L.S., Bailey, J.J., et al.: Recommendations for the standardization and interpretation of the electrocardiogram. Circ. J. Am. Heart Assoc. **115**, 1306–1324 (2007)

Smart Two Level K-Means Algorithm to Generate Dynamic User Pattern Cluster

Dushyantsinh Rathod[1(✉)], Samrat Khanna[2], and Manish Singh[1]

[1] CE/IT Department, Aditya Silver Oak Institute of Technology,
Ahmedabad, India
Dushyantsinh.rathod@gmail.com,
Manishsingh.ce@socet.edu.in
[2] Dean Rai University, Ahmedabad, India
sonu.khanna@gmail.com

Abstract. Data cleaning perform in the Data Preprocessing and Mining. The clean data work of web server logs irrelevant items and useless data can not completely removed and Overlapped data causes difficulty during retrieving data from datasource. Previous paper had given 30% performance of datasource. So We have Implemented Smart Two-level clustering method to get pattern data for mining. This paper presents WebLogCleaner can filter out much irrelevant, inconsistent data based on the common of their URLs and it is going to improving 8% of the data quality, performance, Accuracy and efficiency of any Datasource.

Keywords: Web Usage Mining (WUM) · Data cleaning · Web log mining · Web page mining · Pattern cluster · Preprocessing

1 Introduction

Information mining is the computational procedure of finding examples in huge sum information sets including strategies at the crossing point of manmade brainpower, machine learning of Data System. The WWW is a vast database so this development emerges a the requirement for breaking down more precise the information. The procedure of Extraction and breaking down of Web information is called mining in web page. Web mining is a techniques used to discover patterns from the Web page. WM can be divided in 3 types (1) Web Structure Mining (2) Web Content Mining (3) Web Usage Mining. Web structure mining is the process of discovering the connection between web pages. Web content mining incorporates mining, extraction and joining of valuable information and learning of Web page content. Web Mining is a kind of technique which discover useful information from the Web Page. Mining enterprise manager and employer which is difficult to find out the "right" ands "interesting" data [1]. Web Log are generally noisy and ambiguous. Web applications are expanding at a colossal speed and its clients, are expanding at exponential speed.

There are work done on data cleaning logs irrelevant items and the useless data can not completely removed from log files. When multiple datasources can be combine, data quality problems are present in single data collections, such as files and databases.

© Springer International Publishing AG 2018

S.C. Satapathy and A. Joshi (eds.), *Information and Communication Technology for Intelligent Systems (ICTIS 2017) - Volume 2*, Smart Innovation, Systems and Technologies 84, DOI 10.1007/978-3-319-63645-0_19

2 Web Usage Mining

Web Usage Mining is good technique of discover useful information from the web page and log files. Using usage mining a designer should improve the web site. Web Usage Mining contains three steps [6] (Fig. 1).

(1) Data Preprocessing (2) Pattern Discovery (3) Generate Cluster Pattern

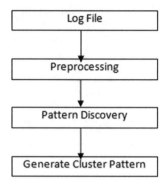

Fig. 1. WebLog.mining process

2.1 Preprocessing of Data

The preprocessing of web logs is intricate and tedious and it is done utilizing the accompanying strides. The principle point of information preprocessing is to choose institutionalized information from the first log records, arranged for client route design disclosure calculation [5].

(1) Data Cleaning
(2) Page view Identification
 3) Path Completion
(4) Formatting

2.1.1 Data Cleaning

Data cleansing is techniques of deleting unnecessary log data from files. Data cleaning contains:-

(a) Global and neighborhood Noise Removal
(b) pictures, video and so forth. Evacuation
(c) Records that fizzled HTTP status code Removal
(d) Robots cleaning Removal
(e) Web commotion can be typically ordered into two gatherings relying upon their granularities.
(f) Global Noise are compares to the superfluous items with gigantic granularities, which are no littler than individual pages.
(g) Local (Intra-Page) Noise are compares to the unessential things inside a Web page (Fig. 2).

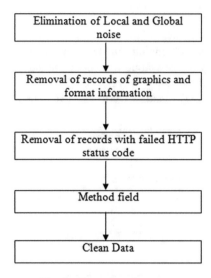

Fig. 2. Data cleansing steps

2.1.2   Page View

Identify Site visit is the accumulation of website page objects. It is the way toward distinguishing page get to documents identified with a solitary online visit. All the online visits has appointed site hit id.

2.2   Pattern Extraction

It is a sort of technique which is utilized as a part of different fields like information mining and example acknowledgment. design disclosure contains an example in which the client utilizes a website pages. There are more calculations accessible for this procedure like the Association Rule mining.

2.3   Design Creation

It contains investigation the example that is removed in example extractions prepare. Significant and fascinating example are kept and rest of the example are evacuated.

3   Problem Statement

Discuss the problem relating to Data cleaning of web log. Web log is generally unwanted Web applications are increasing at an enormous speed and its users are increasing at exponential speed. Difficult to find the "right" or "interesting" information, There are work done on data cleaning of logs irrelevant items and useless data can not completely removed. Overlapped data cause difficulty during Page Ranking, When multiple data sources need to be integrated, data qualities issues are present in single set of datasource, The Standard Log file contains irrelevant inconsistent data. Difficulty of knowledge extraction during Web Log Mining.

4   Two-Level K-Means Clustering

In this paper I Implemented Improved method of clustering, which is used to generate cluster pattern data (Fig. 3).

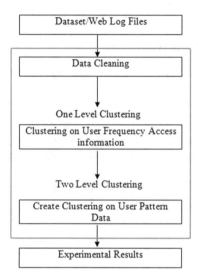

Fig. 3. Two level clustering process

4.1   Two-Level K-Means Clustering Method

The Two-level clustering method is improving the quality of data in the WUM process, which is the two-level clustering. Based on the results of two level clustering method on web log data, it can be concluded that this method can improve the quality of data web log.

- The first level clustering is done in the form of data frequently user access using non-hierarchical clustering method.
- The second level clustering is done by first changing the form of web log data into user access behavior patterns.

5   New Algorithms

5.1   One Level Clustering Algorithm

1. Read *N* no of records from clean data source *DS*
   For i = 1 to i <= N
   Next

2. For each records **R** find frequent access item **F** from data source **DS**
3. Read frequency user access items **F.**
4. If **R = F** frequent records then
5. Save for clustering frequent user access records in frequency access data source **FDS**
6. Make one level cluster from frequency user access records
7. Else not select records
8. End if
9. Next record

5.2   Two Level Clustering Algorithm

1. Read **N** no of records from clean data source **FDS**
   For i = 1 to i <= N
   Next
2. For each records **R** from data source **FDS** find pattern data
3. Read pattern data using specified address from data source **FDS.**
4. If requested records from frequent data source **FDS** with specified pattern then
5. Collect and Save in pattern data source **PDS.**
6. Make two level cluster in pattern data source **PDS .**
7. Else not select that records.
8. End if
9. Next record

6   Results

See Figs. 4, 5, 6, 7 and 8.

Index_No	Date	Client_IP	Server_IP	URI_Stem	Status_Code	Request
0	2015-08-13	10.8.0.15	202.71.129.26	/Papers/SRSExample-webapp.doc	200	/laptops.aspx
1	2015-08-13	10.8.0.13	202.71.129.26	/syllabus.aspx	200	/mobiles.aspx
2	2015-08-13	10.5.0.54	209.85.135.109	/gmail.com	200	/LED.aspx
3	2015-08-13	10.5.0.12	59.162.23.130	/academic/rsrchprgm.html	200	/movies.aspx
4	2015-08-13	10.6.0.20	67.218.96.251	/downloads/index.htm	200	/admission.aspx
5	2015-08-13	10.6.0.22	67.218.96.251	/products/W52XXX-series.aspx	200	/facebook/profile
6	2015-08-13	10.6.0.27	67.218.96.251	/it/experienced/index.htm	200	/powerbank
7	08/13/2015	10.5.0.5	202.71.129.26	http://www.flipkart.com/laptops	200	/Circular.aspx
8	08/13/2015	10.5.0.20	172.30.255.255	http://www.flipkart.com/mobiles	200	/Papers/SRSExample-webapp.doc
9	08/13/2015	10.6.0.26	209.85.135.109	http://www.amazon/Electronics	200	/Drupal-Intro.ppt
10	08/13/2015	10.8.0.15	67.218.96.251	http://in.bookmyshow.com	200	/PMS/PMS.doc
11	08/13/2015	10.8.0.17	202.71.129.26	http://www.ebay.in/laptops	200	/IPL/Schedule.aspx
12	08/13/2015	10.8.0.15	59.162.23.130	/downloads/index.htm	200	/makemytrip/offer.aspx
13	2015-08-13	10.8.0.18	202.71.129.26	/Papers/SRSExample-webapp.doc	200	/laptops.aspx
14	2015-08-13	10.8.0.14	202.71.129.26	/syllabus.aspx	200	/mobiles.aspx
15	2015-08-13	10.5.0.51	209.85.135.109	/gmail.com	200	/LED.aspx
16	2015-08-13	10.5.0.13	59.162.23.130	/academic/rsrchprgm.html	200	/movies.aspx
17	2015-08-13	10.6.0.21	67.218.96.251	/downloads/index.htm	200	/admission.aspx

Fig. 4.   Final clean data

Index_No	Date	Client_IP	Server_IP	URI_Steam	Status_Code	Page_Request	Flag
0	2015-08-13	10.8.0.15	202.71.129.26	/Papers/SRSExample-webapp.doc	404	/samsung.jpg	1
1	2015-08-13	10.8.0.13	202.71.129.26	/syllabus.aspx	404	/LG.jpg	1
2	2015-08-13	10.5.0.54	209.85.135.109	/gmail.com	404	/LED.aspx	1
3	2015-08-13	10.5.0.12	59.162.23.130	/academic/rsrchprgm.html	404	/samsung.jpg	1
4	2015-08-13	10.6.0.20	67.218.96.251	/downloads/index.htm	404	/admission.aspx	1
5	2015-08-13	10.6.0.22	67.218.96.251	/products/W52XXX-series.aspx	404	/facebook/profile	1
6	2015-08-13	10.6.0.27	67.218.96.251	/it/experienced/index.htm	404	/powerbank	1
7	08/13/2015	10.5.0.5	202.71.129.26	http://www.flipkart.com/laptops	404	/Circular.aspx	1
8	08/13/2015	10.5.0.20	172.30.255.255	http://www.flipkart.com/mobiles	404	/Papers/SRSExample-webapp.doc	1
9	08/13/2015	10.6.0.26	209.85.135.109	http://www.amazon/Electronics	404	/Drupal-Intro.ppt	1
10	08/13/2015	10.8.0.15	67.218.96.251	http://in.bookmyshow.com	404	/PMS/PMS.doc	1
11	08/13/2015	10.8.0.17	202.71.129.26	http://www.ebay.in/laptops	404	/IPL/Schedule.aspx	1
12	08/13/2015	10.5.0.59	59.162.23.130	/downloads/index.htm	404	/makemytrip/offer.aspx	1
13	2015-08-13	10.8.0.18	202.71.129.26	/Papers/SRSExample-webapp.doc	404	/laptops.aspx	1
14	2015-08-13	10.8.0.14	202.71.129.26	/syllabus.aspx	404	/mobiles.aspx	1
15	2015-08-13	10.5.0.51	209.85.135.109	/gmail.com	404	/LED.aspx	1
16	2015-08-13	10.5.0.13	59.162.23.130	/academic/rsrchprgm.html	404	/movies.aspx	1
17	2015-08-13	10.6.0.21	67.218.96.251	/downloads/index.htm	404	/admission.aspx	1

Fig. 5. Noisy data with flag storage

Pass No of Cluster	5		Cluster Cration
Cluster No	202.71.129.26		Create

Index_No	Server_IP	Client_IP
0	202.71.129.26	10.8.0.15
1	202.71.129.26	10.8.0.13
7	202.71.129.26	10.5.0.5
11	202.71.129.26	10.8.0.17
13	202.71.129.26	10.8.0.18
14	202.71.129.26	10.8.0.14
20	202.71.129.26	10.5.0.5
24	202.71.129.26	10.8.0.16
26	202.71.129.26	10.8.0.18
27	202.71.129.26	10.8.0.11
33	202.71.129.26	10.5.0.5
37	202.71.129.26	10.8.0.12
39	202.71.129.26	10.8.0.10
40	202.71.129.26	10.8.0.13
46	202.71.129.26	10.5.0.51
50	202.71.129.26	10.8.0.53

Fig. 6. Pattern cluster 1

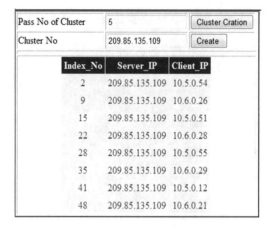

Fig. 7. Pattern cluster 2

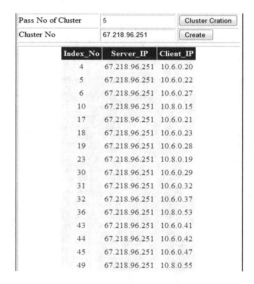

Fig. 8. Pattern cluster 3

7 Comparison Chart

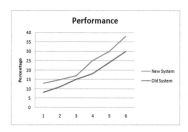

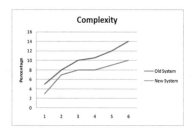

Above performance charts shows that the performance is increasing 8% Quality of data and total is 38% as compared to previous algorithm as 30% quality of data and complexity charts shows that when performance is increasing then by default complexity is decreasing.

8 Conclusion and Future Work

There are many techniques proposed by totally different researchers for the web usage mining. This paper mentioned about Two-level clustering method available for web usage mining. This previous paper has attempted to give EPFLog Miner quality is about 30% Performance of weblog mining. Where these new algorithms gives 8% increase performance and accuracy of WebLogMiner and total quality is 38% and decreasing the complexity of web log data. Web log mining includes of information preprocessing, design extraction and Cluster creation. The aftereffects of Web Log digging can be utilized for different applications like web personalization framework, webpage proposal framework, website change framework, and so on.

In this paper, I describe Two-level Clustering Algorithm for web log preprocessing techniques. In the future work apply this algorithm on Personalize Web recommended system to get high performance, accuracy and efficiency based on different criteria using pattern data clustering mining

References

1. Rathod, D., Khanna, S.: A comparison of k-means clustering and smart two level k-means clustering algorithm. IJSART (2017)
2. Rathod, D., Khanna, S.: A survey on different efficient clustering techniques used in web mining. IJSRD **4**(8), 51–53 (2016)
3. Mengar, K., Rathod, D.: Ant based data reduction in web usage mining using k-means clustering algorithm. IJSDR (2016)
4. Rathod, D., Khanna, S.: Improved two level k-means clustering algorithm to generate user pattern clustering. IJSART (2016)
5. Rathod, D., Khanna, S.: Implemented two level k-means clustering algorithm to improve quality in user pattern mining. IJSART (2016)
6. Rathod, D., Khanna, S.: Improve quality in user pattern mining approach using two level k-means clustering methodology. IJSRD **4**(1), 351–353 (2016)
7. Jeba, J.M.P., Bhuvaneswari, M.S., Muneeswaran, K.: Extracting usage patterns from web server log. IEEE (2016)
8. Sisodia, D.S., Verma, S.: Web usage pattern analysis through web logs: a review. IEEE (2016)
9. Dhanalakshmi, P., Ramani, K., Eswara Reddy, B.: The research of preprocessing and pattern discovery techniques on web log files. IEEE (2016)
10. Mehrotra, S., Kohli, S.: Comparative analysis of k-means with other clustering algorithms to improve search result. IEEE (2015)
11. Shaa, H., Liub, T., Qinb, P., Sunb, Y., Liub, Q.: EPLogCleaner: improving data quality of enterprise proxy logs for efficient web usage mining. Proc. Comput. Sci. **17**, 812–818 (2013). Information Technology and Quantitative Management, ITQM 2013
12. Hussain, T., Asghar, S., Masood, N.: Web usage mining: a survey on preprocessing of web log file. In: Proceedings of the 2010 International Conference on Information and Emerging Technologies (ICIET), pp. 1–6. IEEE (2010)
13. Tyagi, N., Solanki, A., Tyagi, S.: An algorithmic approach to data preprocessing in web usage mining. Int. J. Inf. Technol. Knowl. Manag. **2**(2), 279–283 (2010)
14. Zheng, L., Gui, H., Li, F.: Optimized data preprocessing technology for web log mining. In: International Conference on Computer Design and Applications (ICCDA 2010) (2010)
15. Munk, M., Kapustaa, J., Šveca, P.: Data preprocessing evaluation for web log mining: reconstruction of activities of a web visitor. Proc. Comput. Sci. **1**, 2273–2280 (2012). International Conference on Computational Science, ICCS 2011
16. Nithya, P., Sumathi, P.: Novel pre-processing technique for web log mining by removing global noise and web robots. In: National Conference on Computing and Communication Systems (NCCCS). IEEE (2012)
17. Sujatha, V., Punithavalli: Improved user navigation pattern prediction technique from web log data. Proc. Eng. **30**, 92 (2012). International Conference on Communication Technology and System Design 2011
18. Aye, T.T.: Web log cleaning for mining of web usage patterns. IEEE (2011)
19. Lee, C.-H., Lo, Y., Fu, Y.-H.: A novel prediction model based on hierarchical characteristic of web site. Expert Syst. Appl. **38**, 3422–3430 (2011)
20. Losarwar, V., Joshi, M.: Data preprocessing in web usage mining. In: International Conference on Artificial Intelligence and Embedded Systems (ICAIES 2012), 15–16 July (2012)
21. Agarwal, R., Arya, K.V., Shekhar, S. Kumar, R.: An efficient weighted algorithm for web information retrieval system. IEEE (2011)

Notification of Data Congestion Intimation [NDCI] for IEEE 802.11 Adhoc Network with Power Save Mode

B. Madhuravani[1(✉)], Syed Umar[1], Sheikh Gouse[1], and Natha Deepthi[2]

[1] Department of Computer Science Engineering, MLRIT, Hyderabad, India
madhuravani.peddi@gmail.com, umar332@gmail.com
[2] Department of Computer Science Engineering, CMR ENGG & TECH, Hyderabad, India
priya.natha.28.85@gmail.com

Abstract. IEEE 802.11-power save mode (PSM) independent basic service set (IBSS) Save, the time is divided into intervals of the signals. At the beginning of each interval signal and power saving alarm periodically all open windows (vocals). The station will be in competition with the rest of the frame window frame sent voice data leakage range. Element depends frame transmission IEEE CSMA/CA as defined in 802.11 DCF. A chance of transmit voice frames type of collision energy IBSS success. This article gives an analysis model with a chance of success output transmission window fixed size element. The results of the simulation analysis of the accuracy of the analysis.

Keywords: DCF · Adhoc networks · IEEE 802.11

1 Introduction

IEEE 802.11 wireless LAN MAC Media Access Protocol widely used. This means that both are necessary to find the distribution channel coordination function (DCF) and the coordinates of the optional function (PCF). DCF based on Carrier Sense Multiple Access with Collision (CSMA/CA). CSMA/CA with double exponential regression (DER) [1] algorithm to avoid collisions online. Station-ll space ready to hit where the liquid can be heard in the vicinity of the big lottery DIF (Digital Distribution Frame), delays, otherwise delivery time DIF recorded medium. Then hold the station for calculation. Window minimum and maximum contention dimensions. Physical layer are determined by these values. Backoff counter is lost in the ear canal in the freezing cold and when the line is busy. After each unsuccessful transmission, CW will be doubled with Eq. 1. (CWmin) is called the maximum duration of the suspension. CW is transmitted is reset CWmin. prepared by the IEEE 802.11 standard impact analysis DCF. Bianchi [2] at some point is a Markov model of the IEEE 802.11 DCF appropriate channel modes [3]. Bianchi described a modified version of the business model. Number of documents [4–8], the first model of Bianchi operator error, not the ideal diversion and prisoner treatment. All derived IEEE 802.11 DCF models frame transmission theoretical data. IEEE 802.11 power save mode (PSM) for IBSS, time is divided into intervals, each is divided into two parts, the element and the window data. IEEE 802.11 mode power

© Springer International Publishing AG 2018
S.C. Satapathy and A. Joshi (eds.), *Information and Communication Technology for Intelligent Systems (ICTIS 2017) - Volume 2*, Smart Innovation, Systems and Technologies 84, DOI 10.1007/978-3-319-63645-0_20

saving interval, each node aware of the limited space called first element window DCF. Window element is used to wait for messages to describe food into energy saving mode. If the station has successfully delivered element email frame executives competition in the transfer of relevant data. wireless network services that are required. Wireless devices often rely on batteries. The design of the "energy" and "energy" of the wireless network are the main areas of research. More MAC protocol for the wireless LAN is designed to reduce energy consumption. [9] to improve the economic without powerful wireless MAC protocol selected public collection element of the size of the adjustment window and unusual bleeding after various window sizes. [10] proposed a short window Carrier Sense element component. But best of our knowledge, no model for success.

$$CW_{max} + 1 = 2^m(CW_{min} + 1) \tag{1}$$

IEEE 802.11 power save transmission frame member in IBSS mode. This chapter describes a discrete time Markov model was similar to a part of the element of success. Throughput IEEE 802.11 PSM therefore calculates the targeted element. The simulation tool NS-2 [11] is used to validate the models. The study says. The second part of the first a brief overview of the IEEE 802.11 PSM. Part III, we present a theoretical model to calculate the successful frame element of probability. IV confirms the accuracy of the simulation model. Finally, the fifth section presents conclusions.

2   IEEE 802.11 DCF in Energy Saving Mode

IEEE 802.11 PSM two power settings, power and strength. Power, or to send the player and the current success of all time is. We hope that in the face of the synchronization hub. A saving mode in regular energy station to listen to the message and realized the element of the window period. Transmitter reduces each broadcast/multicast or unicast frame into energy saving mode and the belt element diffusion element. The sending of data packets from the window of the element to go to the station in a power saving mode or staying asleep by determining the transfer member data. Article transmission frame period algorithm CSMA/CA DCF, IEEE 802.11 [1]. A single transmission frame when the base member of the window frame member and sends a confirmation and alert to the end of the next point of the window. If confirmation is received and sent to traditional elements DCF process. If the drive letter of the frame member, can send, for example, when they are based buffer data and after trying to get another member of the frame. When receiving the channel or mail window frame, enter the government at the end of the window element. The product can last ACK frame or sent or received in the mail window. A buffer at the context of energy saving measures after the transfer station frame buffer over time. IEEE 802.11 [1], or the time of delivery, following the post when thrown. Depending on the point size of the transmission window is small, the limit of seven executives is a member is not required [9]. The system and try to get the transmission element and intended to charge two beacon frames with energy saving interval data. The energy saving mode is indicated by a stick. Figure said sending placing executives and managers from one station to post. Ms Atim station B ACK to sleep at the train station and the rest of the series. C center number to save the state at the end part of the window, allowing energy savings.

3   Assumptions and Estimations of Network Model

Modeling and analysis of transmission, try these ideas. Think of the machine. We look forward to the saturation state in which the stations are always sending packets. We thought the fixed voice size window. The channel is very long, that is to say, does not refer to [12]. When the data recorded by the sending of the input image of the police station. No broadcast frame member, but unicasting at Delivery. If the success of the replacement product B, and the pillar of fire units (the population), can send more affordable voice and central B-pillar of fire units. Before the central element, and determining the value of CW + 1 double CWmin unnecessary distractions CWmax CW + 1 and the value of the transmission CWmin CW n + 1. If the successful reset of the main center section B-frames face more police could send part of another object, in this case, part of the host transfer element. Three flagship voice test interval. Atim as ACK after three transmission, the beacon interval, and central data received re-buffered other subsequent headlights minute test. Try sending the voice three times in total center. re-buffered she can continue to save two Beacon intervals. When three beacon interval if the transmission fails the element frame for disposal. The algorithm is derived from the concept [9].

3.1   System Model

(S(t), B(t) (t)) is the stochastic field back-off circle AA, S(t), a backoff B(t) and low suspension (R) (beacon interval) T. It is a particular example, the top of the second row is a drive slot. Follow up co-ginning each goal. Backlog is the part number and try again cube element using a beacon, a beacon of successful federal units and the item number of the track arrested after all. We constructed a three-part system (S (t), B (t) (t)) as indicated in the special time Markov chains Fig. 1.

$$P\{i_1, k_1, a_1 | i_0, k_0, a_0\} = P\{s(t+1) = i_1,$$
$$b(t+1) = k_1, a(t+1) = a_1 | s(t) = i_0,$$
$$b(t) = k_0, a(t) = a_0\}.$$

Let p be the conditional probability of collision, a fixed number of individual regardless of the number of retransmissions bus. The probability P that fights with the frame. Consider probability q window element ends with the existing space. This is regardless of the number of retransmission images. This is not to zero - evidence of a step in the chain of a Markov probability. 2 shows that (1) shows the comparison. 3

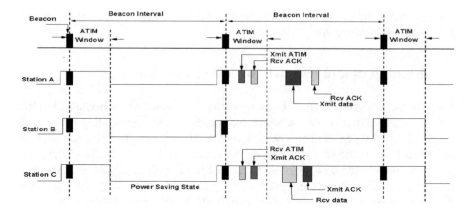

Fig. 1. PSM of IBSS [1]

The first part shows the beginning of each hole and the window of the article to reduce the probability $(1 - q)$. Second part of death does not show table of Article window backlog retaining section, the protocol entry 0 article in the wings to another window. The third part shows the transmission. Good show successful delivery, or to begin sending new framework for the item. Five-part shows that the conflict is the last test interval is a beacon, even attempting to send a frame in the sixth stage of another section 0 window default lighthouse flowers appear and the voice interval, a transmission fails

4   Expected Model

Evaluation of the model in the ns-2 simulator [11]. Simulation of selected areas so that jumps, received intensity of the detection signal and therefore the total station. Assuming that about 20% of the time window position, and that the fire units column. The introduction of the distribution of the performance of IEEE 802.11 DCF access frequencies in direct main power mode (DSSSL) physical layer [1]. The system is used for the parameters listed in Table 1 calculation.

Table 1.   Parameters used for power save mode calculation.

Payload of data packet	1024 bytes
Data	1024 bytes + MAC header + PHY header
ACK	14 bytes + PHY header
PHY header	192 μs
MAC header	28 bytes
Basic rate	1 Mbps
Data rate	2 Mbps
Slot time	20 μs
SIPS	10 μs
DIPS	50 μs

Number of fixed stations, we run 10 different magic and random seed. all symbols that show the results of the simulation. Figure 2, shows that the likelihood of success of the delivery of costly average measuring element. Figure 3 the solid line shows the results of calculation and Markov models broke the mean line of 10 pieces each. Statistics show that the theoretical results and simulation. Figure 4 shows this expensive. Furthermore, according to the results of the simulation results. Note that the transmission of a model Bianchi slightly below the IEEE 802.11 PSM input window.

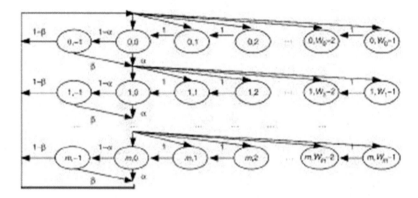

Fig. 2. Provide a framework for the subject Markov model

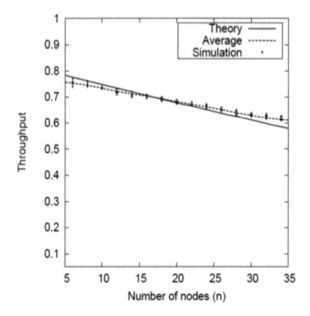

Fig. 3. Notification of data congestion intimation [NDCI] probability

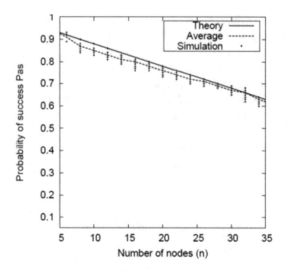

Fig. 4. 802.11 PSM throughput with different node.

5 Conclusion

The article with the model based on Markov analysis of IEEE 802.11 DCF under the heading chain transmit the power saving mode. We are used to calculate the probability of IEEE 802.11 DCF throughput power save mode. The theoretical results are almost identical to the simulation results in terms of normal operation success probability.

References

1. IEEE Std. 802.11: Part 11: Wireless LAN Medium Access Control (MAC) and Physical Layer (PHY) Specifications, Edition 2007, IEEE (2007)
2. Bianchi, G.: Performance analysis of the IEEE 802.11 distributed coordination function. IEEE J. Select. Areas Commun. **18**, 535–547 (2000)
3. Wo, H., Peng, Y., Long, K., Cheng, S., Ma, J.: Performance of reliable transport protocol over IEEE 802.11 wireless LAN: analysis and enhancement. In: INFOCOM (2002)
4. Ergen, M., Varaiya, P.: Throughput analysis and admission control for IEEE 802.11a. Mobile Netw. Appl. **10**, 705–716 (2005)
5. Alshanyour, A., Agarwal, A.: Three-dimensional markov chain model for performance analysis of the IEEE 802.11 DCF. In: IEEE GLOBCOM (2009)
6. Hou, T.C., Tsac, L.F., Lia, H.C.: Throughput analysis of the IEEE 802.11 DCF in multihop ad hoc networks. In: ICWN, pp. 653–659 (2003)
7. Vishnevsky, V.M., Lyakhov, A.I.: IEEE 802.11 LANs: saturation throughput in the presence of noise. In: IFIP Network, Pisa, Italy (2002)
8. Daneshgran, F., Laddomada, M., Mondin, M.: A model of the IEEE 802.11 DCF in presence of non ideal transmission channel and capture effects. In: IEEE GLOBCOM (2007)

9. Jung, E.S., Vaidya, N.H.: Energy efficient MAC protocol for wireless LANs. In: IEEE INFOCOM (2002)

10. Miller, M.J., Vaidya, N.H.: Improving power saving protocols using carrier sensing for dynamic advertisement windows. In: IEEE INFOCOM (2005)

Industrial Internet of Thing Based Smart Process Control Laboratory: A Case Study on Level Control System

Alpesh Patel[✉], Rohit Singh, Jignesh Patel, and Harsh Kapadia

Instrumentation and Control (IC), Institute of Technology,
Nirma University, Ahmedabad 384281, Gujarat, India
{alpesh.patel,14micc24,jbpatel,harsh.kapadia}@nirmauni.ac.in

Abstract. This paper tells us about the smart process control laboratory in which the concept of industrial internet of things (IIOT) is implemented on laboratory-scale trainer kit. A case study of level control trainer is discussed along with implementation and results. PID control algorithm is implemented in order to control level of water in the tank. Today, IIoT is an emerging technology that brought the control and automation on the platform of IoT, i.e., the control and monitoring of sensors and actuator is done from remote location. This example is relatively for home automation where the mobile devices comes into the picture. The device challenges and requirements of the systems are discussed. The software platform chosen is NI LabVIEW through which the application data dashboard is used in mobile devices to communicate with the process computer.

Keywords: Internet of Things · Laboratory scale trainer kit · PID · LabVIEW · Data dashboard

1 Introduction

The embedded platform to communicate and share the information each other, the external environment and with the people is known as Industrial Internet of Things (IIoT). It is also considered as a collection of different objects, platforms, systems and application consist of embedded technology [1]. IIoT is preferable when there is a large number of availability and affordability of sensors, pressure, etc. have helped for the real-time capturing and access of data [2]. It can improve the performance of industrial system as well as lab-scale trainer kit by communicating their data analysis and also take the necessary control action to benefit the society as well. The sensors and actuators are used by the industry to interface the digital world to the physical world [3]. These system are integrated with the big data available to obtain the huge information of the process.

IIoT has a large number of network connected to the industry to increase the performance and also efficiency and help in reducing the downtime [4]. For example - in industry the equipment was placed on the factory floor which detect a minor changes in its operation that will provide information regarding possibility of component failure and then it will take control action before the component failure to reduce the downtime.

© Springer International Publishing AG 2018
S.C. Satapathy and A. Joshi (eds.), *Information and Communication Technology for Intelligent Systems (ICTIS 2017) - Volume 2*, Smart Innovation, Systems and Technologies 84, DOI 10.1007/978-3-319-63645-0_21

Graphically programming is done using the software which works on real-time data is known as LabVIEW [5]. LabVIEW is very easy tool to handle as compared to other because of the available toolbox and icons which will just simply rearrange and combine together to do the programming.

Generally LabVIEW has three main components: Font panel, block diagram and connector panel. In front panel user places the relevant controls and indicators [6]. In block diagram user builds the code by placing different blocks. In connector panel user is allowed to represent single VI as a sub VI icon which can be called in other VI. Control design and simulation toolbox is of a great help in process control applications. It has wide range of controllers from simple PID's and auto tune PID's to advanced controllers like MPC and fuzzy logic [7]. Once can implement a controller of his choice with the help of this toolbox for e.g. Smith Predictor, gain scheduling etc. Real time data logging is also possible in LabVIEW; user can select one of these file formats.xlsx, LVM, TDM, and TDMS. Another important feature of LabVIEW is shared variables, one can share data between loops on single VI or between multiple VI's across the network LabVIEW uses many data sharing methods like UDP/TCP, LabVIEW queues and real time FIFOs. Depending on one's application select any of the above sharing methods [8].

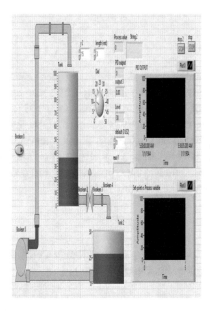

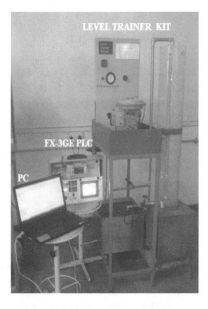

Fig. 1. Front panel of lab-scale trainer in LabVIEW

Fig. 2. Lab scale level control trainer

In many small scale industries and small application of automation, people are using Visual C, C++, and Visual basic for development of GUI [9] and for communication with PLC. It provides satisfactory performance and reduces the cost of software development. Mitsubishi FX-3GE PLC has in-built Ethernet port and also has two analog inputs and one analog output [10].

1.1 Difference Between IIoT, IoT and M2M (Mobile 2 Mobile Communication)

Consumer IoT reflects the benefits of individual consumers whereas Industrial IoT is concentrated on increasing the efficiency, safety and yield of operators with the return of investment done. M2M is a division of Industrial IoT, which drifts to importance very especially on machine 2 machine communication where, IoT expands which include machine to object, people and infrastructure. The IIoT is about creating machine more proficient and easier to monitor.

1.2 Hardware Setup

Figures 1 and 2 gives as idea about the whole process which is going into the level control system. In this process first step is to launch the communication between the Mitsubishi PLC and LabVIEW using the Ethernet TCP/IP. After that build up the shared variables into the LabVIEW and then deploy the shared variables which will give the IP address. Using that IP address connects with available tablet device through which the data could be controlled as well as monitored in the Tablet. Connection between the Tablet and computer is done using hotspot.

2 IIoT Requirements

Industrial automation today is very refined. Manufacturing of computer process is done without human intervention in a "light-out" environment for example if some error occurs no human intervention is responsible for that because automation speed is so rapid. The industrial environment is so accurate and productivity because of the direct interact with other network at speeds without human intervention or direct intervention [11]. Today, this system of industry as well as lab system is connecting to the internet and internal IP network through gateways. These gateways are programing and provisioning in such a way to expose the require data to the enterprise systems.

2.1 Cloud Computing

In lab scale trainer kit it used is limited but in case of industry these will play a vital role in infrastructure (IaaS), platform (PaaS) and application (SaaS).

2.1.1 Software-as-a-Service (SaaS)

It is a platform at which one or more applications and high computational resources are required to run them as per the user demand. In our case to control & monitor more parameters for example- PID values, PV, etc. So to run this services on SaaS Platform.

2.1.2   Platform-as-a-Service (PaaS)

Is a model of service delivery where the computer systems as a computing platform as a lab-scale level trainer kit, controllers, on a demand it is provided to developed and deployed the applications.

2.1.3   Infrastructure-as-a-Service (IaaS)

It is a platform in which the common computing infrastructures of software, sensors and equipment connected to network are provided as per demand service on which the different algorithms are develop and execution takes place. For example- in the developed system various software for e.g. GX-works 2, LabVIEW and hotspot serer, virtual router and equipment like control valve, sensors, wiring, etc. are present.

2.2   Access

In general the industry will occupy the larger space than the Lab-scale trainer kit. So, in that case use the different routers, repeaters, etc. to access the devices to control the parameters at any time or anywhere in the industry. But in lab-scale trainer kit the surrounding area is less so use the hotspot which can work within the range of 10 m or use the virtual router to monitor the data as well as control and access from any place. Hence, the devices should be contact at anytime from anywhere.

2.3   Security

This is very important aspect of the IIoT. How we can secure our data in the cloud? What is the different approaches that the plant should be secure because when IT sector involves it totally comes into the online data storage? The system also contain the smart machines so it can protected using the password [12]. The system should contain the wireless protection as well as virus protection. Finally be sure that the system should not contain SQL injection, Malware injection, DOS (Daniel of service), etc. [13].

2.4   User Experience

The embedded platform to communicate and share the information each other, the external environment and with the people is known as IIoT (Industrial Internet of Things) [14]. The adaption of IIoT is being enabled by the ease of availability of sensors, processors which can be programmed in easy language or high computational power and other technologies which is familiar to the user's that are working on the system [15].

2.4.1   Augmented Operators

To increase productivity the employees which join company use mobile devices, data analysis, augmented readily & also transparent connectivity. As fewer skilled workers are left behind because it will refine and younger replacement worker will need information at their fingertips. That information will be given in a real-time format that is familiar to them, thus the plant becomes more user-centric instead machine-centric or less-machine centric.

2.5 Assets Management

Technology changes after few years so we need to adapt the system which have proactive asset management instead of predictive asset management. Proactive asset management provides IIoT to access more data and process variables that can help them to choose the appropriate system. Multi-variant data analysis can help identify anticipated failures earlier and help to provide more accurate diagnosis of the pending issue [16]. In the lab-scale trainer kit the sensor and actuators are placed in such a way that it can be used in the future also and should be checked at the regular. Buying of the product is according to the proactive asset management.

2.6 Smart Machines

The machine work on the principle of 3 C's Collect, Configure & control. The term smart machines implies a machine that is better connected, more flexible, more efficient and safe. It can quickly responds to demands for example in lab-scale trainer kit which uses the controller as a Programmable logic controller (PLC) Mitsubishi FX-3GE 40 M in which we use Ethernet communication using cat-5 cable instead of RS-485 [17] for the fast communication.

3 IIoT Challenges

3.1 Precision

Precision in IIoT means how your system is more efficient and exactness to the requirement – For example when communicating the PLC with LabVIEW there is a network down in between the process and the whole system will shut down for some time so avoid that error by making the system more accurate for that you have to use NI OPC [18] instead of Ethernet.

3.2 Adaptability and Scalability

IIoT make a great change in the picture of the industrial systems. The conventional and growth of industrial system is described by either (1) end-to-end solutions or registered design or (2) increment the functions at regular interval according to the demand of vendors [19]. This solution is provided but at what cost? In IIoT data can be easily stored and analyzed for the better result. If the vendor defined-monitoring system is available open platform, then there is chances to improve the efficiency as well as prevent system downtime. For example – software like LabVIEW should be updated at regular interval.

3.3 Security

Today, the majority of control network are not secured. Protocol must be contain or support strong encryption, mutual authentication and safety against record/playback attacks. The main challenge of security is comes into the picture when we store the data into the cloud [20]. There are the various type of challenges which occurs for example SQL injection in which the data are attack, cross-sites crypting, Malware-injection and Daniel of service. In this problems the data is stolen or attack by the third party.

3.4 Update and Maintenance

In addition to security, the system should be continuously modified and maintain to meet the requirement functionality and also meet the maintenance requirements. Add to more capabilities the system software is updated at regular interval and the more systems are added to meet the requirements of the industry. The main challenge is comes into the picture when it's time to update the system because the user is depend on the manufacture/vendor to give the updated software so it can be used in the available system.

3.5 Flexibility

The system used in the lab-scale trainer kit or industry should be flexible. In lab-scale trainer kit used the software LabVIEW in which we declare or create the shared variables after that it can deploy then it will communicate with the Tablet. From tablet it can be controlled and monitor but it cannot be controlled from the LabVIEW side in other words it can't be controlled from computer side so if system gets disconnected then the process get stopped and the system get damaged. The network downtime should be less so it can communicate fast with the available system and acquire the data and passed it to the other system.

4   Hardware Implementation

This section tells about the interfacing of sensors and actuators taking the control action through the IIoT technology. The hardware setup is presented in laboratory i.e. laboratory-scale level trainer kit in which the level of the tank is continuously monitoring and controlling by using the IIoT. The software Platform used is LabVIEW in which the shared variables are defined and an application known as dashboard is used to communicate between the software and hardware. The requirements of the smart process control laboratory shown in Figs. 3 and 4 are defined in two parts- (1) Hardware requirements-Level sensor (4–20 mA), Pressure Sensor (4–20 mA), Flow sensor-DP type (4–20 mA), Current to pressure convertor (I/P), Pneumatic Valve (3–15 psig), Router, HUB, PLC (TCP/IP), DAQ card (8-channel 6008/09) with USB/RJ45, NIC, Data server. (2) Software requirements- Basic operating system, LabVIEW, PLC software.

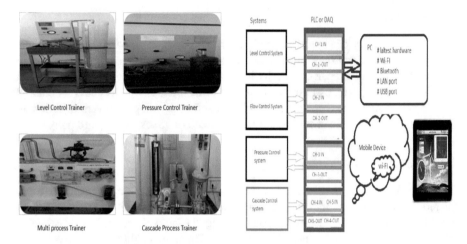

Fig. 3. Different process trainer kit   **Fig. 4.** Smart process control laboratory

The outcome of the smart process control laboratory are - Multi User System, Remote location control, real time data logging, online monitoring, fault detection.

4.1   Implementation of IIoT for Laboratory-Scale Level Trainer Kit

The IIoT will play a vital role in monitoring as well as controlling of the level in the tank. One type of SCADA is developed in both the LabVIEW side as well in mobile devices. First the shared variables defined according to the requirements of the system. In our system defined the controlling variables such as set-point (SP), proportional gain (Kp), integral gain (Ki), differential gain (Kd) and on the other hand monitoring the variables such as process variable (PV), pid output. Assign variable name as per convenient in the data dashboard and

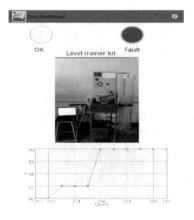

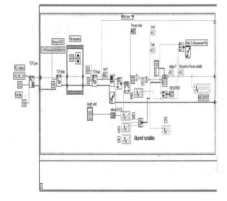

Fig. 5. Monitoring of fault detection   **Fig. 6.** Block diagram of shared variables

connect the device to hotspot. Now, as both computer and mobile devices connected to local hotspot, deploy the program and connect to the ip address monitor on the LabVIEW (Fig. 5).

Figure 6 gives idea about the communication is done using ethernet TCP/IP between the Mitsubishi PLC and LabVIEW. Declared the shared variables in the programming window and data to be stored in the MS-excel. The shared variables then also initial by some value then deploy the variables to obtain the ip. The shared variables can be read as well as write.

4.2   Role of Mobile Devices and Server

The mobile devices is used to monitor and controlling of the data from the remote location. For example if the laboratory-setup is in the department and it's continuously monitor and controlling the level from the other location then laboratory. It's very less time consuming and easily to monitor as a small display in the mobile devices.

Figure 7 shows a graph of PI controller. Data is stored in excel and the same will be sent over some other location using the server.

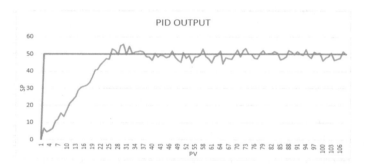

Fig. 7.   Result of PI controller

5   Conclusion

Effectively implementation of IIoT makes the smart process control laboratory very popular among the user. A single mobile device can control and monitor many system at a time in the laboratory and it's a great power of multi-user system. Real time data logging facility is also provided for the analysis purpose in which the detailed data are stored. Online monitoring of the system is done by the developing the SCADA on mobile device using data dashboard application. The user can identify any fault occurring in the laboratory by monitoring the fault in the mobile device.

References

1. Stankovic, J.A.: Research directions for the Internet of Things. IEEE IoT J. 1(1), 3–9 (2014)
2. Abdelzaher, T., Prabh, S., Kiran, R.: On real-time capacity limits of ad hoc wireless sensor networks RTSS, December 2004
3. Bradshaw, V.: The Building Environment: Active and Passive Control Systems. Wiley, New York (2006)
4. Gubbia, J., Buyyab, R., Marusic, S., Palaniswami, M.: Internet of Things (IoT): a vision, architectural elements, and future directions. Future Gener. Comput. Syst. 29, 1645–1660 (2013)
5. Jamal, R., Pichlik, H.: LabVIEW Applications and Solutions. Prentice Hall, Upper Saddle River (1999)
6. Quan, B., Ke-xiang, W., Yu-lan, C., Shi-sha, Z.: Design of force loading monitor and control system based on LabVIEW. In: Fourth International Conference on Digital Manufacturing and Automation (2014)
7. Lakshmi Sangeetha, A., Naveenkumar, B., Balaji Ganesh, A., Bharathi, N.: Experimental validation of PID based cascade control system through SCADA-PLC-OPC interface. In: International Conference on Computer Communication and Informatics (ICCCI 2012), Coimbatore, India, 10–12 January 2012
8. Zanella, A., Bui, N., Castellani, A., Vangelista, L.: Internet of Things for smart cities. IEEE Internet Things J. 1(1), 22–32 (2014)
9. Boyer, S.A.: SCADA: Supervisory Control and Data Acquisition. International Society of Automation (2009)
10. FX-3GE series programmable controller's hardware manual
11. Liu, A., Salvucci, D.: Modeling and prediction of human driver behavior. In: International Conference on HCI (2001)
12. Zunnurhain, K., Vrbsky, S.V.: Security attacks and solutions in clouds. In: 2nd IEEE International Conference on Cloud Computing Technology and Science, Indianapolis, December 2010
13. Wentzlaff, D., Gruenwald III, C., Beckmann, N., Modzelewski, K., Belay, A., Touseff, L., Miller, J., Agarwal, A.: A unified operating system for clouds and manycore: fos. In: Computer Science and Artificial Intelligence Laboratory TR, 20 November 2009
14. Litty, L., Lie, D.: Manitou: a layer-below approach to fighting malware. In: Proceedings of the 1st Workshop on Architectural and System Support for Improving Software Dependability, ASID 2006, New York, NY, USA, pp. 6–11 (2006)
15. Payne, B.D., Carbone, M., Sharif, M., Lee, W.: Architecture for secure active monitoring using virtualization. In: IEEE Symposium on Security and Privacy, pp. 233–247 (2008)
16. IIoT white paper: Requirements for the 'Industrial Internet of Things'
17. FX-Series PLC Training Manual using GX-Developer
18. Sahina, C., Bolatb, E.D.: Development of remote control and monitoring of web-based distributed OPC system. Comput. Stand. Interfaces 31, 984–993 (2009)
19. National instruments. http://ni.com/trend-watch
20. Igure, V.M., Laughter, S.A., Williams, R.D.: Security issues in SCADA networks. Comput. Secur. 25, 498–506 (2006)

Deep Neural Network Based Classification of Tumourous and Non-tumorous Medical Images

Vipin Makde[(✉)], Jenice Bhavsar, Swati Jain, and Priyanka Sharma

Computer Science and Engineering, Nirma University, Institute of Technology,
S.G. Highway, Ahmedabad 382481, India
{15mcec30,15mcec05,swati.jain,priyanka.sharma}@nirmauni.ac.in

Abstract. Tumor identification and classification from various medical images is a very challenging task. Various image processing and pattern identification techniques can be used for tumor identification and classification process. Deep learning is evolving technique under machine learning that provides the advantage for automatically extracting the features from the images. The computer aided diagnosis system proposed in this research work can assist the radiologists in cancer tumor identification based on various facts and studies done previously. The system can expedite the process of identification even in earlier stages by adding up the facility of a second opinion which makes the process simpler and faster. In this paper, we have proposed a framework of convolution neural network (CNN), that is a technique under Deep Learning. The research work implements the framework on AlexNet and ZFNet architectures and have trained the system for tumor detection in lung nodules and well as brain. The accuracy for classification is more than 97% for both the architectures and both the datasets of lung CT and brain MRI images.

Keywords: Tumor identification and classification · Machine learning · Deep learning · Convolution neural network · AlexNet · ZFNet

1 Introduction

Medical imaging have been continuously evolving with the improvement in technology. Over the past years, tumor classification was done manually by the radiologists but over the last couple of years, this process has been changed and has become faster and less erroneous with less or no human intervention. Tumor is an irregular growth of tissues which can be felt in any part of the body within any organ. They may or may not be harmful to the body. Identification and classification of such tumors into benign i.e. not harmful and malignant i.e. harmful is important to do in the earlier stages. Cancer has very large number of mortality rate among the world. Increase in size of the tumor or change in its shape along with the time are the basic symptoms of the cancerous tumors which can

© Springer International Publishing AG 2018
S.C. Satapathy and A. Joshi (eds.), *Information and Communication Technology for Intelligent Systems (ICTIS 2017) - Volume 2*, Smart Innovation, Systems and Technologies 84, DOI 10.1007/978-3-319-63645-0_22

only be identified through the imaging modalities either CT scan or MRI scan. Deep learning is a machine learning techniques that provides the advantages of intrinsic feature extraction using self-learning methods.

Convolutional Neural Network is well known architecture of deep learning. In 1990, a series of LeNet neural network were developed by LeCun et al. [16] and among them the most popular one was LeNet-5. This was a multi-layer artificial neural network which could classify handwritten digits and can be trained using back propagation algorithm. It can gain effective outputs as compared to other, but due to lack of computing power and training dataset it could not perform well on large scale image.

Since 2006, many other CNN architecture were developed to overcome the problems encountered earlier. A new architecture was proposed by Alex Krizhevsky, known as AlexNet [13]. It was deeper and much more wider version of LeNet. After this structure many more models were introduced which were more deeper and can solve complex problems, and they were ZFNet [10], VGGNet, GoogleNet and ResNet. However these networks are more complex which lead to difficulty in optimising and got easily over-fitting state.

2   Related Work

Roth et al. [4] have proposed a method where Medical CT images are classified into five anatomical classes, that are lung, neck, liver, pelvis, legs using deep neural network. The training of 4,298 of 5 classes. Here the data augmentation approach is used to enrich the data-set and improve the performance in classification.

In 2016 Yang et al. [6] have proposed a deep convolutional neural network instead of conventional SVM and ELM algorithm to classify the medical images. They have used DCNN models with 5 layers and 7 layers, consisting multiple layers with convolution, max-pooling, and fully connected layer. The input is reshaped in 32×32 and the n passed on to both deep neural network. Both models are having filters of 3×3 with stride 1×1 for all the convolution layers and in pooling layer they are using 2×2 stride this indicates that the size is reduced to half of its input size.

Amir et al. [1] have proposed a SVM classifier method using random-forest based learning for segmentation of tumor. While training is done using random-forest learning and testing the predicted label are taken as input in SVM classifier. Before training and testing is performed the image pre-processing is done by enhancing the contrast and then filtering by normalizing the intensity. They have used BRATS dataset of 20 patient having high-grade gliomas.

Paulin John [14] has proposed a method of classification of brain tumor in three steps: wavelet decomposition, textural feature extraction and classification. The wavelet decomposing is used to extract the features of the brain MRI images, textural feature extraction is done to differentiate the normal and abnormal tissues which is done by providing a contrast between malignant and normal tissue. And finally the classification is done here by probabilistic neural network.

KL Hua et al. [11] have suggested a framework in their paper which uses the region of interest first of all segmented from the CT images as per the annotations of the radiologists. This is further given to the classifier as a sample to train the DBN and the CNN. Lung nodules larger than 3 mm were taken into account for further consideration. The authors have compared the results for DBN and CNN.

Kumar D. et al. [5] have designed a framework where nodules are extracted using the annotations and then given to the classifier. Here, binary decision tree is used for training. Nodules larger than 3 mm are extracted and given to the classifier as per the ground truth provided along with the data.

Setio A. et al. [2] have discussed a framework of multi-view convolution neural network to reduce the number of false-positive in the CT images for manually extracted nodules which are larger than 3 mm in size. Multiple 2-D streams of ConvNets are used to classify the nodules and finally they are combined using a fusion method for final classification. The authors have said that the described framework is highly suitable for the false positive reduction methods.

Kim B. et al. [3] have designed a framework where they have taken manually segmented pulmonary nodules which were segmented by the well-trained radiologists and feature extraction has been applied here. Features have been extracted from the class labels and then the nodules are passed to the classifier.

3   Architectures

There are various kind of architecture to implement CNN, however the basic modules used are quite similar. The design of all CNN deep learning architecture were designed for the understanding that picture elements are circulated over the whole picture, and convolutions with learn-able parameters are a viable approach to extract similar features at different area with couple of parameters. To classify the data in two classes namely tumorous and non-tumorous, AlexNet [13] and ZFnet [10] architecture are used as shown in Figs. 1 and 2. Both the architecture are having ten layers which are combination of convolution layer, max-pooling layer and fully connected layer as shown in Table 1. The discription of each layer as mentioned in the table and fig are, (i) the depth of each layer indicates that how many kernels are present in that layer, (ii) filter/pooling indicate what is the kernel size and (iii) strides mentioned are indicating that how the kernels are moving. The activation function used is ReLU (Rectified Linear Unit) [15], and in the final layer we have used softmax [8] loss function as the activation function to classify data-set.

1. **Convolution Layer:** Instead of converting image to 1-D array the CNNs perform a convolution operation over the input image. In convolution operation small matrices called filters or kernels are randomly initialised. The filters are moved from left to right, top to bottom and multiplied at each position. Each multiplication gives a single output and hence the resulting output has reduced dimensions. This process of multiplying from left to right and top to bottom with filters is called convolution. There are four main parameters involved in Convolution (i) Number of kernels, (ii) Size of kernel, (iii) Stride and (iv) Padding maintains the size of the image.

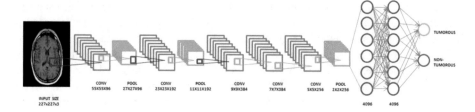

Fig. 1. Alexnet architecture

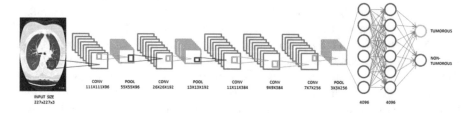

Fig. 2. ZFNet architecture

Table 1. Details of architectures used

Layer	Depth	Filter/Pooling		Stride
		AlexNet	ZFNet	
Input	3	–	–	–
Convolution	96	11×11	7×7	1×1
Max-pooling	96	3×3	3×3	2×2
Convolution	192	5×5	5×5	1×1
Max-pooling	192	3×3	2×2	2×2
Convolution	384	3×3	3×3	1×1
Convolution	384	3×3	3×3	1×1
Convolution	256	3×3	3×3	1×1
Max-pooling	256	3×3	3×3	2×2
Fully connected network	4096	–	–	–
Fully connected network	4096	–	–	–
Output	2	–	–	–

2. **Pooling Layer:** The function of the pooling layer is to reduce the spatial size of the image to represent the reduce amount of parameters and computation in the network, and to control over-fitting. It is often seen that the pooling layer is introduced in between successive convolution layer in an CNN architecture so as to provides a form of translation invariance. The most commonly used one is max pooling, in which the maximum number from sub-matrices

in input image and taking it as the output, while moving left to right, top to bottom, with some stride.

3. **Fully Connected Layer:** Artificial Neural Network (as in Fig. 3) is combination of (i) Input Layer where the image is flattened to a 1-D array, (ii) Hidden Layers which consist activation function and (iii) The outputs in the Output Layer are equal to the number of classes. The output is compares with actual results and compute loss, to modify the elements of matrices such that loss is minimized and then again perform the same things until the network's performance stops improving.

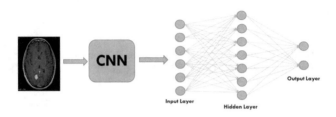

Fig. 3. Artificial neural network

4   Experimental Results

For experimentation we have worked on two datasets, REMBRANDT [9] and 'SPIE-AAPM CT Challenge' [7]. The Data-set REMBRANDT (The Repository of Molecular Brain Neoplasia Data) used has been taken from The Cancer Genome Archive (TCIA) Glioma Phenotype Research Group [12]. It contains 10,10,020 magnetic resonance multi-sequence images of the patients having Malignant Brain Tumor of various types and shapes. 'SPIE-AAPM CT Challenge' dataset has been taken from the public access of The Cancer Imaging Archive (TCIA)[12]. The dataset is a part of 2015 SPIE medical imaging conference. It contains 22,489 lung CT images with the ground truth that is the position of the tumor and the details of which patient has benign tumour and which patient has malignant tumour.

Steps of Classification: The following steps are carried out for classification of brain tumor data set into two classes tumorous and non-tumorous:

1. The Alexnet and ZFNet architecture are created using Tensor-flow and Keras. And the images are labled as non-tumorous(0) tumorous(1).
2. The input images are of 512×512 pixels, and these are re-sized to 227×227 and are stored in batches of training sets and testing set. The data has been split into 80% of training data and 20% of testing data.
3. The training sets are used to train the model by using fit function, which run for a certain number of epochs so as to get correct features to train the data accordingly.

4. The testing is done by passing the test batches to the testing model where it uses softmax loss function as the mode of classification of the dataset.

Performance Analysis: After training and testing was applied on sub-categories of the data-set, the results obtained by AlexNet and ZFNet architecture are mentioned in Table 2 for the REMBRANDT Dataset and it is sub-categorised as (i) Type-A contains frontal MRI, Axial MRI and Co-Axial MRI images. (ii) Type-B contains only Axial MRI images. (iii) Type-C contains few of the top Axial MRI images. (iv) Type-D contains augmented MRI images from case 3. Here the Augmented images means that the same set of images were flipped, rotated to some degree, so as to enrich the dataset and improve the performance in classification.

Table 2. Detailed results of AlexNet and ZFNet architecture for REMBRANDT brain dataset

	Training		Testing	
	Loss	Accuracy	Loss	Accuracy
Type-A^a	0.4237	0.8502	0.4757	0.8175
Type-B^a	0.4544	0.8351	0.4922	0.8064
Type-C^a	0.3986	0.8636	0.3640	0.8802
Type-D^a	0.1083	0.9742	0.1094	0.9766
Type-C^b	0.2802	0.9247	0.3226	0.9028
Type-D^b	0.1210	0.9742	0.1114	0.9766

^{a} AlexNet Architecture
^{b} ZFNet Architecture

Table 3. Detailed results of AlexNet and ZFNet architecture for 'SPIE-AAPM CT Challenge' lungs dataset

	Training		Testing	
	Accuracy	Loss	Accuracy	Loss
Type-I^a	0.48	0.71	0.49	0.69
Type-II^a	0.96	1.49	0.95	1.87
Type-III^a	0.9961	0.0627	0.9962	0.0609
Type-III^b	0.9961	0.0627	0.9962	0.0609

^{a} AlexNet Architecture
^{b} ZFNet Architecture

The results obtained for the AlexNet and ZFNet architectures for the 'SPIE-AAPM CT Challenge' dataset are as mentioned in Table 3. The 'SPIE-AAPM CT Challenge' dataset is sub-categorised as (i) Type-I has dataset divided in two set of Benign and Malignant. (ii) Type-II has dataset divided in two set of Tumorous and Non-Tumorous. Worked on sample set of 2000 images. (iii) Type-III is similar to Type-II but has Complete dataset of 22,489 images.

As referred in Table 2 when the augmented data which has some fixed set of axial brain MRI slices of every patient is used, then maximum accuracy of 97.66% was acquired for both the architectures. And as in Table 3, when the data was divided in two sets of tumorous and non-tumorous images both the architecture accuracy was 99.62% and this result is of testing data.

5   Conclusion and Future Scope

Image processing and analytics using Deep learning techniques has attracted medical imaging researchers the world over. Convolution neural network (CNN) is a well accepted technique under deep learning algorithms. The ConvNet architecture for CNN implementation has been used for object detection in natural images by many researchers. It is also evident that they can be used for medical images as well. In the present work we have focused at using CNN for classification of tumorous and non-tumorous lung nodules and brain medical images. In this paper, the images were manually labeled as no-tumorous(0) and tumorous(1) and the dataset was split into 4:1 ratio of training batches and testing batches respectively. As concerned to brain data set when mixed type of data is considered, the architecture was not able to learn the features of brain tumour for all the three views(frontal, axial, co-axial view of brain MRI). However, when the data was limited to fixed set of axial images, then the learning rate increased, since only axial brim MRI images were used. Similarly we can obtained the same results for frontal and co-axial view too. It is worth mentioning here that since the lung images do not have axial views, hence, maximum accuracy was achieved. As compared to the existing techniques, CNN is auto-configurable and hence, can give accuracy for newer types of tumour categories in future. However, for the network to work that way, it needs a more exhaustive training on categorizing the tumour types. This can be termed as the future extension of this work. The successful implementation of such an architecture will assist the doctors to classify and accordingly determine line of treatment. Hence, reduced human efforts will lead to faster and more accurate analysis for medical diagnostics.

References

1. Amiri, S., Rekik, I., Mahjoub, M.A.: Deep random forest-based learning transfer to SVM for brain tumor segmentation. In: 2016 2nd International Conference on Advanced Technologies for Signal and Image Processing (ATSIP), pp. 297–302, IEEE (2016)
2. Setio, A.A.A., et al.: Pulmonary nodule detection in CT images: false positive reduction using multi-view convolutional networks. IEEE Trans. Med. Imaging **35**(5), 1160–1169 (2016)
3. Kim, B.-C., Sung, Y.S., Suk, H.-I.: Deep feature learning for pulmonary nodule classification in a lung CT. In: 2016 4th International Winter Conference on Brain-Computer Interface (BCI). IEEE (2016)
4. Roth, H.R., Lee, C.T., Shin, H.C., Seff, A., Kim, L., Yao, J., Lu, L., Summers, R.M.: Anatomy-specific classification of medical images using deep convolutional nets. In: 2015 IEEE 12th International Symposium on Biomedical Imaging (ISBI), pp. 101–104. IEEE (2015)
5. Kumar, D., Wong, A., Clausi, D.A.: Lung nodule classification using deep features in CT images. In: 2015 12th Conference on Computer and Robot Vision (CRV). IEEE (2015)
6. Yang, X., et al.: A deep Learning approach for tumor tissue image classification. Biomed. Eng. (2016)
7. Armato III, S.G., Hadjiiski, L., Tourassi, G.D., Drukker, K., Giger, M.L., Li, F., Redmond, G., Farahani, K., Kirby, J.S., Clarke, L.P.: SPIE-AAPM-NCI lung nodule classification challenge dataset. The Cancer Imaging Archive (2015)
8. Gu, J., Wang, Z., Kuen, J., Ma, L., Shahroudy, A., Shuai, B., Liu, T., Wang, X., Wang, G.: Recent advances in convolutional neural networks. arXiv preprint arXiv:1512.07108 (2015)

9. Scarpace, L., Flanders, A.E., Jain, R., Mikkelsen, T., Andrews, D.W.: Data from REMBRANDT. The Cancer Imaging Archive (2015). http://doi.org/10.7937/K9/TCIA.2015.588OZUZB

10. Zeiler, M.D., Fergus, R.: Visualizing and understanding convolutional networks. In: European Conference on Computer Vision, pp. 818–833. Springer, Heidelberg (2014)

11. Hua, K.-L., et al.: Computer-aided classification of lung nodules on computed tomography images via deep learning technique. OncoTargets Ther. (2014)

12. Clark, K., Vendt, B., Smith, K., Freymann, J., Kirby, J., Koppel, P., Moore, S., Phillips, S., Maffitt, D., Pringle, M., et al.: The cancer imaging archive (tcia): maintaining and operating a public information repository. J. Digit. Imaging **26**(6), 1045–1057 (2013)

13. Krizhevsky, A., Sutskever, I., Hinton, G.E.: Imagenet classification with deep convolutional neural networks. In: Advances in Neural Information Processing Systems, pp. 1097–1105 (2012)

14. John, P., et al.: Brain tumor classification using wavelet and texture based neural network. Int. J. Sci. Eng. Res. **3**(10), 1 (2012)

15. Nair, V., Hinton, G.E.: Rectified linear units improve restricted boltzmann machines. In: Proceedings of the 27th International Conference on Machine Learning, ICML 2010, pp. 807–814 (2010)

16. Le Cun, B.B., Denker, J.S., Henderson, D., Howard, R.E., Hubbard, W., Jackel, L.D.: Handwritten digit recognition with a back-propagation network. In: Advances in Neural Information Processing Systems. Citeseer (1990)

Decision Tree Based Intrusion Detection System for NSL-KDD Dataset

Bhupendra Ingre, Anamika Yadav$^{(\boxtimes)}$, and Atul Kumar Soni

Department of Electrical Engineering, NIT Raipur, Raipur, India
ayadav.ele@nitrr.ac.in

Abstract. In this paper, Decision Tree (DT) based IDS is proposed for NSL-KDD dataset. The proposed work uses Correlation Feature Selection (CFS) subset evaluation method for feature selection. Feature selection improves the prediction performance of DT based IDS. Performance is evaluated before feature selection and after feature selection for five class classification (normal and types of attack) and binary class classification (normal and attack). The obtained result is compared and analyzed with the other reported techniques. The analysis shows that the proposed DT based IDS provides high DR and accuracy. The overall result for binary class classification for the dataset is higher than five class classification.

Keywords: Decision Tree · Detection rate · False positive rate · Intrusion Detection System · NSL-KDD datasets

1 Introduction

The use of internet and network based devices are growing rapidly. For many purpose in every day's life internet is the most necessary and sufficient connectivity solution today. Intrusion Detection System can be categorized based on location or processing of data [1] and based on method of detection. According to the method of detection, IDS can be classified into two main subcategories that are: Misuse or Signature based IDS and Anomaly based IDS. Misuse based IDS works by searching for new instances from its previously stored database of signature of known attacks. But these signature based systems will fail when the new instances have unknown (new) attacks [1]. Anomaly based IDS on the other hand captures all the header information of the IP packets running towards the network and checks that those information deviates from normal behavior (i.e., previously studied normal behavior of monitored system), so it can detect new attacks. The paper compared the performance of KDD99 and NSL-KDD dataset [2] using SOM and found that the performance of KDD99 was very higher compared to the latest introduced NSL-KDD dataset i.e. 92.37% accuracy for KDD99 and 75.49% for NSL-KDD dataset [3].

Introduction of the first anomaly based intrusion detection system is proposed by Denning et al. [4] in 1987 and thereafter continuous research work is going on IDS with some new concept and technologies. Denning uses statistical model and rules to get the behavior of pattern from audit data to separate abnormal patterns from normal connection. In 2002, Liao et al. [5] proposed k-nearest neighbour classifier for intrusion

© Springer International Publishing AG 2018
S.C. Satapathy and A. Joshi (eds.), *Information and Communication Technology for Intelligent Systems (ICTIS 2017) - Volume 2*, Smart Innovation, Systems and Technologies 84, DOI 10.1007/978-3-319-63645-0_23

detection using KDDCup dataset. Song et al. [6] in 2005 used sub-sampling to select patterns of KDD Cup'99 training dataset and proposed genetic programming based IDS. In 2007, a novel hybrid method had been developed by Gaddam et al. [7] using cascading k-means clustering and ID3 decision tree algorithm on NAD Dataset for two-class classification and achieved 96.24% and 0.03% DR and FPR respectively. In 2008, Xiang et al. [8] proposed multi level hybrid techniques based on Bayesian clustering and decision tree. Also in 2008, Adetunmbi et al. [9] applied rough set theory for the feature reduction and K-NN classifier is used for pattern classification, this work is done using KDD Cup'99 dataset. In the same year, AdaBoost-based algorithm for network intrusion detection had been proposed by Wei Hu et al. [10]. Author [10] achieves the accuracy of 90–91% with good running speed of proposed algorithm for KDD Cup dataset. Support vector machine (SVM) is supervised learning method and it can be used for classification and regression. Fuzzy association rule is one of the most attractive data mining techniques, fuzzy rule based IDS on KDD Cup'99 dataset is proposed by Tajbakhsh et al. [11] in 2009 and achieved 91% detection rate and 3.34% false positive rate.

In recent 4–5 years, mostly the work is done on network based anomaly detection with some new concepts. Several researches published on IDS during 2000–2008 were analyzed by Thavallaee et al. [12] in 2010. In his survey paper, one can find that the work done on intrusion detection system is more on publicly available dataset (probably 70%) and only 32% researchers use their self-created dataset. Thavallaee also mentioned that only 24% researchers use feature selection method in their work. According to this survey paper, performance of IDS depends mainly on three components, i.e., employed dataset, characteristic of experiments and performance metrics. In 2012, Imran et al. [13] proposed the Linear Discriminate Analysis and Genetic Algorithm for feature selection and used Radial Basis Function (RBF) with cross validation for testing the result. Cross validation is a method to test the performance of classification based problem but it uses same dataset for training and testing. This model is applied by Guo et al. [14] to develop IDS and achieved accuracy of 92.5% and 93.3% for five class classification and binary class classification respectively. In same year, Neethu et al. [15] uses naïve bayes classifier for classification of two class problem on KDDCup dataset and Principal component analysis (PCA) for feature selection. CART 4.5 is applied by Bhoria et al. [16] to detect DoS attack in which 6 fold cross validation method is used to test the system. In aug 2013, Bajaj et al. [17] applied information gain (IG), Gain Ratio (GR) and correlation attribute evaluation algorithm to select optimal feature from the NSL-KDD dataset, further on the reduced set he applied several machine learning algorithm (i.e. naïve bayes, j48, nb tree, lib svm, and simple cart) in WEKA tool. Adaboost machine learning algorithm is proposed by Patil et al. [18] to evaluate the performance of KDDCup'99 and NSL-KDD dataset. The detection rate obtained by the author is 86.27% and 90.00% for KDDCup'99 and NSL-KDD respectively. Bhuyan et al. [19] had done an exhaustive survey on Network Anomaly Detection and presented various detection method, systems and tools. ANN has been applied for intrusion detection and classification in [20].

In this paper, Intrusion Detection System is implemented on NSL KDD dataset, Decision Tree (DT) based CART (Classification and Regression Tree) algorithm is applied for classification of attacks. The benchmark dataset [3] contain 41 features,

CFS attribute evaluation method is used for dimensionality reduction which helps to improve the performance of intrusion detection system in terms of time and space.

2 Proposed Work

Decision tree (DT) based intrusion detection system is implemented to evaluate the performance of the system based on two well known intrusion detection dataset (KDDCup'99 and NSL-KDD). Decision tree is non parametric and non linear data mining approach used for supervised classification learning and regression. The steps involve in proposed method is shown in the Fig. 1. In the first step, dataset is selected from the dataset repository of NSL-KDD dataset. After that preprocessing is applied on

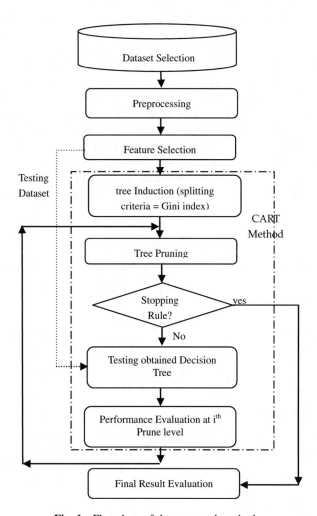

Fig. 1. Flowchart of the proposed method.

selected data to convert non numeric attribute into numeric. In the next step feature selection method is applied for dimensionality reduction. Random sampling is applied to get rid of biased training. Next three steps define CART algorithm, it includes tree induction, tree pruning and testing resultant tree with new test dataset. CART is recursive partitioning method [21], which build binary decision tree by splitting at each node. The root node of the decision tree contains full training samples. In the last step, performance is evaluated using various performance parameters like classification accuracy, detection rate, and false positive rate.

2.1   Dataset Selection and Preprocessing

Above mentioned, NSL-KDD intrusion dataset is used for Decision Tree based Intrusion Detection System (DTIDS), which is publicly available benchmark dataset for intrusion detection. The data repositories of the dataset have various datasets for NSL-KDD, out of which "KDDTrain" and "KDDTest" are chosen for training and testing respectively. The connection patterns of the dataset is described by 41 attributes and one class attribute as normal or attack or the attack types. Both the datasets contain symbolic attribute as well as numeric attribute. The symbolic attribute is converted to numeric attribute and the categorical attribute given separately in the cell array.

The advantage of DT is that it can take categorical attribute, categorical predictor so it doesn't give burden of converting all attributes to numeric. But converting some symbolic attribute to numeric is efficient with respect to time and space when we talk about the complexity of the system. The conversion of symbolic attribute is done as they assign numeric value based on their number of utterances in the feature space, the attribute value which occurs more frequently is assigned as '1' as shown in Table 1.

Table 1. Pre-processing

Categorical attribute	Feature name	Utterances in dataset (NSL-KDD)	Numeric conversion
Binary category	Normal	67343	1
	Attack	58630	2
Types of attack	Normal	67343	1
	DoS	45927	2
	Probe	11656	3
	R2L	995	4
	U2R	52	5
Protocol_Type	TCP	102689	1
	UDP	14993	2
	ICMP	8291	3

(*continued*)

Table 1. (*continued*)

Categorical attribute	Feature name	Utterances in dataset (NSL-KDD)	Numeric conversion
Flag	SF	74945	1
	S0	34851	2
	REJ	11233	3
	RSTR	2421	4
	RSTO	1562	5
	S1	365	6
	SH	271	7
	S2	127	8
	RSTOS0	103	9
	S3	49	10
	OTH	46	11
Service	70 features	High = http (40338) Low = http_2784 (1)	1 to 70

2.2 Feature Selection

The dimensionality of dataset is a big issue in data mining and machine learning, large amount of dataset leads to large amount of storage space and computational time. Other purpose of dimensionality reduction is to advance systems classification performance. Here, both the datasets have large number of pattern with 41 attributes, in which some attributes have no role, some have minimum role and some of the attribute have ambiguous value which leads to misclassification of attacks. To select the optimal feature from the intrusion dataset, Correlation–based Feature Selection (CFS) subset evaluation (cfsSubsetEval) has been applied in WEKA tool. This algorithm uses RankSearch method in which gain ratio attribute evaluation is used to rank all attributes as shown in Fig. 2. Here, gain ratio subset evaluator is specified so forward selection search is used to generate ranked list. From that ranked list, CFS computes the best subset of attributes by considering the individual predictive ability of each feature along with the degree of redundancy between them. At a particular split 'S' gain ratio is define by Eqs. (1 and 2).

$$\text{GainR}(S, A) = {Gain(S, A)}\big/{IntInfo(S, A)} \tag{1}$$

$$IntInfo(S, A) = -\sum_{i \in A} \frac{|S_i|}{|S|} \log\left(\frac{|S_i|}{|S|}\right) \tag{2}$$

CFS subset evaluation processes the training data of NSL-KDD dataset chooses 14 features from each dataset. The selected features are shown in Table 2. Main work of proposed intrusion detection system has been implemented on these selected features.

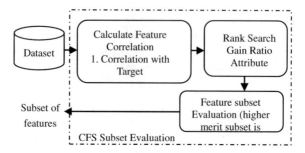

Fig. 2. The feature selection method using cfsSubsetEval.

Table 2. Selected attributes.

Dataset	Selected features	Description
NSL-KDD	1, 2, 3	Duration (cont), protocol_type (disc), service (disc)
	4, 5, 6	Flag (disc), src_bytes (cont), dst_bytes (cont)
	12, 25, 26	Logged_in (disc), serror_rate (cont), srv_serror_rate (cont)
	29, 30, 37	same_srv_rate (cont), diff_srv_rate (cont), dst_host_srv_diff_host_rate (cont)
	38, 39	dst_host_serror_rate (cont), dst_host_srv_serror_rate (cont)

2.3 Random Sampling

In this work, the dataset contains less number of patterns from R2L and U2R category as shown in Table 1. If we supply full training dataset without random sampling to the decision tree classifier then it will get biased training towards normal, dos and probe attack categories which results classification error for unknown attack patterns of testing datasets. So, it is necessary to apply random sampling to choose less number of connection patterns from other categories (normal, dos) and all patterns from R2L and U2R class. From NSL-KDD dataset, 23701 patterns are selected for five class classification and for binary class classification 6808 patterns are selected by random sampling.

2.4 Classification Using CART

2.4.1 Tree Induction

The proposed work is applied on one of the well-known intrusion dataset, i.e., NSL-KDD. The CART (Classification and Regression Tree) decision tree based algorithm is used to evaluate the performance of the dataset. Here, each and every connection patterns have discrete and continuous valued attribute, the proposed method checks for reduction in impurity for each possible splits of individual attribute during tree induction phase. A separate decision tree is created for the dataset, one for five class classification (normal and types of attack) and one for binary class classification (normal and attack). "KDDTrain" dataset of NSL-KDD and intrusion database repository, is used to train the decision tree.

2.4.2   Tree Pruning

The obtained decision tree is pruned to avoid over-fitting of data. Over-fitting occurs due to outliers and noise in the dataset. To overcome this problem, proposed method uses post pruning which uses error rate (pattern misclassified by tree) as pruning criterion. The tree is pruned and after each level of pruning the testing result is obtained.

2.5   Testing the Proposed DT Based IDS

From NSL-KDD dataset repository, "corrected test dataset" dataset is used to test the performance of intrusion detection system. The NSL-KDD testing dataset contain 22544 connection patterns. Each testing dataset have more number of attacks than that of in training dataset. The main aim of proposed IDS is to check for new and unknown attacks, so the system will handle new and unknown attacks. The test result is obtained after each level of pruning and the results are depicted in next section.

3   Performance Evaluation

Result of Decision Tree (DT) based classifier is evaluated using various parameters. The standard parameter includes Classification Accuracy, Detection Rate (DR) of each class, and False Positive Rate (FPR). These performance measures are calculated using Eqs. (3–5). The IDS which have high overall accuracy and detection rate and low false positive rate is considered as a good intrusion detection system.

$$\text{Detection Rate (DR)} = \frac{\text{TP}}{\text{TP} + \text{FN}} \tag{3}$$

$$\text{False Positive Rate (FPR)} = \frac{\text{FP}}{\text{FP} + \text{TN}} \tag{4}$$

$$\text{Accuracy (ACC)} = \frac{\text{TP} + \text{TN}}{\text{TP} + \text{TN} + \text{FP} + \text{FN}} \tag{5}$$

4   Experimental Result and Performance Evaluation

The proposed Decision Tree (DT) based intrusion detection system is implemented using MATLAB. Correlation–based Feature Selection (CFS) subset evaluation (cfsSubsetEval) method is implemented in WEKA for dimensionality reduction. The whole system is implemented using Intel(R) Core(TM)2 Duo CPU T6670@ 2.20 GHz processor system having 4 GB RAM. Here, a publicly available dataset is used to evaluate performance of the intrusion detection system, i.e., NSL-KDD dataset. For each dataset, four confusion matrices are obtained, one for five class classifications and one for binary class classification before feature selection and same two confusion matrices after feature selection. From this dataset repository the proposed scheme collected "KDDTest" for classification of five classes as well as two classes. The test dataset contain 22544 connection patterns and the whole test dataset is used for training and testing.

4.1   Five Class Classifications

The result is tested for five class classification with feature reduction and the result at each pruning level is shown in Table 3. From that table best result is used for confusion plot between testing target and testing predicted output. The best result at 17th pruning level is shown in bold.

Table 3. Testing results at each pruning for NSL-KDD dataset after feature selection (five class).

Pruning level	Overall accuracy (%)	Pruning level	Overall accuracy (%)
1	78.7216110716820	11	80.1366217175302
2	78.7216110716820	12	80.1366217175302
3	78.8014549325763	13	80.4382540809085
4	78.8058907026260	14	80.4382540809085
5	78.7970191625266	15	80.4382540809085
6	78.7970191625266	16	82.9134137686302
7	78.2514194464159	**17**	**83.6674946770759**
8	80.4293825408091	18	83.6630589070263
9	80.4249467707594	19	83.6231369765791
10	80.1321859474805	20	83.5432931156849

Feature selection is applied on the same dataset and 14 attribute is extracted from it. The overall testing accuracy for five class classification with feature selection is 83.7%, shown in Fig. 3. Table 4 shows the confusion matrix for each attack classes with FPR and DR.

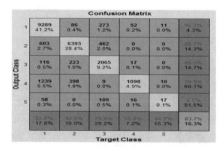

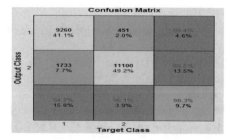

Fig. 3. Confusion plot for NSL-KDD dataset five class classification after feature selection.

Fig. 4. Confusion plot for NSL-KDD dataset binary class classifications after feature selection.

Table 4. Confusion matrix for five class classification.

Actual class		Predicted class	
		Yes	No
DoS	Yes	6393	1065
	No	707	12469
		DR = 85.7	FPR = 5.3
Probe	Yes	2065	356
	No	853	16797
		DR = 85.3	FPR = 4.8
R2L	Yes	1098	1656
	No	85	17764
		DR = 39.9	FPR = 0.4
U2R	Yes	17	183
	No	21	18845
		DR = 8.5	FPR = 0.1

4.2 Binary Class Classification

The result of NSL binary class classification at each pruning level is shown in Table 5 which is obtained after feature reduction. The confusion plot (Fig. 4) is obtained between testing target output and testing predicted output for the best result obtained from the Table 5. After applying feature selection on the same dataset, 90.2% of overall testing accuracy is obtained which is shown in Fig. 4.

Table 5. Testing results at each pruning for NSL-KDD dataset after feature selection (binary class).

Pruning level	Overall accuracy (%)	Pruning level	Overall accuracy (%)
1	86.3112136266856	11	90.1836408800568
2	86.3067778566359	12	90.1703335699077
3	86.3466997870830	13	90.0017743080199
4	86.3644428672818	14	88.1476224272534
5	86.4309794180270	15	88.0855216465578
6	87.3624911284599	16	87.7705819730305
7	89.6336053938964	17	87.8105039034776
8	**90.3122782114975**	18	87.1673172462740
9	90.3034066713982	19	90.3788147622427
10	90.2900993612491	20	86.8301987224982

5 Performance Comparison

The obtained result of Decision Tree based Intrusion Detection System (DTIDS) is compared with other existing technologies that are reported by different authors. As the IDS system is developed for NSL-KDD dataset so its performance is also compared.

The proposed system is also developed for the reduced version of NSL-KDD dataset. For five classes classification the performance result obtained is shown in Table 6. It is justified from Table 6 that applying feature selection is really improves the overall performance of the proposed scheme. It can be found from Tables 6 and 7 and Fig. 5 that the proposed system having high detection rate and overall accuracy with low false positive rate than other reported work.

Table 6. Performance comparison for NSL-KDD dataset (five classes).

Attack	With feature selection	
	DR	FPR
DOS	85.7	5.3
PROBE	85.3	4.8
R2L	39.9	0.1
U2R	8.5	0.1
Accuracy	83.7	

Table 7. Performance comparison with the methods proposed by Bajaj et al. [17]

Classifier method	Detection accuracy	Incorrectly classified
J48	81.94	18.06
Naïve Bayes	75.78	24.21
NB Tree	80.68	19.32
Multilayer perceptron	73.55	26.45
LibSVM	71.02	28.98
Simple CART	82.32	17.68
Proposed method (after feature selection for two class)	**90.3**	**9.7**

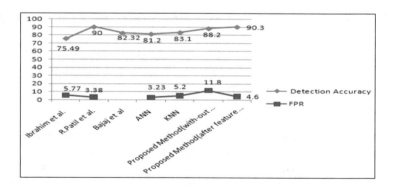

Fig. 5. Comparison of DR and FPR for two class classification (NSL-KDD).

6   Conclusion

In this paper, Intrusion Detection System based on Decision tree has been proposed for NSL-KDD dataset. Four different models have been proposed for the dataset; two model for five class classification (normal and types of attack) in which the model uses feature selection. Likewise a different model is created for binary class classification (normal and attack). The proposed system uses CART algorithm with gini index as splitting criteria for pattern classification and correlation based feature selection (CFS) is used for dimensionality reduction. The proposed system has been tested using the separate testing data of benchmark NSL-KDD dataset. Feature selection method improves the accuracy but more improvement has been found in NSL-KDD dataset as the detection rate of all the attacks has improved. For binary class classification, overall accuracy is higher than five class classification. Feature selection improves the performance of the system; it reduces the FPR and overall performance. The proposed work is also compared with other latest reported techniques and it is found that the proposed system for NSL-KDD dataset has higher overall accuracy and detection rate or low false positive rate. The feature reduction method reduces the time and space complexity as the size is reduces by one third of the original data size.

References

1. Sadek, R.A., Soliman, M.S., Elsayed, H.S.: Effective anomaly intrusion detection system based on neural network with indicator variable and rough set reduction. Int. J. Comput. Sci. Issues (IJCSI) **10**(6), 227–233 (2013)
2. Ibrahim, L.M., Basheer, D.T., Mahamod, M.S.: A comparison study for intrusion database (KDD99, NSL-KDD) based on self organization map (SOM) artificial neural network. J. Eng. Sci. Technol. **8**(1), 107–119 (2013)
3. NSL-KDD dataset. http://nsl.cs.unb.ca/nsl-kdd/. Accessed 21 July 2016
4. Denning, D.E.: An intrusion detection model. IEEE Trans. Softw. Eng. **13**(2), 222–232 (1987)
5. Liao, Y., Vemuri, V.R.: Using K-nearest neighbour classifier for intrusion detection. Comput Secur. **21**, 439–448 (2002)
6. Song, D., et al.: Training genetic programming on half a million patterns: an example from anomaly detection. IEEE Trans. Evolut. Comput. **9**, 225–239 (2005)
7. Gaddam, S.R.: K-Means+ID3: A novel method for supervised anomaly detection by cascading K-means clustering and ID3 decision tree learning methods. IEEE Trans. Knowl. Data Eng. **19**(3), 345–354 (2000)
8. Xiang, C., Yong, P.C., Meng, L.S.: Design of multiple-level hybrid classifier for intrusion detection system using bayesian clustering and decision trees. Patterns Recognit. Lett. **29**, 918–924 (2008)
9. Adetunmbi, A.O.: Network intrusion detection based on rough set and k-nearest neighbour. Int. J. Comput. ICT Res. **2**(1), 60–66 (2008)
10. Hu, W., Maybank, S.: AdaBoost-based algorithm for network intrusion detection. IEEE Trans. Syst. Man Cybernet. B Cybernet. **38**(2), 577–583 (2008)
11. Tajbakhsh, A., Rahmati, M., Mirzaei, A.: Intrusion detection using fuzzy association rule. J. Appl. Soft Comput. **9**, 462–469 (2009)
12. Tavallaee, M., Stakhanova, N., Ghorbani, A.A.: Toward credible evaluation of anomaly-based intrusion-detection methods. IEEE Trans. Syst. Man Cybernet. C. Appl. Rev. **40**(5), 516–524 (2010)
13. Imran, H.M., Abdullah, A.B., Hussain, M., Palaniappan, S., Ahmad, I.: Intrusions detection based on optimum features subset and efficient dataset selection. Int. J. Eng. Innov. Technol. (IJEIT) **2**(6), 265–270 (2012)

14. Guo, C.: A distance sum-based hybrid method for intrusion detection. Appl. Intell. **40**, 178–188 (2013)
15. Neethu, B.: Adaptive intrusion detection using machine learning. Int. J. Comput. Sci. Netw. Secur. **13**(3), 118–124 (2013)
16. Bhoria, P., Garg, K.K.: Determining feature set of DOS attacks. Int. J. Adv. Res. Comput. Sci. Softw. Eng. **3**(5), 875–878 (2013)
17. Bajaj, K., Arora, A.: Improving the intrusion detection using discriminative machine learning approach and improve the time complexity by data mining feature selection methods. Int. J. Comput. Appl. **76**(1), 5–11 (2013)
18. Patil, D.R., Pattewar, T.M.: A comparative performance evaluation of machine learning-based NIDS on benchmark datasets. Int. J. Res. Advent Technol. **2**(2), 101–106 (2014)
19. Bhuyan, M., Bhattacharyya, D.K., Kalita, J.K.: Network anomaly detection: methods, systems and tools. IEEE Commun. Surv. Tutor. **16**(1), 303–336 (2014)
20. Ingre, A., Yadav, A.: Performance analysis of NSL-KDD dataset using ANN. In: International Conference on Signal Processing and Communication System Engineering (SPACES-2015), pp. 92–96
21. Gey, S., Nedelec, E.: Model selection for CART regression trees. IEEE Trans. Inf. Theor. **51**(2), 658–670 (2005)

Univariate Time Series Models for Forecasting Stationary and Non-stationary Data: A Brief Review

Bashirahamad Momin and Gaurav Chavan[✉]

Department of Computer Science and Engineering,
Walchand College of Engineering, Sangli, Mahrashtra, India
bfmomin@yahoo.com, gauravgchavan91@gmail.com

Abstract. Due to advancements in domain of Information processing, huge amount of data gets collected which varies according to different time intervals. Structural models and Time-series models are used for analysing time series data. Time series models are very efficient as compared to structural models because modelling and predictions can be easily done. This paper gives a brief insight into Auto-regressive Models (AR), Moving Average Models (MA), Autoregressive Moving Average model (ARMA) and Autoregressive Integrated Moving Average Model (ARIMA). This paper also helps to understand the characteristics of the data which will be used for Time-series modelling.

Keywords: White noise · Stationarity · Auto Correlation Function (ACF) · Partial Autocorrelation Function (PACF) · Box Jenkins approach

1 Introduction

Data analytics has been a major area of Interest in the industry since the past decade. The amount of data will tend to grow 10-fold in upcoming few years having approximate storage of about 50 zetabytes [1]. Due to advancements in domain of IoT (Internet of Things), many devices send the data to central node or server for processing. These data is being dumped at regular time intervals. Time series data is a collection of observations that are sampled according to time. Some common examples of time series include the hourly readings of air temperature, monitoring of a person's heart rate, and daily closing price of a particular companies stock.

Time series data are analysed in order to identify patterns in data which can be trends or seasonal variations. Hence, understanding the time series mechanism helps to develop a mathematical model which can explain the data in such a way that control, monitoring and prediction can be done with ease. Structural models like regression models can also be used for modelling time series data. But, these models are not necessarily associated with time, for example stock price which varies according to time can be modelled as change in inflation or unemployment rate using structural models like linear regression. In time series model, there is no such concept of dependant or independent variable; there is only one variable which is a model variable which varies according to different periodic trends.

© Springer International Publishing AG 2018
S.C. Satapathy and A. Joshi (eds.), *Information and Communication Technology for Intelligent Systems (ICTIS 2017) - Volume 2*, Smart Innovation, Systems and Technologies 84, DOI 10.1007/978-3-319-63645-0_24

The significance of time series modelling is that same data can be modelled with different periodic trends which in case of structural model are tedious task. Also some factors can be unobservable, hence excluded from regression analysis. In time series analysis, each observation is somewhat dependent upon previous observation, and gets influenced by more than one previous observation. Error term too gets influenced from one observation to another. These influences are mapped using autocorrelation, which is used to either model trend itself or to model the underlying mechanisms. To model using time-series, four common models are being used which are Autoregressive Model (AR), Moving Average Model (MA), Autoregressive Moving Average (ARMA) and Autoregressive Integrated Moving Average (ARIMA). This paper highlights methodologies of each of these time series models and describes their behaviour.

In Sect. 2 a brief review related to time series modelling is discussed. Section 3 focuses on methodologies which are used for analysing time series data. Section 4 highlights observations pertaining to time series modelling. Finally, in Sect. 5, conclusions and future work are discussed.

2   Related Work

In contrast to time series models, great amount of literature is witnessed in the field of neural network approach [2–6]. However, these methods failed to identify the abnormal patterns which can occur in data. Models based on Support Vector Machines (SVM) and stochastic approaches are also utilized for modelling time series data [7, 8]. Genetic Algorithms along with Neural Network are also being used [9–11]. These models were unable to give effective analysis due to complex nature of data, noisy data and high dimensionality of data. Therefore effective Time series models were needed.

The time series models helps to interpret hidden patterns of the data and helps in analysis by fitting a model for forecasting [12, 18]. In this paper, not only time series models are described but also all the way these models are being analysed is also being showcased. But it will take plenty amount of time to any researcher to understand many important concepts relative to time series models. This paper provides in depth knowledge of each of these concepts.

3   Time Series Modelling and Forecasting

Regression without lags fails to account for the relationships through time and overestimates the relationship between dependant and independent variables. To overcome this time series analysis needs to be done. The upcoming sub-sections will help to understand the stationary and non-stationary data, white noise process, AR, MA, ARMA, ARIMA models.

3.1   Stationarity

One of the important properties of time-series process is stationarity of data. So a random process or a stochastic process is known to be stationary when its joint distribution doesn't change over time. Time series data is stochastic or probabilistic in nature because there is no accurate formula when prediction needs to be done. But usually, time series data points are weakly stationary in nature i.e. those data-points which have constant mean μ, constant variance σ^2 and constant auto-covariance i.e. Auto-covariance (Y_t, Y_{t-1}) = Auto-covariance (Y_{t-2}, Y_{t-3}) at regular periodic intervals.

3.2 White Noise

A random process which has expectancy or mean, variance at any time as a constant and auto-covariance is 0. It implies that each observation is uncorrelated with other observations in the sequence (Fig. 1).

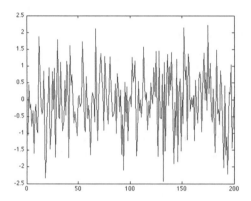

Fig. 1. White noise process [18].

The Ljung-Box statistics [13], which tests the "overall" randomness based on a number of lags, is a standard approach to determine whether the data points exhibit white noise property. Hence if white noise occurs in data, then there is no significance of doing time series modelling or estimation over such data.

3.3 Moving Average Model (MA)

Time series data at time interval t is given as y_t, where y_t which is current value at time t is considered as linear combination of different white noise processes u_t. The model can be represented as follows:

$$y_t = u_t + \theta_1 u_{t-1} + \theta_2 u_{t-2} + \cdots + \theta_q u_{t-q} \tag{1}$$

Where,
$\theta_1, \theta_2, \ldots, \theta_q$ are parameters which are estimated using maximum likelihood function. Lag operator L used to represent same data for previous period.

For example,

$$u_{t-1} = Lu_t \text{ and } u_{t-2} = Lu_{t-1}$$

Therefore,

$$u_{t-2} = L^2 u_t$$

The lag operator is used to make a representation of MA series much better as estimation interpretation becomes much easier. Hence equation is given by:

$$y_t = \sum_{i=1}^{q} L^i \theta_i u_t \tag{2}$$

3.4 Auto Regressive Model (AR)

The AR model depends on the past values and error terms.

$$y_t = \emptyset_1 y_{t-1} + \emptyset_2 y_{t-2} + \cdots + \emptyset_p y_{t-p} + \mu_t \tag{3}$$

Where,
μ_t is a error term given as white noise and p is a lag term.
In terms of Lag operator AR model can be represented as:

$$y_t = \sum_{i=1}^{q} \emptyset_i L_i^i y_t + \mu_t \tag{4}$$

Stationarity condition matters a lot in AR estimation. Non stationary series has a non-declining effect in AR estimation which is undesirable in time series modelling.

3.5 Auto Regressive Moving Average Model (ARMA)

ARMA (p, q) is the process which combines AR series as well as MA series. Mathematically, it can be represented as:

$$y_t = \emptyset_1 y_{t-1} + \cdots + \emptyset_p y_{t-p} + \theta_1 u_{t-1} + \theta_2 u_{t-2} + \cdots + \theta_q u_{t-q} + u_t \tag{5}$$

The ARMA (p, q) process is detected by plotting correlogram of Auto Correlation Function (ACF) and Partial Auto Correlation Function (PACF) against lags, which shows declining curve as lag increases in both ACF and PACF cases as shown in Fig. 2 below:

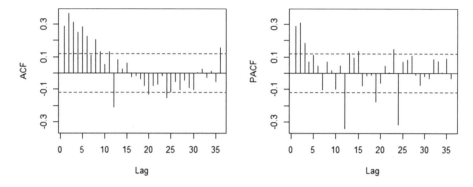

Fig. 2. The ACF and PACF correlogram for ARMA model.

As shown in Fig. 2. The coefficients of acf and pacf values go to negative axis when the coefficient or residual term comes out to be negative.

Box Jenkins Approach for ARMA Model

Box and Jenkins [14] suggested that differencing non-stationary series one or more times can achieve stationarity. Doing so leads to an ARIMA model, with the "I" standing for "Integrated". The approach consists of three main stages which are:

1. Identification.
2. Estimation and
3. Diagnostic checking.

In first stage, the orders of the model are determined. We plot correlogram of ACF and PACF to identify the lag values. In second stage, parameters $\emptyset$ and θ are estimated. The parameters are estimated using Ordinary Least squares [15] and maximum likelihood function [16]. The third stage checks whether the model fit is good. On the other hand, the residual diagnostics, checks whether the residuals have a certain correlation.

Information Criteria for ARMA (p, q) Model Selection

The information criteria for ARMA (p, q) model helps to determine the order of the time series based on Residual Sum of Square (RSS) [17]. The model which gives minimum RSS for value of p and q is best suitable model. When more number of lags are added the one which gives less RSS is best model. But when numbers of lags are added the RSS goes down. Hence, penalty term is considered while finding suitable model which is given as:

$$min(f(RSS) + penalty\ term) \tag{6}$$

Thus adding extra terms will also add penalty while finding the p and q. The standard methods which are used for computing information criteria are:

- Akaike's Information Criteria (AIC).
- Schwarz's Bayesian Information Criteria (SBIC).
- Hannon Quinn Criteria (HQIC).

Where,

$$AIC = \ln\left(\hat{\sigma}^2\right) + \frac{2k}{T},$$

$$SBIC = \ln\left(\hat{\sigma}^2\right) + \frac{k}{T}\ln(T),$$

$$HQIC = \ln\left(\hat{\sigma}^2\right) + \frac{2k}{T}\ln\left(\ln(T)\right)$$

Where,

$\hat{\sigma}^2 = \dfrac{RSS}{T}$, $k = p + q + 1$, and $T = $ *Sample size.*

The model which gives the less value of AIC is best model.

3.6   ARIMA Model

The AR, MA, ARMA models cannot handle non-stationarity that is the series that has trend. In stationary series, the model reverts around its mean value. To make non-stationary series into stationary, we difference it i.e. taking the difference of the series with its own lag. Differencing will give the series leads to a new set of values with constant mean and constant variance. Example,

As shown in table above $\nabla y = y_t - y_{t-1}$ is first order difference, and $\nabla^2 y$ is a second order difference, which gives constant mean and 0 or constant variance. Thus differencing ARMA model makes it integrated, which gives its name ARIMA. It can be represented as ARIMA (p, q, r) (Table 1),

Table 1. First order differencing and second order differencing.

y_t	∇y	$\nabla^2 y$
25		
16	9	2
9	7	2
4	5	

Where,

p = number of AR terms, q = order of differencing and r = number of MA terms.

Also, ARIMA can be represented AR, MA and ARMA model ARIMA(1,0,0) gives AR(1) model, ARIMA(0,0,1) gives MA(1) model and ARIMA(1,0,1) gives ARMA(1,1) model where only differencing is 0.

The main query which arises is to determine the order of differencing. It can be given by observing the following characteristics:

1. Positive auto-correlation at higher lags will need higher order differencing.
2. Zero or negative auto-correlation for first order itself then no differencing is needed.

3. The optimal order of differencing is the one whose prediction gives less value for RMSE (root mean square error) or MAE (Mean Absolute error).

4   Observations

In time series modelling, depending upon characteristics of periodic patterns, predictions are done. Different trends will lead to different periodic patterns which can be classified as hourly, daily, weekly, monthly data. The goal is to find the appropriate statistical relationship of the given series with its own past values. We build the model in training data, and test it on new dataset (holdout) to check whether predicted values match the actual values. Thus, accuracy is computed by computing forecast errors which can be represented as:

$$\text{Forecast error} = \text{Actual value} - \text{Predicted value}.$$

Accuracy can be computed using mean square error (MSE). The model which gives less MSE for Holdout sample is good model for prediction.

The brief behaviours of white noise AR, MA, ARMA and ARIMA for time-series data are shown in Table 2 below:

Table 2. Behaviour of time series models

Model	Characteristic	PACF correlogram	ACF correlogram	Data Characteristic
White noise	Cannot be used for Time series modelling	No spikes	No spikes	Random in nature
AR(p)	1. y_t depends on its own past values 2. P is computed using PACF function	Spikes till p^{th} lag then cuts off to zero	Spikes then decays to zero	Data should be stationary in nature
MA(q)	1. y_t depends on error term which follows a white noise process 2. q is computed using ACF function	Spikes then decays to zero	Spikes till p^{th} lag then cuts off to zero	Data should be stationary in nature
ARMA (p, q)	1. ARMA = AR+MA 2. Value of p and q are determined using AIC criteria	Spikes then decays to zero	Spikes then decays to zero	Data should be stationary in nature
ARIMA (p, d, q)	1. Data is made stationary by differencing it 2. Box-Jenkins approach is used to determine model	Spikes then decays to zero	Spikes then decays to zero	Data should be non-stationary in nature

5   Conclusion

Thus, while doing time series analysis, the data should not be a white noise process or random in nature. Thus only stationary or non-stationary will lead to identification of proper periodic pattern which can be used to determine appropriate forecasting model. The accuracy of forecasting is determined using Root mean square error or Mean Absolute error.

As a part of future work, major focus will be put upon understanding multivariate time series analysis. Also, forecasting over white-noise process can be done by using unsupervised algorithms like Neural Networks. But, it is very complex in nature. Hence, further part of research will be emphasized upon using semi-supervised algorithm like Support Vector Regression.

Acknowledgements. I would like to thank my Guide Dr. B.F. Momin for their continuous help and encouragement. Also thanks to TEQIP-II for financial support in presenting this paper. Also I would like to thank Department of CSE, Walchand College of Engineering, for providing useful resources.

References

1. Barnaghi, P., et al.: Semantics for the Internet of Things: early progress and back to the future. Int. J. Semant. Web Inf. Syst. (IJSWIS) **8**(1), 1–21 (2012)
2. Zhang, C., et al.: A multimodal data mining framework for revealing common sources of spam images. J. Multimed. **4**(5), 313–320 (2009)
3. Zhang, K.: Research on key technologies of college computer room billing management system. In: Key Engineering Materials, vol. 474. Trans Tech Publications (2011)
4. Zhang, M.: Application of data mining technology in digital library. JCP **6**(4), 761–768 (2011)
5. Shen, C.-W., et al.: Data mining the data processing technologies for inventory management. J. Comput. **6**(4), 784–791 (2011)
6. Danping, Z., Deng, J.: The data mining of the human resources data warehouse in university based on association rule. J. Comput. **6**(1), 139–146 (2011)
7. Jiang, J., Long, B.G., Wei, M., Feng, K.F.: Block-based parallel intra prediction scheme for HEVC. J. Multimed. **7**(4), 289–294 (2012)
8. Yang, S.-Y., et al.: Incremental mining of closed sequential patterns in multiple data streams. JNW **6**(5), 728–735 (2011)
9. Fu, Z., Juanjuan, B., Qiang, W.: A novel dynamic bandwidth allocation algorithm with correction-based the multiple traffic prediction in EPON. JNW **7**(10), 1554–1560 (2012)
10. Zhi, Q., Lin, Z.-W., Ma, Y.: Research of Hadoop-based data flow management system. J. China Univ. Posts Telecommun. **18**, 164–168 (2011)
11. Cui, J., Taoshen, L., Hongxing, L.: Design and development of the mass data storage platform based on Hadoop. J. Comput. Res. Dev. **49**(12), 12–18 (2012)
12. Krishna, G.V.: An integrated approach for weather forecasting based on data mining and forecasting analysis. Int. J. Comput. Appl. **120**(11), 0975–8887 (2015)
13. Ljung, G.M., Box, G.E.P.: On a measure of lack of fit in time series models. Biometrika **65**(2), 297–303 (1978)
14. Box, G.E.P., Jenkins, G.M.: Some comments on a paper by Chatfield and Prothero and on a review by Kendall. J. R. Stat. Soc. Ser. A (Gen.) **136**(3), 337–352 (1973)
15. Noreen, E.: An empirical comparison of probit and OLS regression hypothesis tests. J. Account. Res. 119–133 (1988)
16. Pan, J.-X., Fang, K.-T. Maximum likelihood estimation. In: Growth Curve Models and Statistical Diagnostics, pp. 77–158. Springer, New York (2002)
17. Johansen, S.: The Welch-James approximation to the distribution of the residual sum of squares in a weighted linear regression. Biometrika **67**(1), 85–92 (1980)
18. Hamilton, J.D.: Time Series Analysis, vol. 2. Princeton University Press, Princeton (1994)

Skin Detection Based Intelligent Alarm Clock Using YCbCr Model

Mohd. Imran[(✉)], Md. Shadab, Md. Mojahid Islam, and Misbahul Haque

Department of Computer Engineering, Aligarh Muslim University, Aligarh 202002, India
mimran.ce@amu.ac.in, Shadab.md02@gmail.com,
mojahidislam221@gmail.com, misbahul.haque@zhcet.ac.in

Abstract. In this paper, we are implementing an Intelligent Alarm Clock using skin detection approach and YCbCr color space model. This novel application is designed by keeping in mind the busy schedule and fast paced life of people living in this era. Focusing on the data and reports provided by various health organization, total time which is vital for comfortable sleep and peace of mind in an average adult is six or more than six hours on daily basis [1]. In the current scenario of hectic schedule of working in society, spare time of six hours still sound pretty compromising but, in reality, though, it is leading towards chronicle sleep deficiency. Deriving the inspiration from various research references, this alarm clock works on sleeping pattern of a person using Digital Image Processing. This work can be of great help for people suffering from sleeping disorders and hypersomnia. It is implemented using MATLAB. The tool used includes webcam, Image Processing Toolkit, and Image Acquisition Toolbox available in MATLAB.

Keywords: Image processing · Sleep deprivation · Skin detection · Color space

1 Introduction

According to researchers Bonnet and Arand [2] of Dayton Department of Veterans Affairs Medical Center, Wright State University, Kettering Medical Center, and the Wallace Kettering Neuroscience Institute, "There is strong evidence that sufficient shortening or disturbance of the sleep process compromises mood, performance and alertness and can result in injury or death. In this light, the most common-sense 'do no Injury' medical advice would be to avoid sleep deprivation. Data from recent laboratory studies indicate that nocturnal sleep periods reduced by as little as 1.3 to 1.5 h of one night results in reduction of daytime alertness by as much as 32% as measured by the Multiple Sleep Latency Test (MSLT)" [2].

Sleep can be classified mainly in two categories [1, 3]:

a. Non-REM (NREM) This type of sleep generally comprises of three stages of sleep, and having the distinct feature that deepness of sleep is greater than its previous one.

© Springer International Publishing AG 2018
S.C. Satapathy and A. Joshi (eds.), *Information and Communication
Technology for Intelligent Systems (ICTIS 2017) - Volume 2*, Smart Innovation,
Systems and Technologies 84, DOI 10.1007/978-3-319-63645-0_25

b. REM: The REM is the abbreviation for Rapid Eye Movement. Such sleeping gener-
    ally comes during the dreaming when our physical body is in an active state. As the
    name it-self define its basic characteristic of the eye movement during such sleeping
    in a way of back and forth.

As deep sleep recharges the body, paradoxical sleep also called refurbishes the mind
& enhances the capability of human being in learning and keeping sharp memory.
Various researches have been made that proves that night sleep less than six hours results
complexities on health. Some of these issues are well described in [4–10].

This is where our inspiration of this Intelligent Alarm Clock comes. The alarm clock
is designed specifically for the purpose to extend the waking time set in the alarm clock
by certain duration depending upon the disturbances occurred during the sleep.

We present a skin color-based detection algorithm that employs a human skin color
model, which takes into account the luminance Y in classifying skin and non-skin pixels.
In our approach, the distribution of human skin colors is modeled in YCbCr space [11].

2   Related Work

Before introducing our approach, it is necessary to discuss basic principles and related
work in the field of image processing. For skin detection there are different color space
model exists. A color space is a model to quantify color information in its constituent
components. We will focus on two different color space model viz. luminance based
model like YCbCr and RGB based color space which classify the color component in
red, green and blue component which is described below:

2.1   The YCbCr Color Space

This is mostly used color space model. There are various applications like image
processing applications as well as digital video etc., which uses YCbCr technique. The
luminance information which is major component in such format is represented by a
single component, Y, and color information is further divided into two component.
These components acts as two color-difference components also known as chrominance
component Cb & Cr [12] shown in Fig. 1. The YCbCr color space model is a mathematical model which is further explained in [13, 14] for skin detection and segmentation.

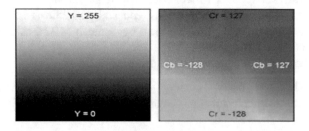

Fig. 1. YCbCr color space

2.2 Color Space for Skin Detection Based on RGB

The mathematical model for RGB color space is explained in [15, 16]. By nature this color model is additive for primary colors red, green, and after mixing together they produce a wide range of color products [17]. The color combination of these color is shown in Fig. 2 where R, G and B represents Red, Green and blue color component of any image. Due to this mixing nature it shows a non-uniform characteristics and failed in the application involving color detection and its analysis [18]. So we have to normalize the RGB model using mathematical expression as below:

$$r = \frac{R}{R + G + B} \tag{1}$$

$$g = \frac{G}{R + G + B} \tag{2}$$

$$b = \frac{B}{R + G + B} \tag{3}$$

$$r + g + b = 1 \tag{4}$$

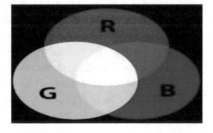

Fig. 2. RGB color space

In this model the R, G and B component is used to describe the color information of an image by quantizing the component value of each varying from zero level to its maximum value for e.g. in integer range from zero to 255 [11] and also dependent on particular application. Despite of various complexities like additive nature and non-uniform characteristics, this color space model is opted in electronic devices for display, analysis and representation [17, 19].

2.3 Skin Detection

The simplest method for separating skin pixels from non-skin pixels is to define thresholds for each channel in a color space. Generally the skin color of people belonging to different geographical region are different, hence it is very difficult process to develop a uniform skin detection and segmentation technique for face recognition. Various studies and research have been done for the characterization of skin pixels and it is clear from study that Cb and Cr entity belonging to skin region exhibit close similarity [20].

Due to this close approximation, the algorithms for skin color model which are based on the Cb and Cr values have a remarkable and wide coverage of human race belonging to different habitat. The thresholds choice should be [Cr1, Cr2] i.e. blue component and [Cb1, Cb2] representing red component, a pixel is classified to have skin tone if the values [Cr, Cb] fall within the thresholds [21].

3   Proposed Technique in Application

The proposed intelligent clock utilizes the concept of skin detection based on YCbCr color model i.e. luminance and chrominance of blue and chrominance of red respectively.

3.1   Implementation Phase

The timely acquisition of the video input frame by frame is applied to various image morphing techniques such as resizing, color model conversion and after series of procedures leading to skin detection. The skin is visible as white pixels on black background. The skin thus detected is again filtered to remove unwanted noise (stray pixels). Then the centroid of the areas where skin is present is calculated and is stored for future reference. Furthermore, another frame is applied to above procedures. Then the difference of the positions of centroids between the successive frames is determined. If the number of centroids is not the same as previous one then it is taken as movement that is accounted as disturbance. Again if the position of centroids of two consecutive frames differs by more than 10 pixels, then it is again taken as disturbance. If these disturbances occur frequently as for e.g. - if disturbances occur from every minute to ten minutes, the waking time is extended by 5 min or so. The proposed system is shown in Fig. 3. The various component or building block of the system are as follows:

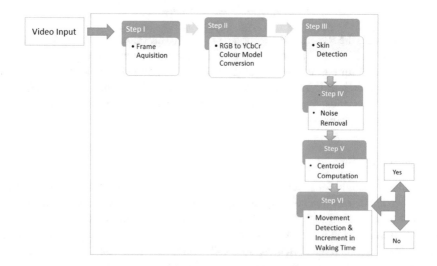

Fig. 3. Module of proposed system

3.1.1   Video Capture
Video is acquired using webcam in YUV color model of size 640 × 480 display resolution. The video is triggered manually. Trigger limit is set to infinity i.e. it will work as long as the webcam is in use. Then frame per trigger is set to 1.

3.1.2   Setting Program Execution till Waking Time
The waking time limit of 7 and ½ hours is set and the main program execution starts for the given time period.

3.1.3   Frame Acquisition
Data from a frame is acquired with the help of getdata (vid, 1) function. The acquired image is of YCbCr color model as shown in Fig. 4.

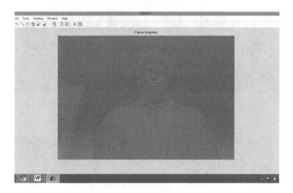

Fig. 4. Acquired frame

3.1.4   Skin Detection
The acquired frame is passed through skin detection algorithm that is based on YCbCr color model i.e. luminance and chrominance of blue and chrominance of red respectively. Luminance corresponds to light intensity that's why the Y component is of great use in our project. The image is then converted to binary image where skin presence corresponds to 1 and absence corresponds to 0 as shown in Fig. 5.

3.1.5   Noise Removal
The binary image thus obtained is passed through several filtration process [22] in which stray pixels are removed and filtered image is obtained as shown in Fig. 6.

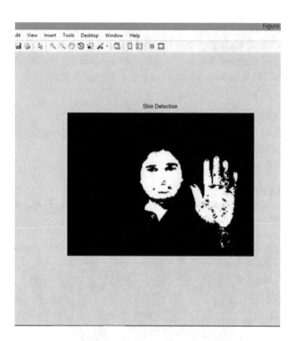

Fig. 5. Detected skin

Fig. 6. Noise removed

3.1.6 Centroid Calculation

For detecting movements and disturbances, centroid of every independent white cluster is calculated through series of procedures involving recursion as shown in Figs. 7 and 8.

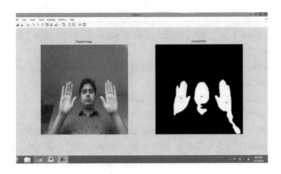

Fig. 7. Centroid plot, sample 1.

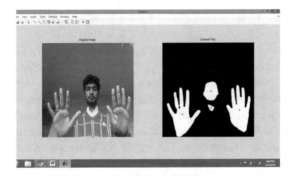

Fig. 8. Centroid plot, sample 2.

3.1.7 Movement Detection and Increment in Waking Time

Disturbances and movements are calculated depending on the change in number of centroids or change in position of centroids for successive frames. If these disturbances occur frequently as for example- if disturbances occur from every minute to ten minutes, the waking time is extended by 5 min. If disturbances occur within gap of hours or half then waking time is not increased and it is assumed that the disturbance is of natural consequence.

4 Limitation and Technical Concerns

During the entire implementation phase and after the completion of all major works, we have faced numerous technical glitches. Some of them that we considered important are illustrated below:

1. Algorithm calibrated for one camera may not work for a different camera, which makes it incompatible for all types of devices
2. Distance between the person/object and camera (detection device) and the pixel/resolution and quality of capturing device, also cause variations in recognition accuracy.
3. Interference of noise also create a negative impact on tracking and face recognition especially when occlusions occur.
4. Items in the background may make recognition difficult. There might be problems of lighting while taking images.
5. Our present project can't work in complete darkness, in case of darkness an infrared night vision camera is a must.
6. Other parameters like ambient environment lighting in poor condition, the physical appearance of person of interest like sunglasses, style & length of hair, or other objects partially covering the subject's face, and low resolution images also causes interference with the accurate skin detection.

5   Conclusion

The primary objective behind this paper is to implement our idea of automated alarm clock using image processing. This paper proposed a system of alarm clock that works on skin detection using YCbCr color space [12] which works very effectively. The YCbCr color space simply converts the image obtained from the input webcam and converts into binary image which further undergoes through various process like filtration of stray pixel, calculation of centroid in pixel for determination and detection of body movement, then noise removal. The output of our algorithm increases the waking time set by the user till the limit. Although there are few minor technical concerns regarding accuracy and efficiency of detected image, but still it fulfill the objective of our proposed system and can be improved in future.

References

1. How Much Sleep Do We Really Need? Signs that You're Not Sleeping Enough. https://www.helpguide.org/articles/sleep/how-much-sleep-do-you-need.htm
2. Bonnet, M.H., Arand, D.L.: We are chronically sleep deprived. Sleep **18**, 908–911 (1995)
3. Antrobus, J.: REM and NREM sleep reports: comparison of word frequencies by cognitive classes. Psychophysiology **20**, 562–568 (1983)
4. Arnal, P.J., Sauvet, F., Leger, D., van Beers, P., Bayon, V., Bougard, C., Rabat, A., Millet, G.Y., Chennaoui, M.: Benefits of sleep extension on sustained attention and sleep pressure before and during total sleep deprivation and recovery. Sleep **38**, 1935–1943 (2015)
5. Boudreau, P., Dumont, G.A., Boivin, D.B.: Circadian adaptation to night shift work influences sleep, performance, mood and the autonomic modulation of the heart. PLoS ONE **8**, e70813 (2013)
6. Devine, J.K., Wolf, J.M.: Integrating nap and night-time sleep into sleep patterns reveals differential links to health-relevant outcomes. J. Sleep Res. **25**, 225–233 (2016)

7. Fukumoto, M., Mochizuki, N., Takeishi, M., Nomura, Y., Segawa, M.: Studies of body movements during night sleep in infancy. Brain Dev. **3**, 37–43 (1981)
8. Fullagar, H.H.K., Skorski, S., Duffield, R., Julian, R., Bartlett, J., Meyer, T.: Impaired sleep and recovery after night matches in elite football players. J. Sports Sci. **34**, 1333–1339 (2016)
9. Goh, V.H., Tong, T.Y., Lim, C.L., Low, E.C., Lee, L.K.: Effects of one night of sleep deprivation on hormone profiles and performance efficiency. Mil. Med. **166**, 427–431 (2001)
10. Harrison, Y., Horne, J.A.: One night of sleep loss impairs innovative thinking and flexible decision making. Organ. Behav. Hum. Decis. Process. **78**, 128–145 (1999)
11. Singh, S.K., Chauhan, D.S., Vatsa, M., Singh, R.: A robust skin color based face detection algorithm. Tamkang J. Sci. Eng. **6**, 227–234 (2003)
12. Basilio, J.A.M., Torres, G.A., Pérez, G.S., Medina, L.K.T., Meana, H.M.P.: Explicit image detection using YCbCr space color model as skin detection. Appl. Math. Comput. Eng. **11**, 123–128 (2011)
13. Phung, S.L., Bouzerdoum, A., Chai, D.: A novel skin color model in YCbCr color space and its application to human face detection. In: IEEE International Conference on Image Processing (ICIP 2002), vol. 1, pp. 289–292 (2002)
14. Menser, B., Brünig, M.: Face Detection and Tracking for Video Coding Applications (2000)
15. El-hafeez, T.A.: A new system for extracting and detecting skin color regions from PDF documents. Int. J. Comput. Sci. Eng. **2**, 2838–2846 (2010)
16. Vadakkepat, P., Lim, P., De Silva, L.C., Jing, L., Ling, L.L.: Multimodal approach to human-face detection and tracking. IEEE Trans. Ind. Electron. **55**, 1385–1393 (2008)
17. Colors. http://www.chai3d.org/download/doc/html/chapter14-colors.html
18. Basha Shaik, K., Mary Jenitha, J.: Comparative study of skin color detection and segmentation in HSV and YCbCr color space. Proc. Comput. Sci. **57**, 41–48 (2015)
19. Google Books: Information Processing and Management of Uncertainty in Knowledge-Based
20. Azad, R., Davami, F.: A robust and adaptable method for face detection based on color probabilistic estimation technique. Int. J. Res. Comput. Sci. **3**, 2249–8265 (2013)
21. Platzer, C., Stuetz, M., Lindorfer, M.: Skin sheriff: a machine learning solution for detecting explicit images. In: Proceedings of the 2nd International Workshop on Security and Forensics in Communication Systems - SFCS 2014, pp. 45–56. ACM Press, New York (2014)
22. Ali, T., Ahmad, T., Imran, M.: UOCR: a ligature based approach for an Urdu OCR system. In: 2016 3rd International Conference on Computing for Sustainable Global Development (INDIACom), New Delhi, pp. 388–394 (2016)

Improved Parallel Rabin-Karp Algorithm Using Compute Unified Device Architecture

Parth Shah[1(✉)] and Rachana Oza[2]

[1] Chhotubhai Gopalbhai Patel Institute of Technology, Bardoli, India
parthpunita@yahoo.in
[2] Sarvajanik College of Engineering and Technology, Surat, India
oza.rachana7@gmail.com

Abstract. String matching algorithms are among one of the most widely used algorithms in computer science. Traditional string matching algorithms are not enough for processing recent growth of data. Increasing efficiency of underlaying string matching algorithm will greatly increase the efficiency of any application. In recent years, Graphics processing units are emerged as highly parallel processor. They out perform best of the central processing units in scientific computation power. By combining recent advancement in graphics processing units with string matching algorithms will allows to speed up process of string matching. In this paper we proposed modified parallel version of Rabin-Karp algorithm using graphics processing unit. Based on that, result of CPU as well as parallel GPU implementations are compared for evaluating effect of varying number of threads, cores, file size as well as pattern size.

Keywords: Rabin-Karp · GPU · String matching · CUDA · Parallel processing · Big Data · Pattern matching

1 Introduction

String matching algorithms are an important part of the string algorithms. Their task is to find all the occurrences of strings (also called patterns) within a larger string or text. These string matching algorithms are widely used in Computational Biology, Signal Processing, Text Retrieval, Computer Security, Text editors and many more applications [1]. In systems like intrusion detection or searching, string matching techniques takes more than half of the total computation time [2]. These string matching algorithms can be divided mainly into two sub category: single string matching algorithms and multiple string matching algorithms. In this paper we focus on mainly multiple string matching algorithms. In multiple string matching algorithm if we want to check only if pattern in text or not then Boyer-Moore performs best, but if we want to find all occurrence of pattern in text then Rabin-Karp provides most optimal results [3].

© Springer International Publishing AG 2018
S.C. Satapathy and A. Joshi (eds.), *Information and Communication Technology for Intelligent Systems (ICTIS 2017) - Volume 2*, Smart Innovation, Systems and Technologies 84, DOI 10.1007/978-3-319-63645-0_26

As the information technology spreads fast, data is growing exponentially. This requires to scale up processing power which is not possible using CPU. CPUs were basically designed for general purpose work. So, it has higher operating frequency and lower number of cores. On the other end Graphics processing units were originally developed for preforming highly parallel operations like graphics rendering. For performing this large amount of computation GPUs have higher number of processing elements (ALUs) and lower operating frequency in compared to normal CPUs. If we able to use these highly parallel processing elements for our pattern matching task, we can greatly improve performance of string matching algorithm.

In this paper, we have implemented modified parallel Rabin-Karp string matching algorithm over GPU where most of computation task is performed on GPU instead of CPU. Previous work by Nayomi et al. uses RB-Marcher technique to find hash while our approach uses Rabin Hashing as hash mechanism. The main contribution of this paper is improved parallel version of Rabin Karp algorithm that utilize inherent parallelism of Rabin Karp algorithm using GPU. In this paper, we have also evaluated performance of our modified approach over different parameters like pattern length, number of threads, filesize as well as number of cores.

Section 2 of this paper describes Related Work in detail. Algorithms and Implementation, Results and Discussion, Conclusion are discussed in Sects. 3, 4 and 5 respectively.

2 Related Work

Accelerating string matching algorithms are one of the major concern in areas where string matching algorithm used heavily. In the era of big data this will greatly speed up the processing task. With the introduction of General Purpose Graphics Processing Units (GPGPUs) we can achieve highly parallel processing capabilities [4]. Using this parallelism we can speedup the task of string matching in efficient manner.

For the task of string matching various algorithms were developed in past like Brute-Force algorithm, Boyer-Moore algorithm, Knuth-Morris-Pratt algorithm, Rabin-Karp algorithm, etc. Blandón and Lombardo [3] had compared all these algorithm and found that Boyer-Moore [5] performs best when there is no pattern matched in text. Knuth-Morris-Pratt [6] algorithm heavily depends on result previous phase's calculation which makes it unsuitable to parallel implementation. That leads us to Rabin-Karp [7] algorithm which is inherent parallel in its design. In Rabin-Karp algorithm, calculation of hash for every substring does not depends on any other information then substring. So we can easily parallelize this operation over GPU which helps us to improve performance of Rabin-Karp algorithm.

For parallel implementation of Rabin-Karp algorithm, Nayomi et al. [8] proposed Rabin-Karp algorithm which uses RB-Matcher [9] for generating hash value. But due to involvement of modular arithmetic in RB-Matcher performance improvement is limited. In our work, we have presented GPU version of

Rabin-Karp algorithm that uses Rabin Hashing as hash function. Our work not only just propose modified approach, but also proves that using experimental results on real data.

3   Algorithm and Implementation

In this section, we discuss about modified Rabin-Karp algorithm and our parallel CUDA implementation details.

3.1   Rabin-Karp Algorithm

Rabin Karp is a string searching algorithm, which supports both single and multiple pattern matching capabilities. It is widely used in applications like plagiarism detection and DNA sequence matching.

In our proposed approach we have used Rabin hashing method to find out pattern strings in a text. In order to compute hash we apply left shift to ASCII value of character and add it to previously calculated hash. We apply this process repeatedly for all character of string. To find out the matching pattern we will compare hash of pattern and hash of each substring of text. If hash of both substring and pattern match then only we will compare string to find match.

Pseudo code for sequential implementation

```
function Rabin_Karp_CPU(result, T, n, P, m)
{
    hx=0,hy= 0;
    for (i = 0;i<m;++i)
    {
        hx = ((hx << 1) + P[i]);
    }
    for (x = 0;x < n - m;x++)
    {
        for (hy = i = 0;i <m; ++i)
        {
            hy = ((hy << 1) + T[i + x]);
        }
        /* Compare hash and match pattern */
        if (hx == hy    &&   compare(P, T + x, m) == 0)
        {
            print("Match at:", x);
            result[x] = true;
        }
    }
    return result;
}
```

In sequential implementation pattern matching function is called for each substring sequentially. First we read text T and pattern P from file. Text T of length n and pattern P of length m is passed as argument to matching function. Function first calculate hash of pattern hx and store it in memory. After that for each substring of length m in text T, hash is calculated. If hash of both pattern and substring matches then only we will compare string. If string is matched we will store its index in result array. And result is returned with containing matching element.

Pseudo code for parallel GPU implementation

```
function Rabin_Karp_GPU(T, P, n, m, hx, result)
{
    blockId = blockIdx.x + blockIdx.y * gridDim.x +
            gridDim.x * gridDim.y * blockIdx.z;
    x = blockId * blockDim.x + threadIdx.x;
    if (x <= n - m)
    {
        hy = 0;
        for (i = 0;i <m; ++i)
        {
            hy = ((hy << 1) + T[i + x]);
        }
        /* Compare hash and match pattern */
        if (hx == hy    &&   compare(P, T + x, m) == 0)
        {
            print("Match at:", x);
            result[x] = true;
        }
    }
}
```

In parallel version, we first load text T and pattern P from file to global memory. Once pattern is loaded in memory we calculate hash hp. Once hp is calculated we pass it as argument to string matching function. At the same time T and P are transferred to GPU memory so that we can access it using GPU's processing elements. String matching function will execute for each substring in parallel on GPU. Processing elements on GPU can be access using grids. Each grid have multiple blocks where each block can execute upto 1024 threads in parallel. We will calculate offset of string index by combining id of thread, block and grid which is stored as x in our pseudo code. Based on offset we calculate hash hy of each substring on individual GPU thread. If hash of both substring and pattern matches then only string is compared. If string matches offset is stored in result and result is transferred back to CPU memory.

3.2   CUDA Architecture

Compute Unified Device Architecture (CUDA) is developed as a parallel computing and application architecture by NVIDIA corporation. It enables user to harness the computing capacity of GPUs. It allows us to send our C/C++ program directly to GPU for execution. Tools for interacting with GPUs having CUDA architecture is available as CUDA Toolkit released by NVIDIA. In this article, CUDA toolkit and CUDA enabled GPUs are used for implementation of parallel GPU version of Rabin-Karp algorithm. Some terminologies used in NVIDIA CUDA programming environment is as follows:

Host: The CPU will be referred to as a host.
Device: An individual GPU will be referred to as a device.
Kernel: Kernel is a function that will be executed in individual block.
Thread: Each thread is an execution of a given kernel with a specific index. Each thread uses its index to access elements in array.
Block: A group of threads that are executed together and form the unit of resource assignment is called block.
Grid: Grid is a group of independently executable blocks.

Our main algorithm for parallel implementation is used as a kernel function. CUDA architectures support maximum of 1024 threads per block. Size of block and grids are selected such that multiplication of number of thread, block and grid will match the total characters in text. This will helps us to effectively optimize all cores of GPU (Fig. 1).

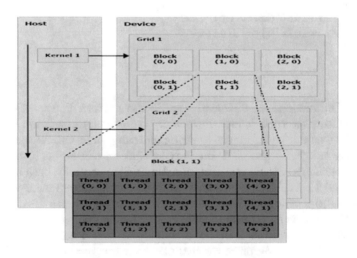

Fig. 1. Architecture of CUDA [10]

3.3   Hardware and Datasets

All the experiment in this paper carried out using Intel Core i7 processor with 4 cores. For GPU implementation we have systems with two different number of cores as increasing number of cores increase total available processing elements. They are as follows: (1) NVIDIA GT920 with 386 CUDA cores and 4 GB Memory and (2) NVIDIA GForce GT 630M with 96 CUDA cores and 2 GB Memory.

For evaluation of algorithm we have used Random DNA sequence generator to generate dna sequence dataset. We have used dataset of different sizes like 2 MB, 10 MB, 20 MB, 40 MB, 80 MB, 160 MB, 360 MB, 640 MB and 1 GB.

4   Result and Discussion

For evaluating performance of our proposed parallel algorithm over serial version we have used speed up ratio as our performance measure. Speed up ratio S is calculated as,

$$S = \frac{T_{GPU}}{T_{CPU}} \tag{1}$$

where T_{GPU} and T_{CPU} are execution time for GPU and CPU respectively.

First we have compared our parallel algorithm over sequential algorithm. For that we have executed our modified algorithm with different number of threads in parallel execution.

Table 1. Performance comparison for different thread size

Number of threads per block	Sequential implementation	Parallel GPU implementation	Speed up ratio
32	6937	4718.750	1.470093
64	6535	2708.375	2.412886
128	6839	1715.790	3.985919
256	6360	1220.096	5.212705
512	6047	971.2466	6.226019
1024	6125	847.1143	7.230430

As we can see in Table 1 as number of threads increases total execution time required for parallel implementation will reduce proportionally.

To check how our algorithm scales when we increase number of cores we have implemented our parallel algorithm on GPUs with 96 and 384 cores. Result of execution on both GPU keeping constant pattern size of 7 over 320 MB filesize is given in Table 2.

As we can see from Fig. 2, increases in speed up ratio is directly proportional to number of cores in GPU. As we increase number of cores total time require for execution will decrease proportionally.

Table 2. Performance comparison for different number of cores

Number of threads per block	NVIDIA GForce GT630M (96 cores) (ms)	NVIDIA GForce 920 (384 cores) (ms)
32	10638.17	4718.75
64	5931.667	2708.375
128	3600.603	1715.79
256	2434.065	1220.096
512	1787.352	971.2466
1024	1495.54	847.1143

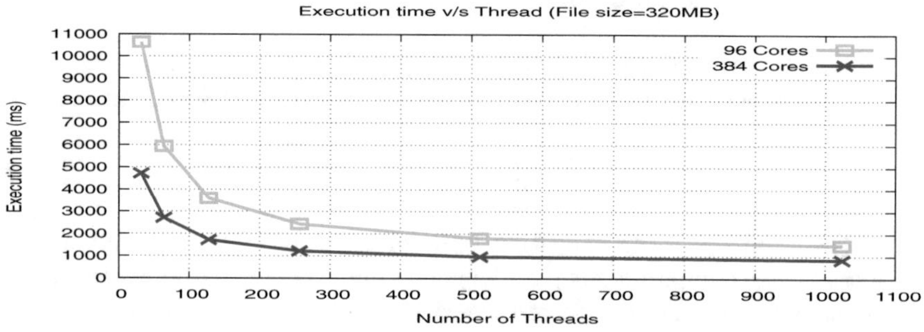

Fig. 2. Performance comparison for different number of cores

In order to check effect of pattern size on execution time, we executed both sequential and parallel version using different pattern size. As we can see from Table 3, increasing input pattern size increases speed up ratio due to fact that GPUs can easily handle computation heavy task with their higher number of cores compared to CPU.

As our main goal is to process and match pattern in real life situations where we encounter different file sizes. So to check performance, we have evaluated performance of our GPU implementation for different filesize in Fig. 3. As we

Table 3. Performance comparison for different pattern length

Pattern length	Intel core i7 (4 cores) (ms)	NVIDIA GForce 920 (384 cores) (ms)	Speed up ratio
25	140	7.51616	18.62653
50	328	13.82605	23.72334
100	578	26.37414	21.9154
200	1187	51.32698	23.12624
800	4578	201.1771	22.75607

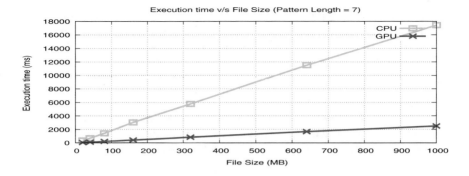

Fig. 3. Performance comparison for different file size

can see, GPU version of parallel algorithm performs better when file size is larger as increasing file size dramatically increase computation which can easily handled by GPU.

5   Conclusion

Comparing implementation of serial and CUDA version of Rabin-Karp string matching algorithm, we can get upto 23× speed-up in CUDA version over serial version. Using Rabin hashing reduces total computation required for generating hash which directly reflect in execution speed. From our experiments we can conclude that maximum speedup is archived when file size is minimum and as file size increase speedup decreases. Similarly by increasing number of cores and execution threads in GPU we can get maximum speedup in task of string matching as it increase total number of tasks that can be executed concurrently.

References

1. Singla, N., Garg, D.: String matching algorithms and their applicability in various applications. Int. J. Soft Comput. Eng. **I**(6), 218–222 (2012)
2. Xu, D., Zhang, H., Fan, Y.: The GPU-based high-performance pattern matching algorithm for intrusion detection. J. Comput. Inf. Syst. **9**(10), 3791–3800 (2013)
3. Gongora-Blandon, M., Vargas-Lombardo, M., et al.: State of the art for string analysis and pattern search using cpu and gpu based programming. J. Inf. Secur. **3**(4), 314 (2012)
4. Ghorpade, J., Parande, J., Kulkarni, M., Bawaskar, A.: GPGPU processing in CUDA architecture. arXiv preprint arXiv:1202.4347 (2012)
5. Boyer, R.S., Moore, J.S.: A fast string searching algorithm. Commun. ACM **20**(10), 762–772 (1977)
6. Knuth, D.E., Morris, J.H., Pratt, V.R.: Fast pattern matching in strings. SIAM J. Comput. **6**(2), 323–350 (1977)
7. Karp, R.M., Rabin, M.O.: Efficient randomized pattern-matching algorithms. IBM J. Res. Dev. **31**(2), 249–260 (1987)

8. Dayarathne, N., Ragel, R.: Accelerating Rabin Karp on a graphics processing unit (GPU) using compute unified device architecture (CUDA). In: 7th International Conference on Information and Automation for Sustainability, pp. 1–6. IEEE, December 2014

9. Chillar, R.S., Kochar, B.: RB-matcher: string matching technique. Int. J. Comput. Electr. Autom. Control Inf. Eng. **2**(6), 2282–2285 (2008). World Academy of Science, Engineering and Technology

10. Ramey, W.: Introduction to CUDA platform. NVIDIA Corporation (2016). http://developer.nvidia.com/compute/developertrainingmaterials/presentations/general/Introduction_to_CUDA_Platform.pptx. Accessed 6 Feb 2016

Privacy-Preserving Associative Classification

Garach Priyanka[✉], Patel Darshana, and Kotecha Radhika

V.V.P. Engineering College, Rajkot, India
priyankait03@gmail.com, darshana.h.patel@gmail.com,
kotecha.radhika7@gmail.com

Abstract. The massive amount of data, if publicly available, can be stored and shared securely for analysis and advancement. Mining of association rule besides classification technique is skilled of discovering useful patterns from big datasets. This technique results in the if-then form of rules and these rules are simple for end users to understand and easy for prediction. But it is apparent that the gathering and analysis of such data causes a serious menace to confidentiality and freedom. Hence, it interprets a field of privacy-preservation of data mining, which deals with efficient conduction and application of data mining without scarifying the privacy of data. This paper puts effort on the construction of class association rules generated by associative classification and applying privacy-preserving techniques on these rules to prevent its disclosure to the uncertified population.

Keywords: Data mining · Classification · Privacy-preservation · Associative classification

1 Introduction

1.1 Classification

The classification method contains two split. Firstly, in the learning phase, classification algorithm constructs a classifier, by learning from, a training set containing database tuples and their corresponding class labels [1]. This step is also called training phase. In the second phase, on basis of the analysis, from training phase which carries connection between class label of the instances and attributes, determination of class of an instance is made. Thus, classification model categorizes data into multiple classes. For example, categorizing news stories as weather, finance, sports, entertainment, etc., the further example includes identification of loan applicants as low, medium, or high credit risks. A new patient's disease can be diagnosed based on the patient's diagnostic data, which includes attributes like sex, age, blood pressure, temperature weight, etc. [2].

Some of the common issues of classification are Imbalance data stream classification [3], Distributed data mining [4], Target detection (multi-class problem) [3], Dealing with unbalanced, cost-sensitive and non-static data [4], Security, data integrity and privacy [4], Distributed data classification [4], Classification of sequence as well as time series data [4].

© Springer International Publishing AG 2018
S.C. Satapathy and A. Joshi (eds.), *Information and Communication
Technology for Intelligent Systems (ICTIS 2017) - Volume 2*, Smart Innovation,
Systems and Technologies 84, DOI 10.1007/978-3-319-63645-0_27

1.2 Paper Organization

This section covers the concept of classification along with its issues and application. Section 2 covers overview and comparison results of different classification techniques. Section 3 introduces privacy-preserving associative classification. Section 4 introduces a proposed framework and its workflow. Section 5 includes a conclusion and future work.

2 Classification Techniques

Major techniques of classification, which are taken into consideration in our work are Decision Tree, Neural Network, Naïve Bayes and Associative Classification. Decision trees consist of tree structures much likely to flowchart, where experiment on an attribute is denoted by every internal node, result of the experiment is denoted by every branch, and a particular class label is carried by every leaf node [5]. Neural network may consists of multiple layers of interconnected nodes, and a non-linear job of neural network's input is produced by this interconnected node and input approach to a node may be directly from the input data or else from various interconnected nodes. Also, with the help of the output result of the network, recognition of several nodes are made [6, 19]. Bayesian Classifier has the skill set of evaluating majority of feasible output on basis of the input. When the class variable is previously known, it takes into consideration, existence (or non-existence) of a meticulous feature of a class, which is not related to the existence (or non-existence) of any additional characteristic" [1, 6, 19]. Associative classification mining is capable of utilizing the rule discovery methods based on association, whose target is to build various classification systems, which is called associative classifiers as well [7].

2.1 Implementation

Data Sets. We are using six real datasets for comparison of different classification techniques. These data sets are available in UCI data repository and its details are described in Table 1.

Table 1. The composition of various datasets.

Dataset	No. of attributes	No. of instances	No. of classes
Census income	48842	14	2
Bank marketing	45211	17	2
Congressional voting record	435	16	2
Shuttle landing	15	6	2
Lenses data	24	4	2

Figure 1, shows the accuracy (in %) of correctly classified instances of different datasets. This calculation of accuracy on the various dataset is carried out using WEKA. It is observed that accuracy of associative classification is better than any other classification method. Neural Network also holds a good result in accuracy, but it has low speed and output of Neural Network is not interpretable, whereas, generated rules of the associative classifier are easy to understand. Thus, due to these advantages, our focus is made on associative classification technique.

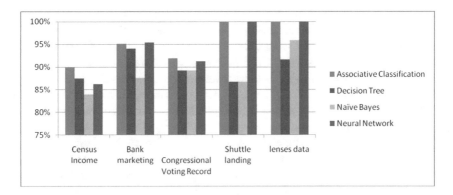

Fig. 1. Accuracy of various classification techniques

3   Privacy-Preserving Associative Classification

3.1   Associative Classification

The main aim of classification is to classify the data into various classes, for e.g. categorizing news stories as weather, finance, sports, entertainment, etc. Thus it is an example of supervised learning. Association rule discovery task [8], aims to find out the correlations among items in the shopping carts, which is a famous example of unsupervised learning. The main target of associative classification, is to search for knowledge rules from the training data set, where the rule consequent (right-hand side) should be either "yes" or "no" [7]. Major techniques are discussed as follows:

CMAR. Weight analysis is performed by CMAR, with the help of multiple strong association rules [9]. This method is bisected as phase one and phase two. The first phase contains rule generation and the second phase consists of class distribution [9]. Advantages: CMAR derives a measure on the strength of the rule, beneath class distribution and conditional support. To efficiently stock up and regain a huge amount of rules for purpose of classification, novel data structure like CR-tree is used by CMAR. Disadvantages: It is slower compared to CBA.

CPAR. Each rule is evaluated by using expected accuracy measure, which helps in avoiding overfitting problem [10]. For rule generation, First Order Inductive Learner (FOIL) [10] is the basic idea, which is used by CPAR. For the target of conducting highest gain among existing instances of dataset, it looks towards finest rule condition [10]. Positive example weights are linked with it, will be decayed by a multiplication factor, after the situation is documented [10]. Advantages [10]: A very small bundle of high-class predictive rules is produced straight from the dataset. Dynamic programming is used to get improved results. Disadvantages [11]. Computational overhead increases during training of dataset due to the usage of greedy algorithm.

CBA. Classification based on association (CBA) composed of three steps. The first step involves discretizing continuous attributes, if present, the second step generates all the class association rules (CAR) using CBA-RG algorithm, and the third step uses CBA-CB algorithm to build a classifier which makes use of generated CARs [12].

The main process of CBA-RG is to discover every ruleitems that comprises support higher than minimum support [12]. The CBA-CB algorithm produces the finest classifier from the entire package of set of laws which contains a calculation of each feasible parts of it from the training data and choosing the part having correct rule sequence whose target is to produce minimum number of errors [12, 20]. There exits 2^m such parts, where m stands for quantity of rules [12]. This algorithm has a heuristic approach [12, 20]. However, the classifier built by CBA performs superior in comparison to C4.5 [12, 20].

The advantages of CBA is that the whole dataset is not required to be fetched into main memory and it is a simple algorithm that finds valuable rules. Also, it has some limitation like redundancy in the form of huge set rules generated during training of dataset and it has the issue of overfitting.

Related Work. Contextualization of relevant papers with their approach, advantages, disadvantages and open issues are shown in Table 2:

Table 2. Literature survey of related work

S. no.	Approach	Advantage	Disadvantage	Open issue
1.	GA-based methodology [13]	Using numeric data, classifier is constructed [13]	In practice, numerical attributes may contain huge set of tricky relations [13]	Identification of new associations among numerical characteristics and encoding structure is required [13]
2.	Multi-label Classifier based Associative Classification (MCAC) [14]	Higher accuracy and good performance [14]	It is estimated that the numbers of phishing website are increased day by day [14]	As long as historical data exists, neat solutions can be merged with the heuristic-based approach [14]
3.	Multi-class multi-label associative classification (MMAC) [15]	Output classifier contains rules with multiple labels [15]	Does not have hyper heuristic approach [15]	Extension of technique to serve continuous data [15]
4.	CBA [12]	More accurate than c4.5 [12]	Training of dataset produces vast bundle of rules resulting in repetition [12]	Issue of overfitting [12]
5.	CMAR [9]	CR Tree, a novel data structure is used to efficiently stock up and recover enormous rules [9]	Slower compared to CBA [9]	Effectiveness should be increased in prediction of fresh class labels with good accuracy of classification [9]
6.	CPAR [10]	Uses dynamic programming to acquire better results [10]	Computational overhead increases during training of dataset due to the usage of greedy algorithm [10]	To further enhance the efficiency and scalability [10]

3.2 Privacy-Preserving Data Mining

Owed to the encouragement of online service usage through campaign like Digital India, collection of digital data increases in online database. These data is very useful for performing various analysis tasks and getting the fruitful patterns from it. Also, getting such fruitful results are not enough because it may disclose the individuals sensitive information which requires to be kept secure. These had lead to the emergence in the field of privacy-preserving techniques, which maintains the privacy of data along with its usefulness. Various available Privacy-Preserving techniques [16–18] are,

Anonymization. It is a method where the individuality or sensitive statistics of documentation holder are to be kept unidentified, parallel assuming that the sensitive data should be conserved for investigation.

Perturbation. In this technique, the original values of data set are changed with some unreal data values so that the result acquired of perturbed dataset does not diverge from arithmetic sequence of the primary dataset.

Randomization. In this technique, the data of available dataset is twisted in a manner that it proves to be improved version compared to pre-defined threshold, to check whether the statistics from users includes accurate information or inaccurate one.

Condensation. This approach forms, unnatural group of data, called clusters, in the dataset and using statistics of these clusters, pseudo-data is generated. The constraints on the cluster are defined in terms of size.

Cryptography. This method uses core functionality of Secure Multiparty Computation (SMC). Encryption is performed on original data values. This approach tends to limitation when more than a few parties are concerned.

4 Proposed Work

4.1 Problem Description

Any unauthorized user must not capture the non-public information. There are problems where the privacy is a major concern. The generated class association rules (CAR) from the associative classification method, are also prospect to redundancy and privacy violation. Hence, there should be some fruitful technique to remove the redundancy, to preserve the privacy of these CAR's as well as maintain the usability of the information. In proposed framework, mainly proposed combination of privacy-preserving technique over the associative classification is used, where particularly k-anonymization technique of privacy-preservation with the classification based on association (CBA) method is used.

4.2   Proposed Framework

To preserve privacy and to construct a global classifier that generates efficient rules
without redundancy, through privacy preserving technique and associative classification
respectively. The proposed work is divided into two phases as shown in Fig. 2.

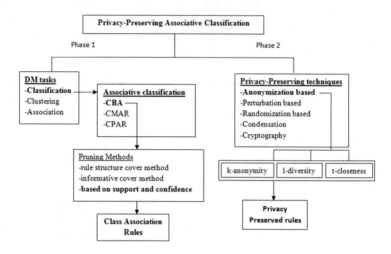

Fig. 2. Proposed framework: privacy-preserving associative classification

In phase 1, among sundry data mining methods, classification strategy is used. Associative classification is one of the categories of classifications. Classification based on
association (CBA) method has several advantages over other associative classification
method, so CAR's are generated using CBA. To remove the redundancy from the
generated rules, pruning method based on support and confidence is taken into consideration which outputs final class association rules.

In phase2, privacy-preserving techniques are applied on the CAR generated in
phase1 through anonymization method. Anonymization has an efficient structure for
handling the categorical and continuous data [18]. k-anonynimization has advantage of
minimal loss of information [17]. Thus using it, we can obtain sanitized CAR's.

5   Conclusion and Future Work

Here, we have made fundamental study of topics like Associative Classification along
with its techniques like CBA, CPAR, CMAR with their pros and cons. Also importance
of Privacy-preserving data mining (PPDM) [18] is discussed along with the brief introduction of PPDM techniques. Using WEKA, we have shown that, among several data
mining techniques of classification, associative classification technique proves to be
better through the experimental result by taking into considerations evaluation parameter
like accuracy. In proposed work, preliminary work on privacy-preserving associative
classification has been studied. The present proposed work can also be extended by

taking into consideration various attacks which violates privacy rules. Also, future analysis can be done by taking concept of distributed database in picture.

References

1. Han, J., Pei, J., Kamber, M.: Data Mining: Concepts and Techniques. Elsevier, Amsterdam (2011)
2. Fu, Y.: Data mining: tasks, techniques, and applications. IEEE Potentials **16**, 18–20 (1997)
3. Krawczyk, B.: Learning from imbalanced data: open challenges and future directions. Prog. Artif. Intell. **5**, 221–232 (2016)
4. Yang, Q., Wu, X.: 10 Challenging problems in data mining research. Int. J. Inf. Technol. Decis. Mak. **5**, 597–604 (2006)
5. Gupta, M., Aggarwal, N.: Classification techniques analysis. In: National Conference on Computational Instrumentation, Chandigarh, pp. 128–131 (2010)
6. Nikam, S.S.: A comparative study of classification techniques in data mining algorithms. Orient. J. Comput. Sci. Technol. **8**, 13–19 (2015)
7. Thabthah, F.: A review of associative classification mining. Knowl. Eng. Rev. **22**, 37–65 (2007)
8. Agrawal, R., Imieliński, T., Swami, A.: Mining association rules between sets of items in large databases. In: SIGMOD Record, vol. 22, pp. 207–216. ACM, New York (1993)
9. Li, W., Han, J., Pei, J.: CMAR: accurate and efficient classification based on multiple class-association rules. In: Proceedings of the 2001 IEEE International Conference on Data Mining, pp. 369–376. IEEE (2001)
10. Yin, X., Han, J.: CPAR: classification based on predictive association rules. In: Proceedings of the SIAM International Conference on Data Mining, pp. 369–376. SIAM, San Francisco (2003)
11. Sasirekha, D., Punitha, A.: A comprehensive analysis on associative classification in medical datasets. Indian J. Sci. Technol. **8**, 1–9 (2015)
12. Liu, B., Hsu, W., Ma, Y.: Integrating classification and association rule mining. In: Proceedings of the Fourth International Conference on Knowledge Discovery and Data Mining, pp. 80–86. AAAI, New York (1998)
13. Chien, Y.W.C.: Mining associative classification rules with stock trading data—a GA-based method. Knowl. Based Syst. **23**, 605–614 (2010)
14. Neda, A., Aladdin, A., Thabthah, F.: Phishing detection based associative classification data mining. Sci. Direct **41**(13), 5948–5959 (2014)
15. Thabthah, F.: Multiple labels associative classification. Knowl. Inf. Syst. **9**(1), 109–129 (2006)
16. Nayak, G., Devi, S.: A survey on privacy preserving data mining: approaches and techniques. Int. J. Eng. Sci. Technol. **3**, 2127–2133 (2011)
17. Vaghashia, H., Ganatra, A.: A survey: privacy preservation techniques in data mining. Int. J. Comput. Appl. **119**, 20–26 (2015)
18. Saranya, K., Premalatha, K., Rajasekar, S.S.: A survey on privacy preserving data mining. In: 2nd IEEE International Conference on Electronics and Communication System, pp. 1740–1744 (2015)
19. Singh, K., Kumar, S., Kaur, P.: Detection of powdery mildew disease of beans in India: a review. Oriental J. Comput. Sci. Technol. http://www.computerscijournal.org/
20. Segrera, S., Moreno, M.: Classification based on association rules for adaptive web systems. Innov. Hybrid Intell. Syst. **44**, 446–453 (2007)

Reverse Engineering of Botnet (APT)

Bhavik Thakar[(⊠)] and Chandresh Parekh

Raksha Shakti University, Ahmedabad, India
br.thakar@outlook.com, cdp_tc@outlook.com

Abstract. Grown internet usage by individual and industries have also increased the attack vector in cyberspace rapidly. Botnet is a digital weapon used by attackers to commit cybercrime in stealthiest way for all type of illegal online activity. Botnet is well articulated attack responsible for many malicious activities in large volume and mass effective against any targeted organization such as confidential data theft, financial loss, distribution of pirated products, e-business extortion and network or service disruption. Because of its global nature of infection and innovative covert techniques of malware development to evade detection, it is also known as advance persistent threat. An analysis of this APT revealed the advancement in sophistication of bot malware by encryption methods, concealed network connections and silent escape as an effective tool for profit-motivated e-crime. Reverse engineering is procedure to analyze malware to classify its type, hazard, impact on machine, information outflow and removal of signature technique. Botnet (APT) detection needs improvised process to identify the channel, architecture and encryption weakness. In bot examination; Programming style, network protocol and behavior analysis can mitigate the APT by creating signature, prototype of behavior based approach and elimination of C&C servers. Reverse engineering is excellent way for defense the modern botnets to immune valuable information by identifying the evidence behavior, log collection and digital forensics. The main aim of study is to determine the most adequate approach to recreate a botnet incident. Network security is prime concern to avoid state sponsored attacks like botnet so security of digital nation and e-governance can be assured.

Keywords: Reverse engineering · BOTNET · Robot network · Malware · Malware analysis · Static analysis · APT · Dynamic analysis · Botware · Network security · Mirai · Cyber security · Cyber forensics

1 Introduction

The internet has evolved global communication distributed medium by various services such as email, social media, web application, mobile internet, e-commerce, data centers, resource sharing, banking, industries data mining, search engine etc. Cyber world broadens the scope of facilitation to human in recent years, it also speeded many criminal activity, malcode development, cryptic currency gambling, corporate espionage, privacy violations, data breeches, illegal resource consumption, channel disruption, unauthorized information access, pirated media market and main importantly nationwide automated critical infrastructure services malfunctioning.

© Springer International Publishing AG 2018
S.C. Satapathy and A. Joshi (eds.), *Information and Communication Technology for Intelligent Systems (ICTIS 2017) - Volume 2*, Smart Innovation, Systems and Technologies 84, DOI 10.1007/978-3-319-63645-0_28

Advanced Persistent Threats (APTs) indicate sophistication of attack over corporate and nationwide critical infrastructure service targets. APTs are specialized in stealthy communication with various stages of intrusion in network, being undetected with predefined attack order for long duration operation to be successful so harvesting of valuable information asset becomes possible. So the Key objective is beyond immediate financial gain but infected machines continue to be in service even after main server breached and initial benefit achieved

Botnet is an emerging threat (APT) against cyber-security as network built from collection of compromised machines (bots or zombies) infected by automated malcode (robots), instructed by an attacker to accomplish distributed malicious task which run remotely, without human intervention for illegal mass financial gain or service/resource consumption, being undetected. Botnet have become online crime vehicle with huge malicious intended network infrastructure which arrange shared distributed environment to launch online criminal activities like corporate espionage, DDOS, click fraud, spamming, phishing, data theft, digital currency gambling, ransomed information and cyber warfare.

Botnets uses command and control server communication over covert channels with infected hosts through which the connected computers can be directed and updated on new target and infection spreading in e-society without victim (compromised host owner)'s knowledge, controlled by botnet master (owner of illegal activity). These covert channels operate on different data communication protocols and topologies such as centralized, distributed hybrid or randomized. As per analysis of C&C architecture, botnets distinguished in IRC, HTTP, DNS and Peer to Peer (P2P) based phenomenon.

Reverse-engineering to botnet attacks defines process used to identify fundamentals of botnet command-and-control protocol reversing, diagnosing and breaking cryptography, as well as reassemble botnet network channels and discovering vulnerability in their architecture. Bot examination explains Programming style, network protocol and behavior analysis can mitigate the APT by creating signature, prototype of behavior based approach and elimination of C&C servers.

2 Methodology

Botnet activate attack by initial exploitation of vulnerability on the target host. It runs remote intrusion code on the machine then use transmission over C&C to inform to Botmaster for further update as Bot, now it could escalate infection to neighbor hosts by same process continual in network. It evades detection by dynamic changing C&C locations through fast flux techniques which usually used for load balancing.

2.1 Botnet Life Cycle

1. Information gathering
2. Vulnerability analysis on target and proposed bots
3. Initial Intrusion

4. Exploitation and connection establishment
5. Securing botnet (maintain conn. Between C&C & bots)
6. Waiting for orders and getting the payload
7. Reporting the results
8. Spread infection in network
9. Intrude or gain credential access
10. Attack
11. Update
12. Remove signature.

2.2   Types

1. Centralized (Client server model)
2. Distributed (peer to peer)
3. Crucial (Randomize, hybrid, hierarchical).

Examining the capabilities of malicious botware help you to better understand the threat ideology, purpose, phenomenon, covert channel, target gain. It can also prevent future attacks.

2.3   Famous Botnets

See Table 1.

Table 1. Famous botnet history details

Year	Botnet	No. of Zombies (observed)	Protocol used	Attack type	OS platform
Centralized architecture (C&C)					
2003	Spybot	50,000	IRC	Keylog	Windows
2004	Bagle	2,30,000	IRC	Spam	Windows
2006	Rustock	1,50,000	HTTP	Spam/DDoS	Windows
2007	Srizibi	4,00,000	HTTP	DDos	Windows
2008	Asprox	15,000	HTTP	SQL Injection	Windows
2008	Kraken	4,00,000	HTTP		Windows
Decentralized (P2P/hybrid) architecture (C&C)					
2008	Conficker	10.5 million	HTTP	Buffer overFlow	Windows
2008	Sality	1 million		Password cracking	Windows
2008	Mariposa	12 million	HTTP	DDoS/Spam	Windows
2009	Waldec	90,000	IRC	Spam	Windows/Linux

(*continued*)

Table 1. (*continued*)

Year	Botnet	No. of Zombies (observed)	Protocol used	Attack type	OS platform
2010	Zeus	3.6 million	HTTP	Form grabbing	Windows
2010	Stuxnet			PLC/SCADA worm	Windows
2010	SymbOS/Zitmo.A/B		SMS	Steal data	Symbian/windows
2011	Android/Geinimi.A		HTTP	Steal data	Android
2011	Kelihos	3 million	HTTP	Spam/Bitcoin theft	Windows
2011	Ramnit	3.2 million	HTTP	Ransomware	Windows
2012	Chameleon	1,20,000	HTTP	Click fraud	Windows
2013	Zer0n3t	200 servers	HTTP	Spam	
2013	Android/AndroRat.A		HTTP	Steal data	Android
2014	Samalt	3,00,000	HTTP	Spam	Windows
2014	Destover		HTTP	Steal & Destroy	Windows
2015	Aethra	8000 server	HTTP	Brute Force	Wordpress portal
2016	Mirai	10 million	IOT	280 Gbps - 1 Tbps DDOS	Linux

2.4   Reverse Engineering

Reverse engineering is procedure to analyze malware to classify its type, hazard, impact on machine, information outflow and removal of signature technique.

Six steps

1. Develop a controlled environment (Creating basic setup for controlled isolated lab most importantly required for analyzing advance persistent threat).
2. Create virtual restoring point
3. Botware sample collection
4. Analysis of findings
5. Recreate incident scenario
6. Documenting the results (Fig. 1).

Set Bot Analysis lab and base lining

1. Allocate physical or virtual systems for the analysis lab
2. Isolate laboratory systems from the real-time environment (sand boxing)
3. Install behavioral analysis tools
4. Install code-analysis tools.

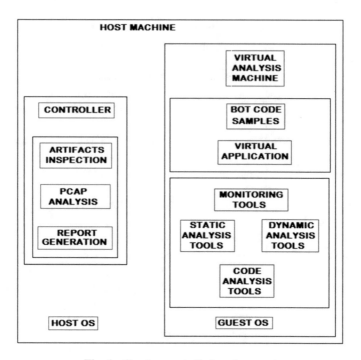

Fig. 1. Creating controlled environment

Utilize online analysis tools (Fig. 2).

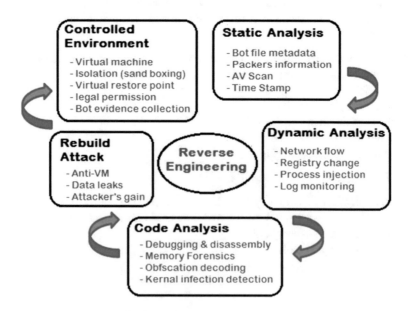

Fig. 2. Reverse engineering process

2.5   Botware Information Collection

Collect Botware evidence traces from infected machines by appropriate digital forensic techniques.

2.6   Information Analysis

Reverse engineering examination stats improvised process by different phase of malware analysis by its various types of approach, findings and results. Static and dynamic analyses are commonly used in studying bot malware.

2.6.1   Methods of Analysis

Static analysis: Static analysis focuses on collected bot malware samples for inspecting its structure and working phenomenon without malcode execution.

Dynamic analysis: Dynamic analysis deals with behavior monitoring during the malware execution.

Code analysis: It is used for investigates botware, which is crucially distributed in the form of binaries and binary code examined by debugging and disassembly of the code extracted from evidence.

All these three analysis explain points given below

1. Identification of threat (APT)
   The first part begins with bots and botnet architectures, fundamental approaches of botnet analysis, both static and dynamic analysis methods to extract relevant pieces of information from malware samples.
2. Communication protocols and cryptography
   Understanding of network protocols used by bots to communicate with their command-and-control servers is crucial for analysis of botnet architecture and the network flow. Bots use cryptic programming to hide their malicious connection from signature based detection and become stealthiest intrusion which make difficult to identify them in known pattern match.
3. Penetrating Botnets
   Botnets is similar to distributed environment operating system; contain loopholes that can be identified from advance reverse engineering techniques. Techniques used to check loose ends in many botnets to perform analysis easier, which includes become part of botnet, communication with other bots, attack botnet hosts and C&C servers with well-crafted packets. Though it is difficult to get architectural information of botnet but weaknesses of communication and formation of infrastructure can be used to control the botnet. Program style and infection of file nature would give information about regulating method any particular botnet and domain generation algorithms(DGA) provides C&C servers network address of infected hosts from backends.
4. Advanced Botnet Invasion
   Reverse engineering used to examine protocol and encryption of particular botware, recreation of attack can be developed to track C&C and architecture of botnet. Preformation of botnet attack in lab with malware to see behavior of stated BOT-NET (APT) will guide to actual information gain about an attacker.

Static and dynamic analyses are commonly used in studying malware. Reverse engineering technique run through these two approaches, this is complex, time-consuming and require deep knowledge of botnets. Code analysis is used for inspecting botware, botnet is found in distributed binaries formation, and that code analyzed by debuggers and disassemblers. Behavior analysis is more concerned with the malicious activity monitoring aspects of the botnet, like changes botcode produce to the operating system environment (registry, boot process, network, file data, autostarts, etc.), its inbound/outbound data transmission from internet, remote devices connection.

3 Analysis

3.1 Static Analysis

1. Identifying common characteristics of botware
2. Running AV scan by multiple vendors for threat observation
3. Signature based identification of bot by hash value
4. Bot API file metadata analysis and artifacts collection
5. first compilation Timestamp of botcode
6. Machine type identification by botcode
7. Packer identification for encryption style of code
8. Bot Portable Executable (PE) File Headers and Sections
9. Research on String and script embedded in botcode
10. Examining Import tables of malcode
11. Investigating Resources used by bot
12. Observation of DLL files embedded in botcode which used for intrusion
13. Finding OEP (the Original Entry Point), OEP is the address of the malware's first instruction before it was packed.

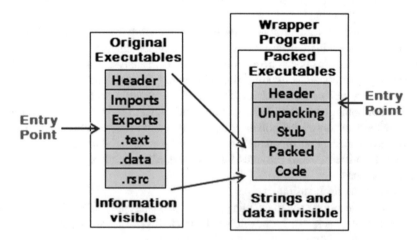

Fig. 3. Packers information

14. MSDN library understanding for windows machine to calculate functions called by botcode
15. Files opened by botware
16. Certificates embedded in files executed by malcode
17. Programming language identification (Fig. 3).

3.2 Dynamic Analysis

1. Unusual system behavior
2. CPU & memory consumption after botware infection
3. Network hosts scanning by infected machine monitoring
4. Communication topology (TCP/UDP) of bot
5. Classifying Registry change and Locating Rogue Service Processes
6. Temp files of the system monitoring
7. Fingerprinting System log
8. Classifying bot family by previously known attack vector scenario
9. Socket, domain and IP address connection triggering artifacts observation
10. URL redirection and browser behavior
11. After infection boot time activity monitoring
12. Examining outbound data leaks after botcode execution
13. Port activity monitoring
14. DNS request based monitoring
15. Parsing user agent from packets transmission occurring in network activity done by victim.

3.3 Code Analysis

1. Debugging and disassembly of code and responsible instruction set cycle
2. Searching of backdoors, downloaders and launchers
3. Viewing executable modules in code of bot
4. Memory forensics, address mapping and process examination
5. Fastflux techniques (Dynamically generation algorithm) identification of C&C servers
6. Comparison of Kernel level and user level infection by exception in protection mechanism of machine
7. Decoding common obfuscations algorithm like exclusive-OR and base64 encoding
8. Scanning for rootkits infiltration after APT code execution
9. Looking for Privilege Escalation by malcode
10. Target identification by C&C communication.

3.4 Reconstructing Attack Scenario

1. Recreate real time scenario of bonet attack in controlled environment to observe botmaster commands, attack gain and evasion from protection mechanism
2. Use anti-virtual machine to inspect real artifacts from bot infection in victim machine (server or zombie)
3. Information leak or destruction cost calculation in bot attack.
4. Identification of physical locations from where bot-attack was launched
5. ISP information analysis for global correlate the C&C connections.

4 Tools

See Table 2.

Table 2. Tools used for bot analysis

S. no.	Static analysis	Dynamic analysis	Code analysis
1	Online AV scanners (VirusTotal, Anubis)	Sysinternal suite	Python
2	VMware, virtualbox	Process explorer	The Sleuth Kit
3	CFF explorer	Process monitor	BinDiff
4	Hex Editors	Tor (for anonymity)	Memoryze
5	PEview	ApateDNS	Import REConstructor
6	PEiD	Wireshark	BinNavi
7	FileAlyzer	Netcat	IDA pro
8	Dependency walker	Regshot	OllyDbg
9	PE explorer	Capture BAT	REMnux
10	YARA	Deep Freeze	The Sleuth Kit
11	Zero Wine	INetSim	BinDiff
12	Hsahmyfile	Snort	
13		TCPView	
14		VERA	

5 Artifacts Concluded in Botnet Analysis

1. IP address, Process Handles, Registry changes, HTTP header packets, memory addressing, DGA algorithms, Encryption, URLs, Ports, Default Credentials
2. Motive (financial gain, information theft, cyber warfare)
3. Channel identification and string parsing
4. Services offered (Key logging, screen reading, privilege escalation, intrusion foothold)
5. Rootkit for evade detection.

6 Case Study

1. Zeus Botnet: Zeus, GOZ, Zbot, Citadel, Ice-IX are different variants in zeus botnet family have compromised over 3.6 million host for Key logging and Banking credential logging. Reverse engineering of zeus states characteristics: RC4 encryption, sysdate as DGA algorithm, HTTP post request from older user agent, URL config, UP time of Host, System Information, FTP/POP3 passwords Stealing
2. Destover Wiper: Wiper Family: Narilam, Dozer, Koredos, Groovemonitor/Maya, Shamoon, Dark Seoul/Jokra, Destover

   Used in Sony attack for Stealing and destruction of data by Guardians of Peace (#GOP) with eight malwares, with capability of upload and download data over C&C. Bot droppers install and run EldoS RawDisk drivers to bypass NTFS security and overwrite disk data/MBR by itself. Seven IP addresses and five backdoors found hardcoded. It uses Backup and Restore Management Windows brmgmtsvc service, adds its own executable malcode and sets a startup '-i'switch with several copies then starts each of them with a -m: MBR Over write, -d: Data Overwrite, and -w: Web Server. Uses hardcoded IP as RDP brute forcing network scanners, Socks proxy server. Cyber terrorism act by stealing data and destruct the host system OS and display GOP wallpaper.
3. Mirai Botnet: Mirai developed in 'C' language formed by Compromised Internet of things Hosts act as IOT botnet to perform DDOS on certain web like Krebs on Security, Dyn, PayPal, Twitter, Reddit, GitHub, Amazon, Netflix, Spotify, RuneScape over huge speed of 250 Gbps to 1 Tbps speed. It intruded to Routers CCTV, DVRs running firmware or NETSurveillance interface with Real Time Streaming Protocol which converted to unique 49,657 zombie hosts and ports 22, 23, 80, 2323, 7547, 48101 with Telnet/SSH. Search engines like Shodan and Censys, provides information for Raspberry Pi available with default ID/Password pair and Mirai generates DDOS with capability of 62 factory default credentials and TCP, UDP, SYN, DNS flood attacks, Application level HTTP flood facility. It has file named killer.c with functions scanner_init, memory_scan_match and string "dvrhelper".

7 Conclusion

Reverse engineering is a crucial process to analyze the botnet by creating signature, collect infected URLs, finding zombies in network, inspecting network intrusion threat, prototyping bot families. By this process network disruption, illegal online activity and organizational data theft can be prevented and even bot specific Intrusion Prevention system can be developed. This also assures the authenticated data flow in digital space by insured e-governance communication for any nation from cyber terrorism.

References

1. Thakar, B., Parekh, C.: Advance persistent threat: botnet. In: Proceedings of the Second International Conference on Information and Communication Technology for Competitive Strategies, ICTCS 2016, Udaipur, India, 4–5 March 2016, Article No. 143. ACM, New York (2016). http://dl.acm.org/citation.cfm?doid=2905055.2905360. Famous Botnet table
2. Sikorski, M., Honig, A.: Practical Malware Analysis: the Hands-on Guide to Dissecting Malicious Software. ISBN 978-1-59327-290-6
3. Ligh, M., Adair, S., Hartstein, B., Richard, M.: Malware Analyst's Cookbook. ISBN 978-0-470-61303-0
4. Ashley, D.: Analysis of a Simple HTTP Bot. SANS Institute whitepapers. https://www.sans.org/reading-room/whitepapers/malicious/analysis-simple-http-bot-33573
5. Satrya, G.B., Cahyani, N.D.W., Andreta, R.F.: The detection of 8 type malware botnet using hybrid malware analysis in executable file windows operating systems. In: Proceedings of the 17th International Conference on Electronic Commerce 2015, ICEC 2015. Informatics, Telkom University, Article No. 5. ACM, New York (2015). doi:10.1145/2781562.2781567. ISBN 978-1-4503-3461-7
6. Pfeffer, A., Call, C., Chamberlain, J., Kellogg, L., Ouellette, J., Patten, T., Zacharias, G., Lakhotia, A., Golconda, S., Bay, J., Hall, R., Scofield, D.: Malware analysis and attribution using genetic information. In: 2012 7th International Conference on Malicious and Unwanted Software (MALWARE). IEEE, Fajardo (2012). ISBN 978-1-4673-4880-5
7. Wu, Y., Zhang, B., Lai, Z., Su, J.: Malware network behavior extraction based on dynamic binary analysis. In: 2012 IEEE International Conference on Computer Science and Automation Engineering, Beijing (2012). ISBN 978-1-4673-2007-8
8. Lastline Whitepaper: The Threat of Evasive Malware, 25 February 2013. https://www.lastline.com/papers/evasive_threats.pdf
9. Microsoft: Understanding anti-malware technologies (2007). http://download.microsoft.com/download/a/b/e/abefdf1c-96bd-40d6-a138-e320b6b25bd3/understandingantimalwaretechnologies.pdf
10. Sanabria, A.: Malware Analysis: Environment Design and Architecture, 18 January 2007. https://www.sans.org/reading-room/whitepapers/threats/malware-analysis-environment-design-artitecture-1841
11. Thapliyal, M., Bijalwan, A., Garg, N., Pilli, E.S.: A generic process model for botnet forensic analysis. In: Conference on Advances in Communication and Control Systems 2013 (CAC2S 2013). Atlantis Press (2013)
12. Cusack, B.: Botnet forensic investigation techniques and cost evaluation. Junewon Park Digital Forensic Research Laboratories. In: ADFSL Conference on Digital Forensics, Security and Law (2014)
13. Zeus Botnet Case Study. https://www.symantec.com/content/en/us/enterprise/media/security_response/whitepapers/zeus_king_of_bots.pdf
14. Zeus Botnet Case Study. https://www.trendmicro.com/vinfo/us/threat-encyclopedia/malware/ZEUS
15. Destover Wiper Case Study. https://securelist.com/blog/research/67985/destover/
16. Mirai Botnet Case Study. https://www.incapsula.com/blog/malware-analysis-mirai-ddos-botnet.html
17. Mirai Botnet Case Study. https://www.malwaretech.com/2016/10/mapping-mirai-a-botnet-case-study.html
18. Mirai Botnet Case Study. https://www.symantec.com/connect/blogs/mirai-what-you-need-know-about-botnet-behind-recent-major-ddos-attacks

Mining Set of Influencers in Signed Social Networks with Maximal Collective Influential Power: A Genetic Algorithm Approach

Gaganmeet Kaur Awal[✉] and K.K. Bharadwaj

School of Computer and Systems Sciences, Jawaharlal Nehru University,
New Delhi 110 067, India
awal.gaganmeet@gmail.com, kbharadwaj@gmail.com

Abstract. The ubiquitous growth of social networks opens a new line of research for developing algorithms and models for influence mining. Determining influential people in the network which consists of both positive and negative links between users is a challenging task. It becomes critical for businesses with fixed budget constraints to identify a group of influential people whose views will influence others' behaviors the most. In this paper, we propose a model that aims to discover a set of influencers in signed social networks with maximal *Collective Influential Power* (*CIP*). We first construct an "influence network" between users and compute the influence strength between each pair of users by utilizing both the explicit trust-distrust information provided by users and the information derived from interactions between them. We then employ an elitist genetic algorithm that discovers a set of influencers with high influence spread as well as maximal enhanced joint influential power over the other users in the network. Experiments are performed on Epinions, a real-world dataset, and the results obtained are quite promising and clearly demonstrate the effectiveness of our proposed model.

Keywords: Influence mining · Signed social networks · Trust-distrust · Genetic algorithm · Collective intelligence

1 Introduction

The proliferation of social networks (SNs) has enabled users to connect with each other, share their views, disseminate information and also get influenced by others who they think are trustworthy or with whom they have frequently interacted in the past. Social influence mining is one such research area of SNs which has recently gained prominence due to its applications in marketing, e.g., advertising, enterprise reputation management and personalized recommendations and viral marketing.

The computational problem of social influence maximization deals with discovering influential users from a social network so that they can exert influence over the other users in the network by word-of-mouth. These helps companies with a small marketing budget reach or target a large section of the population and get maximum return on

© Springer International Publishing AG 2018
S.C. Satapathy and A. Joshi (eds.), *Information and Communication Technology for Intelligent Systems (ICTIS 2017) - Volume 2*, Smart Innovation, Systems and Technologies 84, DOI 10.1007/978-3-319-63645-0_29

marketing investment. With minimal budget requirements, the company can select a small set of users with largest collective influence, which can activate a chain reaction of influence amongst the users in the network.

The above problem has been widely addressed in SNs because of its numerous applications in identifying vital nodes in a social network for information dissemination. Some previous studies focus on using network measures such as degree, betweenness, and closeness centralities, etc. to determine individual node's importance [17]. Kempe et al. [11] presented diffusion models like Linear Threshold (LT) model and Independent Cascade (IC) model inspired from mathematical sociology. They proved influence maximization problem to be NP-hard and provided greedy approximation algorithm to maximize the influence spread of the set of users.

Over the last few years, many studies attempt to improve the efficiency of these greedy algorithms to mine influencers from the social network [4, 5, 12]. These existing works assume that there is a social graph given as input with edges labeled by the strength of influence between the users. However, these influence strengths are not explicitly available in real-life. To tackle this aspect, Goyal et al. [10] proposed an approach to compute the influence strengths by mining action log. Most of the existing works consider only positive influence amongst the users; however, Ahmed et al. [1] presented a new diffusion model that considers positive as well as negative influences between users.

All of the above studies emphasize on the significance of each node without considering the joint influential power (*JIP*) of a group of nodes. The users are inserted into the group one by one as per their features and the parts of the strength of connections between them, without considering the influential power of the group as a whole [18]. Xu et al. [18] proposed a model that considers *JIP* of a group of users and showed that it have a considerable impact on a large number of users. In real-life situations, the influence relationships between users are a function of their trust/distrust due to past behaviors, social actions and the centrality of their connections. Therefore, to address this aspect, we extend the model to consider influence relationships as a combination of both explicit and interactions-based implicit information. Our proposed model aims to identify a set of influencers with the maximal collective influential power which optimizes both influential power of this set of users globally as well as the actual spread of their collective influence.

The problem addressed here can be stated as follows:

Given a signed social network, connecting users with explicit positive (trust) and negative (distrust) links, and rating information of users on other users' articles, how can we leverage this information to discover a set (of size k) of influential users from the network that have maximum collective influential power over the other users in the network?

The main contributions of our work are as follows:

- Determining influence strengths between users of the network through three sources of information, i.e., explicit direct trust-distrust data, interactions-based information, and preference matching between users. We also incorporate the concept of closeness centrality that asserts more importance to the interactions or ratings coming from the central users.

- A model is proposed for **M**ining **I**nfluencers using **G**enetic **A**lgorithm (**MIGA**). We formulate a new fitness function called **C**ollective **I**nfluential **P**ower (***CIP***) which is a harmonic mean of the two objectives: enhanced joint influential power and influence spread.
- Experiments are conducted on Epinions, a real-world dataset.

The rest of the paper is organized as follows: Sect. 2 summarizes the related studies about discovering influential users from the network. The detailed description of our proposed MIGA model is presented in Sect. 3. The experiments performed on the real-world dataset and the results obtained are discussed in Sect. 4. Finally, Sect. 5 concludes the article with few directions for future research.

2   Related Work

The availability of social media serves as a fertile platform to share ideas, views or opinions and to build social connections between users. A significant amount of user-generated content is available in the form of online reviews, ratings, scores, and connections between users, which can be mined to gain deeper insights into the various social phenomena.

The SNs with both positive and negative links between the users are known as *signed social networks* (SSNs) [7]. Users can express their views about other users in the form of like-dislike, trust-distrust, agreement-disagreement, etc. Epinions [14] allows users to rate other users in the form of trust (positive link) or distrust (negative link). The study and analysis of signed links between users can help in unveiling the phenomenon of influence propagation in the network. The presence of negative links between users can emanate negative influence on other users. Thus, both positive and negative influence can be explored and exploited for the influence maximization task. For example, the likelihood of a user performing an action will increase (/decrease) as the number of its trusted (/distrusted) neighbors performing the same action increases.

The phenomenon of collective influence takes inspiration from the concept of collective intelligence (CI) which has recently gained multi-disciplinary importance [2, 15]. Users tend to influence each other's opinions on SNs through their behaviors and various social interactions directly or indirectly. The combined effect that emerges from the associations and as a result of all the interactions and actions of the set of users collectively on the other users is referred to as the collective influence [15].

The influence mining in SNs has gathered a good deal of attention in recent years [3]. The problem of discovering influential users is introduced by Domingos and Richardson [6]. They presented a model based on Markov random field that exploited the network value of customers in the context of a viral marketing application. Kempe et al. [11] address the viral marketing problem and formalize it as influence maximization problem. After this, a significant amount of research has been done considering the spread of influence by improving the efficiency of the greedy algorithms [4, 5, 12] and mining influence strengths [1, 10].

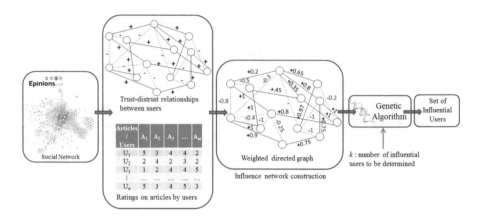

Fig. 1. The framework of our proposed MIGA model

Xu et al. [18] proposed a model to identify influential nodes with maximal joint influential power (*JIP*), where the weights on the edges between users which represent the influence strengths are computed as follows:

- User u trust/distrust user v then, $w_{vu} = +1/-1$.

- User u rates on user v's articles then, $w_{vu} = \dfrac{1 - e^{-(p-n)}}{1 + e^{-(p-n)}}$, where $p(/n)$ is the number of positive (/negative) ratings and supporting (/opposing) interactions from user u to user v.
- Otherwise, $w_{vu} = 0$.

Afterward, the *JIP* of the k influential users is optimized by computing it as:

$$\text{Maximize} \qquad JIP = \sum_{i \in C, j \in \bar{C}} w_{ij} \qquad (1)$$

$$\text{subject to} \qquad C \subseteq V \text{ and } C = k \qquad (2)$$

3   Mining Influencers Using Genetic Algorithm (MIGA)

The main steps of our proposed MIGA model are described below:

1. **Step 1:** The varied information available on the social network from different sources such as trust-distrust relationships between users and the users' rating information on articles are collected.
2. **Step 2:** This information is utilized to determine the strength of influence relationships between users. Both explicit trust-distrust values and implicit past interactions-based behavior of users are combined to construct influence network between users where the weight on edge between two users represents the strength of influence from one user to another user.

3. **Step 3:** The target set of influencers are then discovered based on the constructed influence network. In this work, we employ an elitist genetic algorithm (GA) to identify a set of influential users that have maximal collective influential power in the network.

Figure 1 shows the overall steps in our proposed computational model pictorially. The following sub-sections provide the details for each of these steps.

3.1 Influence Network Construction

The factors that affect the strength of social influence are the strength of relationships between users of the network, the social distance between users, level of concordance or discordance in preferences and their behaviors deduced through direct interactions, e.g., ratings. Influence is usually reflected through various social actions or behavioral patterns in the social network.

We compute the influence relationships by mining frequent patterns of behavioral actions performed by the users. These influence relationships are asymmetric, and the strength varies for different pairs of users. The influence network between users is constructed as a directed weighted graph $G = \{V, E_s, W\}$. The nodes of the graph represent users of SSNs, and the edges correspond to the presence of influence relationship between the users. The weight on edge represents the strength of influence of a user on another user and is adapted from [18]. We consider both explicit as well as implicit information available through the following sources to derive influence strengths on the SSNs.

1. **Explicit information (Trust-Distrust between users):** The trust relationships between users and the distrust information help in measuring the level of positive and negative influence that trusted and distrusted neighbors exert on a user. If a user u trust (/distrust) another user v then $T(u, v) = +1 (/-1)$ otherwise zero. The user tends to get influenced by its trusted neighbors and in a similar way, get negatively influenced by its distrusted neighbors. Both positive and negative influences reflect the differential effects on user's behavior by others in the network. The likelihood of a user performing an action increase if the number of her trusted neighbors performing the same action increases and also the number of her distrusted neighbors performing the same action decreases and vice versa.

2. **Direct interactions-based information (Ratings on article reviews by users):** On various social platforms, users can rate articles written by other users on various scales. Epinions allows users to rate on a five-point scale to express their preference and opinion about the article written by another user. The more the number of high ratings (treated as positive opinion) user u has given to articles A_v written by another user v, the more he/she would tend to get influenced by user v's future articles. The number of positive and negative ratings by user u on articles written by user v is counted up to compute $P_{u \rightarrow A_v}$ and $N_{u \rightarrow A_v}$.

3. **Implicit information through interactions (Concordance or Discordance level):** We measure the level of concordance or discordance between a pair of users based on their shared experiences. We compute the concordance level as the number of

common articles on which the users share similar opinions or views, that is, either both have liked the article, or both have disliked it. For given users u and v, the *Concord(u, v)* is the number of all articles on which both u and v have a common opinion, i.e., both positive or both negative. Similarly, *Discord(u, v)* is the number of all articles on which the users u and v have contradicting opinions.

Another important aspect that determines the influence of others on a user u is CC_u, the closeness centrality of user u [9] which represents how close u is to other users in the network. All the above information on SSNs can be incorporated to measure the influence strength between users as shown in Fig. 2. This influence strength defines the likelihood of user u performing an action under the influence of user v. The final strength of influence w_{vu} of user v on user u is computed as:

$$w_{vu} = \left[\alpha \frac{\left(1 - e^{-\left(P_{u \to v} - N_{u \to v}\right)*CC_u}\right)}{\left(1 + e^{-\left(P_{u \to v} - N_{u \to v}\right)*CC_u}\right)} + (1 - \alpha)T_{(u,v)} \right] \tag{3}$$

where $P_{u \to v}$ and $N_{u \to v}$ are the total positive and negative counts from user u to user v and is given by $P_{u \to v} = P_{u \to A_v} + Concord(u, v)$ and $N_{u \to v} = N_{u \to A_v} + Discord(u, v)$. The weight value w_{vu} lies in the range $[-1, +1]$ where the higher value represents a significant influence of a user on another user. The parameter α can be selected empirically to lay relative importance to explicit and implicit interactions-based information. It is done since some users may choose not to declare their trust or distrust towards another user publicly. But their real behavior can be gauged by analyzing their activities and mining the true dynamics in the form of interactions. Therefore, our model aggregates implicit interactions-based information, trust as well as distrust information to reflect users' real-world behavior towards other users for computing influence rather than using the information individually as done by Xu et al. [18].

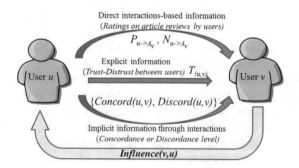

Fig. 2. Influence through various sources of information on SSNs

3.2 Discovering Set of Influential Users Using Genetic Algorithm

The problem of maximizing the collective influence of a set of users (k) is NP-hard [11], and it will be difficult for exhaustive search algorithm to find the optimal solution in a reasonable time given the scale of the real-world SNs is huge. Also, a greedy algorithm may not obtain an optimal solution. Hence, GA seems suitable for addressing the collective influence maximization problem.

GA has been widely used across various combinatorial optimization problems [2, 8] and takes inspiration from the Darwinian biological theory of evolution. The algorithm starts with a random population of potential candidate solutions to a problem which are known as chromosomes. The quality of each chromosome is evaluated in terms of its fitness which is used to determine which chromosomes will be used to generate new ones. The evolution of solutions at each generation of the algorithm is guided by the principle of survival of the fittest to generate a better solution. New chromosome solutions are created using genetic operators like crossover and mutation. The crossover operator generates two new offspring chromosomes from two parent chromosomes by allowing them to exchange some meaningful information and preserve best characteristics to evolve better solutions. The mutation operator is applied to preserve the genetic diversity of the population by introducing changes at the genes level of the individual chromosome. To ensure that the best chromosome solutions are retained from generations to generations, we have employed an elitist approach. This entire process repeats over many generations till a convergence criterion is met.

Chromosome Representation and Genetic Operators

Each chromosome individual is encoded as a fixed-size linear vector which contains k integer-valued identifiers representing the set of influencers. Here k is the number of influencers to be determined. Therefore, each gene of the chromosome represents an influencer.

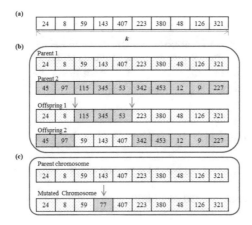

Fig. 3. Chromosome representation and genetic operators (a) chromosome structure (b) two-point crossover operator with crossing points at 2 and 5 (c) mutation operator applied at position 4

In the course of evolution, genetic operators are applied to get better exploration and exploitation in solutions search space. In our model, we have used a two-point crossover in which the segment of chromosome between two crossover points is exchanged between the two parent chromosomes to generate two offspring. A substitution operator is used for mutation where the random gene (/s) in a chromosome is reset to any value from a set of permissible values to obtain a new offspring. These operators may produce infeasible offspring; removal of duplicates is used as a reparation strategy to address this. Figure 3 illustrates an example of encoded chromosome solution and the application of genetic operators.

Fitness Function

A fitness function quantifies the quality of a chromosome and directs the evolutionary process to discover an optimal solution. To solve the collective influence maximization problem in the context of SSNs, we formulate a new fitness function known as *collective influential power* (*CIP*) with the following two objectives, both of which are crucial to finding the set of k influencers with maximal collective influence.

Objective Function 1: Enhanced Joint Influential Power (EJIP). This function is adapted from [18] and is defined as the sum of the influence strengths of a set of k influencers on other users, without considering the inter-influence between them. There-fore, from the constructed weighted influence network, we can find the *EJIP* of the set of target influencers which is a subset of nodes C in the network as summation of the weights on the edges from this subset of nodes to other nodes in the network, that is,

$$\text{Maximize} \qquad Z1 = EJIP_C = \sum_{i \in C, j \in \bar{C}} w_{ij} \tag{4}$$

$$\text{subject to} \qquad C \subseteq V \text{ and } |C| = k \tag{5}$$

where $\bar{C}$ is the complementary set consisting of all nodes of the network that are not in C and k is the size of the subset C.

Objective Function 2: Influence Spread (IS). This function represents the spread of actual influence emanated by the set of k influencers in the network. A user is activated if he/she performs an action (or likes/rates a product) and inactive otherwise. The influ-ence spread is computed as the number of distinct users that get activated under the influence of k influencers, that is, the number of users activated by the set of influencers C and is computed as:

$$\text{Maximize} \qquad Z2 = |IS_C| = \sum_{i \in C} IS_i \tag{6}$$

$$\text{subject to} \qquad C \subseteq V, |C| = k, \text{ and } \forall m, n \in IS_C, m \neq n \tag{7}$$

where IS_i is the influence spread of ith influencer and m, n are the users in the network that become activated.

Fitness Function: Collective Influential Power (CIP). The fitness function is formulated as the harmonic mean of both the *EJIP* and the *IS* of the set of influencers so as to optimize these two objectives simultaneously. It is a measure of collective influence that the group of influencers exerts over the entire network. Hence, to discover the target set of k influencers with maximal *CIP* is to find a subset of users $C \subseteq V$, where the harmonic mean of the sum of influence strengths from the nodes in C to other users in the network and the combined influence spread of C over the entire network is maximal. It is defined as:

$$CIP = \frac{2 * EJIP_C * IS_C}{\left(EJIP_C + IS_C\right)} \tag{8}$$

where $EJIP_C$ (Eqs. 4 and 5) and IS_C (Eqs. 6 and 7) are the enhanced joint influential power and the influence spread of the set of k influencers. This formulation of *CIP* ensures that when both *EJIP* and *IS* are largest, then the *CIP* of the set of k influencers is maximized. It allows easy diffusion of the influence over the network since the power of collective influence, and its actual spread will be well exploited.

4   Experiments and Results

In this section, we present the details of the computational experiments that we have performed to evaluate the effectiveness of our proposed model.

4.1   Dataset and Experimental Setup

We have performed experiments on the real-world dataset of Epinions [14] which is a popular consumer reviews e-commerce website. The dataset used in our experiments consists of 486 users who rated 1,238 articles and have issued 21,725 article ratings. Users can write review articles about various products and can also rate articles written by other on a five-point scale with 1 (minimum) to 5 (maximum). The website also allows users to express their like or dislike towards other users by labeling them as trust $(+1)$ or distrust (-1) worthy.

In our experiments, we employ an elitist strategy which ascertains that the fittest solutions generated so far are retained for the next generations. The termination criterion for the evolutionary algorithm is the number of generations. The GA performance is often sensitive to the choice of suitable values for the evolutionary parameters, e.g.,

Table 1. Genetic algorithm parameters

Parameter	Value
Population size	20
Number of generations	500
Crossover probability P_c	0.8
Mutation probability P_m	0.2

crossover and mutation probabilities, population size, and the number of generations. Table 1 shows the values chosen for these parameters.

4.2   Computational Experiments and Results

To compare the performance of our proposed MIGA model, we consider the following methods for discovering influential users from the SNs: joint influential power (*JIP*) approach (Eqs. 1 and 2) and random method (the users are randomly selected to have k influential users).

For evaluating the effectiveness of our proposed MIGA model, the influence spread is used as the metric to compare the performance of all the methods. We analyze the performance of our proposed MIGA model with the other methods by computing the influence spreads of all methods for varying k, that is, the number of influencers to be determined. A higher value for the influence spread corresponds to a better performance. We observe that our MIGA model considerably outperforms other methods in terms of the number of users activated by the influencers for different sizes of influencers' groups, as depicted in Fig. 4. It is due to the fact our MIGA model discovers influencers with high *CIP* over the network.

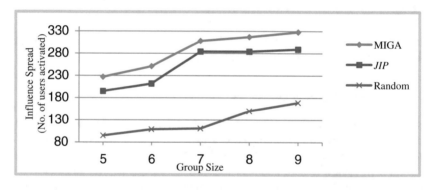

Fig. 4. Performance comparison of MIGA with existing approaches

We also analyze and study the effect of direct influence versus indirect social influence on *CIP* for our proposed MIGA model. Figure 5 shows the variation in *CIP* values for our model at different hops, that is, MIGA1 at hop 1, MIGA2 at hop 2, and MIGA3 at hop 3 with respect to the number of influencers. We observe that as the number of propagation hops increases, the *CIP* values also increases with the number of influencers. It is intuitive since users get indirectly influenced from other users in the network. And the collective abilities of influencers tend to rise considering the extended network of users. We consider the hop count till three that represents "Friend of a FOAF" (where FOAF is a friend of a friend) since after that influence propagation gets diluted in effect.

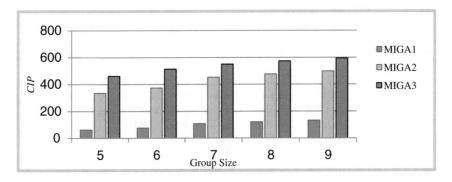

Fig. 5. Effect of direct and indirect influences on *CIP*

5   Conclusions and Future Work

The deep penetration of social networks into the lives of people makes the networks exhibit the various real-world phenomena. Users on social network tend to get influenced by the behavior or actions performed by their neighbors. The presence of this social influence can be mined, measured and analyzed both qualitatively and quantitatively for various real-world business and e-commerce applications.

In our work, we address the problem of maximization of the collective influential power of a set of users in the context of signed social networks for application in viral marketing. We propose a computational model, Mining Influencers using Genetic Algorithm (MIGA) to evolve and discover an optimal set of influencers that have maximal influence collectively over the other users in the network. They have the largest influence spread in terms of activating other potential users in the network and also emanate highest enhanced joint influential power over the other users. Firstly, we mine the explicit signed (trust-distrust) relationships between users as well as their implicit and interactions-based behavioral information to determine the strength of influence relationships between the users of the network. The influence strength takes into account both positive as well as negative influence between users. It also gives importance to the users that are closely connected to more central users so that their influence can be diffused to an enhanced set of users. Our model reflects a holistic view of how real-world relationships exist. Once the influence relationships are determined, an elitist genetic algorithm-based model (MIGA) is employed to discover target group. We have conducted the experimental study to prove the efficacy of our proposed model.

As a future work, exploration and analysis of temporal and topical aspects of influence relationships can be considered [10, 16]. Exploitation of dynamics of the network and the presence of other social-psychological phenomena needs to be investigated [3]. Another interesting future direction would be to study the influence maximization problem in heterogeneous multi-dimensional relational networks [13].

Acknowledgement. This work is, in part, financially supported through the Inspire program by Department of Science and Technology (DST), Government of India.

References

1. Ahmed, S., Ezeife, C.I.: Discovering influential nodes from trust network. In: 28th Annual ACM Symposium on Applied Computing, pp. 121–128. ACM (2013)
2. Awal, G.K., Bharadwaj, K.K.: Team formation in social networks based on collective intelligence-an evolutionary approach. Appl. Intell. **41**(2), 627–648 (2014)
3. Bonchi, F., Castillo, C., Gionis, A., Jaimes, A.: Social network analysis and mining for business applications. ACM Trans. Intell. Syst. Technol. **2**(3), 22 (2011)
4. Chen, W., Wang, Y., Yang, S.: Efficient influence maximization in social networks. In: 15th ACM SIGKDD, pp. 199–208. ACM (2009)
5. Chen, W., Wang, C., Wang, Y.: Scalable influence maximization for prevalent viral marketing in large-scale social networks. In: 16th ACM SIGKDD, pp. 1029–1038. ACM (2010)
6. Domingos, P., Richardson, M.: Mining the network value of customers. In: 7th ACM SIGKDD, pp. 57–66. ACM (2001)
7. Doreian, P., Mrvar, A.: A partitioning approach to structural balance. Soc. Netw. **18**(2), 149–168 (1996)
8. Eiben, A.E., Smith, J.E.: Introduction to Evolutionary Computing, 2nd edn. Springer, Heidelberg (2007)
9. Freeman, L.C.: Centrality in social networks conceptual clarification. Soc. Netw. **1**(3), 215–239 (1978)
10. Goyal, A., Bonchi, F., Lakshmanan, L.V.: Learning influence probabilities in social networks. In: 3rd ACM International Conference on Web Search and Data Mining, pp. 241–250. ACM (2010)
11. Kempe, D., Kleinberg, J., Tardos, É.: Maximizing the spread of influence through a social network. In: 9th ACM SIGKDD, pp. 137–146. ACM (2003)
12. Leskovec, J., Krause, A., Guestrin, C., Faloutsos, C., VanBriesen, J., Glance, N.: Cost-effective outbreak detection in networks. In: 13th ACM SIGKDD, pp. 420–429. ACM (2007)
13. Liu, L., Tang, J., Han, J., Jiang, M., Yang, S.: Mining topic-level influence in heterogeneous networks. In: 19th ACM CIKM, pp. 199–208. ACM (2010)
14. Massa, P., Avesani, P.: Trust-aware bootstrapping of recommender systems. In: ECAI Workshop on Recommender Systems, pp. 29–33 (2006)
15. Schut, M.C.: Scientific Handbook for Simulation of Collective Intelligence. Available under creative commons license version 2 (2007)
16. Tang, J., Sun, J., Wang, C., Yang, Z.: Social influence analysis in large-scale networks. In: 15th ACM SIGKDD, pp. 807–816. ACM (2009)
17. Wasserman, S., Faust, K.: Social Network Analysis: Methods and Applications, vol. 8. Cambridge University Press, Cambridge (1994)
18. Xu, K., Guo, X., Li, J., Lau, R.Y., Liao, S.S.: Discovering target groups in social networking sites: an effective method for maximizing joint influential power. Electron. Commer. Res. Appl. **11**(4), 318–334 (2012)

Face Recognition Across Aging
Using GLBP Features

Mrudula Nimbarte[1](✉) and K.K. Bhoyar[2]

[1] CE Department, BDCE Sevagram, Wardha, India
mrudula_nimbarte@rediffmail.com
[2] IT Department, YCCE Nagpur, Nagpur, MH, India
kkbhoyar@yahoo.com

Abstract. Face recognition over aging is still a difficult but interesting problem in pattern recognition nowadays. It has many real world applications. It is highly affected with many uncontrolled parameters like variations in head pose, expressions and illumination. Aging also varies person to person, thus makes the task more difficult. This paper includes an approach proposed by us for solving this problem. Here, we introduced a novel feature descriptor that is a combination of Gabor and LBP features called as GLBP. We used Principal Component analysis (PCA) for dimensionality reduction and k-NN as a classifier. Proposed approach is experimentally tested on popular aging datasets FGNET and MORPH. It is observed from the experimental results that our approach is better in Rank-1 recognition accuracy as a performance measure.

Keywords: Aging model · Face recognition · LBP · Gabor wavelets · PCA

1 Introduction

Age invariant face recognition is very interesting and challenging area of research in pattern recognition. Apart from routine problems faced by face recognition systems in general (like pose, illumination and expressions variations), aging process in humans makes this task the most difficult. Also the process of aging vary from person to person, making it further difficult. In general face shape and texture are highly affected by aging process [4]. Even it is very difficult to recognize the identity of a person manually over aging [7]. Thus, face recognition across aging is an open challenge in the field Computer Science. There are two types of age invariant face recognition methods: Generative and Discriminative [5]. Generative method includes the simulation of synthetic image at that age and then matching those two images. Whereas, discriminative method involves extraction of age invariant features and classification of that image for recognition [8, 10, 14]. Unavailability of huge dataset having sufficient images of the same person over the long span of time is another issue in this field. It has many interesting real time applications as passport renewal system, driving license renewal system, identification of missing children, providing more security to VIPs, identification of criminals etc. [15]. The basic steps for face recognition across aging are image preprocessing, feature extraction and classification [15]. In this process, researchers either use a combination of new features or choose a new classifier to identify a person over time. In this work, we

© Springer International Publishing AG 2018
S.C. Satapathy and A. Joshi (eds.), *Information and Communication
Technology for Intelligent Systems (ICTIS 2017) - Volume 2*, Smart Innovation,
Systems and Technologies 84, DOI 10.1007/978-3-319-63645-0_30

follow the same process using a novel combination of different feature representations and proposed a new method to solve this problem. To introduce a new simple facial feature called as GLBP is the main contribution of this paper. It is a combination of Gabor wavelets and LBP. Earlier studies also focused their results using such combinations, some of them used a combination of three features also. But our combined feature (using only two features, Gabor wavelets and LBP) is used for solving the problem of face recognition across aging for the first time.

Following is the organization of rest of the paper. Section 1 contains introduction of the topic in the field of image processing and pattern recognition. Section 2 describes adopted approach for solving the problem of age invariant face recognition. Section 3 presents experiments and results while Sect. 4 contains conclusion and future scope.

2   Adopted Approach

We propose a different approach to recognize a face across time lapse. This approach includes of a mixture of Gabor features and LBP features for facial feature representation. For face recognition across aging, there are basically two steps. First, extracting age invariant facial features. Second, selecting a best available or designing a new classifier. The overall model for the proposed system is shown in Fig. 1. Many researchers have shown very good results for both of these steps. Many different facial feature representations are available in this area.

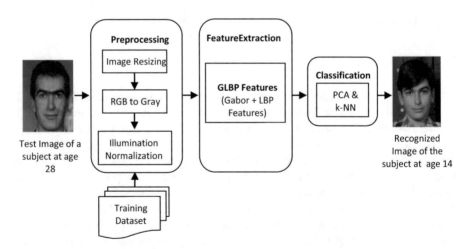

Fig. 1. Proposed model for face recognition across aging

2.1   Image Preprocessing

Image preprocessing helps to improve the quality of the digital image and hence improve the recognition rate of the system. We follow some basic steps as Image Resizing, RGB to gray conversion and Histogram Normalization to adjust the contrast. We resized the images to 240 × 240. Figure 2 shows the Step-by-step preprocessing using sample image.

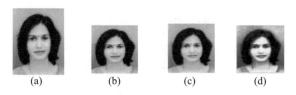

(a)   (b)   (c)   (d)

Fig. 2. Step-by-step preprocessing steps (a) sample original image (b) resized image (c) gray scale image (d) histogram normalized image

2.2 Feature Extraction

Performance of face recognition related applications is highly dependent on the features selected for experimentation. Many of the researchers applied available feature representations and some of them proposed some novel techniques. In our experiments, we used available approaches as Gabor Wavelets and LBP Features. Instead of using them separately we used a combination of them and introduced a new feature referred as GLBP. GLBP contains of both the features combined into it. Some of earlier studies used a combination of Gabor Wavelets, GOP and LBP for feature extraction, but it increases a computational complexity.

LBP Feature. LBP is considered as a good feature descriptor, popularly used as a texture feature as it is found to be effective in object recognition. Many variants of LBP are available nowadays. The basic LBP can be modified as per requirements. For image analysis, the statistics of LBP are generally used. It is an operator that transforms image into array of integer values giving small-scale representation of the image. In original LBP, 3×3 block of image pixels is considered. LBP at a given location of pixel of an image I(x, y), is obtained as the binary number by making a comparison of the intensity values of center pixel and its eight neighboring pixels. As there are 8 neighboring pixels, it has $2^8 = 256$ different labels. This decimal representation is referred as Local Binary Pattern [1, 11]. Figure 3 shows the sample original and LBP image. This decimal code of LBP is computed as follows:

$$LBP(x_i, y_i) = \sum_{n=0}^{7} S(i_n, i_i) 2^n \tag{1}$$

where i_i is gray intensity value of center pixel and i_n is the gray intensity value of eight neighboring pixel [2, 3, 6]. The threshold function S(x) is given as:

$$S(x) = \begin{cases} 1 & if \quad x \geq 0 \\ 0 & if \quad x < 0 \end{cases} \tag{2}$$

Fig. 3. Sample LBP image with original image

Gabor Wavelet Feature. These features are used to represent an image by performing convolution of an image with a set of Gabor kernels. For the input image I(x, y), the output of the convolution can be given as: $O_{\mu,\nu}(Z) = I_z\psi_{\mu,\nu}(Z)$ where

$$\psi_{\mu,\nu}(Z) = \frac{\left\|k_{\mu,\nu}\right\|^2}{\sigma^2} e^{\left(\frac{\|k_{\mu,\nu}\|^2\|z\|^2}{2\sigma^2}\right)} \left[e^{ik_{u,\nu}z} - e^{\frac{\sigma^2}{2}}\right] \tag{3}$$

$Z = (x, y)$, μ, and ν are used to represent the orientation and scales of the Gabor kernels respectively, the norm operation is denoted by$\|\cdot\|$ and the wave vector $k_{\mu,\nu}$ is given as $k_{\mu,\nu} = k_\nu e^{i\phi_\mu}$ where $k_\nu = k_{max}/f^\nu$, and k_{max} is the highest frequency, f is the spacing factor between kernels, and ϕ_μ is equal to $\pi\mu/8$. In most face image processing tasks, Gabor wavelets with five scales and eight orientations are commonly used. The real parts of five scales and eight orientation are given in Fig. 4.

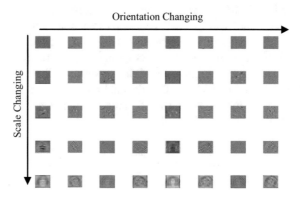

Fig. 4. Real parts of five scales and eight orientation

2.3 Classification Technique

As the feature vectors are available, then it will be checked for dimensionality reduction. It plays very important role in image processing as many images per subject are available. For this purpose we used Principal Component Analysis (PCA) for dimensionality reduction and k-Nearest Neighbor (k-NN) as classifier in our experimentation. There are two types of classifiers: Supervised and Unsupervised. Our approach is an example of unsupervised learning as exact class information is not available for classification. In PCA, face images are represented as "eigenfaces". It is considered as influential tool for solving the problems of face recognition and detection. It includes converting 2-D images to 1-D face vectors. Then obtain a normalized face vector by subtracting average face vector from each face vector. Next, step is to calculate eigenvectors. So, we need to calculate covariance matrix and reduce the dimensionality by selecting k best eigenvectors [1]. In k-NN classifier k lower dimensional face vectors are selected. In our approach we used k as 1. We used Euclidean distance as a distance measure to classify the test vectors for the implementation of eigenfaces [3, 6].

The overall process as shown in Fig. 1, includes training and testing phases. Given input image is first resized as 240 × 240. This resized image is then converted to gray-scale image. It is then applied for histogram normalization to improve the performance against any intensity variations. Next, Gabor wavelet and LBP features are extracted separately. Then, combined feature vector is formed known as GLBP feature vector. PCA is then used for dimensionality reduction. Finally, it is provided for classification using k-NN technique. In training, the overall process is completed on the known images those are available as gallery images. In, testing, the same process is repeated on unknown testing image. Feature vector of testing image is compared one-by-one with the feature vectors of training images. The matched image with lowest Euclidean distance is the output. Figure 5 shows experimental results using some sample images from FGNET dataset in first three columns. The last column shows the recognized image from the gallery. Similarly, Fig. 6 shows sample testing images and correctly recognized images from MORPH dataset. It is observed from these results

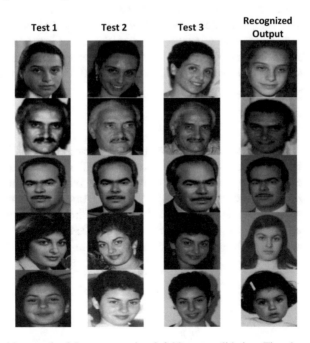

Fig. 5. Recognition result of the system using 3-fold cross validation. First three columns show the testing images at different ages whereas last column shows correctly recognized image from FGNET dataset

Fig. 6. Recognition results of the system. First row shows some sample testing images whereas second row shows correctly recognized image from MORPH (Album 2) dataset

that there are many variations as per aging, hence it is really difficult to recognize manually also. The proposed method also works for all age groups.

3 Experimental Results

3.1 Experimental Setup

Our approach for face recognition across aging is a part of multi-class classification problem. It is also considered as an identification problem. In this technique, the input image of a person at some age is converted into a feature vector set. Here, by using the proposed method final feature vector is obtained. This approach contains a novel combination of Gabor and LBP features used as GLBP. Then this feature vector is matched with the feature vectors of the subjects available in dataset. For this face recognition across aging Principal Component Analysis (PCA) is used for dimensionality reduction and k-NN is used as a classifier. Hence, we used a similarity measure as a Euclidean distance in our approach. These experimentations were performed using MATLAB 2013b (64-bit) version on 2.60 GHz Intel (R) CORE (TM) i-5 CPU and 8 GB of RAM.

3.2 Datasets

The experimentation is done using popular aging datasets FGNET and MORPH. The Face and Gesture recognition research NETwork (FGNET) dataset includes 82 subjects and 1002 images [16]. It contains images of range 0–62 years. In average there are 12 images per subject. Another aging dataset, MORPH (Album II) contains more than 55,000 images of about 13,000 subjects [17]. It contains images of adults. Age ranges from 16–77 years. In our experimentation, we used a training dataset of 40 subjects from FGNET with 280 images whereas 40 subjects from MORPH dataset including 310 images.

3.3   Performance Evaluation

The performance of the proposed system is evaluated using Rank-1 recognition accuracy as a measure. It is considered as the best performance evaluator [9]. In this case the classifier classifies the testing image in only one class on the basis of least Euclidean distance. If it classifies that image into correct class, the person gets recognized properly. For testing, we used LOPO (Leave-One-Person-Out) technique as used by earlier studies. In this testing technique, one person is kept in testing folder while all others are there in training folder. This is mainly because to make sure that the person in training folder is not there in testing folder at the same time. We repeated the same process for three times, every time on new image of the same person at different ages. This 3-Fold Cross validation is done using images from FGNET dataset. For MORPH dataset, we used only one testing folder as there are less number of images per subject as compared with FGNET. Rank-1 recognition accuracy using proposed method for FGNET dataset is 86.66% and for MORPH (Album 2) is 92%. Table 1 shows average Rank-1 recognition rate of all three testing folders.

Table 1. Rank 1 recognition rate using proposed method for both datasets

Images	FGNET (%)	MORPH (%)
Fold 1	88	92
Fold 2	85	
Fold 3	84	
Mean	86.66	92

Table 2 demonstrates the comparison of the Rank-1 recognition rate of the proposed system with some available state-of-arts results. It is found comparable good performance with the existing approaches.

Table 2. Rank-1 recognition rate of state-of-arts with proposed method

Methods	Rank-1 recognition rate (%)	
	FGNET	MORPH
AIFR using NTCA [1] 2015	48.96	83.80
AIFR from graph based view [14] 2014	64.47	–
AIFR using PCA & WLBP [12] 2014	67.30	–
AIFR for cross-age reference coding [18] 2014	–	92.8
AIFR using facial asymmetry [3] 2016	69.40	69.51
AIFR using MEFD [13] 2015	76.2	92.26
AIFR using CNN [15] 2016	85	–
Proposed AIFR using GLBP	**86.66**	**92**

4   Conclusion and Future Scope

In this proposed approach, a novel feature descriptor has been presented for solving the problem of face recognition across aging. It also improves the performance of the system. For experimentation, we used FGNET and MORPH datasets, consist of wide range of age groups. The results obtained from these experimentations show that this approach is better in Rank-1 recognition accuracy as compared with earlier studies. It is also found that we get better results using this approach against the results of the system if these features (Gabor and LBP) used individually. Moreover, using GLBP as combined features, improved the recognition accuracy to 86.66% on FGNET and 92% on MORPH datasets. It is observed that it gives good result on MORPH as compared with FGNET because of less variation in ages. This work can be extended in future using all images of both the datasets and with some other novel techniques.

References

1. Bouchaffra, D.: Nonlinear topological component analysis: application to age-invariant face recognition. IEEE Trans. Neural Netw. Learn. Syst. **26**(7), 1375–1387 (2015)
2. Ali, A.S.O., Sagayan, V., Saeed, A.M., Ameen, H., Aziz, A.: Age-invariant face recognition system using combined shape and texture features. IET Biom. **4**(2), 98–115 (2015)
3. Sajid, M., Taj, I.A., Bajwa, U.I., Ratyal, N.Q.: The role of facial asymmetry in recognizing age-separated face images. J. Comput. Electr. Eng. **54**, 1–12 (2016)
4. Bijarnia, S., Singh, P.: Age invariant face recognition using minimal geometrical facial features. Advanced Computing and Communication Technologies, vol. 452, pp. 71–77. Springer, Singapore (2016)
5. Sungatullina, D., Lu, J., Wang, G., Moulin, P.: Multiview discriminative learning for age-invariant face recognition. In: 10th IEEE International Conference and Workshops on Automatic Face and Gesture Recognition (FG), pp. 1–6 (2013)
6. Karthigayani, P., Sridhar, S.: A novel approach for face recognition and age estimation using local binary pattern, discriminative approach using two layered back propagation network. In: 3rd International Conference on Trendz in Information Sciences and Computing (TISC), pp. 11–16 (2011)
7. Carcagnì, P., Coco, M., Cazzato, D., Leo, M., Distante, C.: A study on different experimental configurations for age, race, and gender estimation problems. EURASIP J. Image Video Process. **2015**, 37 (2015). Springer
8. Li, Z., Park, U., Jain, A.K.: A discriminative model for age invariant face recognition. IEEE Trans. Inf. Forensics Secur. **6**(3), 1028–1037 (2011)
9. Park, U., Tong, Y., Jain, A.K.: Age-invariant face recognition. IEEE Trans. Pattern Anal. Mach. Intell. **32**(5), 947–954 (2010)
10. Ling, H., Soatto, S., Ramanathan, N., Jacobs, D.W.: Face verification across age progression using discriminative methods. IEEE Trans. Inf. Forensics Secur. **5**(1), 82–91 (2010)
11. Bereta, M., Karczmarek, P., Pedrycz, W., Reformat, M.: Local descriptors in application to the aging problem in recognition. J. Pattern Recognit. **46**(10), 2634–2646 (2013). Elsevier
12. Patel, P., Ganatra, A.: Investigate age invariant face recognition using PCA, LBP, Walsh Hadamard transform with neural network. In: International Conference on Signal and Speech Processing (ICSSP-14), pp. 266–274. Elsevier (2014)

13. Gong, D., Li, Z., Tao, D., Li, X.: A maximum entropy feature descriptor for age invariant face recognition. In: Proceedings of IEEE Conference on Computer Vision and Pattern Recognition (2015)
14. Yang, H., Huang, D., Wang, Y.: Age invariant face recognition based on texture embedded discriminative graph model. In: Proceedings of IEEE International Joint Conference on Biometrics (2014)
15. Khiyari, H., Wechsler, H.: Face recognition across time lapse using convolutional neural networks. J. Inf. Secur. **7**(3), 141–151 (2016)
16. The FG-NET Aging Database. http://www.fgnet.rsunit.com
17. MORPH Non-commercial Release Whitepaper. http://www.faceaginggroup.com
18. Chen, B.C., Chen, C.S., Hsu, W.H.: Cross-age reference coding for age-invariant face recognition and retrieval. Computer Vision. Lecture Notes in Computer Science, vol. 8694, pp. 768–783. Springer, Cham (2014)

Semantic Graph Based Automatic Text Summarization for Hindi Documents Using Particle Swarm Optimization

Vipul Dalal[1]([✉]) and Latesh Malik[2]

[1] CSE Department, G.H. Raisoni College of Engineering, Nagpur, India
vipul.dalal@vit.edu.in
[2] Computer Department, Government Engineering College, Nagpur, India
latesh.gagan@gmail.com

Abstract. Automatic text summarization can be defined as a process of extracting and describing important information from given document using computer algorithms. A number of techniques have been proposed by researchers in the past for summarization of English text. Automatic summarization of Indian text has received a very little attention so far. In this paper, we propose an approach for summarizing Hindi text based on semantic graph of the document using Particle Swarm Optimization (PSO) algorithm. PSO is one of the most powerful bio-inspired algorithms used to obtain optimal solution. The subject-object-verb (SOV) triples are extracted from the document. These triples are used to construct semantic graph of the document. A classifier is trained using PSO algorithm which is then used to generate semantic sub-graph and to obtain document summary.

Keywords: Bio-inspired algorithms · Text mining · Text summarization · Semantic graph · PSO

1 Introduction

Hindi is national language of India. It is native language of more than 258 million people in India. The use of Hindi documents in various fields is increasing rapidly. Text summarization allows readers to get a gist of a given document. The process of automatic text summarization consists of two phases. In the first phase, called "preprocessing phase", key textual elements, such as keywords and clauses are extracted from the given text. This requires linguistic and statistical analysis of the text. In the second phase, the extracted text is used as a summary. Such summaries are called "extracts" and this type of technique is called "extractive summarization". Another approach is called "abstractive summarization". In this approach the original text is interpreted and described in fewer sentences. Here linguistic methods are used to examine and interpret the text. The new concepts and expressions are found which can describe the text in a new shorter form such that it conveys the most relevant information from the original text. Such abstracts may or may not contain the sentences from the original document. Extractive summarization is shallow approach and is easy to implement whereas abstractive summarization needs deep understanding and analysis of the document and involves

© Springer International Publishing AG 2018
S.C. Satapathy and A. Joshi (eds.), *Information and Communication Technology for Intelligent Systems (ICTIS 2017) - Volume 2*, Smart Innovation, Systems and Technologies 84, DOI 10.1007/978-3-319-63645-0_31

some elements of Natural Language Generation (NLG), so it is more complex to implement. Our proposed approach extracts summary sentences from the input document only but analyzes semantic relationships of the document elements. In the survey of literature we found very little documented work for summarizing Hindi text [1]. So, in this paper we have proposed a semantic graph based approach for summarizing Hindi text using PSO algorithm. The rest of the paper is organized as follows. In Sect. 2, related work based on bio-inspired techniques is explained. Section 3 explains related work for Indian languages especially for Hindi. Section 4 explains our proposed approach for Hindi document. Experiment and results are discussed in Sect. 5. Finally, Sect. 6 concludes the proposed approach.

2 Summarization Using Bio-Inspired Methods

The extractive automatic text summarization work based on bio-inspired algorithms is as follows.

Binwahlan et al. [2] introduced an approach for feature selection. In their approach five features related to text summarization were used and the PSO was employed to make the system learn to obtain the weights of each feature. These weights are used in their next work [3] to generate the summary. The authors claimed that, their PSO method can generate summaries that are 43% similar to the human generated summaries, whereas summaries generated by MS-WORD are 37% similar.

Abuobieda et al. [4] proposed a feature selection approach based on (pseudo) Genetic probabilistic-based Summarization (PGPSum) model. This model was used for generating extractive summary of single document. Their method was employed as features selection mechanism and was used to obtain the weights of features from texts. These weights were used to obtain tuned scores for the features and to optimize the summarization process. The document summary was represented using these important sentences. The authors claimed that, their PGPSum model is better than Ms-Word benchmarks as the similarity ratio is close to human benchmark summary.

3 Summarization of Indian Text

An approach for generic extractive summarization for single document was proposed by Patel et al. [5]. Various structural and statistical parameters were used in their method. The algorithm was claimed to be language independent and it was applied to generate single-document summary for English, Hindi, Gujarati and Urdu documents. Nagwani et al. [6] developed a frequent term based text summarization algorithm. There are three steps in the algorithm. The first step processes the input document, eliminates the stop words and applies the stemmers to obtain root words. In the second step frequent terms are obtained from the document. These frequent terms are filtered to get the top frequent terms that are further considered. For these terms the semantic equivalent terms are also extracted. A last the third step filters all the sentences in the input document, that contain the frequent and semantic equivalent terms, and generates summary.

Sarkar [7] proposed an extractive approach for Bengali text summarization. A ranking was generated for the sentences based on thematic term and position features. Mishra et al. [8] designed a stemmer named "Maulik" for Hindi Language. This stemmer may be used to obtain root words in the preprocessing phase of summarization. Gupta et al. [9] suggested preprocessing phase for Punjabi text summarization. In this work, they applied stop word removal, noun stemming and cue phrase detection.

4   Proposed Approach

We found from the literature survey that very little work is done for summarization of Hindi text. In this paper, we have proposed an approach based on [10]. Instead of training SVM classifier, we are using Particle Swarm Optimization (PSO) to train the classifier. The PSO approach is well known for its optimization capabilities. Our approach can be outlined as follows:

1. Preprocess a set of training documents as well as the corresponding summaries to extract SOV triples from each sentence.
2. Construct semantic graphs for the training documents and their corresponding summaries using the extracted SOV triples.
3. Train PSO classifier to learn semantic sub-graph structure of the summaries from the semantic graph of the corresponding training documents. This procedure is depicted in Fig. 1.

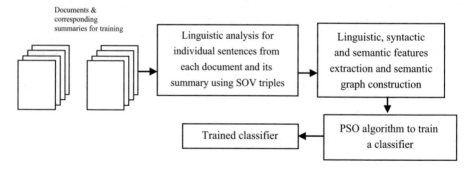

Fig. 1. Offline training phase

4. Preprocess the input document to extract SOV triples and to construct its semantic graph.
5. Use the trained classifier to derive sub-graph structure from the semantic graph of the input document.
6. Generate summary using the sub-graph obtained from the classifier.

This procedure is depicted in Fig. 2.

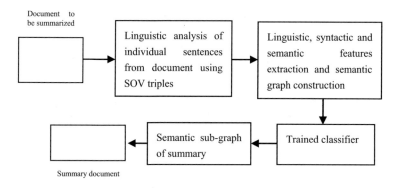

Fig. 2. Real time summarization phase

5 Experiment and Results

The proposed approach is implemented using Java platform. A set of 80 Hindi documents along with their summaries were selected for training purpose. In the preprocessing phase, total 1550 SOV triples were extracted to form the training set. A feature vector comprising of total 144 features was obtained for each SOV triple of each training document. The selected features can be categorized as follows:

1. Linguistic features – This includes POS tags, dependency tags, subject-object-verb tags, word depth in the dependency tree, etc.
2. Semantic Graph features – This includes page rank, hub, authority, number of incoming links, number of out going links, number of direct neighbors, number of indirect neighbors, etc.
3. Document Discourse Structure features – This includes sentence length, word position, word frequency, tf-isf, sentence similarity, etc.

Mainly the graph based features allow our proposed approach to perform deep semantic analysis of document elements. So, our approach gives a good compromise between the simplicity of extractive summarization and the human-like summary generation capability of abstractive summarization.

For the sack of simplicity, co-reference resolution and anaphora resolution were ignored and done manually. After forming the feature set, the PSO algorithm was run to obtain the centroids. The swarm was empirically considered converged if there is no improvement for 10 consecutive iterations or if "swarm size X dimensions" (i.e. 126×144, in this case) number of iterations is executed. The final global best position gives near optimal centroids. The feature vector of each SOV triple in the input document is then compared with these centroids and appropriate label is assigned to each triple. The sentences with at least one SOV triple labeled as part of reduced graph, are then included in the final summary of the document.

The unavailability of benchmark for Hindi summarization makes evaluation of our approach difficult. Therefore, the extracted summary was compared with human extracted summary. The system's performance was measured using precision, recall, F1 score and G score.

$$precision = \frac{no\ of\ summary\ sentences\ extracted\ that\ match\ with\ human\ exctracted\ summary}{total\ number\ of\ sentences\ extracted} \tag{1}$$

$$recall = \frac{no\ of\ summary\ sentences\ extracted\ that\ match\ with\ human\ exctracted\ summary}{no\ of\ actual\ summary\ sentences\ in\ human\ extracted\ summary} \tag{2}$$

$$F1 = 2\frac{precision\ \cdot\ recall}{precision\ +\ recall} \tag{3}$$

$$G = \sqrt{precision\ \cdot\ recall} \tag{4}$$

The performance of the proposed approach is given in Table 1. Higher value of recall indicates more sensitivity of the approach as compared to the accuracy or the precision.

Table 1. Performance metrics for the proposed approach.

Recall	60
Precision	42.86
F1 score	50.01
G score	50.71

6   Conclusion

In this paper we have presented a bio-inspired text summarization approach based on semantic graph of input document for Hindi text. The traditional summarizers rely upon sentence score obtained using various features but do not optimally select the summary sentences. Our proposed approach uses PSO to select the summary sentences optimally. The approach gives reasonably good performance. The adequacy of the approach can be improved if anaphora resolution and co-reference resolution are integrated in the preprocessing phase.

References

1. Dalal, V., Malik, L.: A survey of extractive and abstractive text summarization. In: 6th International Conference on Emerging Trends in Engineering & Tecnology (ICETET) (2013)
2. Binwahlan, M.S., Salim, N., Suanmali, L.: Swarm based features selection for text summarization. Int. J. Comput. Sc. Netw. Secur. IJCSNS **9**, 175–179 (2009)
3. Binwahlan, M.S., Salim, N., Suanmali, L.: Swarm based text summarization. In: International Association of Computer Science and Information Technology–Spring Conference, 2009, IACSITSC 2009, pp. 145–150 (2009)
4. Ali, A.A.M., Salim, N., Ahmed, R.E., Binwahlan, M.S., Sunamali, L., Hamza, A.: Pseudo genetic and probabilistic-based feature selection method for extractive single document summarization. J. Theor. Appl. Inf. Technol. **32**(1), 80–86 (2011). ISSN 1992-8645, E-ISSN 1817-3195

5. Patel, A., Siddiqui, T., Tiwary, U.S.: A language independent approach to multilingual text summarization. In: Conference RIAO2007, Pittsburgh PA, U.S.A. 30 May–1 June 2007. C.I.D., Paris (2007)
6. Nagwani, N.K., Verma, S.: A frequent term and semantic similarity based single document text summarization algorithm. Int. J. Comput. Appl. (0975 – 8887) **17**(2), 36–40 (2011)
7. Sarkar, K.: Bengali Text Summarization by Sentence Extraction (2012)
8. Mishra, U., Prakash, C.: MAULIK: an effective stemmer for Hindi language. Int. J. Comput. Sci. Eng. (IJCSE) **4**(5), 711 (2012)
9. Gupta, V., Lehal, G.S.: Preprocessing Phase of Punjabi Language Text Summarization (2011)
10. Leskovec, J., Milic-Frayling, N., Grobelnik, M.: Extracting Summary Sentences Based on the Document Semantic Graph. Microsoft Research, Microsoft Corporation (2005)
11. Barzilay, R., Elhadad, M.: Using lexical chains for text summarization. In: Proceedings of the Intelligent Scalable Text Summarization Workshop (ISTS 1997), pp. 10–17. ACL, Madrid (1997)
12. Ganesan, K., Zhai, C.X., Han, J.: Opinosis: A Graph-Based Approach to Abstractive Summarization of Highly Redundant Opinions (2010)
13. Hovy, E., Lin, C.-Y.: Automated text summarization in SUMMARIST. In: Mani, I., Maybury, M. (eds.) Advances in Automatic Text Summarization. MIT Press, Cambridge (1999)

Monitoring Real-Time Urban Carbon Monoxide (CO) Emissions Using Wireless Sensor Networks

Movva Pavani[1(✉)] and P. Trinatha Rao[2]

[1] Department of ECE, IcfaiTech School, IFHE University, Hyderabad, India
pavanimovva@ifheindia.org
[2] Department of ECE, GITAM School of Technology, GITAM University, Hyderabad, India
trinath@gitam.in

Abstract. In this paper, we propose a wireless sensor network based portable pollution monitoring system for monitoring the carbon monoxide (CO) concentration levels on the real time basis. Carbon Monoxide which is a critical and primary pollutant in air significantly affects the health of the people. With the rapid industrialisation and the exponential growth of automotive vehicles had led to the deterioration of air quality in the urban areas. Our design consists of Testbed of five nodes with calibrated carbon monoxide sensors for measuring CO concentration levels. By using the multi-hop mesh network, the CO sensors are integrated onto the Waspmote to communicate between the various nodes for the information exchange. The derived concentration levels of carbon monoxide from the different sensors on the board are made available on the internet through the platform which consists of Light Weight Middleware and Net Interface deployed on the server. Designed prototype had been implemented and tested in collecting the emission levels of CO in the Hyderabad city which had shown the consistent results under various circumstances.

Keywords: Wireless sensor networks · Middleware · Calibration · Web interface · Electrochemical sensors · Transducer technology · Respirable Suspended Particulate Matter

1 Introduction

In the last four decades, the world had witnessed an enormous increase in the number of industries and automobile vehicles leading to the deterioration in the air quality. Excessive emission of air pollutants beyond the specific levels had grave consequences on the environment and also on the health of the people. Air pollution monitoring systems [1] are used for measuring the concentration levels of various air pollutants, and the analysis of this data is required for the policy makers to take necessary and appropriate steps to reduce the air pollution for the welfare of their people. Traditional Air Quality Monitoring stations are deployed in limited number due o their large size, power limitations, high maintenance cost and huge initial investment cost involved. Wireless Sensor Network (WSN) presents an attractive and alternate solution for the

© Springer International Publishing AG 2018
S.C. Satapathy and A. Joshi (eds.), *Information and Communication Technology for Intelligent Systems (ICTIS 2017) - Volume 2*, Smart Innovation, Systems and Technologies 84, DOI 10.1007/978-3-319-63645-0_32

large traditional Air monitoring stations as we could able to deploy the sensing stations with smaller size and low-cost sensors [1]. The sensing stations are capable of collecting and transmitting the sensed information to the other sensing stations in the network. In the present work, Gas sensors were used as they are practical in terms price, compactness and robustness. Gas sensors used for air pollution monitoring are classified as electrochemical, photoionization, catalytic bead, solid-state [2] and infrared type. Carbon monoxide (CO) in the ambient atmosphere is a critical pollutant. It is neither photochemical nor photoreactive gas. CO concentration is independent of the intensity of sunlight. High levels of carbon monoxide affects the health of the people causing the breakdown of brain, nervous system and heart due to lack of sufficient supply of oxygen to these parts. Carbon monoxide is safe to humans when its concentration is around 0.2 parts per million (ppm) in air. Volcanoes and bush fires are natural sources of carbon monoxide emissions, whereas automobile exhaust and steel industries are the artificial sources of CO emissions. In this paper, Wireless Sensor Network based Air pollution monitoring [3] of carbon monoxide (CO) on a real time basis is discussed, and the designed prototype is cost-effective, reliable, accurate and scalable.

2   Source Apportionment of Air Pollutants in Hyderabad City

Hyderabad is the capital city of the newborn Telangana state of India which is 400 years old city. It is the fourth most populous city in India and having a density of 19,000 persons/Sq km. Increasing urbanisation and motorization led to deteriorating air quality in the city. It is estimated 3,700 premature deaths and 280,000 additional asthma attacks per year due to the PM pollution in the Hyderabad. The city is now listed in the top ten cities with the worst air quality in India, with significant contributions from transportation and industrial sectors.

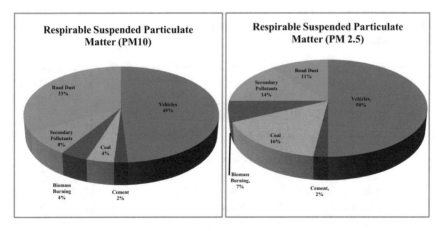

Fig. 1.   (a) Different sources contributing to Respirable Suspended Particulate Matter (PM10). (b) Different sources contributing to Particulate Matter of size less than 2.5 microns (PM2.5)

In Hyderabad city, 82–61% of the respirable suspended Particulate Matter (PM_{10} and $PM_{2.5}$) are the result of an increase in motorised transport due to the constant movement of the vehicles on the roads as shown in Fig. 1. Transport and the Industrial sector remain the primary sources of CO emissions in Hyderabad as shown in Table 1. 2&4 Wheeler Vehicles and Heavy Duty Vehicles (HDVs) are contributing the 77% out of vehicle emissions in Hyderabad as shown in the Table 2.

Table 1. Estimated Carbon monoxide pollutant emissions in Hyderabad by sector wise

Category	Vehicle exhaust	Road dust	Domestic	Industries	Brick Kilns	Construction	Generator sets	Waste burning
CO emissions tons/year	31%	–	20%	32%	12%	0.5%	3%	1.5%

Table 2. Source of Carbon monoxide emissions by vehicle type exhaust in Hyderabad

Vehicle type	2Wheeler vehicle	3Wheeler vehicle	4Wheeler vehicle	Buses	Heavy duty vehicles (HDVs)	Light duty vehicle (LDVs)
% of CO emissions	40%	10%	15%	5%	22%	8%

3   Calibration of Carbon Monoxide (CO) Sensors

To obtain the accurate and precise readings of gases, it is recommended to perform the calibration process for the sensors as per the standards before deploying onto the Waspmote [4]. Gas sensors produce an output voltage of small magnitude equivalent to the concentration of gas (parts per million (ppm)) which is unstable in nature. Signal conditioning circuits are used to stabilise and amplify the level of output voltage obtained from the sensors (Fig. 2).

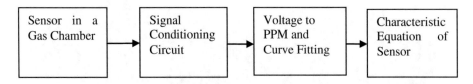

Fig. 2. Block diagram of the gas sensor calibration process

Mass flow controller (MFC) is used for controlling the reference gas concentration during the calibration process [5] of carbon monoxide (CO) Sensors. To obtain the required concentration, the synthetic air is diluted with CO gas (Figs. 3 and 4).

Fig. 3. (a) Experimental setup for calibration process. (b) Mass Flow Controller (MFC). (c) Libelium Waspmote

Fig. 4. Carbon monoxide sensors (a) CO-AX. (b) CO-BX. (c) CO-D4. (d) RCO100F. (e) 3ET1CO1500 (Courtesy: Alphasense Ltd and KWJ Engineering)

In the proposed work for sensing the CO emissions, we had used five electrochemical CO sensors: CO-AX, CO-BX, CO-D4, RCO100F and 3CO1ET1500.

Calibration graphs for the five carbon monoxide sensors are presented in Fig. 5. Calibration graphs of CO sensors are linear and with low noise levels. The concentration of CO sensors is linearly proportional to the output voltage after signal conditioning process. The sensitivity of the sensors is directly proportional to the size of the sensors.

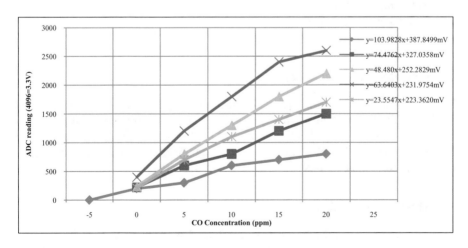

Fig. 5. Calibration curves for five CO sensors

4   Wireless Sensor Based Air Pollution Monitoring System

Air pollution monitoring system used for measuring the concentration of carbon monoxide CO gases on a real-time basis has the following four stages. CO sensors on the Waspmote uses node to node communications to communicate to the backend server through the Base Station (Fig. 6).

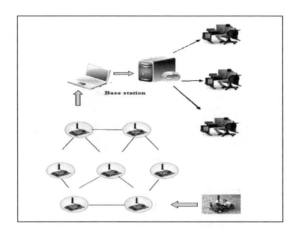

Fig. 6. (a) WSN based pollution monitor system architecture

Measured CO emissions are made available on online through the platform which consists of Light Weight Middleware [6], and Net Interface deployed on the server

1. Calibration of CO sensors
2. Configuring sensor nodes
3. Middleware development
4. Deployment of the prototype

4.1   Configuring Sensor Nodes

A test bed was developed with five nodes for measuring the CO Concentrations in the air. The Calibrated gas sensors are deployed on Waspmotes forming a node, and these nodes communicate with other nodes through the sensors in multihop mesh network [11]. Waspmote consists of wireless communication module which can be configured to establish communication between the nodes in the network.

4.2   Middleware Development

Deployed system [10] is capable of providing the real-time pollution data [7] in the user-friendly formats via the internet and mobile phones through the platform on the server. The platform is the combination of Lightweight middleware which handles data storage and recovery of information and net-based graphical user interface (GUI) for providing data in user-friendly formats.

4.3 Ground Deployment and Observations

A Wireless Sensor [8] based Air pollution monitoring of carbon monoxide (CO) on a real time basis is designed, and the prototype was tested at Patancheru, Hyderabad during the month of May 2016. Our design consists of Test bed with five nodes with calibrated carbon monoxide sensors for measuring CO concentration levels and plotted the concentration of CO with respect to time [9].

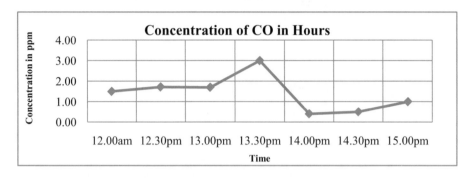

Fig. 7. Concentration of CO in a Hours (12.00 am–15.00 pm)

Figure 7 presents the concentration of CO in the afternoon from 12.00 am–15.00 pm which shows the peak between 1.00 pm–3.30 pm due to increased movement of motor vehicles and peak hours of Industries. Figure 8 presents the measurement of carbon monoxide in a day, and there is an increase in the concentration of CO levels during the peak working hours of Industries. Figure 9 presents the measurement of carbon monoxide in a week days from Monday to Saturday. As the Industrial area Patancheru has power holiday on Monday which had resulted less CO emissions and CO emissions steadily increase from Tuesday to Saturday. Figure 10 presents the measurement of CO

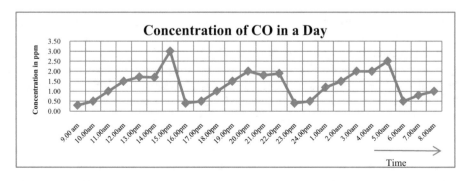

Fig. 8. Concentration of CO in a Day

in a Month showing the concentration of CO is less on Mondays and gradually increases from Tuesday to Saturday which is repetitive in nature for the remaining weeks in a month.

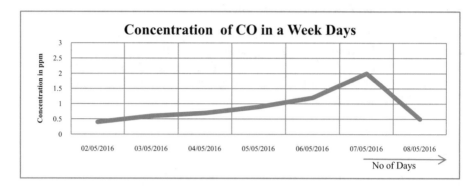

Fig. 9. Concentration of CO in a weekdays

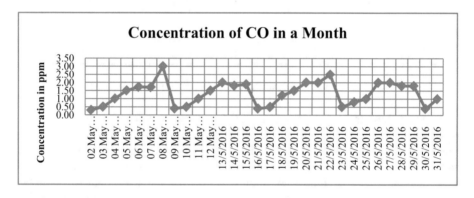

Fig. 10. Concentration of CO in a May month

5   Conclusion

A Wireless sensor-based pollution monitoring system [12] was designed considerating its significance and importance. A comprehensive investigation was done in terms of scientific feasibility and financial practicability in deploying Air quality monitoring systems. A test bed was developed with five nodes for measuring the CO Concentrations in the air. The Calibrated gas sensors are deployed on Wasp motes forming a node, and these nodes communicate [13] with other nodes through the sensors in a multihop mesh network. Deployed system is capable of providing the real-time pollution data in the user-friendly formats via the internet and mobile phones through the platform on the server. The platform is the combination of Lightweight middleware which handles data storage and recovery of information and net-based graphical user interface (GUI) for providing data in user-friendly formats. Designed prototype had been deployed and tested in collecting the emission levels of CO in the Hyderabad city which had shown the consistent results under various circumstances.

References

1. Karl, H., Willig, A.: Protocols and Architectures for Wireless Sensor Networks. Wiley, Chichester (2005)
2. Kularatna, N., Sudantha, B.H.: An environmental air pollution monitoring system based on the IEEE 1451 standard for low-cost requirements. IEEE Sens. J. **8**(4), 415–422 (2008)
3. Jung, Y.J., Lee, Y.K., Lee, D.G., Ryu, K.H., Nittel, S.: Air pollution monitoring system based on geosensor network. In: IEEE International Geoscience and Remote Sensing Symposium, pp. III-1370–III-1373 (2008)
4. Wasp mote technical guide-libelium system. www.libelium.com
5. Wang, W., Kim, T., Lee, K., Oh, H., Yang, S.: Development of a new wireless chemical sensor for CO_2 detection. In: IEEE Sensors, Conference (2007)
6. Prasad, R.V., Baig, M.S.: Real time wireless air pollution monitoring system. ICTACT J. Commun. Technol **2**(2), 370–375 (2011)
7. Shaban, K.B., Kadri, A., Rezk, E.: Urban air pollution monitoring system with forecasting models. IEEE Sens. J. **16**(8), 2598–2606 (2016)
8. Bagde, V.V., Shirbahadurkar, S.D.: Automated system for air pollution detection and control. IJECT **7**(3), 12196–12200 (2016)
9. Anju, D., Jacob, V.: WSN method of pollution monitoring system. IJCSIT Int. J. Comput. Sci. Inf. Technol. **7**(1), 353–359 (2016)
10. Han, G., Zhang, C., Shu, L., Rodrigues, J.J.P.C.: Impacts of deployment strategies on localisation performance in underwater acoustic sensor networks. IEEE Trans. Ind. Electron. **62**(3), 1725–1733 (2015)
11. Boukerche, A.: Algorithms and Protocols for Wireless Sensor Networks. Wiley, Toronto (2009)
12. Pavani, M., Rao, P.T.: Real-time pollution monitoring using Wireless Sensor Networks. In: 2016 IEEE 7th Annual Information Technology, Electronics and Mobile Communication Conference (IEMCON), Vancouver, BC, pp. 1–6 (2016)
13. Pavani, M., Rao, P.T.: End-end delay minimization using real-time routing protocol for Wireless Sensor Networks. In: 2016 IEEE 7th Annual Information Technology, Electronics and Mobile Communication Conference (IEMCON), Vancouver, BC, pp. 1–6 (2016)

Interference Minimization for Hybrid Channel Allocation in Cellular Network with Swarm Intelligence

Dattatraya S. Bormane[1] and Sharada N. Ohatkar[2(✉)]

[1] AISSMS College of Engineering, Savitribai Phule Pune University,
Pune, India
bdattatraya@yahoo.com
[2] MKSSS, Cummins College of Engineering, Savitribai Phule Pune University,
Pune, India
sharada.ok@gmail.com

Abstract. In the wireless cellular networks in order to deal with irregular and expanding demand, channel must be allocated in such a way that spectrum is used efficiently, capacity is maximized with a minimum level of interference; this problem is called Channel/frequency Allocation Problem. The swarm intelligence category of Heuristic technique, i.e. Particle Swarm Optimization and Ant Colony Optimization for Hybrid Channel Allocation is investigated to find the optimal solution to the minimum interference. The fitness function designed is based on Graph Theory in PSO. The designing of fitness function is the probabilistic model with Sequential packing and ordering technique is explored with ACO. The interference level is represented by edges indicating co-channel and co-site. The signal to interference ratio is measured for Kunz benchmarks and the computation time is obtained. The performance of applied PSO and ACO is compared with the literature reported with Genetic algorithm (GA).

Keywords: Hybrid Channel Allocation · Particle Swarm Optimization · Ant Colony Optimization · Signal to interference ratio · Cellular network

1 Introduction

There is an immense growth in population of the mobile user. The CoS, bandwidth requisite and interference can be reduced by efficient reuse of the scarce radio spectrum. The limited frequency spectrum and escalating demand have led to an NP hard problem of channel allocation. The prohibiting factor in radio spectrum reuse is interference caused by the surroundings or by other mobile customer. Interference can be minimized by deploy proficient radio subsystems and by utilization of channel assignment techniques.

Hybrid Channel Assignment (HCA) is best suited for present traffic flow, i.e. some amount of static traffic along with changeable traffic [1]. In this paper Co-Channel (CCI) and Co Site Interference (CSI) [2] are considered to measure the interference level.

The signal to interference ratio (SIR) is the preferred signal power level at the receiver/sum of the CCI level of power and received power. The received power is the

© Springer International Publishing AG 2018
S.C. Satapathy and A. Joshi (eds.), *Information and Communication
Technology for Intelligent Systems (ICTIS 2017) - Volume 2*, Smart Innovation,
Systems and Technologies 84, DOI 10.1007/978-3-319-63645-0_33

transmitted power reduced by attenuation. The propagation model can predict the amount of signal attenuated, HATA propagation model [2] for is applied.

Heuristic methods give optimal solutions at a rational cost of computation for algorithmically complex/time- consuming problems, i.e. NP Hard, such as a channel allocation. The evolutionary computation techniques under swarm intelligence, PSO and ACO are investigated to solve Frequency Assignment Problem (FAP).

The variants of FAP are Min. Span FAP (MS-FAP), Min. Blocking FAP (MB-FAP) and Min. Interference FAP (MI-FAP) [3]. In MS-FAP, the problem is to allocate frequency in a way that the interference is below a definite level, and the span is minimized. MS-FAP is relevant when the network operator wants to deploy new network in a region. In MB-FAP, then by and large blocking probability of the network is minimized. The paper focuses on applying a heuristic algorithm to MI-FAP whose objective is to minimize interference with assigned fixed set of frequencies which is required for operating networks. MI-FAP is more relevant today as there are more operating networks than the new network to be designed.

With latest networks being set up or current networks being expanded, standard GSM is in use as it is cheaper than the most recent technology. Therefore, standard GSM is still relevant and in use in modern networks.

In Sect. 2, the mathematical formulation of MI-FAP is detailed. The benchmarks are detailed in Sect. 3. In Sects. 4 and 5 working of PSO and ACO for MI-FAP is presented. The simulated results in the form of channel allocation matrix, interference graph (CCI and CSI), SIR along with the computation time is calculated and compared with Genetic Algorithm (GA) [14] in Sect. 6. The conclusion and future directions are addressed.

2 MI-FAP Mathematical Formulation

The FAP can be formulated with graph theory-node coloring algorithm. The edge amid base stations indicates interference, i.e. same channel is shared among them [4].

$$G = (V, E) \tag{1}$$

$$V = \{v_0, v_1, \ldots, v_i\} | i \in \mathrm{N} \tag{2}$$

$$E = \{\{v_0, v_1\}, \{v_0, v_2\}, \ldots, \{v_i, v_j\}\} | v \in V, \quad \forall ij \in \mathrm{N}, i \neq j \tag{3}$$

$$D = \{d_{01}, d_{02}, \ldots, d_{ij}\} | \forall \{i, j\} \in E, \exists d_{ij} \in \mathrm{N}^+ \tag{4}$$

$$P = \{\{p\overline{0}0, p\overline{0}1\}.\{p\overline{1}0, p\overline{1}1\}, \ldots, \{p\overline{i}0, p\overline{i}1\}\} | \forall \{i, j\} \in E, \quad \exists p_{ij} \in \mathrm{N}^+ \tag{5}$$

$$F = \{0, 1, 2, 3, \ldots, k\} | \forall k \in \mathrm{N}, \quad \forall v \in V, \exists f \in F \tag{6}$$

$$d_{ij} < |f(i) - f(j)|, \quad \forall ij \in \mathrm{N}, i \neq j \tag{7}$$

Let G (Eq. 1) be the weighted undirected graph, V (Eq. 2) is a set of vertices [5]. Each $v \in V$ (G) is set of base stations. The E (Eq. 3) is set of edges [4]. Forbidden set 'C' is the group of cells, (not use the same channel at the same time). Independent set 'I' is a cluster of cells, having an identical channel at the same time. An edge representation among two vertices v_i and v_j, are the constriction on the frequencies that can be allocated between them [4]. The distance between frequencies assigned to transmitters v_i and v_j is d_{ij} that is set D (Eq. 4). The amount of interference between v_i and v_j is found by (Eq. 7) [4]. Set P (Eq. 5) is the interference matrix, the value of p_{ij}, $\bar{p}i0$ is CCI and $\bar{\bar{p}}i1$ is ACI. The set F (Eq. 6) is frequencies for every transmitters in V [4]. MI-FAP is formulated as $\{V, P, E, F, D\}$, by mapping of $f : V \rightarrow F$.

3   Benchmarks Applied

Kunz put forth the Hopfield model from the actual 24 km × 21 km topographical data from locale around Helsinki, Finland [6]. The solution to FAP was found by applying neural network (NN) technique. The Kunz benchmark problems (Kunz 1, Kunz 2, Kunz 3, Kunz 4) essentially required a huge number of iterations to attain the optimal solution. These Benchmarks (Table 1) are specified by the amount of cells (Nce), the amount of channels (Nch) and a traffic demand vector D [6].

Table 1. Benchmark details

Problem	Nce	Nch	Demand vector D
Kunz 1	10	30	10, 11, 9, 5, 9, 4, 4, 7, 4, 8
Kunz 2	15	44	10, 11, 9, 5, 9, 4, 4, 7, 4, 8, 8, 9, 10, 7, 7
Kunz 3	20	60	10, 11, 9, 5, 9, 4, 5, 7, 4, 8, 8, 9, 10, 7, 7, 6, 4, 5, 5, 7
Kunz 4	25	73	10, 11, 9, 5, 9, 4, 5, 7, 4, 8, 8, 9, 10, 7, 7, 6, 4, 5, 5, 7, 6, 4, 5, 7, 5

4   Channel Allocation with Particle Swarm Optimization (PSO)

The PSO was developed in 1995 by Eberhart & Kennedy. The social behavior of Swarms of Bees, Flocks of Birds, and Schools of Fish, etc. motivated them. They found that the folks in a population turn out to be skilled from earlier experiences and experiences of others. The folks move with the information on its current position, current velocity along with the location of individuals "best" solution along with location of neighbors "best" solution. The individuals and consequently overall population gradually swirl to the "better" area.

In PSO, each and every solution is a "bird"- a "particle" in the search space. The fitness function gives the fittest value, and has velocity, i.e. Channels, indicating flying of the particles. They steer through the problem space by subsequently updating particle's position, i.e. Cells [7].

PSO is initialized with a group of random particles and then searches for minima, the optima by updating the number of iterations. With each iteration, the particle is updated with two "best" values - *pbest* and *gbest*. The *pbest* is the best solution (fitness) the particle has acquired so far. The *gbest* is the best value that is obtained so far by any particle in the population, it is been tracked by the particle swarm optimizer.

After finding the *pbest* and *gbest*, particle revises its velocity and positions with the Eqs. (8) and (9),

$$Vt + 1 = Vt + C1 * r1 * (pbest - X1) + C2 * r2 * (gbest - X1) \qquad (8)$$

$$Xt + 1 = X1 + Vt + 1 \qquad (9)$$

where, Vt: particle velocity, Xt: current particle position, C1, C2 are constant multiplier, and r1, r2 are uniformly distributed random nos. The process is repeated for a certain number of times until the satisfactory solution is achieved.

The PSO is applied to Cellular Network as: The velocity and position relate to channels and cells in cellular network. As in PSO updating velocity and position is done based on the old and new pbest and gbest, in cellular network also each new cell's updating depends on the old cells and its neighboring cells channels.

Table 2 presents the contribution made as compared to reported work with PSO.

Table 2. Contributions for solving CAP with PSO

PSO	Literature survey	Our contribution
Variant	MS-FAP [8–10]	MI-FAP
Assignment	DCA [8–10]	HCA
Fitness function	[8–11]	Hard and soft constraints Graph theory based
Performance measure	Minimized span	Channel assignment matrix
	Calls blocked [9]	Co-channel and co-site edges
	Call rejection ratio [8, 10]	SIR (Hata propagation model)
Benchmarks	21 Cell [9], EX (1, 2), HEX (1, 2), [10] KUNZ (1, 2) [8, 10]	KUNZ (1, 2, 3, 4)
Platform	Not given [9], C programming [8, 10]	MATLAB

5   Channel Assignment with Ant Colony Optimization (ACO)

ACO is a class of productive meta-heuristic algorithm developed by Marco Dorigo who was encouraged by real ant's actions. It was developed as a probabilistic technique that could be used for solving computational problems. The ants are intelligent to build the shortest viable path from their colony to the food source by following pheromone trails. The pheromone is planted in small quantity on different paths by the ant's as they walk, the subsequent ant following senses the pheromone laid on diverse paths and selects the path with more intensity of pheromone, than plants its own pheromone on the path [7].

To evade the MI-FAP problem, we have proposed a fusion structure that encompasses the ordering and sequential packing heuristics into the ACO iterations. We have represented a cell as node and the channel as ant for graph representation. In particular, the ordering technique and the sequential packing technique which is developed for obtaining the solution for FAP are implanted into an ACO to produce an efficient combination. Table 3 presents the comparison of the reported work.

Table 3. Contributions for solving CAP with ACO

ACO	Literature survey	Our contribution
Variant	MS-FAP [12, 13] Reducing span	MI-FAP Reducing interference
Assignment	FCA [12], DCA [13]	HCA
Probabilistic model	τij-pheromone intensity (minimizing span) ηij-minimize interference	τij - ordering technique(δj): CSI ηij - sequential packing (Aij): CCI
Performance measure	Call blocking probability [13] Call dropped [13]	Channel assignment matrix Co-channel and co-site edges SIR (Hata propagation model)
Benchmarks	21 Cell [12] Not given [13]	49 cell KUNZ (1, 2, 3, 4)
Platform	Coded in C++ [12] Not given [13]	MATLAB

Given the demand and considering the interfering constraints, channels are assigned to cells orderly by Ordering Heuristics, which is given as, (Eq. 10)

$$\delta_i = \sum_{j=1}^{n} d_j c_{ij} - c_{ii} \qquad (10)$$

where, d is the traffic demand and c is the compatibility matrix. The cell with a greater demand and subjected to more interference with its adjoining cells has a greater degree and that cell gives priority in assigning channels. Ordering ensures that the CSI is reduced to a large extent.

The cells that are fittest for allocating a specific channel is done sequentially by sequential packing heuristic. Let 'f1' frequency be packed and initially allotted to the cell. The region around the cell with 'f1' will not be allotted 'f1'. The interference free cells is selected which is assigned 'f1' frequency. This process continues till all the cells are occupied. The assignment of all the channels is done until the traffic demand is fulfilled. It ensures CCI is minimized.

The probability equation (Eq. 11) is proportional to the pheromone intensity τ_{ij} and the visibility value η_{ij} connected with the edges. The value of $\tau_{ij} = \delta_j$ is the degree of cell j, the value of $\eta_{ij} = A_{ij}$ is the overlap region indicating ordering and sequential packing heuristics respectively.

$$p_{ij} = \frac{(\tau_{ij})^{\alpha}(\eta_{ij})^{\beta}}{\sum\limits_{h \notin tabu} (\tau_{ij})^{\alpha}(\eta_{ij})^{\beta}} \tag{11}$$

The cells whose demands are fulfilled and cells violating the interference constraints are Tabu. α and β parameter is the weights representing pheromone and visibility respectively. With the assignment of 'f1' the process of finding the solution starts which are continued by assigning the remaining channels simultaneously.

6 Simulation Results

A frequency plan to allocate channels to respective cells is produced with the PSO and ACO on Kunz benchmark. The CCI and CSI are obtained are compared with GA [14]. In [14], GA with the binary string of individuals are taken to find solution for MI-FAP on Kunz benchmarks. The SIR in dBm and the computation time is plotted. The CAM for Kunz 4 with ACO-HCA is in Fig. 1.

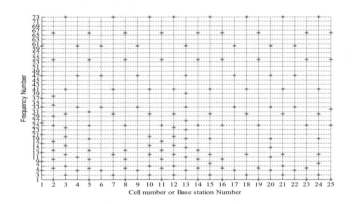

Fig. 1. Channel allocation matrix (KUNZ4) with ACO.

The interference graph is plotted indicating edges - CCI and CSI, Fig. 2. shows a plot of same for [14]. In Figs. 2 and 3, each circle within the cell represents traffic demand. The CCI is indicated by edge between cells if same channel is allotted to respective cells, which are within the reuse distance (3 units), a 'solid line' in Figs. 2 and 3. The CSI, the edge within the cell with frequency separation less than 4 is a 'dashed line' is within the cell. Table 4 depicts that the CCI is absent by applying PSO-HCA and with ACO-HCA the CCI & CSI are removed.

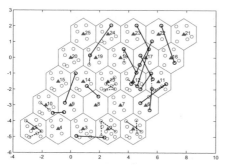

Fig. 2. Interference graph (Kunz 4) [14]

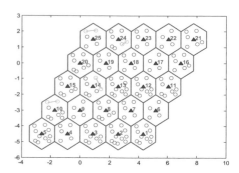

Fig. 3. Interference graph (PSO)

Table 4. Performance of PSO and ACO.

Benchmark		[14] GA	Proposed PSO	Proposed ACO
KUNZ 4	CCI edges	15	0	0
	CSI edges	16	4	0
	Computation time (s)	–	0.5390	0.2541
	Generation	2450	100	–
KUNZ 3	CCI edges	58	0	0
	CSI edges	21	2	0
	Computation time (s)	–	0.4678	0.1706
	Generation	50000 to 100000	100	–
KUNZ 2	CCI edges	32	0	0
	CSI edges	27	4	0
	Computation time (s)	–	0.3196	0.1451
	Generation	–	100	–
KUNZ 1	CCI edges	32	1	0
	CSI edges	21	4	0
	Computation time (s)	–	0.2099	0.1129
	Generation	–	100	–

In Table 4, the results of Kunz's four benchmarks are compared with [14]. The numbers of iterations required, the CCI and CSI edges depicting interference less than [14]. The computation time is least with ACO.

We have worked with OS Windows 7 workstation, a core I3 (IInd generation) processor, having 3.5 GHz speed with 2 GB RAM and is simulated in MATLAB.

The way channel allocation is done influence the SIR of each mobile station. Figure 4 shows plots for Kunz 4 benchmark instance, where x axis: Total demand for all cells, y axis: Range of SIR in dBm. Figure 4 shows that the SIR obtained by proposing PSO and ACO for Kunz 4 benchmark instances is better than [14]. The SIR is also better for Kunz 3, Kunz 2 and Kunz 1 benchmark instances than [14].

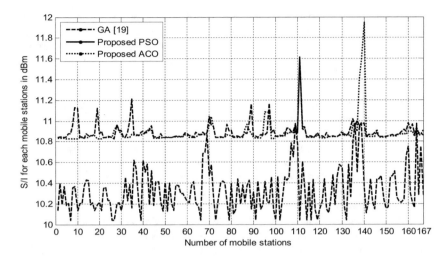

Fig. 4. Comparison of SIR with GA

7 Conclusion

This work presents the HCA technique based on PSO and ACO focusing on MI-FAP. The solution for MI-FAP in the form of SIR and the computation time is obtained for Kunz benchmarks. The SIR is improved than with [14] along with less computation time.

The effectiveness of the swarm intelligence - PSO and ACO can be investigated with the real – life MI-FAP instances and other benchmark instances like COST 256 and CELAR.

References

1. Ohatkar, S.N., Bormane, D.S.: An optimization technique for efficient channel allocation in cellular network. J. Commun. Technol. Electron. **59**(11), 1225–1233 (2014)
2. Theodore, S.R.: Wireless Communications Principles and Practice. Prentice-Hall, Delhi (2006)
3. Aardal, K., van Hoesel, S., Koster, A., Mannino, C., Sassano, A.: Models and solution techniques for frequency assignment problems. Ann. Oper. Res. **153**(1), 79–129 (2007)
4. Eisenbltter, A.: Assigning frequencies in GSM networks. Technical report (ZIB) (2001)
5. Montemanni, R., Smith, D.: Heuristic manipulation, tabu search and frequency assignment. Comput. Oper. Res. **37**(3), 543–551 (2010)
6. Kunz, D.: Channel assignment for cellular radio using neural networks. IEEE Trans. Veh. Technol. **40**(1), 88–193 (1991)
7. Padhy, N.: Artificial Intelligence and Intelligence Systems. Oxford University Press, New York (2013)

8. Chakraborty, M., Chowdhury, R., Basu, J., Janarthanan, R., Konar, A.: A particle swarm optimization-based approach towards the solution of the dynamic channel assignment problem in mobile cellular networks. In: TENCON 2008, Hyderabad, pp. 1–6 (2008)
9. Elkamchouchi, H.M., Elragal, H.M., Makar, M.A.: Channel assignment for cellular radio using particle swarm optimization. In: Proceedings of the 23rd, NRSC 2006, Menoufiya (2006)
10. Ghosh, S., Konar, A., Nagar, A.: Dynamic channel assignment problem in mobile networks using particle swarm optimization. In: Second UKSIM, EMS 2008, Liverpool, pp. 64–69 (2008)
11. Battiti, R., Bertossi, A.A., Brunato, M.: In cellular channel assignment: a new localized and distributed strategy. Mob. Netw. Appl. **6**, 493–500 (2001). Kluwer
12. Yin, P.-Y., Li, S.-C.: Hybrid Ant Colony Optimization for the Channel Assignment Problem in Wireless Communication. I-Tech Education and Publishing, Vienna (2007)
13. Papazoglou, P.M., Karras, D.A., Papademetriou, R.C.: On the implementation of ant colony optimization scheme for improved channel allocation in wireless communications. In: 2008 4th International IEEE Conference Intelligent Systems, Varna, pp. 641–650 (2008)
14. Wang, L.: Genetic algorithms for optimal channel assignments in mobile communication. Ph.D. thesis, School of Electrical & Electronics Engineering, Nanyang Technical University (2006)

Handwritten Numeral Identification System Using Pixel Level Distribution Features

Madhav V. Vaidya[1]($\boxtimes$) and Yashwant V. Joshi[2]

[1] Department Information Technology, Shri Guru Gobind Singhji Institute of Engineering and Technology, Nanded 431606, Maharashtra, India
`mvvaidya@sggs.ac.in`
[2] Department of Electronics and Telecommunication, Shri Guru Gobind Singhji Institute of Engineering and Technology, Nanded 431606, Maharashtra, India
`yashwant.joshi@gmail.com`

Abstract. In this paper, pixel level features of the character are used for Devanagari numeral Recognition. The pixel distribution features for each numeral can be calculated after preprocessing the document image and converting it to binary. Based on these features the numerals are classified into appropriate groups. Histogram feature matching method gives erroneous results for the numbers like one and nine as they are having nearly similar histogram. In the proposed approach pixel distribution features are extracted in four directions. The overall performance of classification can be improved if more number of features is compared. The proposed approach gives improved results as compared to simple histogram matching criteria.

Keywords: Histogram · Numeral · Zig-Zag scanning

1 Introduction

Optical character recognition (OCR) system primarily converts audio or speech into a format which can be stored ahead in digital format. OCR converts scanned text character images into editable text format. Main OCR research application is to reduce human efforts and error-free system running in post office work [1]. Also to reduce the work in different areas like, automation systems [3] and so on. The first most important issue or step is the separation of character. In this stage of implementation any error or inaccuracy will affect the overall accuracy of the approach. For Bengali and Devanagari script line segmentation is performed for machine printed characters PAL and Choudhury [2]. India is having diversification in culture and languages written by different peoples along the separate geographical area who uses different language scripts [4]. Research in off-line character recognition started with printed characters, and then to handwritten numbers and used in many Indian scripts [5], to identify the characters.

Off-line handwritten text or character recognition studied by most researchers [10, 11] in the languages like English, Chinese, Japanese, Latin and Arabic scripts. Very big complications are present in major Indian scripts. For automatic recognition of handwritten text these scripts so far has been studied extensively. Although in the

© Springer International Publishing AG 2018
S.C. Satapathy and A. Joshi (eds.), *Information and Communication Technology for Intelligent Systems (ICTIS 2017) - Volume 2*, Smart Innovation, Systems and Technologies 84, DOI 10.1007/978-3-319-63645-0_34

recent twenty two years, researchers in India showed keen interest in Indian scripts for off-line handwriting recognition. Most researchers such as [13] worked on different images of handwritten characters. Along with the study on off-line handwritten Devanagari word [10–12].

Many of the systems implementation different techniques for separating the text character from background are elaborated but most of them do not separate character correctly causing some unreliable results. Off-line in word-recognition system based on structural information Buse [13]. The word recognition accuracy was improved by two-dimensional fuzzy term classification system. The Park used pattern confidence and lexical confidence, and shape features measurement and evaluation to identify off-line handwritten character or word [14] using adaptive approach. Also Oliveira et al. [15] is a modular system for identifying numeric strings. It is an recognition-based approach which uses an identification and verification strategy. Outputs, such as different levels of identification, and post processing are combined in this approach.

1.1   Creation and Use of Database

Numeric database for experiments can be prepared for Marathi numerals. Marathi is an ancient language mainly spoken in Maharashtra. Marathi numeric representation and of Hindi language numeric representation are quiet the same. They are represented in ten different digits 0 through 9 in Marathi.

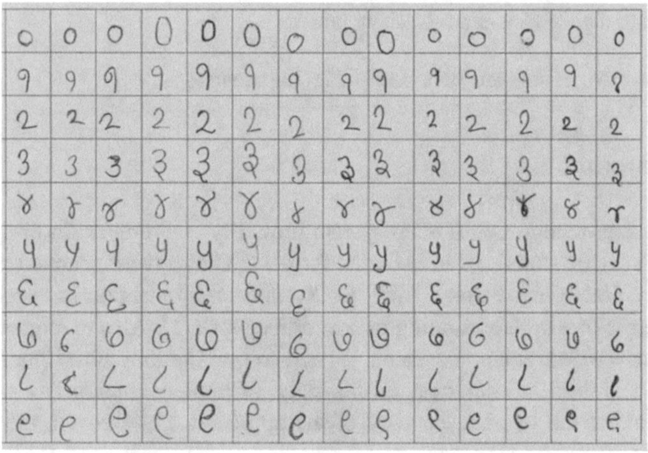

Fig. 1. Sample scanned document image for database creation

A database of almost 6000 sample points of numbers is created by scanning the page documents containing Devanagari Marathi number from zero to nine. The sample document shown in Fig. 1 is scanned by Samsung LaserJet scanners with 600 dpi. Almost every scanned document contains 560 handwritten Marathi sample numeral points. Thus eleven scanned pages containing handwritten numbers written by students of the school are used as a dataset.

2   Numeral Recognition and Feature Extraction

In this paper the results of the proposed approach on numerical classification in Marathi derived features are based on the pixel level. Using the horizontal and vertical launch profile features is removed. Also zig-zag scan vectors are created and Yoga for eight elements unlike each other vectors in two directions is calculated. These features are removed to determine cross-line features. Vectors are drawn and to increase overall performance of the system together. Split and method using conquers the best features or attributes for a given feature is recognized correctly by analyzing the character training phase can be selected. Thus it is less search time and size.

Document image scanners connected to the machine is used to acquire the document. In our experiment image is scanned by a single machine but in general Image captured is coming from different sources, it may have different format and color scheme. Image Enhancement is done to make it easier and better understanding of the preprocessed features. Segmentation algorithms for bounding box can be applied, so that interested part i.e. block from the whole image can be separated. Segmented parts are then normalized to get consistency in the feature dataset in generalized format.

2.1   Preprocessing

After receiving source document image, it has to be preprocessed. In this phase noise reduction and image enhancement algorithms applied to obtain a good quality image. Image may or may not be in the proper required format or color scheme, even it is to be converted to common file format. Any other color scheme in image format is converted to gray scale and then to binary to using binary approach of Otsus (Fig. 2).

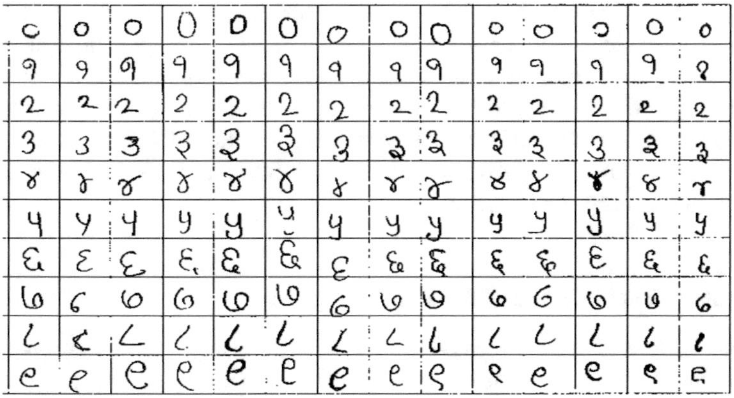

Fig. 2. Document image binarization

2.2   Segmentation

Many algorithms to partition the object from image are available. Here different sample points can be separated using bounding box from scanned database as there are no overlapping numbers. In the process, the individual sample points indicating numerals are cropped and inserted into database. For printed number it is a simple task and requires no additional modifications as the size and orientation of the number is the same, but for other documents every person is having different writing style therefore it can be quite difficult.

The main reason for this type of problem is that handwritten characters or numerals often overlap with each other. In addition, the person with the personal writing style to make radical changes in the normal partition is very cumbersome. In this paper, the basic assumptions used in handwritten numeric recognition are used. Thus using a bounding block the numbers are segmented to generate individual samples as shown in Fig. 3.

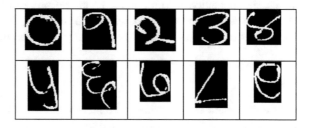

Fig. 3. Segmented individual numbers

2.3   Normalization

These segmented points can vary in either shape or size. As the segmented image size changes, feature related to the number changes which will lead to unexpected results. To solve the problem, segmented characters are scaled to proper dimensions to get standard results in training as well as testing. Thus any size numeric identification is possible with minimal error. Normal size for the selected basic sample points overall greater impact on system performance. Size is bigger, the validation error is low, but over time the complexity of adverse effects. 32×32 pixel sizes as follows to obtain maximum performance are selected.

2.4   Feature Vector Manipulation and Extraction

Traditional feature extraction method uses the simple histogram on only two different colors in a binary image. These features can be used to calculate, and only two bars will be plotted in a graph. A binary image size $m \times n$ ($n \in N$, where m) pixels (or points or status) m rows and n columns, consisting of a rectangular array of elements is called two possible values (black and white) can also move a binary matrix where 1s/0s stand black/matrix Pixel size $m \times n$ can be represented by, respectively [15]. $I_1, I_2, \ldots I_k (k \geq 1)$ components, can be divided in any way set a binary image exclusively.

An image histogram is a graphical representation of the distribution of the pixel data in an image (Fig. 4). The probability distribution of a continuous variable introduced by Karl Pearson estimates Image histogram density and distribution of pixel data which gives a rough estimate, and mainly density probability density function estimation of total pixels [20]. A histogram is used for the probability density of the total area of histogram is forever one (1). Marathi numerals like nine and one may be misclassified as they are having similarity in terms of no of pixel and their distribution.

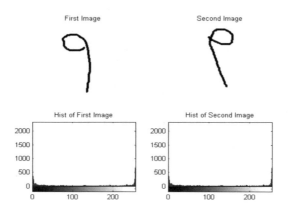

Fig. 4. Simple histogram of 1 and 9

Histogram (Fig. 4) of an image is a graphical representation of the distribution of pixel data in an image. It is an estimate of the probability distribution of a continuous variable introduced by Karl Pearson. Image Histogram gives a rough estimate about the density and distribution of the pixel data, and used mainly for density estimation of total pixels and estimating the probability density function PDF of the underlying variable [20]. The total area of a histogram used for probability density is always normalized to one (1). Using simple histogram features, the Marathi numerals like three and six can be misclassified as they are mirror image of each other having similarity in terms of no of pixel and their distribution.

$$H(j) = Sum\ pixels\ i\ from\ 0\ to\ n-1\ Image(n,n) \tag{1}$$

$$V(i) = Sum\ pixels\ j\ from\ 0\ to\ n-1\ Image(n,n) \tag{2}$$

$$D(i) = Sum\ pixels\ i\ from\ 0\ to\ n-1\ Zigzag(N) \tag{3}$$

$$D'(j) = Sum\ pixels\ i\ from\ 0\ to\ n-1\ Zigzag(N') \tag{4}$$

To overcome this limitation of simple histogram feature and to avoid misclassification, pixels in horizontal vector, vertical vector and zigzag scanned vector by combining eight pixels in a group are calculated. Statistical features of the vectors are calculated and further used for classification (Fig. 5).

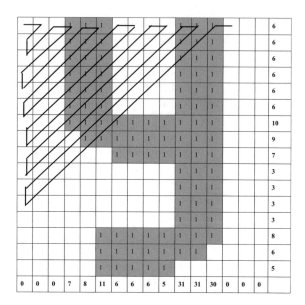

Fig. 5. Zig-zag scanned pixel level features

Horizontal H (i), vertical V(j) Zig-zag Z(k) and Zig-zag Z'(k) scanned vectors can be considered for calculating the statistical features which gives more complexity as each vector gives at least five features. To reduce the dimensionality of features, vectors can be combined and then the statistical features are calculated to reduce the overall complexity. Thus all vectors combinations can be verified to get better classification rate. Thus the concatenated vectors are used to calculate statistical features such as mean, variance, skew, means, average and absolute deviation for clustering and classification.

To calculate Mean using simple equation,

$$\bar{x} = \frac{1}{n}\sum_{i=1}^{n} x_i \quad \text{or} \quad \mu = \frac{\sum x}{N} \tag{5}$$

Calculate Median using,

$$Median = L_1 + \left(\frac{n/2 - (\sum f)l}{f_{median}}\right)c \tag{6}$$

To calculate variance σ2 and Standard deviation,

$$\sigma^2 = \frac{1}{N}\sum_{i=1}^{n}(x_i - \mu)^2 = \frac{1}{N}\sum_{i=1}^{n}x_i^2 - \mu^2 \tag{7}$$

Calculate Skew using,

$$Skew = \frac{(Mean - Median)}{\sigma} \tag{8}$$

2.5 Classification and Recognition

Use the decision tree to categorize Marathi sample numbers. A decision tree is a tree structure, where each interior node that is a test on a non leaf node attribute, each leaf node (or terminal node) holds a class label and is represented in a hierarchical level and each branch represents a result of the trial. Other non-binary tree where two or more nodes [21] can produce each inner node branches, while some decision tree algorithm only produces binary trees. The attribute value of X for which a given tuple related categories label unknown decision tree has been tested against. Decision tree classification rules can be easily changed. Starting from the root node, a path is a leaf node, which predicts that the same holds for features or class tuple direction is detected. Decision tree classifiers great accuracy as they approach the Marathi sample points used in the validation.

All horizontal, vertical Zig-zag scanned vectors obtained from the estimates and the first statistical data is analyzed. Features such as the sum of the mean, variance, skew, and a mean absolute deviation are calculated by statistical measures. These features to create or make a decision tree with various predefined terms to the query parameter are used to compare sample points. Minimum distance with features of the new database features patterns can be calculated by comparison. Decision tree leaf nodes are the classified points of the database. Ten different classes with different patterns are used for classification, decision tree by using binary or multi-ary tree might form for the final result.

3 Experimental Results

Database creation using pages of the Marathi Devanagari numbers written by school children from Zilha Parishad Yelegaon, Nanded, Maharashtra. 600 dots per inch dpi full pages with Samsung SCX3401 machine is used for scanning the documents with 600 dpi. As shown in Fig. 1. Using segmentation the numbers are separated into 6000 sample points. Segmented samples are divided into a 50% training set and 50% the points in test set.

Training of the first scanned 50% samples in a data set using decision tree classification is performed. Then set the performance and accuracy of the test decision tree is used to evaluate. Classification and misclassification rate percentage of tree shown in tabular format is calculated for the decision. Overall classification rates 94.12% is observed on the test set. If the scanned documents are not having proper inclination the system performance will degrade (Table 1).

Table 1. Comparing results for with other methods

Sr. no	Published by	Methodology	Accuracy
1.	B. Shaw, S.K. Parui and M. Shridhar [11]	Segmentation based	84.31
2.	S. Arora and D. Bhatcharjee [22]	Two stage classification	89.12
3.	R.J. Ramteke and S.C. Mehrotra [23]	Moment based	92.28
4.	Hanmandlu and Ramana Murthy [24]	Fuzzy model based	92.65
5.	Proposed approach	Pixel distribution vectors	94.12

4 Comparison with Other DCR Systems

The results of various feature extraction methods are compared with the proposed approach. The work done by different researchers based on different methods in past is compared. Accuracy of the recognition system Using fuzzy model [24] the results of classification of Devanagari characters are better as compared to chain code HMM [12] or two stage classification [22]. It is observed from the results that proposed approach gives better results as compared to the above said methods.

5 Conclusion

In the proposed approach, the various feature of the numerical calculation of binary image. The combination of two or more vectors is used to reduce feature space. The features that received better decision ratios are used to develop tree to reduce the amount of time complexity of the system. The numbers of attributes in this approach are the type that leads to less than fast processing. Numeric taxonomy using the sub images in a hierarchical manner by different sections split points. 94.12% for an overall experiments point's classification rate is observed which was held on the scanned pages. In addition, support vector machine classification rates, genetic algorithm techniques may be used to improve the performance. Other feature extraction methods can be applied for better accuracy. This validation process is very important role as better segmentation methods can be employed. In future the system can be extended to Devanagari connected character recognition.

References

1. Roy, K., Vaidya, S., Pal, U., Chaudhuri, B.B., Belaid, A.: A system for indian postal automation. In: Proceeding of 8th International Conference Document Analysis and Recognition, Seoul, Korea, pp. 1060–1064 (2005)
2. Pal, U., Chaudhari, B.B.: Indian script character recognition: a survey. J. Pattern Recognit. **37**, 1887–1899 (2004)
3. Das, N., Reddy, J.M., Sarkar, R., Basu, S., Kundu, M., Nasipuri, M., Basu, D.K.: A statistical–topological feature combination for recognition of handwritten numerals. J. Appl. Soft Comput. **12**, 2486–2495 (2012)
4. Premaratne, H.L., Jarpe, E., Bigun, J., Parui, S.K.: Lexicon and hidden Markov model-based optimisation of the recognised Sinhala script. J. Pattern Recognit. Lett. **27**, 696–705 (2006)

5. Jayadevan, R., Kolhe, S.R., Patil, P.M., Pal, U.: Offline recognition of Devanagari script: a survey. IEEE Trans. Syst. Man Cybern. C Appl. Rev. **41**(6), 782–796 (2011)
6. Vaidya, M.V., Joshi, Y.V.: Marathi numeral recognition using statistical distribution features. In: IEEE Conference on Information Processing, pp. 586–591 (2015)
7. Plamondon, R., Srihari, S.N.: On-line and off-line hand-writing recognition: a comprehensive survey. IEEE Trans. Pattern Anal. Mach. Intell. **22**(1), 63–84 (2000)
8. Bhattacharya, U., Chaudhuri, B.B.: Handwritten numeral databases of indian scripts and multistage recognition of mixed numerals. IEEE Trans. Pattern Recognit. Mach. Intell. **31**(3), 444–457 (2009)
9. Garain, U., Chaudhuri, B.B.: Segmentation of touching characters in printed Devnagari and Bangla scripts using fuzzy multifactorial analysis. IEEE Trans. Syst. Man Cybern. C Appl Rev. **32**(4), 449–459 (2002)
10. Parui, S.K., Shaw, B.: Offline handwritten Devanagari word recognition: an HMM based approach. In: Proceedings of 2nd International Conference on Pattern Recognition and Machine Intelligence (PReMI), Kolkata, India. LNCS, vol. 4815, pp. 528–535 (2007)
11. Shaw, B., Parui, S.K., Shridhar, M.: Offline handwritten Devanagari word recognition: a segmentation based approach. In: Proceedings of International Conference on Pattern Recognition (ICPR), Tampa, Florida, USA, pp. 1–4 (2008)
12. Lu, Z., Chi, Z., Siu, W.-C.: Extraction and optimization of B-spline PBD templates for recognition of connected handwritten digit strings. IEEE Trans. Pattern Anal. Mach. Intell. **24**(1), 132–139 (2002)
13. Buse, R., Liu, Z.-Q., Bezdek, J.: Word recognition using fuzzy logic. IEEE Trans. Fuzzy Syst. **10**(1), 65–76 (2002)
14. Park, J.: An adaptive approach to offline handwritten word recognition. IEEE Trans. Pattern Anal. Mach. Intell. **24**(7), 920–931 (2002)
15. Young Oliveira, L.S., Sabourin, R., Bortolozzi, F., Suen, C.Y.: Automatic recognition of handwritten numerical strings: a recognition and verification strategy. IEEE Trans. Pattern Anal. Mach. Intell. **24**(11), 1438–1454 (2002)
16. Tan, Y.C.L., Huang, W., Yu, Z., Xu, Y.: Imaged document text retrieval without OCR. IEEE Trans. Pattern Anal. Mach. Intell. **24**(6), 838–844 (2002)
17. Young, C.M., Fugate, M., Hush, D.R., Scovel, C.: Selecting a restoration technique to minimize OCR error. IEEE Trans. Neural Netw. **14**(3), 478–490 (2003)
18. Young Shi, D., Gunn, S.R., Damper, R.I.: Handwritten Chinese radical recognition using nonlinear active shape models. IEEE Trans. Pattern Anal. Mach. Intell. **25**(2), 277–280 (2003)
19. Gonzalez, R.C., Woods, R.E.: Digital Image Processing, 2nd edn. Prentice Hall, Upper Saddle River (2002)
20. Golomb, S.W.: Polyominoes. Charles Scriber's Sons, New York (1965)
21. Han, J., Kamber, M.: Data Mining Concept and Techniques, 2nd edn. Morgan Kaufmann Publishers, Burlington (2006)
22. Arora, S., Bhatcharjee, D., Nasipuri, M., Malik, L.: A two stage classification approach for handwritten Devanagari characters. In: Proceedings of International Conference Computer Intelligence Multimedia Application, pp. 399–403 (2007)
23. Ramteke, R.J., Mehrotra, S.C.: Feature extraction based on moment invariants for handwriting recognition. In: Proceedings of International IEEE Conference of Cybernetics Intelligent System (CIS 2006), Bangkok, pp. 1–6 (2006)
24. Hanmandalu, M., Murthy, O.V.R., Madasu, V.K.: Fuzzy model-based recognition of handwritten Hindi numerals. Pattern Recogn. **40**(6), 1840–1854 (2006)
25. Quinlan, J.R.: Induction of decision trees. Mach. Learn. **1**, 81–106 (1986)

Imbalanced Data Stream Classification: Analysis and Solution

Koringa Anjana$^{(\boxtimes)}$, Kotecha Radhika, and Patel Darshana

V.V.P. Engineering College, Rajkot, India
Koringa.anjana6@gmail.com,
kotecha.radhika7@gmail.com,
darshana.h.patel@gmail.com

Abstract. Through the progress in each hardware and software system technologies, automatic data creation and storage have become quicker than ever. Such data is called as a data stream. Streaming information is present everywhere and it's usually a difficult problem to visualize, collect and examine such huge volumes of information. Data stream mining has become a unique experimental area in information finding because of the large size and rapid speed of data in the data stream, due to this reason conventional classification methods are not effective. In today`s a substantial amount of analysis has been done on this issue whose main aim is to efficiently solve the difficulty of information stream mining with concept drift. Class imbalance is one of the problems of machine learning and data processing fields. Imbalance data sets reduce the performance as well as the overall accuracy of data mining methods. Decision making towards the majority class, which lead to misclassifying the minority class examples or moreover considered them as noise.

Keywords: Data stream · Ensemble method · Hoeffding tree · K-Nearest Neighbor · SMOTE · MSMOTE

1 Introduction

Data stream mining has been appealing considerable attention in today. In data mining, factual induction and machine learning, classification [1] issues have been observed altogether as a chief category of the data analysis tasks. Class imbalance is a most recent topic for research in machine learning and data mining. Class imbalance occurs when the examples of one or more classes are far less than the examples of other classes.

Classification is a type of supervised learning approach, in which estimation of dependent variable's set is based on some other bunch of input attributes [2]. The classification process is made up of two steps: model building and model testing. The various methods used for the classification problem are neural networks, decision trees, rule-based methods, etc. From stationary datasets, various classification models are constructed by plenty of approaches, in which feasibility is based on numerous passes over the stored data. In the case of the data stream, it is not feasible because data streams have infinite length. So it is required that the whole data should be processed in

© Springer International Publishing AG 2018
S.C. Satapathy and A. Joshi (eds.), *Information and Communication Technology for Intelligent Systems (ICTIS 2017) - Volume 2*, Smart Innovation, Systems and Technologies 84, DOI 10.1007/978-3-319-63645-0_35

a single pass. Moreover, considering the perspective of concept drift, the classification issue requires to reconstructed.

Data stream can be conceived as a continuous and changing sequence of data that continuously arrive at a system to store or process [1]. There are various applications of data stream like network traffic, financial fraud detection, telecommunication calling records, remote surveillance system, share market data, remote sensor etc. Also, there are some issues of data stream classification like novel class detection [4], handling concept drifting efficiently [5], scarcity of labeled data [6, 13]
, privacy preserving data stream mining [7], handling data stream with imbalanced classes [3]. Restrictive processing time, limited memory, and one scan of incoming examples are the main threats of data stream classification.

In imbalance data stream, the number of examples of one or more classes is far more than the example of other classes. Imbalance data stream is often utilized in varied real time applications like management of a network traffic, financial fraud detection, credit card transactions etc. In several circumstances, the examples of the class which is in marginal are of more significance. In credit card fraud detection there are several transactions going at a time. Out of them most of the transactions are legitimate only some of the transactions are fraud, so that from millions of transactions find the minority class means fraud transactions may be a troublesome task. If the proportion of legitimate and fraud transaction are same then it's simple to handle, however during this case it's equal to 1:100 therefore from such unbalanced data stream it will be hard to find that fraud transaction. Most of the classifiers can target majority class and it'll take minority class as a noise, but that minority classes are more useful because it contains some valuable information.

2   Related Work

Data streams are vast so that it needs extra space and training time, because of that reasons previous multiphase data processes aren't suitable for classification. Concept-drift happens within the streams once the fundamental idea of the information varies with time. Therefore, the classification model should be restructured endlessly so that it returns the foremost new idea. Yet, one more foremost difficulty is neglected by almost all unconventional data stream classification methods is that the arrival of a new class which is called as "concept-evolution". "The total amount of classes within the data stream is stationary", this standard is followed by most of the present solutions. However, in real-world data stream like intrusion detection, text classification, and fault detection streams etc. have a variable number of classes.

In this section, we present various existing methods of data streams classification.

2.1   Data Stream Classification Methods

In [2, 8], ensemble-based classification approach is used. Data stream have an infinite length so that problem is solved by separating the stream into same sized lumps. Every lump can be put in memory and processed on-line. Every lump is utilized to train single classification model as all the examples within the current lumps are labeled. An unlabeled example is categorized by picking majority vote amongst the classifiers within the ensemble. When a replacement model is prepared, one of the present model within the ensemble is substituted by it, if required.

Bagging is one of the most effective methods in ensemble classification technique. In that, a new dataset is created to train every classifier by arbitrarily selecting examples from the primary dataset. Ultimately, when an unfamiliar instance is given towards every distinct classifier, a weighted or majority vote is used to assume the class.

AdaBoost [16] is the primary illustrative methodology of Boosting. After every iteration, it provides additional attention to tough examples, with the objective of properly classifying samples within the subsequent iteration that were falsely classified throughout this iteration. Therefore, after every round, the weights of falsely classified examples are raised; on the other hand, the weights of properly classified examples are attenuated.

Very Fast Decision Trees (VFDT) [2, 11, 12, 22] is a decision tree knowledge framework created from Hoeffding trees. The tree is divided by utilizing the present finest attribute pleasing into attention that the numerous instances utilized fulfill the Hoeffding bound. There are a few issues of this strategy like ties of properties, limited memory, effectiveness and exactness and so on. The Hoeffding tree creates a decision tree from a data stream by quickly surveying every case in the stream just once, without the need of putting away cases after that they have been utilized to overhaul the tree. Its name is gotten from the Hoeffding bound which is utilized to select a number of examples require to attain the firm level of certainty. Consider, r = real value arbitrary variable with range R, n = number of self-determining observations have been made, r̄ = mean value obtained after n self-determining observations.

$$\varepsilon = \sqrt{\frac{R^2 \ln(1/\delta)}{2n}} \tag{1}$$

Give G a chance to be the heuristic measure used to pick the part quality at a hub. On the off chance that G is data pick up. Subsequent to watching n cases and expecting G is to be augmented, given X_a be the attribute with maximum G, X_b is the attribute with second highest G and $\Delta G = G(X_a) - G(X_b)$ be the distinction between the two qualities. After watching n samples at any hub, if $\Delta G > \varepsilon$, the Hoeffding bound assurances with a likelihood $1 - \delta$ that the genuine $\Delta G \geq 0$ and X_a is in reality the best split attribute at that node.

In [9], Naïve Bayesian classification approach is used. They perform classic Bayesian prediction whereas creating a naive assumption that each one input is independent. Given n totally different classes, the trained Naive Bayes classifier predicts for each unlabeled instance. Bayesian classifiers use a probabilistic approach to deal with classification in light of Bayesian probability standards However, NB makes robust (Naive) assumptions concerning the conditional independence of the options wherever the presence (or absence) of a feature is assumed to be fully unrelated to the presence (or absence) of another feature. Bayesian belief networks address this issue of conditional independence by representing dependencies between options as a directed graph.

Rule-based classifier prompts rules of the shape. IF condition THEN plans that are expressive, littler and effectively address information for gathering. The condition is a conjunction of regular terms of the edge (Attribute = Value) for hard and fast characteristics and (Attribute < Value) or (Attribute $\geq$ Value) for steady data. There are various rule-based classifiers for the data stream and SCALLOP [2] is one of them. In that, when a new record arrives at that time 3 cases are possible either it is positive covering or negative covering or possible expansion. The rules that have less than the minimum optimistic support are dismissed. Moreover, the rules which are not covered by any example of user-defined are discarded. Finally, the vote is taken to assign data instance to unlabeled records.

The on-demand classification [2, 10] technique splits the classification method into 2 parts. First part constantly caches brief summary concerning the data streams and the other part endlessly utilize the brief statistics to do the classification. Micro-clusters define the summary statistics are characterized by the class label type. Every micro-cluster has selected class label that describes the class label of the examples in it. In that, each part of the method is employed in on-line style so that the method is called as an on-demand classification technique.

2.2 Imbalanced Data Management Methods

The random under and oversampling [14, 15, 21] techniques have their numerous limitations. The random undersampling technique will probably eliminate definite essential examples, and random oversampling will result in overfitting

There are 3 strategies [14, 15, 21]: resampling, recognition-based induction scheme and under-sampling. There are two sampling ways for each over and under sampling. Random resampling comprised of oversampling the smaller category arbitrarily till it comprised of as several examples as the majority category and "focused resampling" comprised of oversampling only those minority examples that occurred on the boundary between the minority and majority categories.

SMOTE [16, 17, 20] main goal is to form new marginal class samples by introducing numerous marginal class examples that reside along. SMOTE generate examples by arbitrarily choosing any k nearest neighbor of a minority class example and therefore the creation of the new example values from an arbitrary interpolation of each instance. Thus, the drawback of oversampling is avoided.

With respect to synthetic sampling [15], the synthetic minority oversampling technique (SMOTE) could be a powerful technique that has shown an excellent deal of success in varied applications [15]. The SMOTE algorithm creates artificial information based on the feature area similarities between existing minority examples.

In [18, 19], MSMOTE approach is used which is an upgraded version of SMOTE. Algorithm categorizes the examples of the marginal class into 3 types, latent noise, safe and border examples by the calculation of the distances between all examples. Once MSMOTE creates new examples, the approach to pick out the closest neighbors is modified with regard to SMOTE that depends on the cluster formerly appointed to the example.

2.3   Imbalanced Data Classification Methods

K-nearest neighbor (KNN) [15] classifier is used to attain under-sampling based on the features of the known information distribution. Four KNN under-sampling ways are there that are, namely, NearMiss-1, NearMiss-2, Near- Miss-3 and "most distant" technique. The NearMiss-1 technique picks those majority instances whose mean distance to the 3 closest minority category samples is minimum, whereas the NearMiss-2 technique picks the majority category instances whose mean distance to the 3 farthest minority category examples is minimum. NearMiss-3 picks a specified range of the closest majority examples for every minority instances to ensure that each minority instance is enclosed by some majority instances. Finally, the "most distance" technique picks the bulk category instances whose mean distance to the 3 nearest marginal category examples is that the largest. Experimental results recommended that the NearMiss-2 technique will offer competitive results for imbalanced learning. If the user has chosen K = 1 then it will catch one adjacent neighbor if k = 8 then it will catch eight closest neighbors. If the user needs to pick suitable value for their data set, they have to start from 1 and if the error rate is more than k should be incremented by two and this procedure will be keep going till get the good results.

3   Implementation

We are using 4 different data streams to compare the performance of various classifiers. From that KDDCup99 and Forest Covered Type are available in UCI data Repository and other two are synthetically generated using MOA. Table 1 lists out the data streams used along with their details. Accuracy is a ratio of number of correct prediction to the total of cases to be predicted. Table 2 give the information about the accuracy of various classifiers.

Table 1. Composition of data stream

Data stream	No. of attributes	No. of instances	No. of classes
KDDCup99	42	4,000,000	20
Forest covered Type	54	581,012	7
LED	25	1,000,000	10
Synthetic	10	1,000,000	2

Table 2. Accuracy of classifiers (in %)

Data stream	Naive Bayes	Rule based	KNN	VFDT
KDDCup99	95.809	85.995	84.65	95.699
Forest covered type	59.635	50.356	52.87	61.528
LED	74.066	76.832	66.98	74.111
Synthetic	65.831	63.856	68.75	91.715

Table 3 illustrates time elapsed during execution of this methods.

Table 3. Running time of classifiers (in seconds)

Data stream	Naive Bayes	Rule based	KNN	VFDT
KDDCup99	28.48	33.80	40.95	13.25
Forest covered type	20.31	56.06	50.78	19.19
LED	9.31	15.00	17.42	11.17
Synthetic	31.08	36.65	29.42	28.95

From the result, we can analyze that Hoeffding tree is better than any other classification method in terms of accuracy. If we see the elapsed time, then VFDT also require less time for learning and evaluating the model compare to other method but its accuracies are relatively very less.

4 Proposed Framework

Most real-time applications like credit card transaction, medical diagnosis, telecommunication, surveillance etc. which contain imbalanced data. These data streams are given as input. Mainly there are two types of classes in the data stream. First is majority classes and the second one is minority classes. Classes which have very less number of examples compared to other classes are called minority classes and others are called majority classes. If the data stream is not discretized than first, we discretize data stream. There are various methods available for discretization like histogram, equal-frequency, equal- width and binning. We use binning method for discretization. Once the data stream is discretizing then check the type of class whether it is of a minority class or majority class. After that based on the type of samples, next process will be followed. If the example is of a minority class, then set sampling probability Ps based on the Euclidean distance between samples. If the distance is less, then set probability Ps high because there is more chance to misclassify examples. If distance is more then set sampling probability Ps low because there is less chance to misclassify examples.

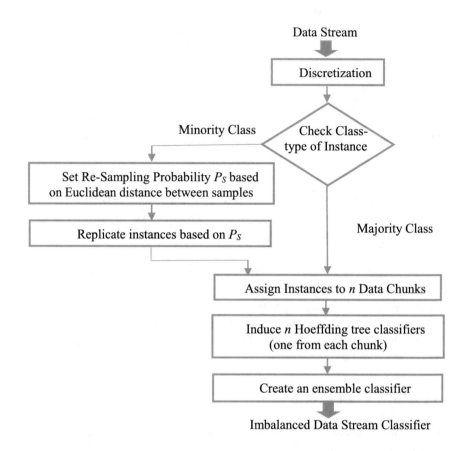

Following equation is proposed to set probability based on Euclidean distance.

$$p_s = 1 - \frac{\sum_{i=1}^{|D|} \left(\sum_{j=1}^{|A|} \varepsilon_{ij} \right)}{|D|} \tag{2}$$

$$\text{Where } \varepsilon_{ij} = \begin{cases} 1 & T_{Aj} \neq D_{Aj} \\ 0 & otherwise \end{cases}$$

In that equation D is training dataset and T is the dataset which we trained using training dataset. $|D|$ is total no of tuples in training dataset while $|A|$ is the total no of attributes in Dataset which we trained. If the training dataset attribute value and an attribute value of dataset which we train are not same then set distance as 1 otherwise set distance as 0. After that based on the probability, we replicate instances if the instance has less probability than it replicates less number of times than instance have more probability. After that assign instances to n chunks. After chunking, n Hoeffding tree classifiers are induced one from each data chunk, as there are many methods available for classification but we use VFDT because it gives better result in case of data stream. Afterward, Ensemble bagging method is used for creating ensemble classifier by using Hoeffding tree classifier which takes the vote from multiple classifiers and then combines output, which results in imbalanced data stream classifier.

5 Conclusion

We have reviewed various problem in data stream classification. From that emerging issue of data stream classification like skewed distribution of classes are discussed in this paper. Several approaches in the literature survey have been summarized with their advantages and disadvantages. We compare various existing methods of data stream classification like Ensemble method, Hoeffding tree, Naïve Bayesian classification and rule-based classification using various data stream. From that, we conclude that Hoeffding tree gives better accuracy than another classifier. We proposed an approach for imbalanced data stream classification. Our future work is to implement this approach and compare the performance of this approach with various other existing methods.

References

1. Mahnoosh, K., Mohammadreza, K.: An analytical framework for data stream mining techniques based on challenges and requirements. Int. J. Eng. Sci. Technol. (IJEST) **3**, 0975–5462 (2011). Mahnoosh, K. et al.
2. Dariusz, B.: Master's thesis: Mining data streams with concept drift. Poznan University of Technology (2010)
3. Mohammad, M., Qing, C., Latifur, K., Charu C., Jing, G., Jiawei, H., Ashok, S., Nikunj, O.: Classification and adaptive novel class detection of feature-evolving data streams. IEEE Trans. Knowl. Data Eng. **25**(7), 1484–1497 (2003)
4. Mohamed, M., Arkady, Z., Shonali, K.: Survey of Classification Methods in Data Streams. Advances in Database Systems. Springer, Berlin (2007)
5. Abhijeet, G., Vahida, A.: Classifier ensemble for imbalanced data stream classification. In: Cube 2012 Proceedings of the Cube International Information Technology Conference. ACM (2012)
6. Abhijeet, B.G.: Classification of Data Streams With Skewed Distribution. Department of Computer Engineering and Information Technology, and Information Technology, College of Engineering, Pune (2012)
7. Mohammad, M., Jing, G., Latifur, K., Jiawei, H., Bhavani, T.: Classification and novel class detection in concept-drifting data streams under time constraints. IEEE Trans. Knowl. Data Eng. **23**(6), 859–874 (2011)
8. Dewan, F., Li, Z., Alamgir, H., Chowdhury, R., Rebecca, S., Graham, S., Keshav, D.: An adaptive ensemble classifier for mining concept-drifting data streams. In: Cube 2012 Proceedings of the Cube International Information Technology Conference. ACM (2012)
9. Charu, A., Jiawei, H., Jianyong, W., Philip, Y.: On demand classification of data streams. In: KDD 2004 Proceedings of the Tenth ACM SIGKDD International Conference on Knowledge Discovery and Data Mining. ACM (2004)

10. Charu, A., Jiawei, H., Jianyong, W., Philip, Y.: A framework for on-demand classification of evolving data streams. IEEE Trans. Knowl. Data Eng. **18**, 577–589 (2006)
11. Pedro, D., Geoff, H.: Mining high-speed data streams. In: KDD 2000 Proceedings of the Sixth ACM SIGKDD International Conference on Knowledge Discovery and Data Mining. ACM (2000)
12. Richard, K.: Improving Hoeffding Trees. University of Waikato, Hamilton (2007)
13. Mohammad, M., Clay, W., Jing, G., Latifur, K., Jiawei, H., Kevin, H., Nikunj, O.: Facing the Reality of Data Stream Classification: Coping with Scarcity of Labeled Data, vol. 33. Springer, Berlin (2012)
14. Yabo, X., Ke, W., Ada, F., Rong, S., Jian, P.: Privacy-Preserving Data Stream Classification, vol. 34. Springer, Berlin (2008)
15. Nitesh, C.: Data Mining for Imbalanced Datasets: An Overview. Data Mining and Knowledge Discovery Handbook. Springer, Berlin (2005)
16. Haibo, H., Edwardo, G.: Learning from imbalanced data. IEEE Trans. Knowl. Data Eng. **21**, 1263–1284 (2009)
17. Chawla, N.V., Bowyer, K.W., Hall, L.O., Kegelmeyer, W.P.: SMOTE: synthetic minority over-sampling technique. J. Artif. Intell. Res. **16**, 321–357 (2002)
18. Shengguo, H., Yanfeng, L., Lintao, M., Ying, H.: MSMOTE: improving classification performance when training data is imbalanced. In: IEEE Second International Workshop on Computer Science and Engineering (2009)
19. Mikel, G., Alberto, F., Edurne, B., Humberto, B.: A review on ensembles for the class imbalance problem: bagging-, boosting-, and hybrid-based approaches. IEEE Trans. Syst. Man Cybernet. Part C (Appl. Rev.) **42**(4), 463–484 (2011)
20. Nitesh, V. C., Aleksandar, L., Kevin, B., Lawrence, H.: SMOTEBoost: improving prediction of the minority class in boosting. In: 7th European Conference on Principles and Practice of Knowledge Discovery in Databases (PKDD), Dubrovnik, Croatia, vol. 107–119. Springer (2003)
21. Xin, Y., Peter, T., Ata, K.: Class Imbalance Learning (2008)
22. Yin, C., Lu, F., Luyu, M.: An improved Hoeffding-ID data stream classification algorithm. J. Supercomput. **1**, 1–12 (2015)

Stabilizing Rough Sets Based Clustering Algorithms Using Firefly Algorithm over Image Datasets

Abhay Jain[1], Srujan Chinta[2], and B.K. Tripathy[2(✉)]

[1] School of Information Technology and Engineering, VIT University,
Vellore 632014, Tamil Nadu, India
abhay.jain2014@vit.ac.in
[2] School of Computing Science and Engineering, VIT University,
Vellore 632014, Tamil Nadu, India
{chintasai.srujan2014, tripathybk}@vit.ac.in

Abstract. Rough Intuitionistic Fuzzy C-Means Algorithm is a combination of Fuzzy Sets, Rough Sets and Intuitionistic Fuzzy Sets. This algorithm provides high quality clustering over numeric datasets. However, RIFCM is highly inconsistent over Image datasets. In this paper, we combine RIFCM with Firefly Algorithm. Firefly algorithm is a meta-heuristic bio-inspired algorithm which mimics the behavior of fireflies. Our experimental results prove that using Firefly algorithm before RIFCM lends stability to the clustering output and considerably reduces the number of iterations required for convergence.

Keywords: Data clustering · Rough set · Firefly algorithm · DB-index · D-index

1 Introduction

Image Segmentation is a process of partitioning an image space and Data Clustering is one of the methods of implementing it. Bezdek et al. [2] introduced Fuzzy C-Means algorithm in 1984. In FCM, every pixel of the image belongs to every cluster with a certain probability. The concept of Rough Sets [8] was first used in data clustering by Lingras et al. [5] in 2004. They proposed the Rough K-Means Algorithm which classifies pixels into the lower or upper approximations of each cluster. In 2006, Mitra et al. [7] combined the concepts of Rough sets and Fuzzy sets [11] to develop Rough Fuzzy C-Means algorithm. Experimental results proved that RFCM produced better results when compared to FCM and HCM (Hard C-Means). Maji et al. [6] added a possibilistic flavor to RFCM in their algorithm RFPCM. Atanassov [1] introduced Intuitionistic Fuzzy Sets in 1986. Chaira et al. [4] created IFCM in 2011 wherein there is a new parameter called the hesitation degree associated with the data clustering. In this paper, the interpretation of fuzzy entropy by Yager et al. [9] is used. In 2013, Tripathy et al. [3] combined the concepts of Fuzzy Sets, Rough Sets and Intuitionistic Fuzzy Sets to develop RIFCM. In this paper, the authors have clearly published results pertaining to Numeric data but were silent about Image datasets. The experimental

© Springer International Publishing AG 2018
S.C. Satapathy and A. Joshi (eds.), *Information and Communication Technology for Intelligent Systems (ICTIS 2017) - Volume 2*, Smart Innovation, Systems and Technologies 84, DOI 10.1007/978-3-319-63645-0_36

results clearly establish the fact that RIFCM is heavily inconsistent when applied to Image datasets. In this paper, we employ Firefly Algorithm [10] to overcome these limitations of RFCM and RIFCM. The rest of the paper is structured as follows: Sect. 2 contains information about Firefly Algorithm; A brief overview of RIFCM is given in Sect. 3; Our methodology is explained in Sect. 4; Sect. 5 contains the Experimental Results and Observations; Sect. 6 contains the conclusion.

2   Firefly Algorithm

Firefly Algorithm was invented by Yang [10] in 2010. This algorithm mimics the behavior of fireflies. Each firefly is attracted to a brighter firefly and the brightest firefly will move randomly. The degree of attractiveness depends on the brightness of the firefly and the distance between the two fireflies. The attractiveness function β is determined by the formula:

$$\beta(r_{i,j}) = \beta_0 e^{-\gamma r_{i,j}^2} \tag{1}$$

Here, β_0 is the default value of attractiveness, γ is the light absorption co-efficient, $r_{i,j}$ is the Euclidian distance between the two fireflies. The equation that governs the movement of a firefly i to a brighter firefly j is:-

$$x_i = x_i + \beta_0 e^{-\gamma r_{i,j}^2}(x_i - x_j) + \alpha\left(rand - \frac{1}{2}\right) \tag{2}$$

The third term in the above equation lends randomness to the fireflies' movement. The major advantage of Firefly Algorithm is that it does not get stuck at the local optima. Another advantage of using Firefly algorithm is that it quickens the convergence rate.

3   Rough Sets Based Clustering Algorithms

A rough set, first described by a Polish computer scientist Zdzisław I. Pawlak [8], is a formal approximation of a crisp set i.e., conventional set in terms of a pair of sets which give the lower and the upper approximation of the original set. In the standard version of rough set theory Pawlak 1991, the lower- and upper-approximation sets are crisp sets, but in other variations, the approximating sets may be fuzzy sets. The lower approximation is a complete set of objects which definitely belong to the solution set. The upper approximation includes all objects which could possibly belong to the solution set. Those objects which are present in the upper approximation but not present in the lower approximation constitute the Boundary region. The first Rough Sets based clustering algorithm is Rough K-Means [5]. In this algorithm, a particular pixel of the image is said to belong to the lower approximation of a cluster if the difference between the distances of the pixel and its two closest centroids is lesser than a threshold value. If the difference is greater than the threshold, then the pixel belongs to the upper approximations of both the clusters. Moreover, every pixel that belongs to the lower approximation also belongs to the upper approximation of that cluster. The cluster centroids are updated after every iteration according to the following formula:

$$
v_i = \begin{cases} w_{low} \dfrac{\sum_{x_k \in \underline{B}U_i} x_k}{|\underline{B}U_i|} + w_{up} \dfrac{\sum_{x_k \in \overline{B}U_i - \underline{B}U_i} x_k}{|\overline{B}U_i - \underline{B}U_i|} & if\ |\underline{B}U_i| \neq \emptyset\ and\ |\overline{B}U_i - \underline{B}U_i| \neq \emptyset \\[4mm] \dfrac{\sum_{x_k \in \overline{B}U_i - \underline{B}U_i} x_k}{|\overline{B}U_i - \underline{B}U_i|} & if\ |\underline{B}U_i| = \emptyset\ and\ |\overline{B}U_i - \underline{B}U_i| \neq \emptyset \\[4mm] \dfrac{\sum_{x_k \in \underline{B}U_i} x_k}{|\underline{B}U_i|} & ELSE \end{cases} \tag{3}
$$

In RFCM [4], the membership matrix is initially calculated using the formula:-

$$
\mu_{ik} = \frac{1}{\sum_{j=1}^{c} \left(\dfrac{d_{ik}}{d_{jk}}\right)^{\frac{2}{m-1}}} \tag{4}
$$

Here, d_{ik} and d_{jk} are the distances of the pixels from the centroids. The difference between the maximum and next to maximum membership values is computed and the rest of the algorithm is the same as RCM. Chaira [4] used the concept of Intuitionistic Fuzzy Sets in her paper. The major difference is that a hesitation degree is added to the membership matrix. This hesitation degree is computed as follows:-

$$
\pi_A(x) = 1 - \mu_A(x)(1 - \mu_A(x)^\alpha)^{\frac{1}{\alpha}} \tag{5}
$$

Tripathy et al. [3] imbibed this Intuitionistic flavor into RFCM and produced RIFCM.

4 Methodology

In this paper, Firefly Algorithm has been used to provide the near-optimum cluster centroids and membership matrix for Rough sets based clustering algorithms. Each firefly is associated with a random value of a cluster centroid. The objective function to be minimized is framed in such a way that the intensity of the brightness of each firefly increases as the cluster centers move towards the global optimum. The cluster centroids and the corresponding membership matrix of the best firefly after convergence are passed to the clustering algorithm. Thus, firefly algorithm is fused with both RFCM and RIFCM. Our experimental data proves that both RFCM and RIFCM are heavily susceptible to the random initialization of the cluster centroids. Thus, Firefly algorithm is used to nullify this vulnerability by initiating both the algorithms with near-optimal cluster centroids and Fuzzy Membership matrix.

5 Results

In order to mathematically compare the clustering quality of the algorithms, two indices have been used: DB index and Dunn index.

$$DB = \frac{1}{c}\sum_{i=1}^{c} \max_{k\neq i}\left\{\frac{s(v_i) + s(v_k)}{d(v_i, v_k)}\right\} \text{ for } 1 < i, k < c \qquad (6)$$

$$Dunn = \min_i\left\{\min_{k\neq i}\left\{\frac{d(v_i, v_k)}{\max_l s(v_l)}\right\}\right\} \text{ for } 1 < k, i, l < c \qquad (7)$$

Lower the value of DB index and higher the value of D index, the better the clustering quality. RFCM, RFCMFA, RIFCM and RIFCMFA have been tested on three images: Brain MRI, Rice Copy and Satellite images. Each of these four algorithms was run 50 times for each image and the results have been presented in the form of graphs.

For all the images, two cluster centers were considered. The X-axis of each graph represents the iterations and the Y-axis represents the DB/Dunn index values. The maximum and minimum index values are explicitly mentioned for each image and the performance of each algorithm is discussed and visualized.

5.1 Brain MRI

From the graphs, it is evident that Firefly is stabilizing the performance of both RFCM and RIFCM. The average DB value of RFCM is 1.6, RIFCM is 1.4, RFCMFA is 0.93 and RIFCMFA is 0.91. The Dunn values are also very consistent for RFCMFA and RIFCMFA. Both the hybrid algorithms converged within 7 iterations whereas RIFCM and RFCM took 15 iterations on an average. Hence, it can be clearly observed that Firefly algorithm has eliminated the randomness associated with RFCM and RIFCM (Figs. 1, 2 and 3).

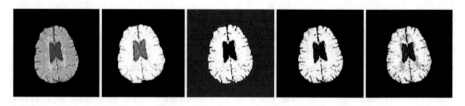

Fig. 1. The first image is the input image and the following segmented images are the outputs after applying RFCM, RFCMFA, RIFCM, and RIFCMFA respectively.

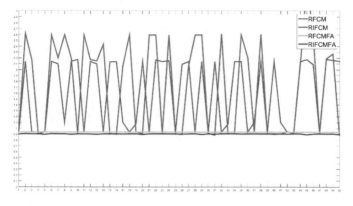

Fig. 2. DB values for Brain MRI image

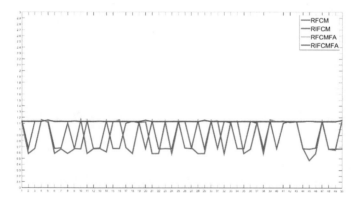

Fig. 3. Dunn values for Brain MRI Image

5.2   Rice Image

The best clustering is achieved by RIFCM. However, its DB values range from 0.5 to 3.6. On the other hand, RIFCMFA's DB values range between 0.9 to 1.0. On this image, RFCM is also stable but RFCMFA results in a nearly straight line. The average number of iterations for the hybrid algorithms is 9. RFCM and RIFCM took 17 iterations on an average to converge (Figs. 4, 5 and 6).

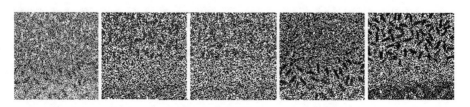

Fig. 4. The first image is the input image and the following segmented images are the outputs after applying RFCM, RFCMFA, RIFCM, and RIFCMFA respectively.

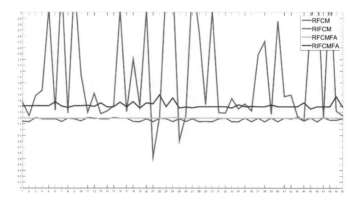

Fig. 5. DB values for Rice Image

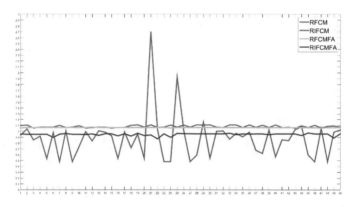

Fig. 6. Dunn values for Rice Image

5.3 Satellite Image

Depending upon the initial random cluster centers, the DB and Dunn values of RIFCM and RFCM are either very good or very bad. The average number of iterations for RFCM and RIFCM are 34. For RIFCMFA and RFCMFA, the average number of iterations is 19 (Figs. 7, 8 and 9).

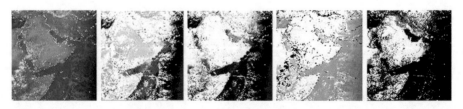

Fig. 7. The first image is the input image and the following segmented images are the outputs after applying RFCM, RFCMFA, RIFCM, and RIFCMFA respectively.

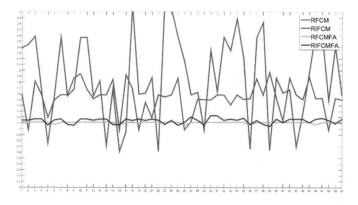

Fig. 8. DB values for Satellite Image

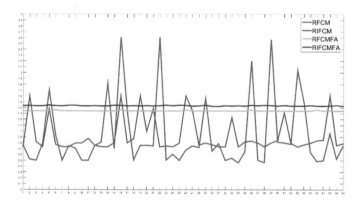

Fig. 9. Dunn values for Satellite Image

6 Conclusion

In this paper, we propose a novel algorithm to stabilize the two rough sets based clustering algorithms: RFCM and RIFCM. Our experimental results prove that using Firefly algorithm to provide the near optimal cluster centroids and fuzzy membership matrix eliminates the susceptibility of RFCM and RIFCM to random initialization of data. This results in a stable and reliable clustering output. Moreover, the number of iterations required for convergence has also reduced considerably. Our future works include analyzing image characteristics such as the optimal number of clusters and trying to use Firefly algorithm for finding the appropriate number of reducts.

References

1. Atanassov, K.T.: Intuitionistic fuzzy sets. Fuzzy Sets Syst. **20**(1), 87–96 (1986)
2. Bezdek, J.C., Ehrlich, R., Full, W.: FCM: The fuzzy C-means clustering algorithm. Comput. Geosci. **10**(2–3), 191–203 (1984)
3. Bhargava, R., Tripathy, B.K., Tripathy, A., Dhull, R., Verma, E., Swarnalatha, P.: Rough intuitionistic fuzzy C-means algorithm and a comparative analysis. In: Proceedings of the 6th ACM India Computing Convention, p. 23. ACM, August 2013
4. Chaira, T.: A novel intuitionistic fuzzy C means clustering algorithm and its application to medical images. Appl. Soft Comput. **11**(2), 1711–1717 (2011)
5. Lingras, P., West, C.: Interval set clustering of web users with rough k-means. J. Intell. Inf. Syst. **23**(1), 5–16 (2004)
6. Maji, P., Pal, S.K.: Rough set based generalized fuzzy C-means algorithm and quantitative indices. IEEE Trans. Syst. Man Cybernet. B (Cybernet.) **37**(6), 1529–1540 (2007)
7. Mitra, S., Banka, H., Pedrycz, W.: Rough–fuzzy collaborative clustering. IEEE Trans. Syst. Man Cybernet. B (Cybernet.) **36**(4), 795–805 (2006)
8. Pawlak, Z.: Rough sets. Int. J. Parallel Prog. **11**(5), 341–356 (1982)
9. Yager, R.R.: Some aspects of intuitionistic fuzzy sets. Fuzzy Optim. Decis. Mak. **8**(1), 67–90 (2009)
10. Yang, X.S.: Firefly algorithms for multimodal optimization. In: International Symposium on Stochastic Algorithms, pp. 169–178. Springer, Berlin, October 2009
11. Zadeh, L.A.: Fuzzy sets. Inf. Control **8**(3), 338–353 (1965)

Analyzing the Stemming Paradigm

Rupam Gupta[1(✉)] and Anjali G. Jivani[2]

[1] Sardar Vallabhbhai Patel Institute of Technology, Vasad, India
shibi_sengupta@yahoo.com
[2] The Maharaja Sayajjirao University of Baroda, Vadodara, India
anjali.jivani-cse@msubaroda.ac.in

Abstract. This paper discusses affix removal and statistical based Stemming algorithms in detail with stemmer-generated output from some Standard English text and dictionary. Comparative empirical studies of all these stemmers are also discussed here with respect to number of stem token generation from single root morphed word variants and computation time. First part of the paper deals with introductory discussion of stemming and lemmatization. Second part of the paper focuses on algorithms of affix and statistical based stemmers with their empirical output. Last part describes the steps of the comparative tool for the same. Finally conclusion section wraps up whole discussion about stemming. This paper can assist researchers working in the field of text mining.

Keywords: Stemming · Lemmatizing · Text mining · Index Compression Factor (ICF)

1 Introduction

Stemming algorithms are applied as a prerequisite task of any text mining application or natural language text analysis for extracting root words or stemmed token from different semantically similar morphed words. Morphological variants of a single root word carry similar semantic meaning in a particular text. So with the help of stemming algorithm, single token is generated from different morphological variants of a word, derived from a single root word, present within a text. This will help to extract and to retrieve core information from any un-structured and semi-structured text.

Modified version of stemming technique is called *lemmatizing* which mostly applies morphological finite state transducer method to generate dictionary root word from different morphological variants.

W.B Frake categorized the stemming process into four major categories i.e. affix, successor, table lookup and N-gram [1].

Affix removal stemmers apply a set of pre-defined conditional and transformation rules on textual word to generate single token from different inflectional variants after chopping off suffixes and prefixes without exploiting linguistic basis (e.g. schools to school/woman to wom/believe to belief/central, center to cent). Lovins developed first affix removal stemming algorithm in 1968 [3]. After Lovins' stemming, Porter and Paice

© Springer International Publishing AG 2018
S.C. Satapathy and A. Joshi (eds.), *Information and Communication
Technology for Intelligent Systems (ICTIS 2017) - Volume 2*, Smart Innovation,
Systems and Technologies 84, DOI 10.1007/978-3-319-63645-0_37

developed their own affix removal algorithm in 1980 [2, 4]. N-gram stemming is totally different, it is a method of conflating terms based on a joint diagram. Though, it is called "stemming method", but stem token is not produced here. This concept was conceived by George W. Adamson and Jillian Boreham in 1974 [10]. N-gram stemming is executed based on some statistical computation. Successor Variety stemmer had aimed to generate segment from word based on structural linguistic. This concept was introduced by Margaret A. Hafer and Stephen F. Weiss in 1974 [11]. Successor Variety decomposes a word across its' structure to generate segment.

2   In Depth Analysis of Stemming Algorithms

2.1   Lovins' Stemmer

Lovins' stemming algorithm is developed based on two major concepts: iteration and longest-match. Iteration is handling suffixes of any morphed English word to generate dictionary root word or stem token [3]. It uses 294 terminated character sequences, 29 predicates and 35 conversion rules to generate stem token form different morphed word variants.

Suffixes which are attached at the end of the words (e.g. bus, school, add, accept) such as -s, -es and -ed, are called inflectional suffixes (e.g. buses, schools, added, accepted). Lovins' algorithm handles inflectional suffices as well as irregular singular-plural suffices. Below mentioned steps broadly describes Lovins' stemming algorithm [3].

1. Algorithm starts with "Longest–match" technique. It determines the position from where the terminating list to start probing for a matching sequence, so that the stem is a minimum of two characters length. If match is identified then the end part of the word is stemmed (e.g. metallic to metall, metal to met, metallically to metall, magnetizable to magnet). From amongst all the matches, the one with the maximum length is identified with its related condition and then it is eliminated.
2. If ending suffix is found from given 294 suffices ending list, that suffix is checked as to whether the pre-defined context sensitive rule is satisfied or not (e.g. induction to induct).
    2.1. If, suffix present in the context of that word, is satisfying the condition, suffix will be removed to generate 1st phase stem (e.g. in the word "believed", context sensitive rule "ed E" is found and then condition rule "E" is applied to generate "believ").
3. Now, 1st phase stem will be re-coded for generating final one.
    3.1. Final consonant will be undoubled, if it exists within stem.
    3.2. Stem is recoded based on matched transformation rule retrieved from 35 word ending transformation rules and then final stem token will be created (e.g. induct to induc, metal to metal, magnet to magnet).
    3.3. If, no transformation rule is matched, previously generated stem will become final stem.

Features: Most of the irregular singular-plural morphed words are correctly stemmed here. "Index" and "indices" are stemmed into "indic" single stem token. Stem token "forml" is generated from both morphed words "formulae" and "formula". Based on execution of this algorithm on some DUC .doc files [7], average computational time (in milliseconds) to generate stem token is lowest with compare to other affix removal stemmers.

2.2 Porter's Stemmer

This stemmer checks the presence of suffices within a word and then matches that suffix with already stored suffices. Algorithm's six steps are listed below:

1. This step identifies suffices and recodes them accordingly.
    1.1. It identifies and removes suffix ("-sses", "-ies", "-s") from words having plural part-of-speech and recodes to generate stem token (e.g. actresses to actress, abilities to ability, ponies to pony & pennies to penni).
    1.2. It also removes ending "-ed" used in past tense and ending "-ing" used in present continuous tense from words and recodes them. (e.g. "abbreviated" to "abbreviate" & "abbreviate" to "abbreviate" and then, applying recode rule to convert "abbreviat" into "abbreviate").
2. Second step recodes terminal 'y' into 'i' in presence of another vowel within stem or original word (e.g. "penny" to "penni", " ability" to "abiliti").
3. Third step eliminates double suffices and also does recoding based on given list of suffices (e.g. from "running", -"ing" suffix is removed and then "runn" is recoded into "run").
    3.1. It indexes 2nd last character of stem token. It maps double suffices to single suffix.
    3.2. This step also recodes "-ational", "-tional", "-tion", "-izer", "-ation" and "-fullness" into "-ate", "-tion", "-ize", "-ate" and "-full" respectively (e.g. "application" into "applicate"). It removes "-ic", "-icate", "full" and "-ness" endings from stem word and recodes accordingly (e.g. "applicate" to "applic" using recoding rule "-icate>"-ic").
4. It eliminates "-ance", "-er", "-able", "-ant", "-iti" endings from stem word using <c>vcvc<v> (e.g. "acceptance" and "acceptable" into "accept").
5. Final step removes "-e" ending from input stem token (e.g. "assemble" to "assembl").

In this algorithm, "generalizations" is stemmed into "generalization" in step-1 and then into "generalize" in step-2. Step-3 coverts it into "general" and then final stem "gener" will be generated by step-5 [2, 8].

Features: Distinct 6370 stem tokens are generated with the help of this algorithm. Size of the vocabulary is reduced by one third with the help of this suffix stripping process. This Algorithm is quite alert at the time of stripping of a word, if the size of the stemmed token is too short, stripping will be avoided.

Average computational time (in milliseconds) to generate a set of stem token from DUC .doc file is quite low [7].

2.3 Paice Stemmer

Paice algorithm is working based on rule table. Each rule directs either for deletion or for replacement of endings from a word to generate stem token. The rules are clustered into subdivisions corresponding to the last letter of the suffix. The rule table is examined rapidly by viewing the concluding letter of the present word or shortened word. Inside each subdivision, the sequencing of rules plays a meaningful role. Each rule comprises of five elements out of which two are non-compulsory: [4]

(a) "An ending of one or more characters, held in reverse order;" [4]
(b) "An optional intact counter "*";" [4]
(c) "A digit specifying the remove total (may be zero);" [4]
(d) "An optional append string of one or more characters;" [4]
(e) "A continuation symbol, ">" or "."." [4]

e.g. the rule "sei3y>" indicates that last three characters (here, 3 is "remove total") of a word ends with "-ies" ("sei" in reverse order) are replaced by "-y" (append string) and then stemmer again will apply its' rule.

Paice stemming algorithm is given below: [4, 5]

1. "Selection of relevant subdivision: Initially, the last letter of the token is inspected (e.g. "center" has final letter 'r'); if there does not exist any subdivision matching to that letter, then process will be terminated; otherwise, first rule in the relevant subdivision is applied (e.g. applying rule "re2 {-er>-}" for "center"/ "hsiug5ct. {-guish>-ct}" for "distinguish")." [4]
2. "Verify suitability of rule: If the terminating letters of the token are not matched with the reversed ending string in the rule, then process will jump into step 4; if the terminating sequenceis matched, and the intact counter is initialized, then process will jump to 4; if the suitability conditions are not fulfilled, then process will move to 4." [4]
3. "Apply rule: The sum of characters indicated in the "remove total" (mentioned above in rule elements) are deleted from right hand of the form; if there exists an "affix string", then it is added to the token; (e.g. "-er" is removed from "center"/"-guish" is removed from "distinguish"). If the continuation symbol "." exists into rule element, then process will be terminated; (e.g. "-guish" is replaced by "-ct" in "distinguish" and final stem will be "distinct" due to presence of "." into rule "hsiug5ct."), otherwise, if continuation symbol ">" exists within rule element, process will be continued and will move to 1." [4]
4. "Inspect for another rule: Process will move to the next rule in the table; if the subdivision letter has been altered, then process will be terminated; otherwise process will move to 2." [4]

   Note: sample values of stem rule table which is hold by Paice algorithm, is shown below:

"[ai*2. { -ia > - if intact }; a*1. { -a > - if intact }; bb1. { -bb > -b }; city3s. { -ytic > -ys }; ci2> { -ic > - }; cn1t> { -nc > -nt } dd1. { -dd > -d }; dei3y> { -ied > -y };]"

Based on generation of stem token with the help of Paice algorithm, calculated ICF is quite high with compare to other affix removal. Based on ICF value, Paice is stronger stemmer with compare to others.

For measuring the strength of a stemmer, Index Compression Factor (ICF) is calculated [6].

"N = Number of unique words before stemming
S = Number of unique stems after stemming
ICF = Index Compression factor
ICF = (N − S)/N

Greater the value of ICF, greater will be strength of the stemmer."

Features: Based on generation of stem token with the help of this algorithm, calculated ICF is quite high with compare to other affix removal stemmers. So, Paice algorithm has exhaustive list of suffices as well as rule. Average computational time to execute this algorithm (in milliseconds) to generate stem token from DUC .doc file is quite less with compare to Porter and K-Stemmer affix removal stemmers [2, 9].

2.4 N-gram Stemming

This process is applied for classification of text. The number of shared digram within a pair of character strings is used to compute Dice's Similarity Coefficient which helps to develop cluster from sets of character strings within text [10].

A digram is a pair of successive letters. There is a little bit perplexity in using "stemming" term here because this method is not at all generating stem [1]. Selection of single N-gram from a word is a language independent approach for some languages. N-gram pseudo stemmer decomposes a word form into its digrams. "Similarity measure" of two word forms is calculated based on the number of shared unique digram of a word with another word's unique digram which are generated in this method.

Association measure used in N-gram stemming, is calculated from co-occurrence of words in the text. Association measures are measured between pairs of words based on common distinctived digrams [10].

(i) Association measure between pair of terms "phosphorus" & "phosphate" are calculated here [10].
Phosphorus =>ph, ho, os, sp, ph, ho, or, ru, us
Unique digrams = ph, ho, os, sp, or, ru, us
Phosphate =>ph, ho, os, sp, ph ha, at, te
Unique digrams = ph, ho, os, sp, ha, at, te

Dice's coefficient (similarity)

$$S = \frac{2C}{A + B} = \frac{2 * 4}{7 + 7} = .57 \tag{1}$$

A = number of unique digrams in the 1st word
B = number of unique digrams in the 2nd word
C = number of unique digrams shared by A, B

Table 1. Similarity matrix based on Dice similarity

	Statistics	Statistic	Statistical	Statistician
Statistics	.00	1.08	.93	.87
Statistic	.00	.00	1.00	.93
Statistical	.00	.00	.00	.81
Statistician	.00	.00	.00	.00

Above mentioned matrix is justifying the observation of Adamson and Boreham that the similarity matrix is parsed. Cluster is generated based on cut-off value 0.6 in similarity matrix [16]. So, this process helps to cluster the documents based on similarity measure in between adjacent words by creating virtual digram (Table 1).

Features: N-gram stemming technique is used for statistical natural language processing (e.g. Clustering, Speech Recognition, and Phonemes modeling).

2.5   Successor Variety Stemming

This stemmer is processed by decomposing word into segment and is developed based on structural morphology to identify the word and morpheme boundaries through phonemes' distribution in a large body of utterances [1]. The word is segmented into "stems and affixes" by using certain statistical properties of a corpus (successor and predecessor variety counts) which decides the index from where word should be parsed. Hafer's and Weiss's three approaches to generate stem [11] from a word, are discussed here.

1. *Cut off method:* The simplest method to decompose a test word, is first to choose some cut off "k" and then decompose the word at a particular index where its' successor (or predecessor or both) variety meets or over reaches k. This method was adopted by Haris in his phonetic analysis. In this method, careful decision to select of 'k' is required [11].

   To demonstrate the usage of successor variety stemming, the below mentioned example is self-explanatory [11].
   "Test Word: READABLE
   Corpus: ABLE, APE, BEATABLE, FIXABLE, READ, READABLE, READING, READS, RED, ROPE, RIPE.

Threshold = 2 is considered and segmentation will be done when successor variety > = threshold" [11].

In "Readable" word, "Read" segment has 3 successor variety (A, I, S) and next all segments (reada, readab, readabl, readable) has only one successor variety. So, from "readable", "read" segment is generated which will become stem token.

2. *Complete Word Method:* A word will be decomposed at a particular index before that, if that word-prefix or suffix is observed to be a whole word in the corpus. If a shortest substring "w'' of a word "w" exists as a word in a document, is considered as a stem of that particular word (w) [12].

   "READ" is a broken part (segment) from "READABLE" word which is identified as a complete word in the corpus [11, 13].

3. *Entropy method:* [11, 13]

   It exploits the benefits of the dissemination of successor variety characters.

The method's working process is as follows:

Let $|D\alpha i|$ be the number of words in a text body beginning with the i length sequence of letters of test word α.

Let $|D\alpha ij|$ be the number of words in $|D\alpha i|$ has the successor j.

Then probability that the successor letter of αi is the jth letter of the alphabet is given

by $\dfrac{|D\alpha ij|}{|D\alpha i|}$

The entropy of successor system for a test word prefix αi is

$$H\alpha i = \sum_{p=1}^{26} -\frac{|D\alpha ij|}{|D\alpha i|} \cdot \log_2 \frac{|D\alpha ij|}{|D\alpha i|} \qquad (2)$$

Importance of each successor letter is measured by entropy calculation within segmentation process. Entropy is calculated by successor letter's probability of occurrence. Rare successor letters will have a smaller impact to take decisions in segmentation with compare to highly probable successor variety.

Determination of stems: After segmentation of a word, which segment will be considered as stem, will be determined. In most cases the first segment is chosen as the stem. When initial segment appears in many different words, it is probably a prefix. Then, the second segment should be considered as a stem.

Hafer and Weiss used the following rule:

if (first segment occurs in <= 12 words in corpus)
first segment is considered as stem
else (second segment will become stem) [11]

Briefly, successor variety stemming can be stepped down into following operations:

(1) Identify the successor varieties for a word,
(2) Utilize this information to segment the word using one of the above mentioned methods,

(3) Select one of the segments as a stem.

Feature: The objective of Hafer and Weiss was to develop a stemmer that requires little or no human intervention. They mentioned that affix removal stemmers uses human preparation of suffix lists to work well, whereas successor variety stemming requires no such preparation.

3   Comparative Study

A tool is developed for analyzing comparative performance in between Porter, Lovins and Paice. The steps are given below: [14, 15]

1. Punctuation, numeric values, stop words and proper nouns are removed after tagging by Stanford Part-of-Speech software tool and then filtered collection of words are used as an input of this comparative tool.
2. Filtered sorted unique words are used for execution of Porter, Lovins, and Paice algorithm after removing duplicate words. Filtration process is only applicable for text.
3. Computational time taken by Porter, Lovins, Paice algorithm on given words, are also calculated using java thread.

Based on some standard set of words and for DUC text [7], this comparative tool generates stem token for three affix-removal stemmers' algorithm (Tables 2, 3 and 4).

Table 2. Affix removal stemming output given below:

Word	Porter	Lovins	Paice
center	center	center	cent
central	central	centr	cent
woman	woman	woman	wom
women	women	wom	wom
believe	believ	belief	believ
belief	belief	belief	believ
state	state	st	stat
statement	statement	stat	stat
station	station	stat	stat

Table 3. Number of stem word generation for a set of wordlist [17]

Morphed words	Porter	Lovins	Paice
Abilities, ability	1	1	1
Abnormal, abnormally	1	1	1
Abolish, abolished	2	3	2

Table 4. Comparison between affix removal stemmers

Text	Actual size	Effective size	ICF	Time	ICF	Time	ICF	Time
AP880911-0016 [7]	322	103	Porter		Lovins		Paice	
			0.029	21.6	0.067	11.6	0.048	10.6
AP880912-0095 [7]	782	256	0.066	23.2	0.07	19.8	0.085	13
AP880912-0137 [7]	666	217	0.059	26.2	0.073	12.8	0.078	13.4

4   Conclusion

Affix removal stemming, Successor variety and N-gram stemming algorithms are elaborately discussed here with the help of output generated from our comparative tool. Approaches of developing affix removal stemming with Successor and N-gram stemming are quite different. Affix removal stemming focus on inflection part of any morphed word without considering word meaning and text's context where it resides. But, Successor variety and N-gram exploits the structure of a word.

References

1. Chapter 8: Stemming Algorithms: W. B. Frakes Software Engineering Guild, Sterling, VA 22170
2. Porter, M.F.: An algorithm for suffix stripping. Program **14**, 130–137 (1980)
3. Lovins, J.B.: Development of a stemming algorithm. Mech. Transl. Comput. Linguist. **11**(1/2), 22–31 (1968)
4. Paice, C.D.: Another stemmer. ACM SIGIR Forum **24**(3), 56–61 (1990)
5. Paice, C.D.: An evaluation method for stemming algorithms. In: Proceedings of the 17th Annual International ACM SIGIR Conference on Research and Development in Information Retrieval, pp. 42–50 (1994)
6. Frakes, W.B.: Strength and similarity of affix removal stemming algorithms. ACM SIGIR Forum **37**(1), 26–30 (2003)
7. DUC text. http://www-nlpir.nist.gov/projects/duc/data.html
8. Porter, M.F.: Snowball: a language for stemming algorithms (2001)
9. Krovetz, R.: Viewing morphology as an inference process. In: Proceedings of the 16th Annual International ACM SIGIR Conference on Research and Development in Information Retrieval, pp. 191–202 (1993)
10. Adamson, G.W., Boreham, J.: The use of an association measure based on character structure to identify semantically related pairs of words and document titles
11. Hafer, M., Weiss S.F.: Word Segmentation by Letter Successor Varieties Margaret. Department of Computer Science, University of North Carolina, Chapel Hill, North Carolina 27514, USA. Information Storage and Retrieval, vol. 10, pp. 371–385. Pergamon Press (1974). Printed in Great Britain
12. Stein, B., Potthast, M.: Putting Successor Variety Stemming to Work. Faculty of Media, Media Systems Bauhaus University Weimar, Weimar
13. Al-Shalabi, R., Kannan, G., Hilat, I.: Experiments with the successor variety algorithm using cutoff and entropy methods. Inf. Technol. J. **4**(1), 55–62 (2005). ISSN 1812-5638, 2005 Asian Network for Scientific Information

14. Jivani, A.G., et al.: A comparative study of stemming algorithms. Int. J. Comp. Tech. Appl. IJCTA **2**(6), 1930–1938 (2011). www.ijcta.com, ISSN 2229-6093
15. Jivani, A.G., Gupta, R., et al.: Empirical analysis of affix removal stemmers. Int. J. Comput. Technol. Appl. IJCTA **5**(2), 393–399 (2014). www.ijcta.com, 393 I
16. Chapter 16: Clustering Algorithms. http://orion.lcg.ufrj.br/Dr.Dobbs/books/book5/chap16.htm
17. http://www.comp.lancs.ac.uk/computing/research/stemming/Links/resources.htm (Grouped WordList A)

ABC Based Neural Network Approach for Churn Prediction in Telecommunication Sector

Priyanka Paliwal$^{(\boxtimes)}$ and Divya Kumar

Computer Science and Engineering Department, Motilal Nehru National Institute
of Technology Allahabad, Allahabad, UP, India
priyanka07mnnit@gmail.com, divyak@mnnit.ac.in

Abstract. Customer churn prediction has always been an important aspect of every business. Most of the companies have dedicated churn management teams which work for both churn prevention and churn avoidance. In both of the scenarios it is highly required to identify customers who may change their service providers. In this paper we have tried to propose a neural network based model to predict customer churn in telecommunication industry. We have than used Artificial Bee Colony (ABC) algorithm for neural network training and observed a substantial improvement in accuracy. To prove the efficacy of our model we have compared it against Genetic Algorithm (GA), Particle Swarm Optimization (PSO) and Ant Colony Optimization algorithm (ACO). Simulation result shows that ABC trained neural network is more accurate than others in predicting customer churn in telecommunication sector.

Keywords: Customer churn · Neural network · Evolutionary algorithms · Swarm intelligence · Artificial bee colony

1 Introduction

In the present scenario, retaining old customers is as important as attracting the new ones. Total revenues of every industry is majorly dependent on their customer base. This situation is much realistic in telecommunication sector. In today's telecommunication market there is a good availability of service providers, each one trying to offer more and more services to its customers that too at a very reasonable price. Because of this competition, the customers of this sector are more susceptible to churn out. Thus an efficient churn prediction model is always a prime requirement of this sector [1]. To fulfill this purpose the customer relations team always maintain and use behavioral data base of customers for corporate decision making processes. Data-mining techniques and analytical algorithms are applied over this data-base to identify the customers who are at the risk of churning out [2,3]. Churn prediction is thus an interdisciplinary research area which encompasses a through data-base management, data-mining and model development for churn prediction. This problem is further magnified due to the high dimentionality of customer data-base, containing a lot of information about customer demographics, plans availed, complains

© Springer International Publishing AG 2018
S.C. Satapathy and A. Joshi (eds.), *Information and Communication Technology for Intelligent Systems (ICTIS 2017) - Volume 2*, Smart Innovation, Systems and Technologies 84, DOI 10.1007/978-3-319-63645-0_38

raised, usage and billing etc. Although the problem is very complex but keeping in view the loss that may incur due to retention, a good number of statistical as well as meta-heuristic churn prediction models are proposed in the past.

In literature a lot of work is available on Logistic Regression [4,5], Decision Trees [6,7], Artificial Neural Network [3,8] and Support Vector Machine [9–11] models for churn prediction. These models are particularly successful and are most focused by the research scholars working in the domains of churn identification [12]. Artifical Neural Network (ANN) [13] is a simple and robust technique which is based on the structure and working of human brain. Its utility in machine learning, classification and pattern recognition is already well-proven [14,15]. A feed-forward neural network works with the help of neurons organized in a group of layers. These neurons are connected through weights which are to be optimized at the time of learning/training phase. Once the neural network is trained it can be used for classification. Tsai et al. in [3], Sharma et al. in [8], Parag in [16] and Song et al. in [17] and many others have used artificial neural networks in their research regarding churn prediction. In all the cases, authors have claimed that their ANN based models exhibits efficiency and substantial accuracy in churn case identifications.

Usually an ANN is trained through back-propagation algorithms [18]. However, latest trends of research shows that more sophisticated training can be provided to an ANN via Evolutionary and Swarm Intelligence algorithms [19–21]. The set of Evolutionary Algorithms (EA) majorly includes Genetic Algorithms (GA) [22], Evolutionary Strategies [23], Evolutionary Programming [24] and Genetic Programming (GP) [25] on the other hand Particle Swarm Optimization (PSO) [26], Ant Colony Optimization (ACO) [27] and Artificial Bee Colony (ABC) [28] are the most studied Swarm Intelligence (SI) based algorithms. ABC is a relatively new Swarm Intelligence optimization algorithm which based on the foraging behavior of honey bees swarms. It has been extensively used in the past as an efficient optimization technique [12]. Optimizing the neuron's weight in neural networks is one among various applications of ABC [29].

In this manuscript we have tried to show that ABC trained neural network can be efficiently used for customer churn prediction in telecommunication industry. To meet our objective the rest of the paper is organised as follows: In Sect. 2, we have described the churn prediction problem with its parameters and then have proposed a simple neural network based model to predict customer churn in telecommunication industry in Sect. 3. In the last sections we have proved the efficacy of our algorithm by comparing the results of ABC trained ANN with GA, PSO, ACO. It has been observed that the proposed ABC based ANN training algorithm considerably improves the performance of traditional back-propagation trained ANN.

2   Problem Statement

Customer churn prediction is the term used to determine the customers who can churn in near future, from a given service provider. Customer retention is one of the elementary characteristic of Customer Relationship Management (CRM) because it is always profitable to keep existing customers besides attracting the new one.

In the churn prediction problem, for some n customers associated with a telecom company within last month, we have been provided with their *churn status* (whether they have left the company or not) and a data set containing the information regarding their *age, gender, marital status, dependents status, tenure, payment method, services availed, technical support, type of contract* and *monthly charges*. We have used the data-set provided by IBM Watson Analytics and for more information regarding this data-set we redirect the readers to [30]. From this data base we have to discover the patterns or associations for identifying the customers who may churn out. For this purpose a neural network based model is required to be proposed. *error ratio*, as defined in Eq. (1) is used to predict the accuracy of the model. It measures the number of falsely classified customers out of all n predictions. A customer is falsely classified if he/she is a churner classified as non-churner (*Case B*) or vice-versa (*Case C*). The values of $|A|$, $|B|$, $|C|$ and $|D|$ can be found using confusion matrix [3] as shown in Table 1.

$$error\ ratio = \frac{|B| + |C|}{|A| + |B| + |C| + |D|} \tag{1}$$

Table 1. Confusion matrix

		Actual	
		Churner	Non-churner
Predicted	Churner	*Case A*	*Case B*
	Non-churner	*Case C*	*Case D*

3   Proposed Methodology

Inspite of having various mathematical methods for classification like Support Vector Machines, Bayesian Networks, Adaboost and Decision Trees, we have used evolutionary approach. This is because of the motivation gained from the success of evolutionary approaches on other classification problems [31,32]. To tackle this problem we have used two Artificial Neural Network based approaches. The first approach uses the traditional ANN with back-propagated error based learning. The second approach uses an Artificial Bee Colony tuned ANN. The details of both the techniques are presented in the next sub-sections.

3.1   Back-Propagation Trained Neural Network

In this approach, Back-propogation algorithm [18] is used to train the neural network. The neural network which is trained, acts on 10 inputs for 1 output.

It contains 2 hidden layers, each with 5 neurons, thus there exists a total of 80 neuron weights in this network. For each sample input 20 epochs are used for training with log-sigmoid activation function.

3.2 ABC Trained Neural Network

For training a neural network with ABC an initial population P of size 200 is formed first. Each population member P_i $i \in \{1, \ldots 200\}$ is a 80 dimensional vector of real numbers in the range $(0, 1)$ i.e. $P_i[p_{i1}, p_{i2}, \ldots, p_{i80}]$, $j \in \{1, \ldots, 80\}$, $p_{ij} \in \mathbb{R}$ and $0 < p_{ij} < 1$. In this 80 dimensional vector, each dimension represents a corresponding neural network weight between two neurons.

Once the initial population is formed ABC algorithm come into action. In our experiments we have used ABC algorithm as described in [28] to evolve the neural network weights. This ABC model consists of three types of bees: employed, onlookers and scouts. Employed and onlooker bees represents exploration and exploitation of the search space. The timbal bees turns into scouts which are randomly re-initialized. This model delineates two leading modes of behavior of bees which are sufficient for self-organization and coordinated decision making, *(1)* Recruitment of more foragers towards rich food sources searched by forager bees (positive feedback). *(2)* Probabilistic desolation of forager bees which have found poor food sources (negative feedback). Fitness assignment to artificial bee is an important issue in ABC algorithm. In the proposed approach fitness of a chromosome P_i is directly proportional to the number of cases correctly classified when neural network weights are according to the chromosome values. For n trials, i.e. while using a data-set of n different customers, over a chromosome P_i, Eq. (3) is used for fitness evaluation of P_i.

$$raw\_fitness(P_i) = \sum_{i=1}^{n} \begin{cases} 1, & for\ Case\ A\ and\ D, \\ -1, & for\ Case\ B\ and\ C, \end{cases} \tag{2}$$

$$fitness(P_i) = \frac{raw\_fitness(P_i)}{n} \tag{3}$$

4 Results

All the experiments are done on machines running $Matlab\,2014a$ over $Windows\,7$, $4\,GB\,RAM$ and $corei5$ processor. *error ratio*, as defined in Eq. (1) is used to compare different algorithms. The available data-set [30] contains the information base of 7000 customers. Out of these 7000 entries, 5600 are used for training and 1400 are used for testing purposes. Each algorithm (Back-propagation [18], ABC [28], GA [22], PSO [26] or ACO [27]) is executed 5 times, each time with a fresh start. The best reading is chosen to be recorded. From our simulation results, we observed that ABC trained neural network is most accurate with an *error ratio* of 0.11 and Back-propagation is found to be least accurate with 0.19 *error ratio*. PSO came out to be second accurate algorithm with

0.13 *error ratio*. ACO and GA trained neural network exhibits an *error ratio* of 0.18 and 0.16 respectively (Fig. 1).

For further in-depth analysis of ABC algorithm, we have plotted the accuracy trends of ABC algorithm in Fig. 2. This figure shows the variations in the *error ratio* readings taken at different iterations. The *error ratio* apparently decreases from 0.75, as the ABC algorithms converges iteration by iteration and a minimum *error ratio* of 0.11 is obtained at iteration number 180. Thus *error ratio* has decreased nearly 7 times. An increase in the *error ratio* can also be observed after iteration number 180. This is due to the notification of the

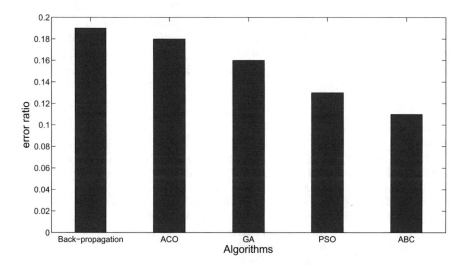

Fig. 1. *error ratio* obtained from different neural networks, trained through Back-propagation, ACO, GA, PSO and ABC.

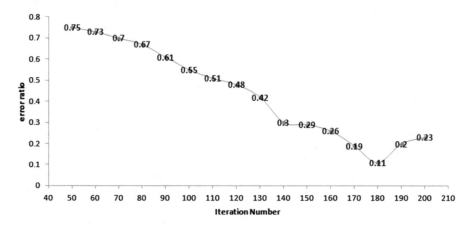

Fig. 2. *error ratio* obtained at different iterations, while running ABC algorithm. *error ratio* is minimum (0.11) at iteration number 180.

occurrence of scout bees, to be reinitialized, who had already converged in the previous iterations.

5 Conclusion

This manuscript has focused on the churn prediction aspect of telecommunication industry. We have first explained the importance of churn prediction for telecommunication and then described the role of Artificial Neural Networks (ANNs) on churn prediction. We have then proposed a Artificial Bee Colony (ABC) based model for training ANN. The ABC trained neural network is then used to predict the customers who are at a risk of churn in near future. To prove the validity of out model, we have compared the accuracy of ABC trained neural network with the neural networks trained through Back-propagation, ACO, GA and PSO. The experimental results clearly validate that ABC trained neural network out performs others.

References

1. Idris, A., Rizwan, M., Khan, A.: Churn prediction in telecom using random forest and PSO based data balancing in combination with various feature selection strategies. Comput. Elect. Eng. **38**(6), 1808–1819 (2012)
2. Au, W.-H., Chan, K.C., Yao, X.: A novel evolutionary data mining algorithm with applications to churn prediction. IEEE Trans. Evol. Comput. **7**(6), 532–545 (2003)
3. Tsai, C.-F., Lu, Y.-H.: Customer churn prediction by hybrid neural networks. Expert Syst. Appl. **36**(10), 12547–12553 (2009)
4. Nie, G., Rowe, W., Zhang, L., Tian, Y., Shi, Y.: Credit card churn forecasting by logistic regression and decision tree. Expert Syst. Appl. **38**(12), 15273–15285 (2011)
5. Hosmer Jr., D.W., Lemeshow, S., Sturdivant, R.X.: Applied Logistic Regression, vol. 398. Wiley, Hoboken (2013)
6. Bin, L., Peiji, S., Juan, L.: Customer churn prediction based on the decision tree in personal handyphone system service. In: 2007 International Conference on Service Systems and Service Management, pp. 1–5. IEEE (2007)
7. Quinlan, J.R.: Induction of decision trees. Mach. Learn. **1**(1), 81–106 (1986)
8. Sharma, A., Panigrahi, D., Kumar, P.: A neural network based approach for predicting customer churn in cellular network services. arXiv preprint arXiv:1309.3945 (2013)
9. Coussement, K., Van den Poel, D.: Churn prediction in subscription services: an application of support vector machines while comparing two parameter-selection techniques. Expert Syst. Appl. **34**(1), 313–327 (2008)
10. Zhao, Y., Li, B., Li, X., Liu, W., Ren, S.: Customer churn prediction using improved one-class support vector machine. In: International Conference on Advanced Data Mining and Applications, pp. 300–306. Springer, Heidelberg (2005)
11. Vladimir, V.N., Vapnik, V.: The Nature of Statistical Learning Theory (1995)
12. Kamalraj, N., Malathi, A.: A survey on churn prediction techniques in communication sector. Int. J. Comput. Appl. **64**(5) (2013)

13. Hagan, M.T., Demuth, H.B., Beale, M.H., De Jesús, O.: Neural Network Design, vol. 20. PWS publishing company, Boston (1996)
14. Basheer, I., Hajmeer, M.: Artificial neural networks: fundamentals, computing, design, and application. J. Microbiol. Meth. **43**(1), 3–31 (2000)
15. Zhang, G.P.: Neural networks for classification: a survey. IEEE Trans. Syst. Man Cybern. Part C (Appl. Rev.) **30**(4), 451–462 (2000)
16. Pendharkar, P.C.: Genetic algorithm based neural network approaches for predicting churn in cellular wireless network services. Expert Syst. Appl. **36**(3), 6714–6720 (2009)
17. Song, G., Yang, D., Wu, L., Wang, T., Tang, S.: A mixed process neural network and its application to churn prediction in mobile communications. In: Sixth IEEE International Conference on Data Mining-Workshops, ICDMW 2006, pp. 798–802. IEEE (2006)
18. Hecht-Nielsen, R.: Theory of the backpropagation neural network. In: International Joint Conference on Neural Networks, IJCNN 1989, pp. 593–605. IEEE (1989)
19. Bullinaria, J.A., AlYahya, K.: Artificial bee colony training of neural networks. In: Nature Inspired Cooperative Strategies for Optimization, NICSO 2013, pp. 191–201. Springer, Heidelberg (2014)
20. Karaboga, D., Kaya, E.: An adaptive and hybrid artificial bee colony algorithm (aABC) for ANFIS training. Appl. Soft Comput. **49**, 423–436 (2016)
21. Branke, J.: Evolutionary algorithms for neural network design and training. In: Proceedings of the First Nordic Workshop on Genetic Algorithms and its Applications. Citeseer (1995)
22. Goldberg, D.E.: Genetic Algorithms. Pearson Education, India (2006)
23. Rechenberg, I.: Evolution strategy. In: Computational Intelligence: Imitating Life, vol. 1, pp. 147–159 (1994)
24. Fogel, L.J.: Intelligence Through Simulated Evolution: Forty Years of Evolutionary Programming. Wiley, Haboken (1999)
25. Koza, J.R.: Genetic Programming II: Automatic Discovery of Reusable Subprograms. MIT Press, Cambridge, MA, USA (1994)
26. Kennedy, J.: Particle swarm optimization. In: Encyclopedia of Machine Learning, pp. 760–766. Springer, Heidelberg (2011)
27. Dorigo, M., Birattari, M., Stutzle, T.: Ant colony optimization. IEEE Comput. Intell. Mag. **1**(4), 28–39 (2006)
28. Karaboga, D., Basturk, B.: A powerful and efficient algorithm for numerical function optimization: artificial bee colony (ABC) algorithm. J. Global Optim. **39**(3), 459–471 (2007)
29. Karaboga, D., Gorkemli, B., Ozturk, C., Karaboga, N.: A comprehensive survey: artificial bee colony (ABC) algorithm and applications. Artif. Intell. Rev. **42**(1), 21–57 (2014)
30. Analytics, I.W.: Using customer behavior data to improve customer retention. https://www.ibm.com/communities/analytics/watson-analytics-blog/predictive-insights-in-the-telco-customer-churn-data-set/
31. Dasgupta, D., Michalewicz, Z.: Evolutionary Algorithms in Engineering Applications. Springer, Heidelberg (2013)
32. Freitas, A.A.: Data Mining and Knowledge Discovery With Evolutionary Algorithms. Springer, Heidelberg (2013)

Expertise Based Cooperative Reinforcement Learning Methods (ECRLM) for Dynamic Decision Making in Retail Shop Application

Deepak A. Vidhate[1]([⊠]) and Parag Kulkarni[2]

[1] Department of Computer Engineering, College of Engineering,
Pune, Maharashtra, India
dvidhate@yahoo.com
[2] iKnowlation Research Labs Pvt. Ltd., Pune, Maharashtra, India
parag.india@gmail.com

Abstract. A novel approach for dynamic decision making in retail application by expertise based cooperative reinforcement learning methods (ECRLM) is proposed in this paper. Different cooperation schemes for cooperative reinforcement learning i.e. EGroup scheme, EDynamic scheme, EGoal-oriented scheme proposed here. Implementation outcome includes demonstration of recommended cooperation schemes that are competent enough to speed up the collection of agents that achieves excellent action policies. This approach is developed for a three retailer shops in the retail market. Retailers be able to help with each other and can obtain profit from cooperation knowledge through learning their own strategies that exactly stand for their aims and benefit. The retailers are the knowledge agents in the hypothesis and employ reinforcement learning to learn cooperatively in situation. Assuming significant hypothesis on the dealer's stock policy, refill period, and arrival process of the consumers, the approach is modeled as Markov decision process model thus making it possible to apply learning algorithms.

Keywords: Cooperation schemes · Multi-agent learning · Reinforcement learning

1 Introduction

The retail store sells the household items and gain profit by that. Retailers are interested about their selling, their profit. By accepting certain steps, the portion that can reason break or decrease the revenue can be prohibited. The aim of predicting the sales business is to collect data from various shops and analyze it by machine learning algorithms. The proficient significance of the practical information by ordinary ways is not practically achievable because the information is extremely vast [1]. The ample of information of a group are casted in such a way that it does include sense. By learning thoroughly the appropriate measures can be taken. Retail shops example is considered

© Springer International Publishing AG 2018
S.C. Satapathy and A. Joshi (eds.), *Information and Communication
Technology for Intelligent Systems (ICTIS 2017) - Volume 2*, Smart Innovation,
Systems and Technologies 84, DOI 10.1007/978-3-319-63645-0_39

here. Walmart is an example for huge shops, big bazaars etc. Most of the time retailers will not be doing well in getting the consumer's requests because they will be unable in the estimation of marketplace prospective. In some particular occurrences the speed of sale or shopping is more. Sometimes it might reason insufficiency of the items. The relationship between the consumers and the shops is evaluated and the modifications that require to gain extra yield are prepared. The history of buy of each item in each shop and department is maintained. By examining these, the sales are predicted that facilitate the understanding of yield and loss happened throughout the year [1, 2]. Let us consider example Christmas in some branch for the period of the specific session. In Christmas celebration the sales is more in shops like clothing, foot wears, jewelry etc. Throughout summertime the purchase of cotton clothing is more; in winter the purchase for sweaters is more. The purchase of items alters as indicated by the season. By examining this past record of purchases, the sales can be forecasted for the future [2]. That discovers the result to predict highest revenue in the industry of retail shop market. The retailers monitor the consumers and they intended them by several beautiful schemes. In order they will be back to the shop and pay for more time and money. The major target of retail shop market preparation is to acquire highest revenue by significant the knowledge and where to provide gainfully and in which shops [2, 3].

There are many challenges in the retail shop forecasting. Some of them are retailers be unsuccessful in the estimating the possibility of the market. Retailers disregard the seasonal chances. The human resources are insufficient and the workers do not exist as and when required. The retailers experience the complexity in storage management system [3]. The retailers sometimes pay no attention to the competition or cooperation in the market. Retailers build the strategies that encourage the success and the extremely target plan. The strategies should be such that they facilitate to achieve the highest revenue.

Generally income of the sale of a specific product are kept which is the result of forecasting the maximum potential of quantity of sales in given period of time and under uncertain environment. Market sale determined by the customer's behavior, the cooperation, facilities support etc. These make effect on the sales of future of a particular shop. Shop and inventory scheduling is significant and is organized policy method in individual shop level [3, 4]. Goods to be buy and sale, store management and space management are the major work in planning of a shop. By monitoring the past history of the shop it helps to put up a scheme of sales of the shop and build any changes in the idea so that it can be highest cost-effective. The fundamental information presented by the existing shop is extremely useful in the forecasting of sales [4].

A novel move toward dynamic decision making in retail shop application by expertise based cooperative reinforcement learning methods (ECRLM) is projected here [4, 5]. Section 2, illustrates novel approach for dynamic decision making by expertise based cooperative reinforcement learning methods (ECRLM). Section 3, presented expertise based cooperation schemes, Sect. 4 demonstrate the system of retail shops designed by Markov decision method and Sect. 5 express concluding remark.

2   Expertise Based Cooperative Reinforcement Learning Methods (ECRLM)

The communication in multiagent reinforcement can build a sophisticated collection of accomplishments achieved from the agents' proceedings. The part of accomplishments set (i.e. a complete action plan) is allocated to the agents via a *Incomplete Action Plan* (Q_i). Normally such incomplete policies maintain incomplete information about the state. These strategies can be incorporated to improve the sum of the partial rewards received using satisfactory association model. The action plans are generated by the way of multiagent Q-learning algorithm by gathering the rewards and constructing the agents to go nearer to the excellent policy $Q*$. When strategies Q_1,........,Q_x are incorporated, it is possible to build up new strategy that is *Complete Action Plan* ($CAP = \{CAP_1, ..., CAP_x\}$), in which CAP_i indicates the **best rewards** received by agent i all over the learning method [4, 5].

Algorithm 1 illustrates the *Splan* algorithm which gives out the agents' learning particulars. The strategies are considered by the Q-learning algorithm for each model. The best reinforcements are distributed to *CAP* that forms a gathering of the best collected rewards by the agents. These rewards will be again given by the way of the extra agents [4–6]. Cooperation is implemented by the changing of incomplete rewards as *CAP* is forecasted by the way of the best reinforcements. A *value* function is applied to discover the excellent strategy among the previous states and last state for a specified plan that calculates *CAP* with the excellent reinforcements. The value function is found out by the adding of stages the agent demand to reach at the final-state and the total of the acquired values in the strategy among each initial state and the final state.

```
Algorithm 1: Multiagent RL Algorithm
Algorithm Splan (I, technique)
1. Initialization Qᵢ(s, a) and CAPᵢ(s, a)
2. Communication by the way of the agents i ∈ I;
3. Agents cooperate till the target state is found;
episode ← episode +1
4. Revise policy which determines the reward value;
Q(s, a)← Q(s, a) +α (r + γ Q(s', a') - Q(s, a))
5. Fco-op (epi, scheme, s, a, i);
The cooperate task decide a cooperation scheme. epi,
scheme, s, a, I are the factors, in which epi is
current iteration, coordination scheme is {group,
dynamic, goal-oriented}, s and a are state and action
chosen accordingly;
6.  Qᵢ←CAP that is Qᵢ of agent i ∈ I is customized by
the way of CAPᵢ.
```

2.1   Expertise Rewards

More expert agents ʾdiscover additional rewards and punishments of the set. As an effect, if the set achieves reinforcement then expertise agents will obtain additional rewards as compare to other agent. On the opposite, further agents receive more punishments as measured to expert agents when the group gets punishment. Expert agents normally execute better than other agents [6, 7]. They find extra chance to conduct correct action as measured apart from less expert agents. Agents acquire rewards (rewards and penalty) as follows [7]:

$$r_i = Rx\frac{e_i}{\sum_{j=1}^{N} e_j} \tag{1}$$

2.2 Expertise Criteria

Expertise criteria consider both reinforcements and penalty as a symbol of being knowledgeable. It indicates that negative and positive results, calculated based upon the cost of reinforcement and penalty signals, are both important for the agent. This is the addition of the absolute value of the reinforcement signals [6–8].

$$e_i = \sum_{t=1}^{now} |r_i(t)| \tag{2}$$

3 Cooperation Schemes

Various cooperation schemes for cooperative reinforcement learning are as [7]:

(i)   E*Group scheme* – reinforcements are issued in a sequence of steps.
(ii) E*Dynamic scheme* – reinforcements are issued in each action.
(iii) E*Goal-oriented scheme* – issuing the addition of reinforcements when the agent reaches the goal-state (S_{goal}).

```
Algorithm 2 Cooperation schemes
Fco-op (epi, scheme,s,a,i)
q : count of agents
   switch between schemes
   In case of EGroup scheme
            if episode mod q = 0 then
            get_Policy(Qᵢ, Q*,CAPᵢ);
   In case of EDynamic scheme
            r ← Σˣⱼ₌₁ Qj(s,a);
            Qᵢ(s,a)← r;
            get_Policy(Qᵢ, Q*,CAPᵢ);
   In case of EGoal-oriented scheme
            if S = Sgoal then
            r ← Σˣⱼ₌₁ Qj(s,a);
            Qᵢ(s,a)← r;
            get_Policy(Qᵢ,Q*,CAPᵢ);
```

```
Algorithm 3 get_Policy
Function get_Policy(Q_i, Q*,CAP_i)
    while each agent i ∈ I
    while each state s ∈ S
      if value(Q_i, s) ≤ value(Q*,s) then
              CAP_i(s,a) ← Q_i(s,a);
    done
```

```
Algorithm 4 get_Ereward
Function get_Ereward(e_i, e_j, N, R)
while agent i ∈ I do
        while state s ∈ S do
        get_Expertise(e_i);
```
$$r_i = Rx \frac{e_i}{\sum_{j=1}^{N} e_j}$$
```
        done
done
return r_i
```

```
Algorithm 5 get_Expertise
Function get_Expertise(r_i)
while agent i ∈ I do
        while state s ∈ S do
```
$$e_i = \sum_{t=1}^{now} |r_i(t)|$$
```
        done
done
return e_i
```

```
Algorithm 6 get_ExpertAgent
Function get_ExpertAgent(e_i)
while agent i ∈ I do
        while state s ∈ S do
                get_expertise(ei);
                If ei > ej then
                ea← ei;
        done
done
return ea;
```

The *EGroup scheme* appears to be extremely strong meeting extremely quick to a best action plan Q*. Reinforcements obtained by the agents are produced in series of pre identified stages. They gather reasonable reward values that cause a good convergence. In the *EGroup scheme* the global policy converges to a best action strategy as there is an intermission of series necessary to gather good reinforcements [8, 9]. The global action policy of the *EDynamic scheme* is able to gather excellent reward values in small learning series. It is observed that after some series, the performance of global strategy reduces. This takes place because the states neighboring to the goal state begin to gather much advanced reward values giving to a local maximum. It punishes the agent as it will no longer stay at the other states. In the **EDynamic scheme** as the reinforcement learning algorithm renews learning values, actions with higher gathered reinforcements are chosen by the top possibility than actions with small gathered reinforcements. Such a policy is recognized as *greedy* search [9, 10]. In the **EGoal-driven scheme** the agent distribute its learning in a changeable number of sequences and the cooperation acquired when the agent arrives at the goal-state. The global action strategy of the *EGoal-driven* scheme is capable to gather excellent reward values, agreed that there is a sum of iteration series to gather values of acceptable rewards. The execution of the cooperative learning algorithms is generally small in the early series of the learning process with the *EGoal-driven* scheme [10, 11].

Expertise agents scheme: The only expertise agents learning is shared in this scheme [10].

4   Model Design

The case with wedding season is considered for the development. Beginning from choosing site, invitation cards, decoration, booking the caterers, shopping of clothes, gifts, jewelry and additional accessories for bride and groom, so many actions are concerned [11, 12]. Such seasonable conditions can be practically executed as follows: Consumer who would purchase in clothing shop surely go for purchase of jewelry, footwear, and further related items [12, 13]. Retailers of various items can come jointly and in cooperation fulfill consumer demands and can acquire the profit by an enhancing in the item sale [13]. Below are mathematical notations for above model.

- Consumers enter at the market by following a Poisson process with rate λ.
- The seller has limited stock capacity I_{max} and follows a fixed reorder policy

States: Assume maximum stock level at each shop $= I_{max} = i1, i2, i3 = 20$.
    State for agent 1 become (x_1, i_1) e.g. (5, 0) that means 5 consumers requests with 0 stock in shop 1. State for agent 2 become (x_2, i_2), State for agent 3 become (x_3, i_3),
    State of the system become **Input** as (x_i, i_i). **Actions:** Assume set of possible actions i.e. action set for agent 1 is (that means Price of products in shop 1), A1 = Price $p = \{8 \text{ to } 14\} = \{8.0; 9.0; 10.0; 10.5; 11.0; 11.5; 12.0; 12.5; 13.0; 13.5\}$. Set of possible actions i.e. action set for agent 2 is A2 = Price $p = \{5 \text{ to } 9\} = \{5.0; 6.0; 7.0; 7.5; 8.0; 8.5; 9.0\}$. Set of possible actions i.e. action set for agent 3 is A3 = Price $p = \{10 \text{ to } 13\} = \{10.0; 10.5; 11.0; 11.5; 12.0; 12.5; 13.0\}$. **Output** is the possible action taken i.e. price in this case. It is now the state-action pair system can be easily modeled using Q learning i.e. **Q(s, a)**. There is need to define the reward calculation [13]. **Rewards:** Reward is calculated in the system as R(s, a) during Q update function.

5   Results

Algorithms are tested on one year's transaction dataset of three different retail shops and results are observed. The experimentation was conducted into situation having dimensions among 120 to 350 states. Result of shop agent 1 for the period of one year sell duration using proposed cooperative expertise methods is given below.

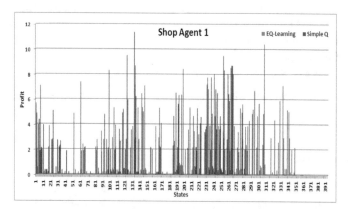

Fig. 1. Profit obtained by shop agent 1 using two learning schemes

For Shop agent 1, above graph (Fig. 1) describes comparison between simple Q learning and proposed expertise based Q learning (EQ-Learning) algorithms. It shows that expertise based Q learning algorithm gives better results in terms of profit vs states as compared to simple Q learning algorithm.

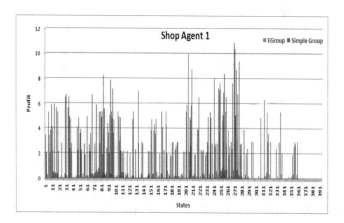

Fig. 2. Profit obtained by shop agent 1 using two learning schemes

For Shop agent 1, above graph (Fig. 2) describes comparison between simple group learning and proposed expertise based group learning (EGroup) scheme. It shows that expertise based group learning algorithm gives better results in terms of profit vs states as compared to simple group scheme. Result of shop agent 2 for the period of one year sell duration using proposed cooperative expertise schemes is given below.

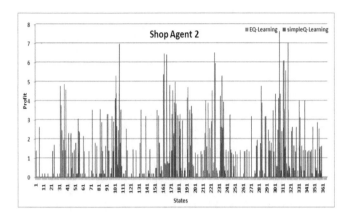

Fig. 3. Profit obtained by shop agent 2 using two learning schemes

For Shop agent 2, above graph (Fig. 3) describes comparison between simple Q learning and proposed expertise based Q learning (EQ-Learning) algorithms. It shows that expertise based Q learning algorithm gives better results in terms of profit vs states as compared to simple Q learning algorithm.

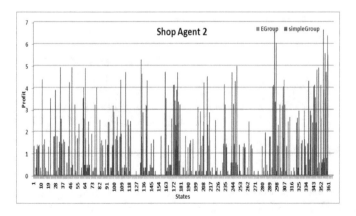

Fig. 4. Profit obtained by shop agent 2 using two learning schemes

For Shop agent 2, above graph (Fig. 4) describes comparison between simple group learning and proposed expertise based group learning (EGroup) scheme. It shows that expertise based group learning algorithm gives better results in terms of profit vs states as compared to simple group scheme. Result of shop agent 3 for the period of one year sell duration using proposed cooperative expertise schemes is given below.

For Shop agent 3, below given graph in Fig. 5 describes that expertise based Q learning algorithm gives better results in terms of profit vs states as compared to simple Q learning algorithm.

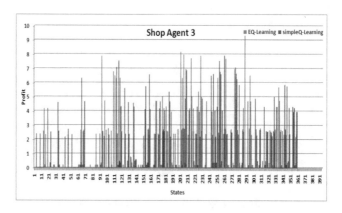

Fig. 5. Profit obtained by shop agent 3 using two learning schemes

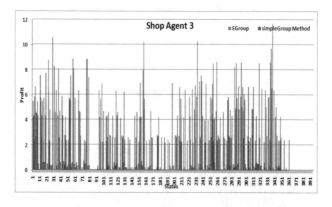

Fig. 6. Profit obtained by shop agent 3 using two learning schemes

For Shop agent 3, above graph (Fig. 6) describes that expertise based group learning algorithm gives better results in terms of profit vs states as compared to simple group scheme.

6 Conclusion

Most of the shopping retailers prefer to draw the consumer's attention to the shop and make more yields. Consumers buy more products by the particular discounts and get the preferred products which can be obtained in the complimentary price and make them happy. The aim of the retailer shops that reasons less failure and effective techniques can be implemented to increase more revenue. All cooperation schemes are able to ensure excellent reinforcements which were obtained during the learning process and modify by means of a collection of excellent reinforcements acquired in partial action policies. By replacing the Q function using four different expertises based reinforcement schemes i.e. EGroup, EDynamic and EGoal oriented, the shop agent evaluate excellent likely items that provides more yields to it. Dynamic decision making by expertise based cooperative reinforcement learning methods (ECRLM) for retail shop application demonstrate that such techniques can present to a quick involvement of agents that restore reinforcements.

References

1. Vidhate, D.A., Kulkarni, P.: New approach for advanced cooperative learning algorithms using RL methods (ACLA). In: VisionNet 2016 Proceedings of the Third International Symposium on Computer Vision and the Internet, pp. 12–20. ACM DL (2016)
2. Vidhate, D.A., Kulkarni, P.: Enhancement in decision making with improved performance by multiagent learning algorithms. IOSR J. Comput. Eng. **1**(18), 18–25 (2016)
3. Raju Chinthalapati, V.L., Yadati, N., Karumanchi, R.: Learning dynamic prices in multi-seller electronic retail markets with price sensitive customers, stochastic demands, and inventory replenishments. IEEE Trans. Syst. Man Cybern. C Appl. Rev. **36**(1), 92–106 (2008)
4. Choi, Y.-C., Ahn H.-S.: A survey on multi-agent reinforcement learning: coordination problems. In: IEEE/ASME International Conference on Mechatronics and Embedded Systems, pp. 81–86 (2010)
5. Abbasi, Z., Abbasi, M.A.: Reinforcement distribution in a team of cooperative Q-learning agent. In: Proceedings of the 9th ACIS International Conference on Software Engineering, Artificial Intelligence, and Parallel/Distributed Computing, pp. 154–160. IEEE (2008)
6. Vidhate, D.A., Kulkarni, P.: Multilevel relationship algorithm for association rule mining used for cooperative learning. Int. J. Comput. Appl. (IJCA) **86**, 20–27 (2014)
7. Gao, L.-M., Zeng, J., Wu, J., Li, M.: Cooperative reinforcement learning algorithm to distributed power system based on multi-agent. In: 3rd International Conference on Power Electronics Systems and Applications Digital Reference: K210509035 (2009)
8. Al-Khatib, A.M.: Cooperative machine learning method. World Comput. Sci. Inf. Technol. J. (WCSIT) **1**, 380–383 (2011). ISSN 2221-0741
9. Vidhate, D.A., Kulkarni, P.: Improvement in association rule mining by multilevel relationship algorithm. Int. J. Res. Advent Technol. (IJRAT) **2**, 366–373 (2014)
10. Panait, L., Luke, S.: Cooperative multi-agent learning: the state of the art. J. Auton. Agents Multi-Agents **11**, 387–434 (2005)

11. Tao, J.-Y., Li, D.-S.: Cooperative strategy learning in multi-agent environment with continuous state space. In: IEEE International Conference on Machine Learning (2006)
12. Berenji, H.R., Vengerov, D.: Learning, cooperation, and coordination in multi-agent systems. In: Intelligent Inference Systems Corporation, Technical report, October 2000
13. Vidhate, D.A., Kulkarni, P.: Innovative approach towards cooperation models for multi-agent reinforcement learning (CMMARL). In: International Conference on Smart Trends for Information Technology and Computer Communications, pp. 468–478. Springer, Singapore (2016)

Investigating the Effect of Varying Window Sizes in Speaker Diarization for Meetings Domain

Nirali Naik, Sapan H. Mankad$^{(\boxtimes)}$, and Priyank Thakkar

Institute of Technology, Nirma University, Ahmedabad, India
{14mcec16,sapanmankad,priyank.thakkar}@nirmauni.ac.in

Abstract. Speaker Diarization deals with determining "who spoke when?" with the help of computers. It is extremely useful in speech transcription, subtitle generation and extracting opinions among others. Diarization is a process in which a relatively long audio recording is processed and speech segments are labeled using respective speaker identities. Such systems are also helpful in determining number of speakers in any conversation or meeting. This field of research is active since long and researchers have been successful to improve the system over time. In this paper, we have made an attempt to study the effect of varying window sizes and threshold criteria on performance of speaker diarization system. Experiments are conducted using LIUM (Laboratoire d'Informatique de l'Universite du Maine) toolkit on Augmented Multiparty Interaction (AMI) meeting data corpus. The proposed approach has shown promising results in case of significantly less number of frames per window.

Keywords: Speaker diarization · Diarization error rate · Bayesian information criterion

1   Introduction

Speaker Diarization is a problem of determining "who spoke when?" from speech utterances. These utterances may have originated from broadcast news room, or telephonic conversations, or audio chat, or meeting room, or press conference, or sports commentary and so on. Speaker diarization focuses on dividing the input audio file into speaker specific segments and labelling them based on the time duration accordingly. An audio may also include non-speech parts such as silence, laughter, claps, etc. including Lombard effect. A primitive and abstract level diarization is to segment an audio into speech and non-speech part. In this paper, we have taken into consideration a more generalized approach in which speech segments are labelled based on different speakers. Speaker Diarization is useful to index an audio file in a sense that if one needs to refer a specific portion

© Springer International Publishing AG 2018
S.C. Satapathy and A. Joshi (eds.), *Information and Communication Technology for Intelligent Systems (ICTIS 2017) - Volume 2*, Smart Innovation, Systems and Technologies 84, DOI 10.1007/978-3-319-63645-0_40

of the audio later, then there is no need to traverse the entire file and can jump directly to the particular time stamp.

The paper is organized as follows. Stages of a typical speaker diarization system are described in Sect. 2. Section 3 discusses a review of earlier work in the field of speaker diarization. Section 4 discusses the proposed approach. Scenario for experiment and result analysis are presented in Sect. 5. Section 6 completes the paper with concluding remarks and future scope.

2   Stages of Diarization System

The basic block diagram of Speaker Diarization is shown in Fig. 1. Since the speaker identity is not known, they are labelled rather than named explicitly.

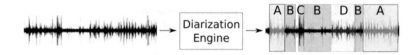

Fig. 1. Speaker diarization [9]

The basic flow of Speaker diarization is illustrated in Fig. 2. A robust audio processing system must handle noise as a part of pre-processing for better results. During *Feature Extraction*, features are extracted from the signal which exhibit speaker-specific characteristics so that it is easier to distinguish among them. Short-term spectral features and voice source features give better performance due to the fact that they are easy to extract [8] in contrast to high level features such as prosody. Moreover, they represent speaker's physiological aspects such as vocal tract length, shape and size. Long-term features (prosodic features) are more robust, but they are difficult to extract. Fusion of short-term cepstral features and long-term prosodic features may result into better efficiency. Speaker specific features help in distinguishing segments pertaining to different speakers during the diarization. *Speech Activity Detection* is basically aimed at separating the speech and non-speech regions from the audio file. Silence can be identified and eliminated using energy level of the speech sample. This stage is also referred to as voice activity detection (VAD).

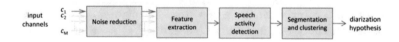

Fig. 2. Block diagram of speaker diarization [4]

During *segmentation*, audio file is divided into multiple, speaker-specific segments so that every segment consists only of one single speaker's speech. During the process of segmentation, a speaker change detection is marked. For each segment, the model for the concerned speaker is built using a mixture of Gaussian distributions. Segmentation detects the points where speakers are changing. A window is used to traverse through the file and detect these changes. Speakers are identified according to models generated for them. Generally, two approaches are used for speaker segmentation: one is single pass speaker change detection, and second is multiple pass re-segmentation.

A typical and more general approach is to use two windows and comparing speaker models for both windows. The segmentation output is decided using two hypotheses. One states that both windows have the same acoustic information and hence belong to same speaker and a single speaker segment model can represent them, while the second hypothesis states that both the windows have different acoustic information and therefore two speaker segment models are needed. Threshold or distance measures such as Bayesian information criterion (BIC), Generalized Likelihood Ratio (GLR) [6] or Kullback Leibler (KL) divergence [14] are used to determine whether or not two consecutive segments belong to the same speaker.

The next task is to perform *clustering* by joining different segments for a common speaker. A window is used to traverse through the entire file once again to measure the speaker change. There are two possibilities: (1) two consecutive segments belong to the same speaker, or (2) they belong to different speakers. The speaker change detection is done using some predefined threshold criterion. An issue that can arise is that if a single segment consists of multiple speakers (overlapped region), where it is not possible to separate speakers anyhow; this may lead to high error rate. The greedy approach to deal with this issue is to remove the overlapped regions and perform the segmentation and clustering. Another, more sophisticated method found in literature is that a separate model for overlapped regions can be built and handled accordingly. However, dealing with overlapped regions in speaker diarization is still a big challenge [2]. The output of the segmentation phase is homogeneous segments belonging to single speaker, located anywhere in the audio. For this purpose, hierarchical agglomerative clustering is the

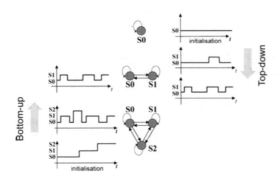

Fig. 3. Hierarchical speaker clustering [4]

most suitable algorithm for forming speaker clusters. In this process, it is assumed initially that all the segments belong to separate individual cluster each, and iteratively we keep on merging the segments for the same speaker into a single cluster. This is referred to as bottom-up approach. Figure 3 depicts a typical flow of both bottom-up and top-down approaches for speaker clustering.

3 Related Work

A survey on different approaches for performing diarization was carried out and various scenarios have been presented in subsequent subsections.

During a meeting room scenario where multiple people are participating and speaking, it is obvious that more than one microphones (microphone arrays) are placed. Each microphone has its own recorded signal, and multiple channels are needed to handle them. Moreover, the type and location of microphones also vary. This gives rise to the cocktail party problem. One solution proposed in [5] is to perform separate diarization on each channel, and then grouping the labels of multiple segments and re-segmenting them. An alternate and more realistic technique is to pick the most dominant signal from multiple channels, preserving it and discarding others [7]. However, the most effective approach of solving this issue is Acoustic Beamforming [3]. In this approach, a single channel is derived through the weighted sum of microphone channels. Other approaches in literature involve selecting the reference channel automatically, computing the channel delays between various microphone channels, selecting the optimal delay and so on.

Two main approaches for performing diarization are (1) *Step-by-step approach*, which is conventional approach wherein clustering is carried out only after segmentation is completely done. CLIPS [13] is an algorithm which employs this approach. Speaker change detection is done using GLR, which gives the relation or association between two speech sequences. If we have two acoustic sequences, we can infer using GLR whether those two sequences belong to the same speaker or no. (2) *Integrated approach* which can rectify or minimize the error occured during the previous step. The output of the clustering phase is again given to the segmentation module of re-segmentation phase. The LIA [13] algorithm offers this flexibility. A hybrid approach which combines both these approaches produces better results compared to individual implementations when LIA, the integrated approach is applied after the CLIPS step-by-step segmentation [12].

Significant amount of work has been done to improve the existing scenario in diarization systems by addressing issues such as handling overlapping regions, detecting dominant speakers, employing hybrid of step-by-step and integrated approaches and so on. However, it was felt that there is a need to identify a better representation of the data which may lead to improvement in error rate.

4   Proposed Work

In this paper, we propose following idea. We made an attempt to observe the impact of dividing the audio file into different number of frames on the performance of the system. Figure 4 depicts the architecture of our proposed workflow.

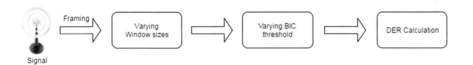

Fig. 4. Proposed approach

The experiments were performed by changing the window size (from 350 frames to 50 frames, in multiples of 50) and segment length repeatedly, with varying BIC threshold for both segmentation and clustering. The typical length of a frame is kept 20 ms. Hence, a window of size 350 frames carries data of $350 \times 20 = 7000$ ms duration. The Diarization Error Rate (DER) is calculated for all these combinations.

5   Experimental Evaluation

This section describes the dataset used for the experiments. Standard evaluation measures to determine the efficiency of diarization systems are discussed at the end of the section.

5.1   Dataset

The implementation was carried out using AMI meeting corpus [10]. This dataset is built for speaker diarization and rich transcript evaluations. Various scenario based and non-scenario based recorded meeting speech audio files are provided with this dataset. Each meeting folder contains 16 files. Files are recorded with 2 microphone arrays and each array has 8 microphones. Hence, each file corresponds to each microphone.

Since the file was already partially preprocessed, acoustic beamforming step was skipped during the implementation. LIUM toolkit [11] based on Java was used for implementation. The 13-dimensional MFCC features were extracted. To remove silence and other non-speech regions, Viterbi decoding with 8 one-state HMMs was used. These eight models consisted of two silence models (wide and narrow band), three models of wide band speech (clean, over noise or over music), and one model each for narrow band speech, jingles and music. One Gaussian was initially built, followed by splitting it till the number of segments were reached. GMMs were built using the Expectation Maximization (EM) algorithm. For boundary detection, the instantaneous change points were detected by traversing through the entire audio file. This was done using GLR wherein change point was considered when the GLR distance within the window reached the threshold. The second traversal in the audio file was carried out using BIC based clustering to merge subsequent segments generated from the same speaker.

A segment object was identified by name of the audio file, a starting point and duration of the segment. The BIC acts as a terminating criterion for clustering. In agglomerative, bottom-up clustering, closest clusters are merged as per this BIC threshold.

5.2   Evaluation Measures

In order to measure the performance of speaker diarization system, the most commonly used evaluation parameter is Diarization Error Rate (DER)[1]. The DER is computed using Eq. 1.

$$DER = \frac{Miss + FalseAlarm + Confusion}{TotalReferenceSpeech} \tag{1}$$

where $Miss$ is a speech segment labelled as non-speech, $FalseAlarm$ is labelling non speech region as speech, and $Confusion$ is the duration in which the system gets confused between different speakers.

5.3   Results and Discussion

Table 1 summarizes the DERs for various window sizes. It is evident from the results that the window size significantly affects the performance. Two important observations are discussed below.

First Observation. If window size is too large, then for that region, the acoustic features will be used to model the speaker. If a small speech part is present and most of the portion is non-speech, then entire region characteristics will be modelled as nonspeech, which would cause a Miss scenario.

Second Observation. If window size is too small, then it may be difficult to extract features properly, and in turn model may not exhibit speaker specific characteristics.

From both observations, it can be seen that it leads to degradation in system performance and increase in DER. Further, it can be observed that the best performance (DER = 10.43%) is achieved for a window size of 100 frames, segmentation BIC threshold = 2, and clustering threshold = 3.5. It is interesting to note that reducing the window size below 100 frames leads to degraded performance. For instance, a window of size 50 frames increases DER back to 15.53% due to the reason stated in the second observation mentioned earlier.

Table 1. Diarization Error Rates (DER) for varying window sizes and BIC threshold values

Window size (no. of frames)	Segment length (ms)	Segmentation BIC	Clustering BIC	DER (%)
350	7000	1	2	14.1
		1.5	3.5	14.0
		2	3.5	13.6
		2.5	3	14.75
300	6000	1	2	14.2
		1.5	3.5	14.09
		2	3.5	13.68
		2.5	3	14.85
250	5000	1	2	13.84
		1.5	3.5	13.75
		2	3.5	13.35
		2.5	3	14.47
200	4000	1	2	13.29
		1.5	3.5	13.22
		2	3.5	12.83
		2.5	3	13.90
150	3000	1	2	11.76
		1.5	3.5	11.69
		2	3.5	11.34
		2.5	3	12.3
100	2000	1	2	10.82
		1.5	3.5	10.75
		2	3.5	**10.43**
		2.5	3	11.31
50	1000	1	2	16.04
		1.5	3.5	15.97
		2	3.5	15.53
		2.5	3	16.77

6   Conclusion and Future Scope

This paper describes the effect of windows having different sizes on diarization system performance on standard AMI corpus for meeting room scenario having multi-distant microphones. Experiments were carried out by changing the window size varying from 350 frames to 50 frames with different BIC-based threshold values for segmentation and clustering. It is seen that reducing the window size to a certain limit gives better performance. From the results, it can be noted that there is a trade-off between window size and performance of the system as discussed in Sect. 5.3. One should be careful in speech-nonspeech discrimination, otherwise miss rates may increase. Further, due to more number of participants present in meeting room conversations compared to telephone or broadcast news scenario, confusion error has to be dealt with properly in case of overlapping.

The major limitation in diarization is that, in most of the cases, neither the number of speakers nor any speaker-specific feature or identity in the audio files is available in advance. This makes the task difficult and more challenging. Moreover, dealing with overlapped regions is also an interesting research area. Further, an analysis of the speaker specific speech can be made and his or her preferences or opinion can be detected on the agenda of meeting. Semi-supervised and Deep Learning based approaches also can be investigated for more improvements. This problem can also be attempted on group songs to extract information about singers from songs.

References

1. NIST 2000 speaker recognition evaluation. http://www.nist.gov/speech/tests/spk/2000/index.htm
2. Anguera, X., Bozonnet, S., Evans, N., Fredouille, C., Friedland, G., Vinyals, O.: Speaker diarization: a review of recent research. IEEE Trans. Audio Speech Lang. Process. **20**(2), 356–370 (2012)
3. Anguera, X., Wooters, C., Hernando, J.: Acoustic beamforming for speaker diarization of meetings. IEEE Trans. Audio Speech Lang. Process. **15**(7), 2011–2022 (2007)
4. Evans, N., Bozonnet, S., Wang, D., Fredouille, C., Troncy, R.: A comparative study of bottom-up and top-down approaches to speaker diarization. IEEE Trans. Audio Speech Lang. Process. **20**(2), 382–392 (2012)
5. Fredouille, C., Moraru, D., Meignier, S., Besacier, L., Bonastre, J.F.: The NIST 2004 spring rich transcription evaluation: two-axis merging strategy in the context of multiple distant microphone based meeting speaker segmentation (2004)
6. Gish, H., Siu, M.H., Rohlicek, R.: Segregation of speakers for speech recognition and speaker identification. In: Proceedings of 1991 International Conference on Acoustics, Speech, and Signal Processing, ICASSP 1991, pp. 873–876, vol. 2, April 1991
7. Jin, Q., Schultz, T.: Speaker segmentation and clustering in meetings. INTER-SPEECH **4**, 597–600 (2004)
8. Kinnunen, T., Li, H.: An overview of text-independent speaker recognition: from features to supervectors. Speech Commun. **52**(1), 12–40 (2010)
9. Knox, M.T.: Speaker Diarization: current Limitations and new directions. Ph.D. thesis, EECS Department, University of California, Berkeley, May 2013
10. Carletta, J., Ashby, S., Bourban, S., Flynn, M., Guillemot, M., Hain, T., Kadlec, J., Karaiskos, V., Kraaij, W., Kronenthal, M., Lathoud, G., Lincoln, M., Lisowska, A., McCowan, I., Post, W., Reidsma, D., Wellner, P.: The AMI Meeting Corpus: A Pre-announcement. In: Renals, S., Bengio, S. (eds.) Proceedings of the Second international conference on Machine Learning for Multimodal Interaction,

MLMI 2005. LNCS, vol. 3869, pp. 28–39. Springer, Heidelberg (2006). doi:10.1007/11677482_3

11. Meignier, S., Merlin, T.: Lium spkdiarization: an open source toolkit for diarization. In: CMU SPUD Workshop, vol. 2010 (2010)
12. Meignier, S., Moraru, D., Fredouille, C., Bonastre, J.F., Besacier, L.: Step-by-step and integrated approaches in broadcast news speaker diarization. Comput. Speech Lang. **20**(2), 303–330 (2006)
13. Moraru, D., Besacier, L., Meignier, S., Fredouille, C., Bonastre, J.F.: Speaker diarization in the ELISA consortium over the last 4 years (2004)
14. Siegler, M.A., Jain, U., Raj, B., Stern, R.M.: Automatic segmentation, classification and clustering of broadcast news audio. In: Proceedings of DARPA Speech Recognition Workshop, pp. 97–99 (1997)

DEAL: Distance and Energy Based Advanced LEACH Protocol

Ankit Thakkar[✉]

Department of Information Technology, Institute of Technology,
Nirma University, Ahmedabad 382 481, Gujarat, India
ankit.thakkar@nirmauni.ac.in

Abstract. Wireless Sensor Network is made of tiny energy-constrained nodes with a limited amount of communication, computation and storage capabilities. Also, it is inconvenient to change batteries of the sensor nodes due to large-scale deployment in hostile environments. Hence, network longevity becomes the prime concern for WSNs. This paper presents Distance and Energy based Advanced LEACH protocol named DEAL. DEAL considers energy and distance of a node during the cluster head election process. Simulation results show that DEAL enhances the stability period and slashes the instability period as compared to ALEACH protocol.

Keywords: Distance · Energy-efficient · Clustering · DEAL · ALEACH

1 Introduction

Wireless Sensor Networks (WSNs) is made of tiny nodes that have sensing capabilities including a limited amount of communication and computation capabilities. WSNs used in the wide variety of applications such as defense applications, Ecology, human-centric applications, robotics, and many more [3]. Energy-efficient protocol design is the prime concern for all types of WSNs irrespective of the application being served by the underlying WSN. This is because WSN nodes are battery operated and large-scale deployment of the nodes makes it inconvenient to change batteries of the nodes. Some application demands deployment of the nodes in hostile environments which also makes it inconvenient to change batteries of the nodes. This leads to conserving the energy of the sensor nodes by making optimal usage of the available energy and at the same time WSN should satisfy the requirements of the underlying application.

There are different techniques to conserve energy of the sensor nodes such as power-aware scheduling [14], use of routing protocols and data aggregation techniques, energy-efficient MAC protocol design, topological control etc. [8]. This paper focuses on the design of energy-efficient routing protocol for WSNs. A classification of WSN routing protocols depend upon the network structure used by the underlying WSN, and according to network structure usage, it can

© Springer International Publishing AG 2018
S.C. Satapathy and A. Joshi (eds.), *Information and Communication
Technology for Intelligent Systems (ICTIS 2017) - Volume 2*, Smart Innovation,
Systems and Technologies 84, DOI 10.1007/978-3-319-63645-0_41

be classified into three categories: flat, hierarchical, and location-based routing [1]. This paper focuses on single-hop hierarchical-based routing schemes as it has dual advantages: scalability and energy-efficiency [1]. The hierarchical routing protocol provides energy-efficiency by reducing communication distance between sender and base station (BS) through the election of cluster heads (CHs). In this scheme, each node elects itself as CH and serves to a group of nodes for a specific duration called round. A group of rounds forms an epoch.

LEACH [6] is a prominent clustering scheme that uses a stochastic approach to elect CHs. It provides network longevity compared to direct communication between nodes and BS. However, LEACH suffers from the following drawbacks: (i) it neither consider the remaining energy of nodes nor their distance from BS during the CH election process and (ii) it assumes that CHs distribution is uniform in the network.

The problem of LEACH protocol had been overcome by many protocols by considering the unconsumed energy of the nodes during the CH election process. In [5], authors have proposed CH election scheme that considers the ratio of the unconsumed energy of the node and initial energy of the node along with the stochastic approach. In [4], authors have proposed stochastic CH election scheme that also considers the ratio of the unconsumed energy of the node and maximum energy of the network. EECS [17] selects the highest energy node as CH with low control overhead. In addition to that, BS needs to broadcast SYNCH messages to provide synchronization between each phase. In [7], a stochastic approach based CH election scheme is proposed that considers the remaining energy of the node and energy consumption of the node during data transfer. E-LEACH [16] improves the LEACH protocol by considering the unconsumed energy of the node and energy required for the data transfer. In [11], authors have proposed a novel self-guided adaptive clustering approach to prolong the lifetime of wireless sensor networks by allowing/disallowing nodes to take part in the clustering process.

ALEACH [2] defines threshold value $Th(n)$ of a node by considering general probability(G_p) and current state probability (CS_p). G_p is defined using stochastic approach while CS_p considers the current energy of the nodes. In [12], authors have extended ALEACH by assigning weight to G_p and CS_p. LEACH-B [15] finds a set of tentative CHs from which final CHs are selected by considering nodes residual energy. These protocols have shown improvement over LEACH protocol and its variants by considering the current energy of a node, but these protocols did not consider the distance of CH from BS which is also an important parameter that needs attention during CH election process [10]. This paper has extended ALEACH protocol and proposed a new protocol named Distance and Energy based Advanced LEACH protocol (DEAL).

The remaining paper is structured as follows: DEAL approach is explained in Sect. 2, Experimental parameters and result analysis are given in Sect. 3, and Sect. 4 covers the concluding remarks.

2   DEAL: Distance and Energy Based Advanced LEACH Protocol

Like ALEACH, each round of DEAL protocol is also divided into two phases. The CH election phase of DEAL is different from ALEACH while the steady-state phase of DEAL is same as the ALEACH. ALEACH has only considered the energy of the nodes during the CH election process. On the other side, DEAL considers the mean value of the energy levels of the node and its standard deviation, the distance of a node from BS and distance of the farthest alive node from BS, and node density to elect CHs. In addition to that, DEAL assigns weight to G_p and CS_p. The threshold value $Thr(n)$ for DEAL is given by Eq. 1.

$$Thr(n) = (1 - p) * G_p + p * CS_p \tag{1}$$

Here, p denotes the optimal percentage of CH, G_p and CS_p denote the general probability and current state probability respectively. The G_p and CS_p are weighted by $(1 - p)$ and p respectively. In Eq. 1, the value of G_p and CS_p is prescribed by Eqs. 2 and 3 respectively.

$$G_p = \frac{k}{N - k * (r \bmod \frac{N}{k})} \tag{2}$$

Here, k denotes the desired number of CHs/Round, N denotes the number of nodes, and current round number is denoted by r.

$$CS_p = \left(\frac{std(E_i)}{\frac{mean(E_i)}{Eo_i}} \right) - \left(p * \frac{Dist2BS_i}{FarthestAliveNodefrmBS} \right) + \frac{N}{x_m * y_m} \tag{3}$$

Here, $std(E_i)$, $mean(E_i)$ and Eo_i denote the standard deviation of the node's residual energy of last twenty rounds, mean value of energy levels of last twenty rounds and initial energy of node i respectively; $Dist2BS_i$ denotes distance of the node from BS, $FarthestAliveNodefrmBS$ denotes the distance of the farthest alive node from BS, N denotes the total nodes deployed in the node deployment area $x_m * y_m \, m^2$. Each node is capable of storing its residual energy for the last twenty rounds [13]. The energy parameter ensures that the node with higher energy and far from the farthest alive node with respect to BS should be elected as CH.

Each node elects itself as CH once per epoch. In CH election process, a node assumes a random number $rnd_n \in [0, 1]$ and compares with the $Thr(n)$. A node is elected as CH only if $rnd_n < Thr(n)$, otherwise the node act as a member for the ongoing round. Once a node is elected as CH for the ongoing round, it will act as a member node for the remaining round of the same epoch, where epoch is given by $\frac{1}{p}$. The steady state phase of DEAL is same as ALEACH.

3   Experimental Parameters and Result Analysis

3.1   Assumptions About the Network Model

The proposed approach DEAL is tested under following assumptions as these assumptions are also used to test ALEACH.

- Random deployment technique is used to deploy nodes in two-dimensional space
- BS usually placed in the center of the node deployment area
- Each node has some data to be sent to BS
- Location of the BS is known to each sensor node
- Sensor nodes are energy constrained
- There is no energy constraint for the BS
- Each node is capable of transmitting data at different energy-levels depending upon the communication distance between sender and receiver
- The proposed approach DEAL has been tested using first order radio energy model presented in [2,6]

3.2   Experimental Parameters and Result Analysis

Extensive simulations are carried out in MATLAB. The experimental parameters are given in the Table 1. The performance of DEAL is compared to ALEACH on the basis of stability period (time between network setup to the death of the first node [9]) and instability period (time between the death of the first node to death of the last node [9]). The larger stability period and shorter instability period indicates better clustering process [9].

Table 1: Experimental parameters

Parameter name	Value
Node deployment area	$100\,\text{m} \times 100\,\text{m}$
Number of nodes (N)	200 and 500
Initial Energy/Node (Eo)	0.5 J and 1 J
Simulation time	Till death of a last node
Sink position	at (50 m, 50 m)
Packet size	4000 bits
E_{elec}	50 nJ/bit
ϵ_{fs}	$10\,\text{pJ/bit/m}^2$
ϵ_{mp}	$0.0013\,\text{pJ/bit/}m^4$
Optimal CH election probability	5%
Simulation runs	Each protocol run sixty-five times with a given N and Eo

The simulation results of stability period and instability period with an initial energy of 0.5 J/Node and 1 J/Node is shown in the Figs. 1 and 2 respectively. It can be evident from the figures that the median value of stability period for the proposed approach is larger than the ALEACH protocol. It can also be seen that the median value of instability period is shorter for the proposed approach

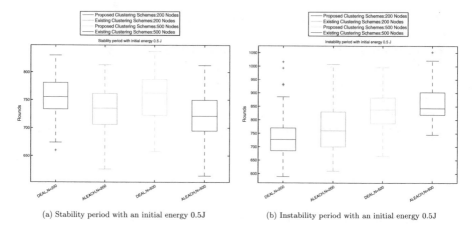

(a) Stability period with an initial energy 0.5J   (b) Instability period with an initial energy 0.5J

Fig. 1: Stability period and instability period for ALEACH and DEAL with different node densities with initial energy 0.5 J/Node

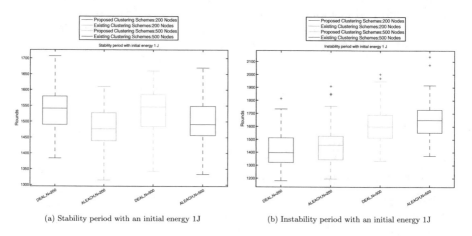

(a) Stability period with an initial energy 1J   (b) Instability period with an initial energy 1J

Fig. 2: Stability period and instability period for ALEACH and DEAL with different node densities with initial energy 1J/Node

as compared to ALEACH. This is because of proposed approach has considered standard deviation of the node's energy level for last twenty rounds along with the mean value. The proposed approach also considers the distance of the node from BS as well as the farthest alive node in the network. In addition to that, node density also affects the network performance. This can be evident through comparing the results obtain for a network of 200 nodes with the results obtained for a network of 500 nodes. To see the effect of node density on the proposed approaches, number of nodes are changed in the experiments and rest all other parameters are kept same.

The simulation results have been also statistically verified using Wilcoxon signed-rank. It is a nonparametric test to test for the median differences for two populations when the observations are paired. The result of this test concludes that median of stability period improves with the proposed approach with 95% probability.

4 Concluding Remarks

The paper presents and discusses Distance and Energy based Advanced LEACH (DEAL) protocol. DEAL considers mean and standard deviation of the node's unconsumed energy for the last twenty rounds. It also considers node density as well as the distance of nodes from BS during the CH election process. The proposed approach DEAL enhances stability period and slashes instability period to provide better clustering scheme.

References

1. Al-Karaki, J.N., Kamal, A.E.: Routing techniques in wireless sensor networks: a survey. IEEE Wirel. Commun. **11**(6), 6–28 (2004)
2. Ali, M.S., Dey, T., Biswas, R.: ALEACH: advanced LEACH routing protocol for wireless microsensor networks. In: International Conference on Electrical and Computer Engineering, 2008, ICECE 2008, pp. 909–914. IEEE (2008)
3. Arampatzis, T., Lygeros, J., Manesis, S.: A survey of applications of wireless sensors and wireless sensor networks. In: Proceedings of the 2005 IEEE International Symposium on Intelligent Control 2005, Mediterrean Conference on Control and Automation, pp. 719–724. IEEE (2005)
4. Han, U.P., Park, S.E., Kim, S.N., Chung, Y.J.: An enhanced cluster based routing algorithm for wireless sensor networks. In: International Conference on Parallel and Distributed Processing Techniques and Applications. vol. 1 (2006)
5. Handy, M., Haase, M., Timmermann, D.: Low energy adaptive clustering hierarchy with deterministic cluster-head selection. In: 4th International Workshop on Mobile and Wireless Communications Network, 2002, pp. 368–372. IEEE (2002)
6. Heinzelman, W.R., Chandrakasan, A., Balakrishnan, H.: Energy-efficient communication protocol for wireless microsensor networks. In: Proceedings of the 33rd Annual Hawaii International Conference on System Sciences, 2000, p. 10. IEEE (2000)
7. Hu, G., Xie, D.M., Wu, Y.Z., et al.: Research and improvement of LEACH for wireless sensor networks. Chin. J. Sens. Actuators **6**(20), 1391–1396 (2007)
8. Li, W., Bandai, M., Watanabe, T.: Tradeoffs among delay, energy and accuracy of partial data aggregation in wireless sensor networks. In: 2010 24th IEEE International Conference on Advanced Information Networking and Applications (AINA), pp. 917–924. IEEE (2010)
9. Smaragdakis, G., Matta, I., Bestavros, A.: SEP: a stable election protocol for clustered heterogeneous wireless sensor networks. Technical report, Boston University Computer Science Department (2004)
10. Thakkar, A.: Cluster head election techniques for energy-efficient routing in wireless sensor networks-an updated survey. Int. J. Comput. Sci. Commun. **7**(2), 218–245 (2016)

11. Thakkar, A.: SKIP: a novel self-guided adaptive clustering approach to prolong lifetime of wireless sensor networks. In: Proceedings of the International Conference on Communication and Computing Systems (ICCCS 2016), Gurgaon, India, 9–11 September 2016, p. 335. CRC Press (2017)
12. Thakkar, A., Kotecha, K.: WALEACH: weight based energy efficient advanced LEACH algorithm. Comput. Sci. Inf. Technol. (CS & IT) **2**(4), 117–130 (2012)
13. Thakkar, A., Kotecha, K.: A new bollinger band based energy efficient routing for clustered wireless sensor network. Appl. Soft Comput. **32**, 144–153 (2015)
14. Thakkar, A., Pradhan, S.: Power aware scheduling for adhoc sensor network nodes. In: 3rd International Conference on Signal Processing and Communication Systems, 2009, ICSPCS 2009, pp. 1–7. IEEE (2009)
15. Tong, M., Tang, M.: LEACH-B: an improved LEACH protocol for wireless sensor network. In: 2010 6th International Conference on Wireless Communications Networking and Mobile Computing (WiCOM), pp. 1–4. IEEE (2010)
16. Xu, J., Jin, N., Lou, X., Peng, T., Zhou, Q., Chen, Y.: Improvement of LEACH protocol for WSN. In: 2012 9th International Conference on Fuzzy Systems and Knowledge Discovery (FSKD), pp. 2174–2177. IEEE (2012)
17. Ye, M., Li, C., Chen, G., Wu, J.: EECS: an energy efficient clustering scheme in wireless sensor networks. In: 24th IEEE International Performance, Computing, and Communications Conference, 2005, IPCCC 2005, pp. 535–540. IEEE (2005)

Comparative Study of DCT and DWT Techniques of Digital Image Watermarking

Azmat Rana[1]([⊠]) and N.K. Pareek[2]

[1] Department of Computer Science, New Look Girls PG College,
Banswara, Rajasthan, India
azmat.rana1@gmail.com
[2] University Computer Centre, Vigyan Bhawan, Block-A,
Mohanlal Sukhadia University, Udaipur, Rajasthan, India

Abstract. This paper presents a comparative analysis of Digital Cosine Transformation (DCT) technique and Discrete Wavelet Transformation (DWT) technique in digital image watermarking. We have used standard digital images for analysing of Watermarked Images and applied standard attacks on the watermarked image. An experimental comparison was made using Matlab.

Keywords: Watermarking · Discrete Wavelet Transformation · Discrete Cosine Transformation

1 Introduction

In today's era, we work in the environment of digital information processing using network based techniques. Various digital watermarking techniques are available and they are being extensively used to make identification and authentication of digital data over network. Digital watermarking techniques are considered as part of cryptography and steganography [1]. Both, cryptography and steganography are used to protect multimedia data against illegal distribution with others. But the watermarking includes the concept of hiding digital image as hidden background of original image or any other kind of data [2]. To maintain watermarking, different techniques are available in literature and they can be classified into two broad categories named as frequency domain [3, 4] and spatial domain [5]. Frequency domain uses different techniques like Discrete Fourier Technique (DFT), Discrete Cosine Transform (DCT), Discrete Wavelet Transform (DWT) etc. whereas in spatial domain techniques Least Significant Bit (LSB) [6], patchwork [7] etc. are used.

On comparison of spatial domain techniques with frequency domain techniques, it was found in literature that frequency domain techniques are more suitable because they are more robust and maintain fidelity with reduction of computational cost and false positive rate [8, 9].

© Springer International Publishing AG 2018
S.C. Satapathy and A. Joshi (eds.), *Information and Communication
Technology for Intelligent Systems (ICTIS 2017) - Volume 2*, Smart Innovation,
Systems and Technologies 84, DOI 10.1007/978-3-319-63645-0_42

2   DCT Watermarking Technique

Discrete Cosine Transform is a technique which converts digital data into cosine frequency component which further used for processing of data [10, 11]. It converts X Image of $M \times N$ into frequency domain by Eq. (1).

$$x(u, v) = \sqrt{\frac{2}{M}}\sqrt{\frac{2}{N}} a_u\, a_v \sum_{u=0}^{M-1} \sum_{v=0}^{N-1} X(M,N) \cos\frac{(2M+1)u\pi}{2M} \cos\frac{(2N+1)v\pi}{2N}$$

(1)

where x(u,v) is DCT coefficient with row u and column v. Values of both a_u and a_v are set to $1/\sqrt{2}$ when u. v = 0, otherwise 1. The converted image again reconverted using the inverse of it by Eq. 2.

$$y(M, N) = \sqrt{\frac{2}{u}}\sqrt{\frac{2}{v}} \sum_{u=0}^{M-1} \sum_{v=0}^{N-1} a_u a_v\, x(u, v) \cos\frac{(2M+1)u\pi}{2M} \cos\frac{(2N+1)v\pi}{2N}$$

(2)

3   DWT Technique

In this technique, we use wavelets which are special functions based on basal functions to represent digital signals. They applied on two dimensional images for processing of images by 2D filters of each dimension [15]. These filters can be divide into four sub bands CA, CH, CV and CD where CA represents approximation coefficient, CH represents horizontal, CV vertical and CD represents diagonal coefficients of the DWT.

4   Algorithm Used for Embedding and Extracting Watermark

Step 1: Input the cover and watermark Images.
Step 2: Convert both images into frequency domain (either use DCT or DWT as we require analyzing).
Step 3: Embed the watermark image with the cover image with the specific α Time.
Step 4: Apply standard attacks on the watermarked image.
Step 5: Extract embedded watermark and compare it with original watermark image.

5   Experimentation and Results

In order to analyse both the techniques of watermarking, we choose two standard images with size of 256×256. Images shown in Fig. 1(a) and (b) are used as cover and watermark image respectively in our experiment.

(a)

(b)

Fig. 1. (a) Cover image and (b) Watermark image.

We create watermark using DCT technique on the basis of different levels of watermark image with cover image and extract the watermark again from the cover image and analyze them on different parameters like Correlation Coefficient, Mean Square Error and Peak Signal Noise Ratio. Result obtained using DCT are shown in Table 1.

Table 1. DCT based watermarking

Watermark (α Times)	PSNR	MSE	Correlation coefficient
0.01	44.7611	1.9507	0.9999
0.02	39.2344	7.0584	0.9997
0.03	35.8003	15.7731	0.9994
0.04	33.3808	27.9014	0.9990
0.05	31.5428	43.1670	0.9984
0.06	30.0037	62.3370	0.9977
0.07	28.7218	84.8378	0.9969
0.08	27.6560	109.8444	0.9959
0.09	26.9066	132.2196	0.9949

We repeat the same procedure using DWT technique and create the Watermark with different levels of Watermark Image. Result obtained using DWT are shown in Table 2.

From Tables 1 and 2, one cannot have any idea about trend of MSE. For this purpose, we plot trends for both DCT and DWT based watermarking using standard tool Matlab. Figure 2 shows plotting of DCT based watermark and original watermark

Table 2. DWT based watermarking

Watermark (α Times)	PSNR	MSE	Correlation coefficient
0.01	4.7804	0.2622	0.0554
0.02	4.8676	0.2570	0.1111
0.03	4.9558	0.2518	0.1657
0.04	5.0448	0.2467	0.2467
0.05	5.1347	0.2416	0.2700
0.06	5.2255	0.2366	0.3189
0.07	5.3174	0.2317	0.3652
0.08	5.4101	0.2268	0.4089
0.09	5.5039	0.2219	0.4499

comparison on the basis of calculated Mean Square Error (MSE) of them. Similarly, Fig. 3 shows plotting of DWT based watermark and original watermark comparison on the basis of calculated Mean Square Error (MSE) of them.

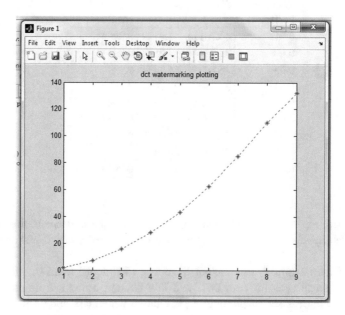

Fig. 2. DCT Watermark plotting on the basis of MSE

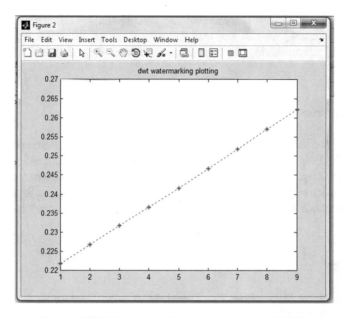

Fig. 3. DWT Watermark plotting on the basis of MSE.

Finally, we apply standard attacks like poisson, salt and paper, Gaussian, Rotation and Cropping on watermarked images to check their robustness. Further, we analyze original watermark and the extracted watermark by calculating again Correlation Coefficient, Mean Square Error and Peak Signal Noise Ratio respectively. On the basis of DCT based watermarking, we obtain results shown in Tables 3.

Table 3. Attacks applied on DCT based watermark.

Types of attack	PSNR	MSE	Correlation coefficient
Poisson	9.5700	0.005589e+004	0.0939
Salt & Pepper	26.4322	0.000135e+004	0.1668
Gaussian	6.1196	1.2371e+004	0.2051
Rotate (30°)	8.9481	6.4499e+003	0.0242
Crop (100,200)	7.6449	8.7071e+003	0.1956

Similarly, on the basis of DWT based watermarking, we obtain results shown in Tables 4.

Table 4. Attacks applied on DWT based watermark.

Types of attack	PSNR	MSE	Correlation coefficient
Poisson	5.5039	0.2219	0.4499
Salt & Pepper	5.4127	0.2267	0.2497
Gaussian	5.4556	0.2242	0.2842
Rotate (30°)	1.0615	0.6173	0.0372
Crop (100, 200)	4.0834	0.3078	0.1932

After analysis of results generated by comparison of original watermark and extracted watermark, we can say that the robustness of watermark based on DWT technique of watermarking is comparatively better then the DCT watermarking technique.

6 Conclusion

In this paper, we have analysed the DCT and DWT based image watermarking. We have taken standard digital images for the processing and applied different attacks on the watermarked image to check their robustness against them. We prepared watermark with different density of the watermark image. From obtained results, we found that DWT based watermarking technique is more robust as compare to watermark based on DCT.

References

1. Cox, I.J., Miller, M.L., Bloom, J.A.: Digital Watermarking. Morgan Kaufmann Publishers, San Francisco (2002)
2. Hartung, F., Kutter, M.: Multimedia watermarking techniques. Proc. IEEE **87**(7), 1079–1107 (1999)
3. Cheddad, A., Condell, J., Curran, K., Mc Kevitt, P.: Digital image steganography: survey and analysis of current methods. Sig. Process. **90**(3), 727–752 (2010)
4. Mohanty, S.P., Ramakrishnan, K.R., Kankanhalli, M.S.: A DCT domain visible watermarking technique for images. In: 2000 IEEE International Conference on Multimedia and Expo, ICME 2000, vol. 2, pp. 1029–1032. IEEE (2000)
5. Podilchuk, C.I., Delp, E.J.: Digital watermarking: algorithms and applications. IEEE Sig. Process. Mag. **18**(4), 33–46 (2001)
6. Meerwald, P.: Digital image watermarking in the wavelet transform domain. Master's thesis, Department of Scientific Computing, University of Salzburg (2001)
7. Potdar, V.M., Han, S., Chang, E.: A survey of digital image watermarking techniques. In: 2005 3rd IEEE International Conference on Industrial Informatics, INDIN 2005, pp. 709–716. IEEE, August 2005
8. Cox, I.J., Miller, M.L., Bloom, J.A.: Watermarking applications and their properties. In: Proceedings of the International Conference on Information Technology: Coding and Computing. IEEE (2000)
9. Lai, C.C., Tsai, C.C.: Digital image watermarking using discrete wavelet transform and singular value decomposition. IEEE Trans. Instrum. Meas. **59**(11), 3060–3063 (2010)
10. Suhail, M.A., Obaidat, M.S.: Digital watermarking-based DCT and JPEG model. IEEE Trans. Instrum. Meas. **52**(5), 1640–1647 (2003)
11. Singh, A.K., Dave, M., Mohan, A.: Hybrid technique for Robust and imperceptible image watermarking in DWT–DCT–SVD domain. Natl. Acad. Sci. Lett. **37**(4), 351–358 (2014)

Comprehensive and Evolution Study Focusing on Comparative Analysis of Automatic Text Summarization

Rima Patel[(✉)], Amit Thakkar, Kamlesh Makwana, and Jay Patel

Charotar University of Science and Technology, Changa, Gujarat, India
patelrima94@gmail.com,
{amitthakkar.it,kamleshmakvana.it,jaypatel.it}@charusat.ac.in

Abstract. In the escalating trend of atomization and online information, text summarization bolster in perceiving textual information in the form of summary. It's highly tedious for human beings to manually summarize large documents of text. In this paper, a study on abstractive and extractive content rundown strategies has been displayed. In Extractive Text Summarization it talk about TF-IDF, Cluster based, Graph theory, Machine learning, Latent Semantic Analysis (LSA) and Fuzzy logic approaches. Abstractive rundown techniques are ordered into two classes i.e. Structured based approach and Semantic based approach. In Structure Based approach it talk about Tree based, Template based, Ontology based, Lead & Phase based and Rule based method. In Semantic Based Approach it talks about Multimodal semantic, Informative item based and Semantic graph based method. The central idea of this method has been elaborated further, apart from idea, the advantages and disadvantages of these methods have been procured.

Keywords: Abstractive Text Summarization · Extractive Text Summarization

1 Introduction

The substantial and superfluous amount of information depicting on World Wide Web (www), the area of Text Summarization is critical in the field of information retrieval. Nowadays, people are used to with the web to discover data through [1] data recovery instruments, for example, Google, Yahoo, Bing etcetera. The way toward consolidating a source message into a feasible form safeguarding its data substance is called outline.

Summarization is a tedious and erroneous job. As consequences, summarization has become the pioneer need for the technical world. Illustrating a logical summarization of the documents which bolster up the most viable information.

As technological depended life lead summarization for various purpose and in many domain for example, news articles outline, email synopsis, short message of news on portable, and data [2] rundown for businessperson, government authorities, specialists include with internet searchers to get the synopsis of significant pages.

On the internet, there is numerous such examples like Text Compacter, Simplify [2], Tools4Noobs, FreeSummarizer, WikiSummarizer and SummarizeTool are online

© Springer International Publishing AG 2018
S.C. Satapathy and A. Joshi (eds.), *Information and Communication Technology for Intelligent Systems (ICTIS 2017) - Volume 2*, Smart Innovation, Systems and Technologies 84, DOI 10.1007/978-3-319-63645-0_43

rundown instruments. Open Text summarizer [2], Classifier4J, NClassifier, CNGL Summarizer are few generally utilized open source outline instruments.

The technological boon of text summarization is encouraging archive choice and writing seeks, improvement of record ordering effectiveness, free from predisposition and they are valuable being referred to noting systems [3] where they give customized data.

The aim of Text Summarization is to provide a well-managed and critical information [2] from authorized documents.

In this paper, first section gives introduction of text summarization. Second section gives details of text summarization features. Third section gives information about text summarization techniques. Forth section gives brief idea about extractive techniques. Fifth section gives brief idea of abstractive techniques. Sixth section gives strengths and weakness of these techniques. Seventh section gives conclusion of this paper.

2   Text Summarization Features

For producing summary, sentences required a few catchphrases, express or certain components to chose where it is imperative for definite synopsis. Here are list of features, can be used for selection of key points in the summary.

- Location Method:- It is depend on the location of the sentences in the text. The beginning or the end of the text is considered to be important. In the document, leading more sentences or last few sentences or conclusion are considered as greater chance to be added in the summary [4, 5].
- Cue Method:- It depends on the supposition that such expressions give a "logical" setting for distinguishing imperative sentences [5]. In this method, positive words like "confirmed, huge, best, this paper" and negative words like "scarcely, outlandish" are assigned weight to text.
- Title/Heading Word:- In this method, title, heading and sub headings are considered to be more important sentences [4]. These sentences contain words, which are added in the summary.
- Sentence Length:- It shows the size of the sentences in the summary. Long and short sentences are not reasonable for outline in general [2].
- Proper Noun:- Text Summarization provides proper nouns are more important. Proper noun means name of people, name of place or organization. Containing proper nouns are greater chance to choose in the summary [2, 4].

3   Text Summarization Techniques

In text summarization has two approaches: Extractive Text Summarization and Abstractive Text Summarization.

Extractive Text Summarization:- In this approach, it extricate the words [5], sentences or expressions from the first content. Extractive Approach depends on measurable investigation of individual or blended elements, for e.g. area, prompt, title or proper noun to extract the sentences. Search engine also generate extractive summary [4] from web pages.

Abstractive Text Summarization:- In this approach, it comprise of comprehension and examination of source report and after that make rundown. Its plan [6] to give a summed up rundown, data in compact way, and required propelled dialect era and pressure procedures.

4   Extractive Text Summarization

There are many techniques in extractive text summarization:

1. Term Frequency - Inverse Document Frequency (TF-IDF):- In this method, words are weighted based on TF-IDF techniques. Term frequency is used to count the no. of terms in the document. Inverse document frequency is used for unseen words which are not in the document. Most astounding weight of words or most elevated scoring sentences are picked to be a bit of the rundown [2]. Wi = {w1, w2, w3,..} is the collection of extract words. TF(wi) = wi/N, where N is the aggregate no. of words in the collection. IDF(wi) = log10(N/wi), where N is the total no. of documents.

2. Cluster Based Method:- There are different "themes" appearing in the document. Document clustering is most important to generate a meaningful summary. Sentences to the subject of the cluster (Ci) [6]. Second variable that is area of the sentence in the document (Li). Third factor that increase the score of the sentence in its similarity to the first sentence in the document to which it belongs (Fi). The overall score (Si) of a sentence i is a weighted [2] sum of the above three factors: Si = w1 * Ci + w2 * Fi + w3 * Li, where w1, w2, w3 are weighted age of inclusion of summary.

3. Graph Theory Method:- Document is designed as a graph, in which graph vertices denotes the sentences. If similarity occurs [4] between the sentences than edges do take place. The similarity between corresponding sentences are analyzed on many factors like content overlap and term frequency. Query specific summaries and generic summaries [7] sentences are both selected from their one sub graph and from different sub graph respectively. Vitality is shown in nodes due to high cardinality number.

4. Machine Learning Method:- This strategy depends on the principle of Bayes Theorem of inverse probability. In this process, document and their extractive summaries are given. Sentences [4] are included or excluded in the summary by calculating the probabilities of their relevance, using Bayes Theorems rule: $P(s \cdot < s | F1, F2, \ldots, FN) = P(F1, F2, \ldots, FN | s \cdot S) * P(F1, F2, \ldots, FN)$. where F1, F2, ..., FN are features which are used for classification. s is a sentence from the document collection and S is summary to be generated and $P(s \cdot < s | F1, F2, \ldots, FN)$ is the probability that sentence s will be chosen to form the summary given that it possesses features F1, F2,..., FN.

5. Latent Semantic Analysis (LSA):- Single Value Decomposition (SVD) connected to archive word grids that are semantically identified with each other, despite the

fact that they don't share basic words. The basic idea behind the use of LSA is that words that for the most part happen in related settings are connected in a similar solitary space. This method can extract the content – sentences and topic-words from document. LSA method consists of three steps: Input Matrix Generation, Singular Value Decomposition, Sentence Selection [2, 4].

6. Fuzzy Logic Method:- Fuzzy logic is the branch of mathematic in which each element in a set possesses a 'degree' of membership to the set. Every normal for a content like sentence length, likeness to pretty much nothing, comparability to watchword are consider in this technique. This all are input in the fuzzy system. In the yield in view of sentence qualities and the accessible standards in the learning base, every sentence is gotten esteem from zero to one. In the yield, the esteem demonstrates the level of the significance of the sentence in the last outline. In fluffy rationale technique, every sentence of the report is spoken to by sentence score. At that point apply plummeting request to that sentence score [4]. An arrangement of most astounding score sentences are extricated as report synopsis.

5   Abstractive Text Summarization

There are two approaches in the abstractive text summarization: Structure based approach & Semantic based Approach.

A.   Structure Based Approach

Through subjective blueprints, for example, layouts, extraction guidelines and structures like lead and phrase, tree, ontology structure, structure based approach encodes most basic information from the document(s) [1]. These are methods of structure based approach.

1. Tree Based Method:- For representing the text/content of a document this techniques uses a dependency tree [6]. Theme intersection algorithm used for content selection summary. The procedure utilizes either a dialect generator o a calculation for era of rundown.

2. Template Based Method:- This system utilizes a format to speak to an entire record. To recognize content bits semantic examples or extraction tenets are coordinated. Bits are the markers and are extricated by Information Extraction Systems. This system is works when the data is available in the record [7].

3. Ontology Based Method:- For same topic having same knowledge, ontology techniques is used. Ontology represents the domain. On the web, the greater part of records are space related in light of the fact that they examine a similar subject or occasion. Next significant terms are created by preprocessing and classifier orders [7] those terms. At that point enrollment degree is made different occasions are connected with these.

4. Lead & Phrase Method:- This technique is concentrate on the expressions that have same syntactic head irregularity in lead and body sentences. Same lumps are looked in lead and body sentences. At that point utilizing closeness metric, these expressions are adjusted. In the event that the body expression has rich data and has same

comparing phrase then substitution happens. However, in the event that body phrases has no partner then inclusion occur [7].

5. Rule Based Method:- This strategy create rundown in light of terms of classes and a rundown of angles. This plan utilizes an administer based data extraction module, content choice heuristics and at least one examples to produce a sentence [7]. Content determination module chooses the best applicant among the ones created by data extraction standards to reply at least one parts of a classification. Toward the end, era examples are utilized for era of outline sentences.

B. Semantic Based Approach

In Semantic based technique, semantic portrayal of document(s) is utilized to encourage into nature dialect generation (NLG) system [1]. The primary concentrates on recognizing thing and verb by handling phonetic information. These are techniques for semantic based approach.

1. Multimodal Semantic Model:- The abstract summary is generated base on semantic model. Normally, the document has text and images. First step is to build semantic model using knowledge. Utilizing data thickness metric the data is connected which checks the fulfillment, association with other and number of event of an expression. The expression give the connections and the ideas [7].
2. Informative Item Based Method:- In this strategy, the substance of outline are created from dynamic portrayal of score documents [1]. It creates short, intelligent, data rich and less repetitive rundown.
3. Semantic Graph Based Method:- A semantic diagram called Rich Semantic Graph (RSG) is manufacture where the hubs speaks to verb and things while edge give semantic and topological connections [1, 7]. It utilizes heuristic tenets to decrease the created rich semantic chart to more lessened diagram and in this manner abstractive outline is delivered. It takes a shot at the semantic and gives less excess and very much organized outline.

6   Pros and Cons of Techniques

See Tables 1, 2 and 3.

Table 1. Pros & Cons of extractive text summarization techniques [4]

Techniques	Advantages	Disadvantages
TF-IDF	Good heuristic for determining keywords	No semantic relation mapping
Graph theoretic approach	Can generate query specific summaries	Accuracy will depend upon selection of affinity function
Latent semantic analysis approach	Semantic relations are captured	Polysemy issues (Inability to capture multiple meanings of a word)
Fuzzy logic approach	Compression ratio is as low as 20%	Overhead of designing membership function
Machine learning approach	Simple	Statistical data is required

Table 2. Pros & Cons of structure based approach

Techniques	Advantages	Disadvantages
Tree based method	-It walks on units of the given document read and easy to summary	-It does not have an entire model which would [2] incorporate a dynamic representation for substance determination
Template based method	-It creates outline is profoundly cognizant in light of the fact that it depends on pertinent data [2] recognized by IE framework	-Requires outlining of layouts and speculation of format is to troublesome
Ontology based method	-Drawing connection or setting is simple because of cosmology -Handles instability at sensible sum	-This technique is limited only for Chinese news -Handling uncertainty is a tedious task for creating rule based system
Lead & phrase method	-It is helpful for semantically fitting redresses for changing a lead sentence [2]	-Due to parsing errors it deplete sentence with grammatical mistakes and with repetitions -It will accentuation on revamping strategies with unsatisfied model which would incorporate a conceptual for substance determination
Rule based method	-It has a potential for creating [2] outlines with more noteworthy data thickness than current condition of workmanship	-In this approach, every one of the tenets and example are physically composed [2], which is furious and tedious

Table 3. Pros & Cons of semantic based approach

Techniques	Advantages	Disadvantages
Multimodal semantic model	-Most pioneer favorable position of this structure is that it gives conceptual synopsis, whose points of confinement [2] are astounding on the grounds that it involves notable literary and graphical substance all through the record	-The restriction of this structure is that it is physically assessed [2] by people
Informative item based method	-The key quality of this strategy is that it creates [2] short, cognizant data with rich and less repetitive synopsis	-It dismisses because of the trouble of making significant and linguistic sentences from them -Phonetic nature of outlines is low because of mistaken parses
Semantic graph based method	-It produces compact, lucid and less excess and linguistically remedy sentences [2]	-This strategy is constrained to single archive abstractive synopsis

7 Conclusion

Content outline is advancing sub – branch of Natural Language processing as the need for compressive, sensible, and unique of subject because of huge measure of data accessible on net. Content rundown is exceedingly requirement for investigator, advertising official, improvement, scientists, government associations, understudies and educators moreover. This paper notifies about the information of both extractive and abstractive methodologies alongside systems utilized, its execution accomplished, alongside favorable circumstances and drawbacks of each approach. Content synopsis assumes a huge part in both business and additionally investigate group. As abstractive synopsis requires greater contribution in learning, it is bit complex than extractive approach at the same time; abstractive outline gives relevant and careful rundown contrast with extractive.

References

1. Khan, A., Salim, N.: A review on abstractive summarization methods. J. Theor. Appl. Inf. Technol. **59**, 64–72 (2014)
2. Gaikwad, D.K., Mahender, C.N.: A review paper on text summarization. Int. J. Adv. Res. Comput. Commun. Eng. **5**(3), 22–28 (2016)
3. Ranjith, S.R.: A survey on sentence similarity based automatic text summarization techniques. Technical Research Organization India
4. Oak, R.: Extractive techniques for automatic document summarization: a survey. Int. J. Innov. Res. Comput. Commun. Eng. **4**(3) (2016)
5. Munot, N., Govilkar, S.S.: Comparative study of text summarization methods. Int. J. Comput. Appl. **102**(12), 0975–8887 (2014)
6. Saranyamol, C.S., Sindhu, L.: A survey on automatic text summarization. Int. J. Comput. Sci. Inf. Technol. **5**(6), 7889–7893 (2016)
7. Bhatia, N., Jaiswal, V.: Trends in extractive and abstractive techniques in text summarization. Int. J. Comput. Appl. **117**(6), 0975–8887 (2015)
8. Murty, M.R., et al.: A survey of cross-domain text categorization techniques. In: 2012 1st International Conference on Recent Advances in Information Technology (RAIT). IEEE (2012)
9. Kasture, N.R., Yargal, N., Singh, N.N., Kulkarni, N., Mathur, V.: A survey on method of abstractive text summarization. Int. J. Res. Emerg. Sci. Technol. **1**(6), 53–57 (2014)

An Intelligent Real Time IoT Based System (IRTBS) for Monitoring ICU Patient

Bharat Prajapati[1]($\boxtimes$), Satyen Parikh[1], and Jignesh Patel[2]

[1] Acharya Motibhai Patel Institute of Computer Studies, Ganpat Vidyanagar,
Mehsana-Gozaria Highway, Kherva 384012, North Gujarat, India
{bbp03, satyen.parikh}@ganpatuniversity.ac.in
[2] Department of Computer Science, Ganpat Vidyanagar, Mehsana-Gozaria
Highway, Kherva 384012, North Gujarat, India
jmp03@ganpatuniversity.ac.in

Abstract. Internet of Things (IoT) enable humans to get higher level of automate by developing system using sensors, interconnected devices and Internet. In ICU, patient monitoring is critical and most important activity, as small delay in decision related to patients' treatment may cause permanent disability or even death. Most of ICU devices are equipped with various sensors to measure health parameters, but to monitor it all the time is still challenging job. We are proposing IOT based system, which can help to fast communication and identifying emergency and initiate communication with healthcare staff and also helps to initiate proactive and quick treatment. This health care system reduces possibility of human errors, delay in communication and helps doctor to spare more time in decision with accurate observations.

Keywords: Internet of Things (IoT) · Intelligent software agent · ICU · Real time · Health parameters

1 Introduction

IoT applications are changing businesses and bringing more efficiency by analyzing relevant data across the sectors like Home Automation, Smart Cities, Environment, Energy, Retail, Logistics, Agriculture, Health and Life Style. Along with development of IoT technology, opportunity to address upcoming challenges and research directions created [1, 11].

This paper proposed architecture of IoT system for healthcare sector especially useful in ICU, CCU, and Ambulances etc. Efficient monitoring in ICU, CCU or ICU on wheel is indispensable need in healthcare. Doctors always prefer to have precise information in marginal time about the patients under treatment. Presently nurses do continuously monitoring for such critical cares but availability of qualified nurses and other healthcare staff is big concern particularly most of developing country like India, China [10]. Another advantage of proposed system is to reduce the chances of human errors significantly, as in one of the finding by US institute of Medicine argues that medical errors persist as the number 3 killer claiming lives of some 400,000 people each year [4].

© Springer International Publishing AG 2018
S.C. Satapathy and A. Joshi (eds.), *Information and Communication*
Technology for Intelligent Systems (ICTIS 2017) - Volume 2, Smart Innovation,
Systems and Technologies 84, DOI 10.1007/978-3-319-63645-0_44

In nutshell the proposed intelligent real time IoT based system for monitoring ICU Patient will prevent from human errors and allow to continuous patient monitoring with less support staff; also provide efficient communication for precise information. Real time patient monitoring system collect data through bed side patient monitors. Inter-communication network system uploads this data on cloud for further processing. Intelligent software agent process this data further and sending notification to special monitoring cell and doctor.

2 Related Work

The proposed work is inspired by the couple of researchers, R. Kumar; M. Pallikonda Rajasekaran proposed Raspberry Pi board to makes monitoring of patients' health parameters like temperature of body, respiration-rate, heart beats and body movement [2].

Similar kind of cheaper modular monitoring system prototype was proposed for mobile support. This support brings faster and better medical involvements in emergency medical cases. In this model low power sensors measures Electrocardiogram (EKG), Oxygen Saturation at the peak of blood pulsation (SpO_2), temperature and movement [3]. Researchers have suggested similar conceptual frame work to deploy on cloud [4]. There are standard telemedicine and medical data base exists as per IEEE 11073, which are compatible to proposed model [5]. Also there are few health systems are available, which uses mail notification for communications and provides features like remote monitoring and controlling [6]. There are development in the medical devices to support such framework and algorithm has been developed for wearable devices for get health data [7, 8, 14]. Mobile applications are available for monitor infant & old age people's health and helping in decision making [9, 12, 15]. To associate doctors, staff and patients, IoT based communication is also proposed for homecare of the Non-Communicable Disease (NCD) patients [10, 17]. Researchers have proposed systems to monitor data for critical deceases like heart, which can be further transmitted to practitioners and alarm system in the event of arrhythmia episode [13, 16].

3 System Components and Architecture

The proposed system is designed to get health information of bedside patients. Devices are available, which can gather the data from patients' body and display it. We are extending facility to pass this information to communicate further and process it in desired way. Bedside monitors are devices which continuously reading patient data. Bedside patient monitors measures capnography, conventional diagnostic, 12-led ECG, non-invasive and invasive blood pressure, respiration, FAST-SpO_2, temperature, BIS and cardiac output. To achieve objective to store, communicate data from bedside monitors or other wearable devices, the following components would be needed.

Sensor: Sensors are preliminary responsible to capture continuously patients' health data. It can be part of wearable devices or bedside monitor system. Typical sensors used in bedside monitor measured following parameters (Fig. 1 and Table 1).

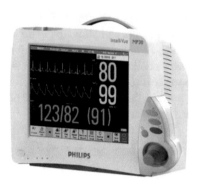

Fig. 1. Bedside patient monitor

Table 1. Typical health parameters

Arterial Carbon Dioxide Partial Pressure (PaCO$_2$)
Arterial Oxygen Partial Pressure (PaO$_2$)
Arterial Oxygen Saturation (SaO$_2$)
Central Venous Pressure (CVP)
Central Venous Oxygen Saturation (ScvO$_2$)
End Tidal Carbon Dioxide (ETCO$_2$)
Mean Arterial Pressure (MAP)
Mixed Venous Oxygen Saturation (SvO$_2$)
Oxygen Saturation at The Peak Of Blood Pulsation (SpO$_2$)
Positive End Expiratory Pressure (PEEP)
Pulmonary Artery Catheter (PAC)
Systemic Vascular Resistance (SVR)

Interconnection Networks: Interconnection networking allows sensor to transmit captured health data to the other systems' component like server etc.

Server and Database: Serves need to manage all the received real time patient data. Also server has database for standard health parameters. The system should be able to update standard health parameters as per WHO indicators. The following tables give an example of sample of standard health parameters (Tables 2 and 3).

In healthcare, critical patient are treating in ICU. Due to critical, health doctors needs to continuously monitor the patient. Doctor use bed side monitors for observing the patient. Bedside monitors continuously transmitting patient data i.e. Heart Rate, Blood Pressure, Oxygen Level, Pressure in Brain etc. to local server for the analyzing purpose. An Intelligent real time monitor system analyze the human body parameters. Following are the different standards recommended by the American Heart Association for blood pressure

In recent development, we have blood oxygen monitors, which are able to check blood and oxygen level from the human body, So it is easy to connect output of this monitor to IRTBS.

Table 2. Standard health parameters (Blood Pressure)

Stages of blood pressure	Systolic mm Hg (upper #)		Diastolic mm Hg (lower #)
Acceptable	<120	&	<80
Pre-hypertension	120 to 39	or	80 to 89
Stage 1-Hypertension High Blood Pressure	140 to 159	or	90 to 99
Stage 2-Hypertension High Blood Pressure	160 or higher	or	100 or higher

Table 3. Standard health parameters (Heart Rate for men and women)

Standard heart-rate for men						
Age-Group	18 to 25	26 to 35	36 to 45	46 to 55	56 to 65	>65
Sportsperson	49 to 55	49 to 55	50 to 56	50 to 57	51 to 56	50 to 55
Outstanding	56 to 61	55 to 61	57 to 62	58 to 63	57 to 61	56 to 61
Good	62 to 65	62 to 65	63 to 66	65 to 67	62 to 67	66 to 69
Above-Average	66 to 69	66 to 70	67 to 70	68 to 71	68 to 71	66 to 69
Below-Average	74 to 81	75 to 81	76 to 82	77 to 83	76 to 81	74 to 79
Poor	≥ 82	≥ 82	≥ 83	≥ 84	≥ 82	≥ 80
Standard heart-rate for women						
Age-Group	18 to 25	26 to 35	36 to 45	46 to 55	56 to 65	>65
Sportsperson	54 to 60	54 to 59	54 to 59	54 to 60	54 to 59	54 to 59
Outstanding	61 to 65	60 to 64	60 to 64	61 to 65	60 to 64	60 to 64
Good	66 to 69	65 to 68	65 to 69	66 to 69	65 to 68	65 to 68
Above-Average	70 to 73	69 to 72	70 to 73	70 to 73	69 to 73	69 to 72
Below-Average	79 to 84	77 to 82	79 to 84	78 to 83	78 to 83	77 to 84
Poor	≥ 85	≥ 83	≥ 85	≥ 84	≥ 84	≥ 84

PRESSURE IN THE BRAIN (INTRACRANIAL PRESURE) - The pressure inside the head may rise in patients with head injuries or after a stroke. The kind of brain pressure is known as intracranial pressure, It may block the blood flow in brain. A probe can be inserted in the brain to measure and help doctors provide therapies to reduce it. Value up to 20 mm Hg are acceptable, but the team may decide to use higher values on some patients.

Cellular Phone or Personal Digital Assistant (PDA): People from the different location, communicate which each other through cellar phone. Communication through mobile phones not only reduces the cost but it also provides faster way of communication between peoples. Doctors can immediate take decision and provide quickly medical treatment to critical patient admitted in ICU.

Intelligent Software Agent: Intelligent software agent are automated code preliminary responsible to do essential processing on health data captured by the sensors. The processing may have different level. At very basic level the captured data stored in database and compared with the standard health parameters. Incase if any parameters crossed the boundary (lower limit and upper limit) defined by the standards. Immediately software agent generates notification and sends it to Emergency Care Unit and concerned practitioners. The intelligent automated software agents' help for following activities, in case of causality observed based on captured health data and standards.

(1) It confirms doctor responsible to provide treatment to patient.
(2) Sending notifications to Emergency Care Unit (ECU) and appointed doctor for concerned patient.
(3) Confirms preparedness of emergency care unit resources Staff and Medical Equipment and allow communicating delegated person of ECU and responsible doctor.
(4) Provide required patients' details to concerned doctor(s).
(5) Helps doctor(s) to refer similar historical cases and consult related other medical practitioners.

System Architecture: The following diagram represents the architecture of IRTBS for typical patient monitoring (Fig. 2).

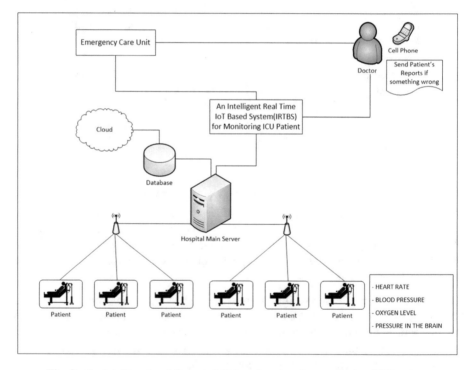

Fig. 2. An intelligent real time an IoT based system for monitoring ICU patient

4   Conclusion and Future Work

Real time IoT based system for monitoring ICU Patient reduces chance of human errors significantly as patient admitted in hospital require 24×7 continuous monitoring. As it is normal scenario to have much more patients as compared to attendees doctors. IoT based systems automate the critical observations in ICU and facilities doctors to spare more time towards decision. Future work can extend by connecting and coordinating doctors for their availability for treatment and contribute/serve to the hospitals and helps in balancing the load. The proposed architecture can be tested and compared with existing system.

Acknowledgements. We are thankful to U N Mehta Institute of Cardiology and Research Centre, Ahmedabad, India, Sterling Hospital, Ahmedabad, India to extend their support to pursue research work and providing access to their knowledge of ICU monitors system. We are also thankful to below mentioned researcher in references, who has inspired us by their research work and publications.

References

1. Stankovic, J.A.: Research directions for the Internet of Things. Internet Things J. **1**, 3–9 (2014)
2. Kumar, R., Pallikonda Rajasekaran, M.: An IoT based patient monitoring system using raspberry Pi. Int. Conf. Comput. Technol. Intell. Data **35**(2), 1–4 (2016)
3. Archip, A., Botezatu, N., Serban, E., Herghelegiu, P.-C., Zal, A.: An IoT based system for remote patient monitoring. In: 17th International Carpathian Control Conference (ICCC), pp. 1–6 (2016)
4. Tyagi, S., Agarwal, A., Maheshwari, P.: A conceptual framework for IoT-based healthcare system using cloud computing. In: 6th International Conference - Cloud System and Big Data Engineering, pp. 503–507 (2016)
5. Blumrosen, G., Avisdris, N., Kupfer, R., Rubinsky, B.: C-SMART: efficient seamless cellular phone based patient monitoring system. In: IEEE International Symposium on a World of Wireless, Mobile and Multimedia Networks, pp. 1–6 (2011)
6. Dhanaliya, U., Devani, A.: Implementation of e-health care system using web services and cloud computing. In: International Conference on Communication and Signal Processing - (ICCSP), pp. 1034–1036 (2016)
7. Azariadi, D., Tsoutsouras, V., Xydis, S., Soudris, D.: ECG signal analysis and arrhythmia detection on IoT wearable medical devices. In: 5th International Conference on Modern Circuits and Systems Technologies (MOCAST), pp. 1–4 (2016)
8. Biswas, S., Misra, S.: Designing of a prototype of e-health monitoring system. In: IEEE International Conference on Research in Computational Intelligence and Communication Networks (ICRCICN), pp. 267–272 (2015)
9. Stutzel, M.C., Fillipo, M., Sztajnberg, A., Brittes, A., da Motta, L.B.: Smai mobile system for elderly monitoring. In: IEEE International Conference on Serious Games and Applications for Health (SeGAH), pp. 1–8 (2016)
10. Liu, Y., Niu, J., Yang, L., Shu, L.: eBPlatform an IoT-based system for NCD patients homecare in China. In: IEEE Global Communications Conference, pp. 2448–2453 (2014)

11. Hassanalieragh, M., Page, A., Soyata, T., Sharma, G., Aktas, M., Mateos, G., Kantarci, B., Andreescu, S.: Health monitoring and management using Internet-of-Things (IoT) sensing with cloud-based processing: opportunities and challenges. In: IEEE International Conference on Services Computing, pp. 285–292 (2015)
12. Al-Adhab, A., Altmimi, H., Alhawashi, M., Alabduljabbar, H., Harrathi, F., ALmubarek, H.: IoT for remote elderly patient care based on fuzzy logic. In: International Symposium on Networks, Computers and Communications (ISNCC), pp. 1–5 (2016)
13. Nigam, K.U., Chavan, A.A., Ghatule, S.S., Barkade, V.M.: IoT-beat an intelligent nurse for the cardiac patient. In: International Conference on Communication and Signal Processing (ICCSP), pp. 0976–0982 (2016)
14. Mun, L.B., Jinsong, O.: Intelligent healthcare service by using collaborations between IoT personal health devices. Int. J. Bio Sci. Bio Technol. **6**(1), 155–164 (2014)
15. Wasnik, P., Jeyakumar, A.: Monitoring stress level parameters of frequent computer users. In: International Conference on Communication and Signal Processing (ICCSP), pp. 1753–1757 (2016)
16. Mohammed, J., Lung, C.-H., Ocneanu, A., Thakral, A., Jones, C., Adler, A.: Internet of Things remote patient monitoring using web services and cloud computing. In: IEEE International Conference on Internet of Things (iThings), pp. 256–263. IEEE (2014)
17. Ghosh, A.M., Halder, D., Alamgir Hossain, S.K.: Remote health monitoring system through IoT. In: 5th International Conference on Informatics, Electronics and Vision (ICIEV), pp. 921–926 (2016)

SLA Management in Cloud Federation

Vipul Chudasama[(⊠)], Dhaval Tilala, and Madhuri Bhavsar

Computer Science and Engineering, Institute of Technology,
Nirma University, S.G. Highway, Ahmedabad 382481, India
{vipul.chudasama,15mcen28,madhuri.bhavsar}@nirmauni.ac.in

Abstract. Now a days cloud computing is the major area of research Because cloud computing has own many benefits. Cloud computing also provides cost effective Resources so that it can become more and more helpful to IT trends. Distributed computing is an extensive arrangement that conveys IT as an administration. It is an Internet-based registering arrangement where shared assets are given like power disseminated on the electrical grid. Cloud suggest to a particular IT environment that is expected with the end goal of remotely provisioning versatile and measured IT resources. Whereas the Internet gives open access to many Web-based IT assets, a cloud is commonly exclusive and offers access to IT assets that is metered following SLAs implementation depends on guidelines that are redesigned in runtime so as to proactively recognize conceivable SLA Violations and handle them in a proper manner. Our proposed framework allows the creation and implementation of effective SLA for provisioning of service. SLA management are one kind of common comprehension between CSP (Cloud Service Provider) and customer.

Keywords: Cloud federation · SLA life cycle · SLA management architechure · SLA violation

1 Introduction

Mostly customers save their data and can retrieve them anytime and from anywhere without having care about backup and data recovery. Cloud Computing has major its own benefits so all the technology trends are the move towards it. Commonly understanding of cloud computing is continuously expanding and concepts used to define it and often need clarifying. The information sources can incorporate databases, information distribution centers, the Web, other data storehouses, or information that are pushed into the framework dynamically. It has been said that Cloud is another innovation in PC time yet its market execution demonstrates a very surprising picture. Organizations are currently leasing assets for capacity and other computational purposes frame cloud so that the framework cost can be diminished.

© Springer International Publishing AG 2018
S.C. Satapathy and A. Joshi (eds.), *Information and Communication*
Technology for Intelligent Systems (ICTIS 2017) - Volume 2, Smart Innovation,
Systems and Technologies 84, DOI 10.1007/978-3-319-63645-0_45

Cloud Computing mainly refers to provide on-demand services with the better utilization of the resources which is used by the each and every host in the cluster. It also provides the flexible manner to the provisioning of the resource for giving faster and efficient result to the user. Resource scheduling and allocation are mainly considered to be a critical issue in Cloud computing. Generally, it is difficult or hard to find and optimal resource allocation which minimizes the execution and effectively utilized the resource which is available.

2   Related Work

R. Buyya [2] contributed that Existing system for Resource administration frameworks in Data server center Service Level Agreement arranged Resource allocation. Currently no work has been accomplished for client driven administration, computational hazard management, into Market-based asset provisioning framework for powerfully changing ventures necessity of cloud computing. So paper presents Architecture of SLA-oriented Resource Allocation and vision and Challenges. According to architecture combination of market-based resource provisioning policies and technologies for dynamically allocation of resources.

Adil Maarouf [5] provides SLA lifecycle with practical modeling in cloud computing. Because Research in service level Agreement has been increased. It is Because of the main role of solution the trade-off between Cloud Service Provider and Customer. Each and every SLA has a different Lifecycle to be operated and life cycle of SLA include different stages according to their requirement and domain. Here the given SLA lifecycle steps for changes of the point of view of all customers required in the agreement. For illustrations are demonstrates that fundamental phases, structures, processes and substances, UML displaying graphs.

S.K. Garg [8] investigated that for hosted any type of application on cloud and giving services to customer software company has to keep their own hardware or rent it from framework providers. Software as a service providers company incurs extra cost. The paper shows the proposed algorithm shows the and designed in a way that guaranteed that SaaS suppliers are capable of maintaining dynamic change of clients and their demand to framework level parameters. In this paper also considering QoS are like response time, service initiation time.

Torsten Braun [1] introduce the objective for the manages Enterprise For Distributed application in cloud computing environments Under Service level Agreement maintains optimize Resource control. Dissertation mainly focuses on how SLA can be used as the input of cloud management system. Which goal is to increasing efficiency for allocating resources to start with creator presents SLA semantic model. Also, characterize Multi-Purpose VM Allocation calculation for productive asset designation in Infrastructure clouds.

Kuan C. Lai [3] proposed resource allocation enable and reduces energy consumption in cloud services providers and also get more profit. Some times client submits more jobs to the server and cloud provider has no enough resources for completing jobs at that time cloud provider has to lease resources from others.

So for solving this problem introduce a combinatorial auction-based problem for different resource allocation and this system concludes that more profit to cloud customer. A. Voulodimos [6] defines customized SLAs in cloud computing environments because now days for the demand of online storage are increasing rapidly. Cloud customer can store and get back data without having no caring about backup and data recovery. In this system customer can choose their Requirement and creates SLA according to their needs. SLA are automatically managed according to their rules and updated in runtime and also manages SLA violation in proper manner.

3   Existing Scenario

– Currently, there are various Types of work done regarding SLA management out of which most of the work done is in the area of Single cloud means only for the public or private cloud.
– But there is some work are done for SLA management for public cloud or Private cloud and some method is developed for SLA monitoring.
– Existing algorithms for SLA Violation are not too much efficient and there is consider only one or two Parameters.
– There is a need for Modify the SLA management algorithm and expand successfully for cloud Federation and also develop an algorithm for SLA Monitoring system with the violation.

3.1   SLA Management in Cloud

1. Discover service providers: where specialist organizations are situated by necessities.
2. Define SLA: incorporates meaning of administrations, parties, penalty approaches and QoS parameters. In this progression it is possible for gatherings to consult to achieve an understanding.
3. Establish agreement: an SLA format is built up and filled in by particular understanding, and gatherings are beginning to focus on the assention.
4. Monitor SLA violation: the supplier's conveyance execution is measured against to the agreement.
5. Terminate SLA: SLA ends due to timeout or any party's violation.
6. Enforce penalties for SLA violation: in cloud services that there are any gathering and damaging contract terms, the relating punishment and conditions are executed [5].

4   Proposed System

4.1   Proposed System Architecture

Here we are having private clouds and public clouds. There is one central broker at the top level who are receiving the user request after its transfers to the

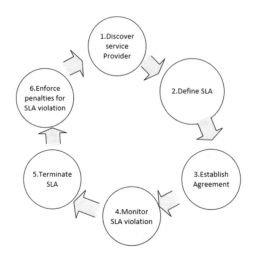

Fig. 1. SLA (Service Level Agreement) life cycle [7]

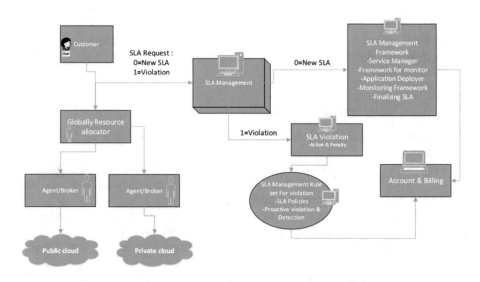

Fig. 2. Proposed block diagram of SLA management in cloud federation

broker of public and private cloud agent. So its checks resource catalogue and take appropriate action on it.

We assume that in our federated Cloud the public cloud will have a large pool of resources like mips, ram, bandwidth, storage etc. Also we assume that private cloud will have a smaller pool of resources as compare to the public cloud. The clouds are connected with the help of broker also known as Cloud Broker [4].

Algorithm 1. Algorithm for User Requirement

I/p: get user input.
O/p: create Federation (allocate resources)
User gives requirement
No of Vm
Vm Configuration (Mips, Bandwidth, RAM, Storage)
Analyze the requirement
Check Resources
Then
Classify the public and private cloud
if resources are available(for user requirement) **then**
    Create VM();
end if
Do Federation ();

Now New User Request is come for the resource allocation and SLA Management to Centrally located Agent. After Analyze the request agent decides according to parameters it decided that Federation are needed or not. If Federation are Needed than Agent Contact to Public and Private Cloud Broker For allocating Resource.

After Completing this task Request Transfer to the SLA Management System and according to their given id (0 = New SLA, 1 = Violation) transfer the Request to that Department. After Completing this it will Transfer to the Account and billing Section and generate the bill according to their User ids.

One time SLA Will be generated After that SLA management system are monitoring the Whole System. If violation are occurred due to any given threshold limit user will not getting service at that time violation are detected.

5   Proposed Algorithm

Basic terms: Mi = Mips of ith cloud, RAi = RAM of ith cloud, Bi = Bandwidth of ith cloud, Ri = total Resources in ith cloud, Ri = Mi + RAi + Bi.

Parameters: T.M = Threshold of MIPS, TRA = Threshold of RAM, TB = Threshold of Bandwidth, TU = Threshold of Uptime, TS = Threshold of storage, TC = Threshold of Max Cost, Sid = Id of SLA, Uid = User ID, Rid = Request ID, minCPU = minimum CPU percentage provided, max-CPU = maximum CPU percentage provided, minMEM = minimum Memory amount provided, maxMEM = maximum Memory amount provided.

Algorithm 2. Algorithm for User Requirement

/* Create VM */
Create VM (Vid, User id, Mips, RAM, BW, Size, Vmm, Cloud Schedular)
{
Set(Vid, UserID, Mips, Ram, BW, Size, VMM, Cloudlet Schedular);
}
Do Federation ();
Allocate resources to client;
Identify the user requirement
ID1 = New SLA
ID2 = Violation
if ID == 1 **then**
  Create New SLA ();
else if ID == 2 **then**
  Violation ();
end if

Algorithm 3. Create New SLA Algorithm:

1: i/p: Used Id
2: o/p: Generate New SLA
3: User Entry Requirement
4: Client Analysis Requirement
5: Create SLA();
6: /*Create New SLA */
7: Create SLA
8: Get (Request type,Account type,Contract length,Response time,Vm types,Service
   initiation time,Vm price,Data transfer time,Data transfer speed);

Algorithm 1 defines user requirement, in this algorithm first check the user requirement after analyzing the requirement algorithm create new VM if needed. And if the federation is needed it create a federation.

Algorithm 2 shows the create VM according to the requirement and it allocates to Customer as resource services. After that, it again checks the request type and takes appropriate action on it.

After creating federation in Algorithm 3 it creates SLA for the appropriate user according to their IDs. For creating SLA following parameters are considered(request type, Account type, contract length, response time) etc.

Define Penalty: $P1 =$ Penalty for Mips/unit $= 0.01$, $P2 =$ Penalty for Ram/unit $= 0.03$, $P3 =$ Penalty for Bandwidth/unit $= 0.05$, $P4 =$ Penalty for Uptime/unit $= 500$, $P5 =$ Penalty for Storage/unit $= 0.07$, $P6 =$ Penalty for Cost/unit $= 100$. for (i=1;i<7;i++) {Total Penalty $= \Sigma_{i=1}^{7} |ProvidedResource(i) - RequestedResource(i) * Pi;$}

Algorithm 4. Algorithm for Violation

```
 1: i/p: Violation ID
 2: o/p: Identify Violation and Take appropriate action
 3: Monitor SLA
 4: Detect Violation(Parameters like MIPS, RAM, BANDWIDTH, UPTIME )
 5: if  (T.RA < 50GB) then
 6:    Violation occurred due to insufficient RAM
 7: else if (T.M < 1500000) then
 8:    Violation occurred due to MIPS then
 9: else if (T.B < 10 GB)  then
10:    Violation occurred due to insufficient Bandwidth
11: else if (T.U < 97%) then
12:    Violation occurred due to insufficient UPTIME
13: else if (T.R < 2000GB) then
14:    Violation occurred due to insufficient STORAGE
15: else if (TC > User Defined) then
16:    Violation occurred due to insufficient Cost
17:    Identify Violation Parameters
18:    Take Appropriate action (Penalty, termination)
19: end if
```

Here, in this Algorithm 4 is a sample algorithm for detecting the violation of services. After violation detection, it defines appropriate penalty, count the total amount of penalty.

6   Conclusion and Future Work

Proposed model is beneficial as per user Requirement which can be managed by CSP (Cloud Service Provider). This model provides transparency to Customer Relationship with the cloud provider. In future proposed model can be executed for billing information for better quality of SLA and customer satisfaction.

References

1. Antonescu, A.-F., Braun, T.: Service level agreements-driven management of distributed applications. In: Cloud Computing Environments, pp. 1122–1128. IEEE, April 2015
2. Buyya, R., Garg, S.K., Calheiros, R.N.: SLA-oriented resource provisioning for cloud computing: challenges, architecture and solutions. In: 2011 International Conference on Cloud and Service Computing (CSC), pp. 1–10. IEEE (2011)
3. Chang, C.C., Lai, K.C., Yang, C.T.: Auction-based resource provisioning with sla consideration on multi-cloud systems. In: IEEE 37th Annual Computer Software and Applications Conference Workshops (COMPSACW 2013), pp. 445–450. IEEE, July 2013

4. Iqbal, W., Dailey, M.N., Carrera, D.: SLA-driven dynamic resource management for multi-tier web applications in a cloud. In: 2010 10th IEEE/ACM International Conference on Cluster, Cloud and Grid Computing (CCGrid), pp. 832–837. IEEE (2010)

5. Maarouf, A., Marzouk, A., Haqiq, A.: Practical modeling of the SLA life cycle in cloud computing. In: 2015 15th International Conference on Intelligent Systems Design and Applications (ISDA), pp. 52–58. IEEE (2015)

6. Mavrogeorgi, N., Alexandrou, V., Voulodioms, A.: Customized SLAs. In: Cloud Environments, pp. 262–269. IEEE (2013)

7. Patel, K.S., Sarje, A.K.: VM provisioning method to improve the profit and SLA violation of cloud services providers. In: 2012 IEEE International Conference on Cloud Computing in Emerging Markets (CCEM), IEEE (2012)

8. Wu, L., Garg, S.K., Buyya, R.: SLA-based resource allocation for software as a service provider (SaaS) in a cloud Computing Environment. In: 2011 11th IEEE/ACM International Symposium on Cluster, Cloud and Grid Computing (CCGrid), IEEE (2011)

Weight Based Workflow Scheduling in Cloud Federation

Vipul Chudasama$^{(\boxtimes)}$, Jinesh Shah$^{(\boxtimes)}$, and Madhuri Bhavsar

Institute of Technology, Nirma University, S.G. Highway, Ahmedabad 382481, India
{vipul.chudasama,15mcec23,madhuri.bhavsar}@nirmauni.ac.in

Abstract. To cater the need of a user as the requirement of infrastructure, application building environment or contemporary software private and public clouds are exiting for provisioning of such services. But cloud providers are facing the problem of how to deploy their applications over different clouds keeping in mind their different requirements in terms of QoS (Cost, resource utilization, execution time). Different clouds have different advantages such as one cloud will be more reliable and efficient whereas, private cloud will be more secure or less expensive. In order to get the benefits of both clouds, we can use the concept of cloud federation. When using cloud federation it becomes important how to schedule large workflows over federated clouds. Proposed work addresses the issue of scheduling large workflow over federated clouds. SMARTFED is used for the cloud federation and our algorithm is used to schedule different workflows according to the QoS parameters over the federation.

Keywords: Cloud computing · Workflow scheduling · Cloud federation · QoS

1 Introduction

1.1 Cloud Technology

Cloud computing means storing and accessing resources over the Internet which are physically not present. We can say that the cloud is in fact Internet. With the development of computing and internet technologies, every day new challenges are arising. One of those challenges is increasing in demand of connectivity and resource handling. In the current trend of IT infrastructure, dynamic resource and access become compulsory. To overcome these challenges, a very elastic infrastructure is needed that can be scaled according to increasing or decreasing demand. These factors are responsible for the concept of Cloud. It has been said that Cloud is a new technology in computer era but its market performance show a totally different picture. Companies are now renting resources for storage and other computational purposes from the cloud so that the infrastructure cost can be decreased. Service model of cloud (SaaS, PaaS, and IaaS) based on a pay-as-you-go model.

© Springer International Publishing AG 2018
S.C. Satapathy and A. Joshi (eds.), *Information and Communication Technology for Intelligent Systems (ICTIS 2017) - Volume 2*, Smart Innovation, Systems and Technologies 84, DOI 10.1007/978-3-319-63645-0_46

1.2   Workflow Scheduling in Cloud

A workflow can be defined as a set of tasks or steps which are used to complete any process. The workflow management describes task synchronization, task invocation, and the way information flow and these are done in a fixed manner. Workflow scheduling maps and manages workflow tasks execution on shared resources. Thus the workflow scheduling algorithm finds a correct order in which the task execution takes place also obeying the users' constraint. Workflow scheduling will schedule jobs in such a way that all the jobs will get executed taking minimal time while maintaining the QoS parameters with satisfying clients needs (Fig. 1).

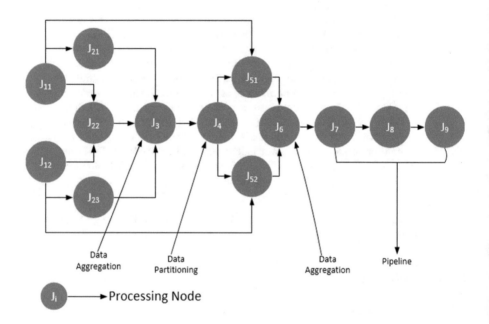

Fig. 1. Example- Montage workflow [1]

The Montage [1] is an open source toolkit developed by the NASA/Infrared Science Archive which was developed to process input images in the Flexible Image Transport System (FITS) format to generate custom mosaics of the sky. Montage application can be represented as a workflow which will be executed in cloud environments.

2   Related Work

Luiz et al. [4] in which they proposed HCOC algorithm. In HCOC algorithm a decision is made which resources should we leased from the public cloud and aggregated to the private cloud to execute a workflow within a given execution time while providing sufficient processing power. The limitation of HCOC is that the scheduling is only limited to single cloud environment.

Diaz-Montes et al. [8] in which they suggested a service framework enable autonomic execution of dynamic workflows for the multi-cloud environment. It allows the customer to customize scheduling policies to use those resources according to their needs. The limitation of this algorithm is that it incurred excessive data transfer cost because data locality was not properly exploited and too much weight is assigned to VM in decision making.

Raffaelli et al. [5] discussed an approach named GraspCC-fed which estimates the optimally the amount of VM to allocate for each workflow. The problem with this approach is that it is limited to the single cloud environment while scheduling of SciCumulus-fed.

Wen et al. [9] proposed an approach which quantifies the most reliable workflow deployments based on entropy. Also, its monetary cost which considers the data storage, the price of computing power, and inter-cloud communication is considered.

Diaz-Montes et al. [6] introduced a framework which allows the users to have customized scheduling policies which drive the way resources are federated and used. In dynamic software-defined resource federation, it also manages end to end execution of the data intensive application workflows.

Jhu et al. [7] proposed an algorithm in which a set of agents that can federate multiple clouds and automatically monitor the entire system, and manage the job if it's a workflow job. Also, an algorithm which dispatches jobs dynamically is proposed in order to reduce the VM usage.

3   Proposed System

3.1   Proposed Architecture

Here we propose private and public cloud environment in which there is a central cloud where the user submits the workflow and the resources from different clouds are allocated for the workflows. The scheduler schedules the workflow according to the weights of the QoS parameters. We assume that in our federated architecture the public cloud will have a large pool of resources like CPU, RAM, bandwidth, storage. Also, we assume that private cloud will have a smaller pool of resources as compare to the public cloud. The clouds are connected with the help of broker also known as Cloud Broker. The workflow provided by user will be an XML file in which dependencies between nodes will be defined in form of edges between them and the running time of each node.

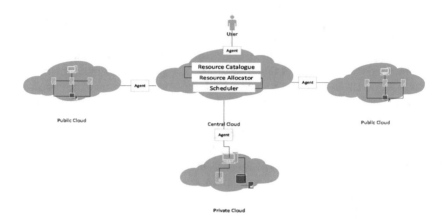

Fig. 2. System architecture

Here we assume that the public cloud has more resources and it's cost is higher than private cloud. The private cloud has limited resources and nearly zero cost.

3.2   Proposed Algorithm

Folllowing terms are used in the algorithm

Basic terms	Scheduling parameters	Weights
Ni = ith node of a Workflow	Pi1 = Cost parameter in ith workflow	Wc = Weight of cost parameter in ith workflow
CCi = Cost per unit resource of Ci cloud	Pi2 = Resource utilization parameter in ith workflow	Wr = Weight of resource utilization parameter in ith workflow
n = total node in workflow	Pi3 = Execution time parameter in ith workflow	We = Weight of exe-cution time parameter in ith workflow

Algorithm 1. Algorithm for Workflow Scheduling

   Input   : Workflow XML file, parameters weight, parameters value, resource catalog
   Output: Properly scheduled Workflow in federated cloud

```
 1  ReadWi() //Read workflow
 2  {
 3         Read → Xml file
 4         Detect dependencies in the workflow using Edges and the level
 5         Read the nodes in the workflow
 6         Assign cloudlet for each node
 7  }
 8  //Calculate Parameter
 9  Pi(Wi)
10  {
11         return ← max(Wc,Wr,We)
12  }
13
14  //Classify Workflow
15  for i=0;i<n;i++ do
16  |   Classify Ni()
17  |   {
18            Switch(Pi)
19            {
20            Case P1:
21            if C3|C5=available then
22      |          Assign Ni to C3|C5
23            else
24      |          Assign Ni to minCost(C1,C2,C4)

25            Case P2:
26            Assign Ni to Ci having
27            max(Resource Utilization)
28
```

$$ResourceUtilization = \frac{no.\ of\ freehosts}{no.\ of\ total\ available\ hosts} \quad (1)$$

```
29            Case P3:
30            Assign Ni to C1 | C2 | C4 having max(mips)
31  end
32            }
33  }
```

As per Algorithm 1, the workflow is in form of XML file in which a DAG is given having nodes and dependencies. Each node in the workflow is assigned a cloudlet for execution. For each workflow, the user will assign weights to each scheduling parameters. As per table, no the parameter with the highest weight will be taken as scheduling parameter. If cost is the scheduling parameter then the private cloud will be preferred as they have low cost, if a private cloud is unavailable the public cloud having the least cost will be chosen.

If Resource utilization is the parameter the ratio is computed and the cloud having the highest value of the ratio is selected. If Execution time is the parameter then the cloud having the highest available MIPS is selected.

Algorithm 2. Algorithm for Federation

---

1 //**Federate Workflow**
2 Federation()
3 {
4 FindFedCloud()
5 {
6 Sort Weights(Wi) in descending order
7 **if** cloud Ci with Pi= Wc == available **then**
8 | federate Wi to Ci;
9 **else if** cloud Ci with Pi= Wr == available **then**
10 | federate Wi to Ci;
11 **else if** cloud Ci with Pi= We == available **then**
12 | federate Ni to Ci;
13 }
14 }

During federation, it is necessary to find out on which cloud the workflow needs to be scheduled. To find out this we sort the weight of workflow in descending order and check one by one which cloud is available with the parameter having maximum weight, if it is not available then we check the parameter with the next highest weight and so on until the workflow is assigned to a different cloud.

4   Tools

SmartFed [2] is a software simulator for Cloud Federations, as modeled in the Contrail European Project. It has built on top of CloudSim and requires such library for running.

The SmartFed extends libraries of CloudSim and workflowsim provides methods and packages for cloud federation. The SmartFed has added packages and methods for Federation Datacenter, allocator, creating applications from DAG using vertex and edges, SLA, Monitoring, Queue management, and storage while it directly imports methods and packages for Datacenter, broker and cloudlets from CloudSim.

First federated datacenters are created and then the required host are created in each datacenters. CloudSim is initialised and then federation process is initiated. For allocating the VMs in feredated datacenters roundrobin algorithm is used. The resource is estimated, the monitoring hub moniters the federation process. The application which is to be run in VMs of federated datacenters are added to a list and from list the applications are allocated to VMs. After the successful execution of applications it's execution time and cost of execution is displayed.

For scheduling of workflow in federated cloud WorkflowSim [3] is used. WorkflowSim is a open source workflow simulator. It models workflows with a DAG model in form of xml files which are known as dax files in WorkflowSim.

The process of executing and scheduling workflow is as below.

- Workflow Mapper It's use is to import the Directed Acyclic Xml file and other metadata info like file size etc. from the Workflow Generator. It then creates a list of tasks and assigns those tasks to an execution node.
- Workflow Engine Based on the dependencies, the tasks are managed by Workflow Engine. The free tasks are released to the Clustering Engine by Workflow Engine.
- Clustering engine To reduced the scheduling overhead it merges the tasks into jobs.
- Workflow Scheduler It matchs the jobs to resources and submits them to a worker node for its execution.

5 Conclusion and Future Work

The proposed algorithms can be used for scheduling the workflows on federated environment. Performance of the algorithm can be analysed as per QoS parameter (Cost, Resource utilization, Execution time) with well known cloud providers. It can be extended for multiple workflows.

References

1. Bharathi, S., et al.: Characterization of scientific workflows. In: Third Workshop on Workflows in Support of Large-Scale Science, 2008. WORKS 2008. IEEE (2008)
2. https://github.com/ecarlini/smartfed
3. http://www.workflowsim.org/
4. Bittencourt, L.F., Madeira, E.R.M.: HCOC: a cost optimization algorithm for workflow scheduling in hybrid clouds. J. Internet Serv. Appl. **2**(3), 207–227 (2011)
5. Coutinho, R.C., et al.: Optimizing virtual machine allocation for parallel scientific workflows in federated clouds. Future Gener. Comput. Syst. **46**, 51–68 (2015)
6. Diaz-Montes, J., et al.: Supporting data-intensive workflows in software-defined federated multi-clouds. In: IEEE Transactions on Cloud Computing (2015)
7. Jhu, S.-R., et al.: Implementing a workflow agent on federated cloud. In: 2015 IEEE International Conference on Smart City/SocialCom/SustainCom (SmartCity), pp. 915–920. IEEE (2015)
8. Montes, J.D., et al.: Data-driven work flows in multi-cloud market- places. In: 2014 IEEE 7th International Conference on Cloud Computing (CLOUD), pp. 168–175. IEEE (2014)
9. Wen, Z., et al.: Cost effective, reliable and secure workflow deployment over federated clouds. IEEE Trans. Serv. Comput. (2016)

Change Detection in Remotely Sensed Images Based on Modified Log Ratio and Fuzzy Clustering

Abhishek Sharma[(✉)] and Tarun Gulati

Electronics and Communication Department,
Maharishi Markandeshwar University, Mullana, Ambala, India
abhishek.kaushik1@gmail.com, gulati_tarun@mmumullana.org

Abstract. This paper proposes a method for change detection based on Modified log ratio. The ratio of variance and mean has been considered along with the logarithmic ratio of the pixels in the images. The ratio of variance and mean played an important role in balancing the preservation of details and robustness to noise in difference image. The multi-temporal images are compared through the proposed method and a difference images has been generated. Fuzzy c means clustering (FCM) has been used to classify the changed and unchanged areas. The results are compared based upon various parameters like false negative (F_n), false positive (F_p), Kappa coefficient (K_c) and percentage correct classification (PCC). The qualitative and quantitative results show that the proposed method offers less speckle noise, higher accuracy and Kappa coefficient value as compare to the other direct comparison based algorithms.

Keywords: Change detection · Image differencing · Image rationing · Image regression · Fuzzy clustering

1 Introduction

Change detection is a technique in which two or more images are compared to detect the pixels that have undergone change with time [1]. The important challenge that change detection faces is the preservation of detail in the difference image and robustness to errors. [2]. Many change detection algorithms have been mentioned in the literature [3–7]. Some of the popular direct comparison based methods are ratioing, regression and differencing [8, 9]. Image ratioing is the technique which is used most widely for change detection [10]. Log ratio technique has been found very effective in the presence of speckle noise because of its ability to convert the multiplicative speckle noise into additive noise. The changed and unchanged pixels are classified into different clusters through fuzzy c means clustering. The changed areas belong to one cluster while the unchanged areas belong to the other. There are many clustering techniques in the literature like Fuzzy c means clustering [11, 12], k means clustering [13] and Nonsubsampled contourlet transform (NSCT) based clustering [14].

This paper proposes a modified log ratio method in which ratio of variance and mean of the pixels is considered along with the logarithmic ratio of the pixels. This enhances

© Springer International Publishing AG 2018
S.C. Satapathy and A. Joshi (eds.), *Information and Communication
Technology for Intelligent Systems (ICTIS 2017) - Volume 2*, Smart Innovation,
Systems and Technologies 84, DOI 10.1007/978-3-319-63645-0_47

the features by preserving the details of the pixels along with making the change detection process more robust to speckle noise.

The organization of this paper is as follows: The Modified change detection approach has been proposed in the next section. Third section introduces the datasets and parameters used for analysis. The results and analysis has been introduced in fourth section. The conclusion has been presented in the fifth section.

2   Proposed Method

Consider two images $I_{m1} = \{I_{m1}(i,j), 1 < i < r, 1 < j < c\}$ and $I_{m2} = \{I_{m2}(i,j), 1 < i < r, 1 < j < c\}$ of size $r \times c$, i.e., of a scene taken at different times. The images are must be co-registered before applying any change detection algorithm. The proposed approach involves the two main steps. In the first step, a difference image is generated and in the next step, the difference image is classified into changed and unchanged areas by fuzzy c means clustering algorithm.

2.1   Difference Image Generation

The proposed method generates a difference image by modifying the existing log ratio method as given in Eq. 1.

$$X = \delta \left| \frac{\log I_{m2} - \log I_{m1}}{\log I_{m2} + \log I_{m1}} \right| \tag{1}$$

where $\delta = \dfrac{\sigma}{\mu}$; μ and σ represents the total mean and total variance of the difference image respectively. X denotes the difference image.

2.2   Clustering

In fuzzy clustering, the pixels under consideration are allocated to the desired number of clusters as per the degree of belongingness. Fuzzy C Means (FCM) is the basic among all the clustering algorithms [11].

FCM is based upon the minimization of objective function J_m as mentioned in Eq. 2.

$$J_m = \sum_{i=1}^{p} \sum_{k=1}^{q} u_{ik}^n \|y_i - r_k\|^2, \ 1 \leq n \leq \infty \tag{2}$$

Where n is a positive real number. The value of membership of y_i which belongs to the cluster k is represented by u_{ik}, y_j denotes the ith value of measured data, the centre of cluster is denoted by r_k, and $\|y_i - r_k\|$ represents the similarity between the centre of cluster and measure data. The objective function is iteratively optimized to carry out the fuzzy partitioning based upon updating $r_{k\ and}$ u_{ik} as given in Eqs. 3 and 4.

$$u_{ik} = \cfrac{1}{\sum_{k=1}^{q} \left(\cfrac{\|y_i - r_k\|}{\|y_i - r_j\|} \right)^{\frac{2}{n-1}}} \qquad (3)$$

$$r_k = \frac{\sum_{1}^{p} u_{ik}^n \cdot y_i}{\sum_{1}^{q} u_{ik}^n} \qquad (4)$$

This iteration will continue till

$$\max_{ik} \left\{ |u_{ik}^{(j+1)} - u_{ik}^{(j)}| \right\} > \varepsilon \qquad (5)$$

Where ε is any positive value less than 1 and j is the number of steps for which the iteration run (Fig. 1).

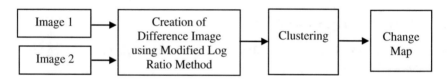

Fig. 1. Process of change detection

Process of change detection takes two multi-temporal images as input and generates a difference image in output. Fuzzy c means clustering is applied on the difference image which classifies the changed and unchanged areas into different clusters. The clustering generates a binary change map in which the changed and unchanged regions are given by pixel value as 1 and 0 respectively.

3   Dataset and Parameters

To compute the effectiveness of the proposed algorithms, two multi-temporal image datasets has been used. One dataset belongs to the Reno Lake Tahoe areas with pixel size 200×200 captured on 5th August, 1986 and 5th August, 1992 [15]. The images show the effect of draught on Reno Lake. The ground truth has been generated by manual analysis.

The effectiveness of the algorithms is computed based upon percentage correct classification (PCC) and Kappa Coefficient (K_c) [17].

$$PCC = \frac{(T_p + T_n)}{(T_p + T_n + F_p + F_n)} \qquad (6)$$

$$\text{If A} = \frac{((Tp + Fn) \times (Tp + Fp) + (Fp + Tn) \times (Tn + Fn))}{(Tp + Tn + Fp + Tn)^2} \qquad (7)$$

$$K_c = \frac{PCC - A}{1 - A} \qquad (8)$$

True positive (T_p) are the changed pixels which has been identified correctly as changed pixels. The value of (T_p) is 1 if the value of corresponding pixels in output of proposed algorithm and ground truth are both 1. Otherwise (T_p) will be zero. True negative (T_n) are the unchanged pixels which have been correctly identified as unchanged. The value of (T_n) is 1 if the corresponding pixels value in output of proposed algorithm and ground truth are both 0. Otherwise (T_n) will be zero. False positive (F_p) are those pixels which are actually changed but identified as unchanged pixels. The value of (F_p) is 1 if the pixels value in output of algorithm is 1 and the value of corresponding pixel in ground truth is 0. Otherwise (F_p) will be zero. False negative (F_n) are those unchanged pixels which have been identified wrongly as changed. The value of (F_n) is 1 if the pixels value in the output of algorithm is 0 and the value of corresponding pixel in ground truth is 1. Otherwise (F_n) will be zero (Fig. 2).

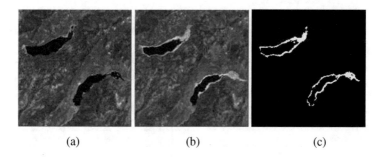

(a)   (b)   (c)

Fig. 2. Image dataset belonging to Reno Lake Tahoe area. (a) Image captured in August, 1986. (b) Image captured in August, 1992. (c) Ground truth

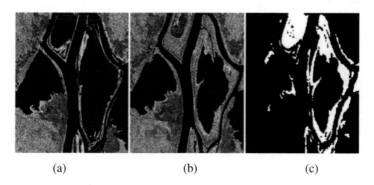

(a)   (b)   (c)

Fig. 3. Image dataset belonging to Ottawa area. (a) Image captured in July 1997. (b) Image captured in August 1997. (c) Ground truth

The second dataset belong to Ottawa city with pixel size 373×451 captured in July 1997 and August 1997 before and after summer flooding respectively [16] (Fig. 3).

4 Results and Analysis

The results of change detection process has been analyzed by applying the proposed algorithms on images datasets of Reno Lake and Ottawa city.

4.1 Results of Reno Lake Dataset

As shown in Table 1, Image regression method has yielded PCC equal to 96.3% and the value of Kappa coefficient is 0.95, Mean ratio operator has yielded PCC equal to 97.9% and Kappa coefficient value is 0.97, Direct differencing method has yielded PCC equal to 99.2% and Kappa coefficient value is 0.98. The PCC of Log ratio as obtained is equal to 99.4% and Kappa coefficient value is 0.99. The proposed method has yielded highest accuracy equal to 99.8% and the Kappa coefficient value is also highest equal to 0.997 which proves that the proposed algorithm offers highest accuracy and less speckle noise as compare to its preexistences.

Table 1. The results on the data set of Reno Lake obtained by different change detection methods.

Method	F_p	F_n	PCC (%)	(K_c)
Image regression	1343	138	96.3	0.95
Mean ratio	852	1	97.8	0.97
Direct differencing	133	183	99.2	0.98
Log ratio	0	211	99.5	0.993
Proposed method	32	42	99.8	0.997

The proposed algorithm offers the maximum resemblance with the ground truth as given in Fig. 4(e). If the output of proposed method is compared with the other methods then it is understood that the proposed method has recovered the maximum pixels with least speckle noise.

4.2 Results of Ottawa Dataset

The quantitative results of proposed algorithm and other existing algorithms for Ottawa dataset has been given in Table 2.

The proposed method offers the highest PCC and Kappa coefficient value equal to 95.52% and 0.9421 respectively. Change map obtained by proposed algorithm and other existing algorithms has been given in Fig. 5(a)–(e). The visual analysis shows that the change map obtained by the proposed algorithm has least spots as shown in Fig. 5(e).

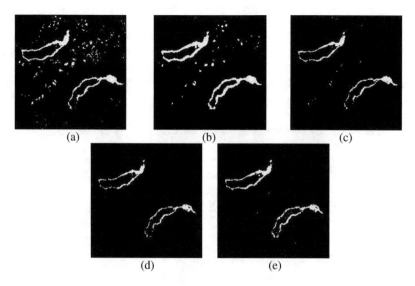

Fig. 4. The change map for Reno Lake dataset obtained by (a) Image regression. (b) Mean ratio. (c) Direct differencing. (d) Log ratio. (e) Proposed algorithm

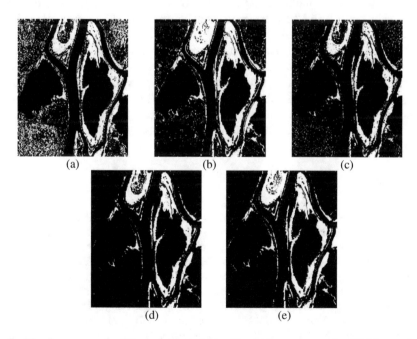

Fig. 5. The change map for Ottawa data set obtained by (a) Image regression (b) Mean ratio (c) Direct differencing (d) Log ratio (e) Proposed method

The change map for Ottawa dataset obtained by proposed method has least spots as compare the other methods presented in this paper. The change map given in Fig. 5(a–c) presents more spots while lot of information has been lost in case of log ratio method shown in Fig. 5(d) although it contain less spots. So, based upon the quantitative and visual analysis of results, it is proved that the proposed method is most effective in terms of preservation of details and removal of speckle noise.

Table 2. The results on the dataset of Ottawa data obtained by different change detection methods.

Method	F_p	F_n	PCC (%)	(K_c)
Image regression	51410	12961	83.00	0.7770
Mean ratio	21918	4105	93.13	0.9115
Direct differencing	13497	16086	92.19	0.8975
Log ratio	5069	13075	95.21	0.9376
Proposed method	9095	7869	95.52	0.9421

5   Conclusion

The paper proposed a modified log ratio approach in which the difference image is generated by considering the variance to mean ratio and logarithmic ratio of the image. The ratio of variance to mean preserve the pixel details and reduces the speckle noise up to large extent. The qualitative analysis proves that the output obtained through proposed method has been affected less by speckle noise as compare to other methods presented for comparison. Also, the proposed methods offer the least spots in the output change map. The quantitative analysis also proves that the accuracy and kappa value of the proposed approach is better than the other methods. So, it is concluded that the presented approach produce the best results for change detection over other direct comparison based methods.

References

1. Radke, R.J., Andra, S., Al-Kofahi, O., Roysam, B.: Image change detection algorithms: a systematic survey. IEEE Trans. Image Process. **14**(3), 294–307 (2005)
2. Bovolo, F., Bruzzone, L.: A detail-preserving scale-driven approach to change detection in multi-temporal SAR images. IEEE Trans. Geosci. Remote Sens. **43**(12), 2963–2972 (2005)
3. Celik, T.: Unsupervised multiscale change detection in multitemporal synthetic aperture radar images. In: 17th European Signal Processing Conference, pp. 1547–1551 (2009)
4. Dianat, R.: Change detection in optical remote sensing images using difference-based methods and spatial information. IEEE Geosci. Remote Sens. Lett. **7**(1), 215–219 (2010)
5. Sumaiya, M.N., Shantha, R.: Logarithmic mean-based thresholding for SAR image change detection. IEEE Geosci. Remote Sens. Lett. **13**(11), 1726–1728 (2016)
6. Bovolo, F.: A framework for automatic and unsupervised detection of multiple changes in multitemporal images. IEEE Trans. Geosci. Remote Sens. **50**(6), 2196–2212 (2012)

7. Rafael, W., Speck, A., Kulbach, D., Spitzer, H., Bienlein, J.: Unsupervised robust change detection on multispectral imagery using spectral and spatial features. In: 3rd International Airborne Remote Sensing Conference and Exhibition, Copenhagen, Denmark (1997)

8. Singh, A.: Digital change detection techniques using remotely sensed data. Int. J. Remote Sens. 10(6), 989–1003 (1989)

9. Nordberg, M.L., Evertson, J.: Vegetation index differencing and linear regression for change detection in a swedish mountain range using landsat TM_ and ETM+_ Imagery. Land Degrad. Dev. 16, 139–149 (2005)

10. Oliver, C., Quegan, S.: Understanding Synthetic Aperture Radar Images. Artech House, Norwood (1998)

11. Bezdek, J.C.: Pattern Recognition with Fuzzy Objective Function. Plenum, New York (1981)

12. Dunn, J.C.: A fuzzy relative of the ISODATA process and its use in detecting compact well-separated clusters. J. Cybern. 3, 32–57 (1973)

13. Celik, T.: Unsupervised change detection in satellite images using principal component analysis and k means clustering. IEEE Geosci. Remote Sens. Lett. 6(4), 772–776 (2009)

14. Da Cunha, A.L., Zhou, J., Do, M.N.: The nonsubsampled contourlet transform: theory, design and application. IEEE Trans. Image Process. 15(10), 3089–3101 (2006)

15. U.S. Geological Survey. http://geochange.er.usgs.gov

16. Gong, M., Zhou, Z., Ma, J.: Change detection in synthetic aperture radar images based on image fusion and fuzzy clustering. IEEE Trans. Image Process. 21(4), 2141–2151 (2012)

17. Rosin, P.L., Ioannidis, E.: Evaluation of global image thresholding for change detection. Pattern Recogn. Lett. 24(14), 2345–2356 (2003)

Topic Detection and Tracking in News Articles

Sagar Patel[(✉)], Sanket Suthar, Sandip Patel, Nehal Patel,
and Arpita Patel

Chandubhai S. Patel Institute of Technology, CHARUSAT, Changa, India
{sagarpatel.it, sanketsuthar.it, sandippatel.it,
nehalpatel.it}@charusat.ac.in,
arpitapatel_29@yahoo.com

Abstract. We have presented an idea in this paper for detecting and tracking topics from news articles. Topic detection and tracking are used in text mining process. From data which are unstructured in text mining we pluck out information which are previously unknown. The objective of this paper is to recognize tasks occurred in different news sources. We are going to use agglomerative clustering based on average linkage for detecting the topics, calculate the similarity of topics using cosine similarity and KNN classifier for tracking the topics.

Keywords: Detecting · Tracking · Article · Text mining · Extract · Information · Unstructured · Agglomerative · Similarity · KNN classifier · Vector space model (VSM)

1 Introduction

Topic detection and tracking is challenging topic in information retrieval technology that can be used in the text mining. In topic detection we finding the most important topics in a collection of news articles [1]. Our approach combines a variety of learning techniques. Topic detection is an unsupervised task and topic tracking is supervised task. We are going to use agglomerative clustering to create topic clusters and KNN classifier for tracking topics. To identify the serious news, we identify the clusters that fall into same category [11].

The corpus considered the news from various news sites over the world like Times of India and CNN, and of various subscription news wires. Thus the collection of different news from different source have same events. Newspapers normally receive the news from various news agencies with very few changes. Thus the corpus of news articles contains the same events written by different journalists which must be eliminated from the collection [2, 12].

The objective of research is to recognize interesting events happens in the world. Analysts are continuously trying to identify latest news and story from a very large resources of information that arrives daily. So for journalist it is easy to understand and identify actual events. The objective is to decompose topics is to extract the events which never seen before and combine them which represents same news stories. The scope of the research is text in news articles obtained from the various newspaper websites [13].

© Springer International Publishing AG 2018
S.C. Satapathy and A. Joshi (eds.), *Information and Communication Technology for Intelligent Systems (ICTIS 2017) - Volume 2*, Smart Innovation, Systems and Technologies 84, DOI 10.1007/978-3-319-63645-0_48

Text mining utilizes techniques from the field of data mining, combines methodologies from various other areas such as categorization, information retrieval, clustering, summarization, information extraction, computational linguistics, concept linkage and topic tracking [8].

1.1 Literature Review

To extract new tasks and then find the topic alike as an information streaming task which was done by querying different article versus the profile of newly identified topic. Topics representation was done using vector of stemmed words and their (term frequency–inverse document frequency) values, considering nouns, adjectives, verbs etc. In the results they produced ten and twenty features unique results. Friburger and Maurel (2002) observed that the identification and usage of proper geographical references, names, significantly improves document similarity calculation and clustering [3]. Hyland et al. (1999) put news in clusters and detected topics manipulating the different combinations of various named entities to link related articles [4]. At the end topic score is assigned to each word. The main problem in this technique is the processing time, as they have pointed out, and there is some need for optimization for some of their calculations [14, 15].

Topic detection is an unsupervised and topic tracking is supervised. In our approach, we are going to use hierarchical agglomerative clustering for topic detection based on average linkage using document similarity vector. For document similarity, we have to use cosine similarity based on TF-IDF [5]. For topic tracking, KNN classifier will be used. It will judge incoming news is on topic or off topic [16, 17].

1.2 Proposed Approach

Regarding the current events, a system is required to detect and track topics within news articles. We would be choosing any one or two domain from politics, sports, science and discovery etc. Focusing on chose domain goal is to implement a system that gives quite satisfying results about current events with all related stories using the optimal approach. We would be using clustering for topic detection and classifier for topic tracking [18] (Fig. 1).

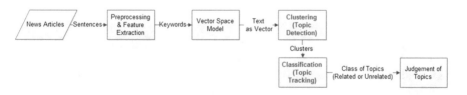

Fig. 1. System flow

In our approach first step is preprocessing on collected text news. A corpus consists of a large contents are gathered from various sources. Due to heterogeneity, data becomes noisy and inconsistent. Due to inconsistency in data, mining process can lead to confusion and hence result becomes inaccurate. So to extract consistence and accurate data pre-processing is applied to the data. Sagar et al. (2016) describes all available pre-processing which can be applicable for text mining [6]. In this work, we are going to use Tokenization, Stop word removal, Stemming methods [19, 20].

Tokenization: Tokenization extracts words from sentences. The example of tokenization is given bellow in which it gives the tokens, the words extracted from given sentence, separated by comma [6].

> **Input**: Ravi plays volleyball. **Output**: [[Ravi], [plays], [volleyball], [.]]

Stop-Word Elimination: Sometimes words are not used in proper way in articles. These stop words are not needed for text mining applications. Stop word elimination reduce the text data like "the", "in", "a", "an", "with" etc. Thus the system performance improves.

Stemming: Stemming feature allows the system to reduce different grammatical forms/word forms of a word like its noun, adjective, verb, adverb etc. to its root form. Its aim is to obtaining the stem of a word, means its morphological root, by removing the affixes combination of prefix and suffix that carry grammatical or lexical information about the word. Stemming is widely uses in Information Retrieval system as stemming having the feature of stems of the words and that allows some phases of the information retrieval process to be improved, and reduces the size of index files. Many Indian languages like Tamil, Malayalam, Gujarati, Marathi, Hindi, Bengali etc. in highly inflected form [7, 21].

Following are the steps of pre-processing:

- First of all, tokenization will be applied on texts of news articles. Here in tokenization sentences will be broken into words. Example: Text Mining is used to extract knowledge from unstructured data. After applying Tokenization it will be like Text, Mining, is, used, too, etc.
- Then from the set of all words stop words will be removed. Here it will remove noninformative words like the, more, and, when, etc.
- Then stemming will be applied on the words to get root word. Here it will remove suffix to generate word stem. Example: walking, walked, walks will become walk (Fig. 2).

Vector Space Model. In this research work, we are dealing with texts. For finding similarity between texts we need to convert texts into vectors, so we are going to use VSM model. We are using the SVM model as the baseline system because it is easy to implement, robust and more competitive compared to other the additional elaborate systems [8]. For weighting the term we will use TFIDF and for similarity, we will use cosine similarity measure.

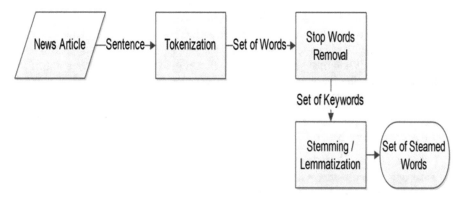

Fig. 2. Preprocessing

TFIDF, a Term Weighting Approach. Weights of the words occurred in documents are assigned by TFIDF weight. The word with higher tf*idf score has more important in the document. TF represents that how many times that term occurs in an article. IDF represent that how many documents contain that term [9, 10].

$$idf(t) = \log\left(1 + \frac{m}{dt}\right) \tag{1}$$

Where,

m = Total number of documents

dt = Number of documents in which term t occurs

TFIDF combines TF and IDF by multiplication [19].

$$tfidf_d(t) = tf_d(t) \cdot idf(t) \tag{2}$$

Cosine Similarity. To measure similarity of two articles cosine similarity used in the VSM. It measures the cosine of the angle between two vectors in N-dimensional space [5]. Cosine similarity between two documents d_i and d_j can be defined as:

$$sim(d_i, d_j) = \left[\sum_{k=1}^{n} w_{ik} \cdot w_{jk}\right]\left[\sum_{k=1}^{n} w_{ik}^2 \sum_{k=1}^{n} w_{jk}^2\right]^{-\frac{1}{2}} \tag{3}$$

Where,

n = Document vector size,

w_{ik} and w_{jk} = tf.idf weights of term-k in document di and dj respectively.

Topic Detection. Agglomerative clustering has been used successfully for topic detection. It is a sequence of nested partitions. It is defined by disjointed clustering, which individualizes each of the N documents within a cluster. This process is repeated and when the sequence progresses cluster containing all N documents the number of clusters decreases [22].

An important feature of creating topic clusters based on keywords is the occurrence of data overlap between clusters. Let's take one example of single story which contains seven different keywords which describes its content, so the story will appear in seven different clusters. When using agglomerative clustering to create a topic tree, the effects of data overlap on the measure of cluster similarity need to be considered. We will be using agglomerative clustering with average distance measures [23].

Algorithm for Topic Detection. Algorithm for Agglomerative Hierarchical Clustering (average linkage base):

1. Initially, consider each keyword as cluster itself.
2. Take only two clusters with the smallest distance from all current clusters, Here for the distance, we would be using average distance:

$$\frac{1}{|A||B|} \sum_{a \in A} \sum_{b \in B} d(a, b) \tag{4}$$

Where A and B are two sets of observation and A and B are two points.
3. Find two clusters that are nearest to each other.
4. Merge two original cluster by replacing these two clusters with a new one.
5. Until one remaining cluster present in the pool repeat Steps 3 and 4.

Topic Tracking. KNN is an instance-based classification method. The system converts the incoming document into a vector and compares it to the training stories. Based on the cosine similarity between them, the k nearest neighbors is selected. The score for this document is calculated by subtracting the similarity coefficients of the negative stories from the positive ones. If it's high then the story is related to the topic. The threshold for determining whether the score is high or not differs from a topic to another. So it's hard to apply the same threshold to all topics.

Algorithm for Topic Tracking

1. News texts di is given in the training set.
2. News in the training and testing sets are in vectors.
3. Calculate the cosine similarity values $\cos(d_i, d_o)$ between new text d_i and training text set d_j and select k texts most similar to new text

$$sim(d_i, d_j) = \left[\sum_{k=1}^{n} w_{ik} \cdot w_{jk} \right] \left[\sum_{k=1}^{n} w_{ik}^2 \sum_{k=1}^{n} w_{jk}^2 \right]^{-1/2} \tag{5}$$

Where,
n = Document vector size
w_{ik} and w_{jk} = The tf.idf weights of term-k (t_k) in document d_i and d_j respectively
4. If the news is more similar then they are related otherwise not related.
5. When p (d_i, d_j) > O, it suggest that new text di belongs to this topic; when p (d_i, d_j) < 0, it's decided that new text "di" does not belong to this topic.

2   Conclusion

At the end conclusion of this paper is that, we have combined machine learning approaches. We would be applying system for sports domain, if time permits would check our approach on politics, entertainment, science and discovery, etc. We have used Agglomerative hierarchical clustering using average distance measure for topic detection and K-nearest neighbour classifier for topic Tracking. We select K Nearest Neighbor classifier for tracking because it gives better performance. As well as it makes the fewest assumptions of about terms, stories and efficient decisions surface for the tracking task. For future work we will detect and track broadcast news.

References

1. Perez-Tellez, F., Pinto, D., Cardiff, J., Rosso, P.: Clustering weblogs on the basis of a topic detection method. In: Mexican Conference on Pattern Recognition, pp. 342–351. Springer, Berlin (2010)
2. Pouliquen, B., Steinberger, R., Ignat, C., Käsper, E., Temnikova, I.: Multilingual and cross-lingual news topic tracking (1998)
3. Friburger, N., Maurel, D., Giacometti, A.: Textual similarity based on proper names. In: Proceedings of the Workshop Mathematical/Formal Methods in Information Retrieval, pp. 155–167 (2002)
4. Hyland, K.: Disciplinary discourses: writer stance in research articles. In: Writing: Texts, Processes and Practices, pp. 99–121 (1999)
5. Schultz, J.M., Liberman, M.: Topic detection and tracking using idf-weighted cosine coefficient. In: Proceedings of the DARPA Broadcast News Workshop, pp. 189–192. Morgan Kaufmann, San Francisco (1999)
6. Patel, S.M., Dabhi, V.K., Prajapati, H.B.: Extractive based automatic text summarization. J. Comput. 12(6), 550–563 (2017)
7. Bijal, D., Sanket, S.: Overview of stemming algorithms for Indian and non-Indian languages. arXiv preprint arXiv:1404.2878 (2014)
8. Makkonen, J.: Semantic classes in topic detection and tracking (2009)
9. De, I.: Experiments in first story detection, pp. 1–8 (2005)
10. Bigi, B., Brun, A., Haton, J.-P., Smaïli, K., Zitouni, I.: Dynamic topic identification: towards combination of methods (2001)
11. Kumar, A.A.: Text data pre-processing and dimensionality reduction techniques for document clustering, vol. 1, no. 5, pp. 1–6. Sri Sivani College of Engineering (2012)
12. Saha, A., Sindhwani, V.: Learning evolving and emerging topics in social media: a dynamic NMF approach with temporal regularization. In: Proceedings of the Fifth ACM International Conference on Web Search and Data Mining, pp. 693–702. ACM (2012)
13. Acun, B., Ba, A., Ekin, O., Saraç, Mİ., Can, F.: Topic Tracking Using Chronological Term Ranking, vol. 25. Springer, London (2011)
14. Hoogma, N.: The modules and methods of topic detection and tracking. In: 2nd Twente Student Conference on IT (2005)
15. Can, A.F., Kocberber, S.: Novelty detection for topic tracking, vol. 63, no. 4, pp. 777–795 (2012)
16. Kaur, K.: A survey of topic tracking techniques. Int. J. Adv. Res. 2(5), 384–393 (2012)
17. Elkan, C.: Text mining and topic models the multinomial distribution (2013)

18. Fukumoto, F., Yamaji, Y.: Topic tracking based on linguistic features. In: LNAI, vol. 3651, pp. 10–21 (2005)
19. Cieri, C., Graff, D., Liberman, M., Martey, N., Strassel, S.: Large, multilingual, broadcast news corpora for cooperative research in topic detection and tracking: the TDT-2 and TDT-3 corpus efforts, January 1998 (1999)
20. Eichmann, D., Ruiz, M., Srinivasan, P., Street, N., Culy, C., Menczer, F.: A cluster-based approach to tracking, detection and segmentation of broadcast news. In: Proceedings of the DARPA Broadcast News Workshop, pp. 69–76 (1999)
21. Kaur, K., Gupta, V.: Tracking for Punjabi language. Comput. Sci. Eng. Int. J. **1**(3), 37–49 (2011)
22. Allan, J., Harding, S., Fisher, D., Bolivar, A., Guzman-lara, S., Amstutz, P.: Taking topic detection from evaluation to practice, pp. 1–10 (2004)
23. Mohd, M.: Design and evaluation of an interactive topic detection and tracking interface (2010)

Efficient Dimensioning and Deployment Criteria for Next Generation Wireless Mobile Network

Sarosh Dastoor[1], Upena Dalal[2(✉)], and Jignesh Sarvaiya[2]

[1] Electronics and Communication Department, SCET, Surat, India
sarosh.dastoor@scet.ac.in
[2] Electronics Department, SVNIT, Surat, India
{udd,jns}@eced.svnit.ac.in

Abstract. The journey of Cellular Mobile Network could be considered from rags to riches. The base-band of Cellular network and its working lies in its proper planning, dimensioning and deployment. Some of the dimensioning techniques for the next generation mobile wireless network have been discussed in this paper. The research has been focused for (Next Generation) LTE-A (Long Term Evolution-Advanced) Network Planning, and dimensioning of the Heterogeneous Network. It is divided into three categories. Firstly, planning based on various radio network parameters (like Quality of Service, Cost, power and path-loss) under consideration. Second part comprises of network planning based on various issues and corresponding solution. Various Deployment strategies using Voronoi Tessellation, AHC (Agglomerative Hierarchical Clustering) and Traditional K-means Deployment (TKD) methods have been discussed. Third part shows the importance of planning for deployment of the network with demography. Macro network, urban scenario with moderate density has been considered for dimensioning.

Keywords: Dimensioning · Deployment · LTE-A · Heterogeneous network · Voronoi tessellation · Agglomerative hierarchical clustering · Traditional K-means

1 Introduction

Radio Network Planning (RNP) process is the pre-operational phase of the Mobile Communication consisting of dimensioning (also known as initial planning), comprehensive planning (detail planning) and network operation and finally optimization. Radio Network Planning process is described in [1]. Radio planning process is quite crucial as it is the basic seed for the development of the Strategy to be used for the deployment of a cellular network in a given area.

Efficient eNB deployment strategy for heterogeneous cells in 4G LTE (Long Term Evolution) systems has been discussed in [2]. It uses Modified AHC (Agglomerative Hierarchical Clustering), Weighted K-means approach and Modified Geometric Disc Cover (to find the location of pico-cell). Energy efficient scheme for heterogeneous cellular networks from deployment to working has been considered in [3]. LTE radio

© Springer International Publishing AG 2018
S.C. Satapathy and A. Joshi (eds.), *Information and Communication Technology for Intelligent Systems (ICTIS 2017) - Volume 2*, Smart Innovation, Systems and Technologies 84, DOI 10.1007/978-3-319-63645-0_49

network planning with HetNets (heterogeneous network) is shown in [4]. Base station placement optimization is shown using Simulated Annealing algorithm. Technical challenges and issues for each of the main areas of study for LTE-Advanced (LTE-A) and the evolution Beyond 4G (B4G) systems has been considered in [5]. Planning, minimization and optimization of the location of BS in a given area (case study) has been shown in [6] using heuristic algorithm. In [7], three diverse schemes for deployment were developed with respect to energy efficiency for different traffic states. The major requirement of next generation mobile network lies in the support of extremely high data rates ranging from 1 Gbps to 10 Gbps [10].

The objective of this paper is to highlight the planning of the wireless network with various strategies adopted and the deployment patterns of the heterogeneous network in a region for its roll-out. Proper planning with reference to clutter and demography helps the service provider to enhance the quality of service and reduce the cost (installation and maintenance cost). Formation of micro and pico cell along with the macro cell, gives rise to heterogeneous network [12]. It helps to enhance the utility of the wireless network in terms of QoS (Quality of Service), throughput and hand-off.

Here the paper considers a particular region of Surat city, which is dimensioned by wireless technology, so we start to plan out first and then obtain network parameters. Subsequently nominal design is put on paper and the process of site acquisition gains momentum. After the review of site, comes the important parameter like coverage and capacity design for the deployment of the network. Initial optimization is required to find the optimum input parameters required by the network to provide highest efficiency.

Organization of the rest of the paper is as follows. Section 2 covers the radio network planning model; the practical approach is shown in Sect. 3. The results and analysis of dimensioning using K-means and AHC is shown in Sect. 4, followed by the concluding comments in Sect. 5.

2   Radio Network Planning Model

Planning of next generation mobile network is a NP Hard (Non-deterministic Polynomial-time) Problem as the network is Heterogeneous in nature. To handle NP-Hard problem we consider one factor at a time with its influence on other and move in the descending order of the hierarchy (with priority). The approach for radio network planning is described with the following steps:

A. Information Collection

Information related to LTE-A network proposal is collected with Operator's network sites density and distribution. The frequency band, environmental information, subscribers region of operation, traffic model forecast and Digital map is collected from the service provider or governing agency.

B. Deployment Strategy

Large cell (dense/urban) for coverage and small cell for capacity is considered with reference to deployment. Clustering of UEs (User Equipments) in a cell could be made using K-means Clustering and AHC (Agglomerative Hierarchical Clustering).

C. Requirements and Target

Coverage requirements include coverage area (obtained using digital map) and the required coverage QoS (Quality of Service). Capacity requirements include subscriber target and density and path loss models associated with the demographical terrain.

D. Radio Network Design

It includes Link Budget Analysis and Capacity analysis. The input parameter for the Radio Network Design includes the area to be covered (in a region), target service at cell edge, LTE frequency, associated Bandwidth and the number of subscribers. The corresponding output obtained from Radio Network Design is Cell range, number of sites, sectors, centroid for BS deployment, SINR, throughput etc.

E. Site Selection and Survey

Cellular range or inter-site distance is obtained from existent network site distribution information, which results in the network design.

F. Coverage Prediction and Simulation

Setting of input parameters along with the coverage prediction is considered. The Cellular network design starts with the collection of data which includes the position of

Table 1. Architecture for simulation environment

Sr no.	Parameters	Specified value used
1.	Number of user equipments	150–300
2.	Area under consideration	10 km^2 (Surat region)
3.	Clustering algorithms used	K-means and AHC
4.	System bandwidth	2600 MHz, FDD
5.	Number PRB	50
6.	Max eNodeB transmission power	40 W per PRB
7.	Path loss model	NLOS modified Hata, Walfisch Ikegami and Modified Hata model
8.	Downlink frequency	10 MHz
9.	Path loss exponent	2–5
10.	The receiver noise figure	7 dB
11.	Shadow fading standard deviation	10 dB
12.	eNodeB antenna height (h_b)	12 m
13.	UE height (h_r)	1.5 m
14.	Deployment strategy	Roof-top pole (geometrical strategy)

UEs in the Cartesian plane of a region. For the same an Excel Sheet has been obtained from Teleysia Network (Ahmedabad) that contains positional co-ordinates of UEs and three BS (Base Stations) locations. Network operators existing sites density and distribution could also be obtained. A particular architecture has been defined for simulation and dimensioning purpose as shown in Table 1. Around 150–300 UEs were considered to be distributed in the Surat region from Athwalines to City-light Road.

Currently large cell deployment strategy with roof-top mechanism has been considered with small cell capacity. Path loss has been found out using Okumura Hata model, Walfisch Ikegami Model and Modified Hata Model.

2.1 Path Loss Propagation Model

Here fast fading mechanism is neglected as the characteristics of problem in small-scale deviations are fairly rapid in space. The operating frequency is considered as $f = 2{,}600$ MHz, $h_b = 12$ m, 10 m or 8 m in accordance to the BS configuration and User Equipment height $h_r = 1.5$ m. The median path loss considered at a generic distance d is calculated by using the COST-231 Hata model which is given by [8, 11]:

$$PL(d)\ [dB] = 46.3 + 33.9 \log(f) - 13.82 \log(h_b) - a(h_r) + (44.9 - 6.55 \log(h_b)) \log(d) + C_m \tag{1}$$

The parameter C_m denotes the gain of antenna and is equal to zero (dB) for suburban areas. Here it is considered that cable losses are mitigated and output power radiated by antenna is same as the input provided. The function $a(h_r)$ is the gain function at receiver which is defined as [8]:

$$a(h_r) = (1.1 \log(f) - 0.7)h_r - (1.56 \log(f) - 0.8) \tag{2}$$

Figure 1 shows the LTE-A network planning, dimensioning and the deployment strategy used for simulation.

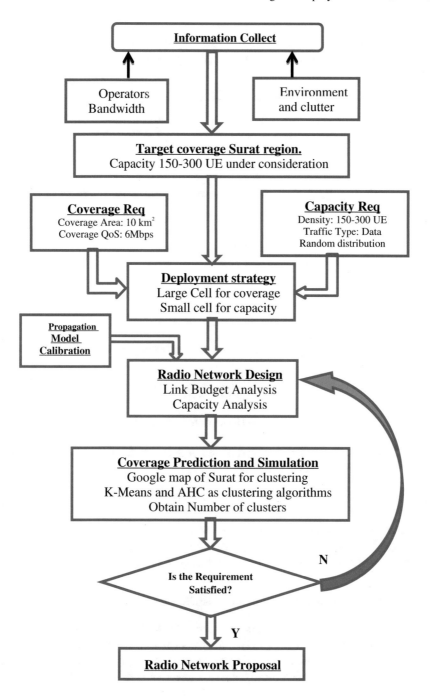

Fig. 1. LTE-A network planning and deployment strategy

Formation of clusters lead to the number of eNBs required in that area. The response of various path loss models is shown in Fig. 2. Modified Hata model has been designed to extend the range upto 10 km and 4 GHz of frequency.

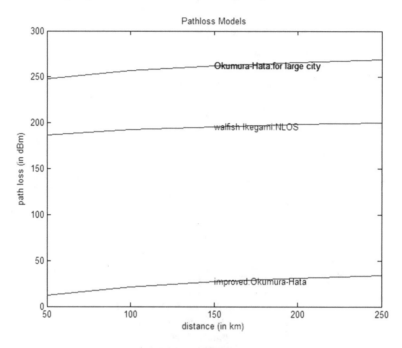

Fig. 2. Path-loss models: Okumura Hata, Walfisch Ikegami and improved Hata model

Throughput associated with a tessellation under consideration could be calculated using Shannon's Channel capacity theorem given by

$$T_h = BW\big[log_2(1 + SINR)\big] \tag{3}$$

Where, BW is the channel bandwidth and SINR stands for Signal to Interference plus Noise Ratio. SINR is calculated as the ratio of power received and the interference power (10–20 dB) with AWGN (Additive White Gaussian Noise) considered as −90 dB.

2.2 Voronoi Tessellation for Random Distribution

It mimics the modeling of random coverage area of small cells in a densely populated urban environment filling the finite two dimensional area without any gaps and overlying. The geometry could be created by taking pairs of neighbor points and drawing a line that is perpendicular bisector to the line connecting both points [9].

2.3 Estimate the Number of eNBs

All the UEs are arranged in the Cartesian plane according to their co-ordinates. AHC (Agglomerative Hierarchical Clustering) scheme is used to find the number of clusters and hence the number of eNBs. Using K-means Clustering Algorithm, preliminary deployment of large cells is done and cluster centre is obtained (centroid). All the scattered UEs are checked with the availability of resources like signal strength in the coverage area. If any uncovered UE is found then based on minimum SINR required, pico-cells could be added to cover such UEs and hence the range is adjusted accordingly. We can thus dimension macro and micro/pico cells.

2.4 Coverage Prediction Using K-means and AHC

K-means: It is a well-known algorithm used to obtain the cluster from the given data group and position of the objects in the plane. Assuming K number of clusters, the K-means algorithm chooses seed as follows: [2].

An observation is chosen uniformly at random from the data set, X. The chosen observation is considered to be the first centroid, symbolized as c_1. Calculate the distance from each observation to the first symbol c_1. 'Let the distance between c_j and the observation m be denoted as $d(x_m, c_j)$. The next centroid, c_2 could be selected at random from X with probability' [2, 4]

$$\frac{d^2(x_m, c_1)}{\sum_{j=1}^{n} d^2(x_j, c_1)} \tag{4}$$

Repeat the above procedure until k centroids are chosen.

3 Practical Cellular Network Design

Practical design of the cellular network requires a thorough insight and it is sectored into the following flow as provided by Teleysia Networks:

(i) Cell Planning Review (Obtained through Simulation)
    This stage of Cellular planning, results review before optimization. At this stage, the location of eNB and signal environment is analyzed. Key Performance Indicators include SINR, RSRP (Reference Signal Received Power), throughput, bandwidth etc. It is adopted by the service providers before roll-out of the network. A data-base is collected containing the location co-ordinates of the UE.

(ii) Single Cell Function Test (Single Site Verification)
    eNB's Initial Parameters, Installation status and Cell design values are compared using Single Cell Function Test. We can also check the performance by eNB test document. Initial parameters like Antenna gain-tilt, transmission power of the BS, VSWR (Voltage Standing Wave Ratio), path-loss etc. are checked through test procedure and are tuned accordingly if required.

(iii) Cluster Optimization (Cluster Quality Verification)

Total number of clusters is reduced to optimize the Base Stations required. At cell-outage where the signal quality drains, femto cell or pico cell could be deployed. RSSP (Reference Signal Received Power) is measured at the cell outage.

4   Results and Analysis

After getting the data about User Equipments and its location in the Cartesian plane, Surat region was considered and using Google maps, the UEs were distributed based on their location as shown in Fig. 3(a). K-means clustering of the UEs have been shown in Fig. 3(b). The clusters so formed are assigned with the centroid. The latitude and longitude of 21° and 72° respectively indicates (part of) Surat region.

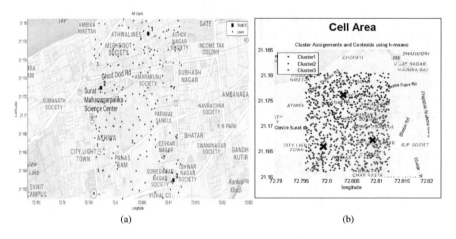

(a)   (b)

Fig. 3. (a) Map of Surat representing the user equipment by blue dots and eNodeB using black dots, (b) K-means clustering on Surat region

Figure 4(a) shows the representation of four clusters using AHC. The corresponding dendogram is shown on the left hand side of the AHC clustering. It shows that when the dendogram is sliced at a vertical height of slightly greater than 0.04, three clusters could be obtained. Voronoi Tessellation provides the rough estimation of the number of active users/customers in that area (which is modeled by a point in our city). The geometry stability of it is quite good as a change caused by distortion of cell shape results in the corresponding change in the shape of the Voronoi Cell.

Figure 4(b) shows the Voronoi tessellation generated with 15 cells. Point P is considered as a reference point from which the cell centre is measured. Table 2 shows the position of UE from the central BS with the corresponding throughput associated.

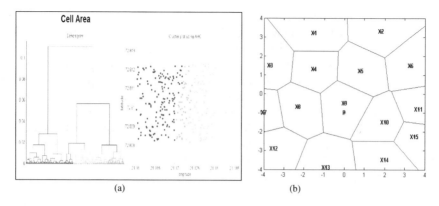

Fig. 4. (a) Dendogram AHC clustering (4 clusters), (b) Voronoi tessellation for 15 cells

Table 2. Position of UEs from central BS and corresponding throughput

Cell no.	X-coordinate	Y-coordinate	Throughput (Mbps)
1	−1.50	3.20	4.8150
2	1.80	3.30	4.7636
3	−3.70	1.50	4.8136
4	−1.50	1.30	5.3064
5	0.80	1.20	5.4450
6	3.30	1.50	4.8986
7	−4	−1	4.9361
8	−2.30	−0.70	5.4527
Ref	0	−0.50	6.4463
9	2	−1.50	5.5490
10	3.70	−0.80	5.0171
11	−3.50	−2.90	4.9408
12	−0.90	−3.90	5.2138
13	2	−3.50	5.1633
14	3.50	−2.25	5.0140
15	−1.50	3.20	4.8150

Throughput has been calculated using Eq. (3). If the corresponding network is expanded with more number of cells then, at one stage the throughput and the corresponding SINR will fall below a threshold value.

5 Summary

Geometrical and stochastic strategy is a relevant approach for the dimensioning and deployment of the heterogeneous network. AHC provides the approximate number of clusters required for the deployment of eNB with the proper selection of its threshold.

Voronoi tessellation provides good distribution of users in the given geometry with non-uniform areas. If SINR and throughput drastically falls at the cell boundaries, it implies the requirement of a pico-cell deployment. Simulation results substantiate the requirement of heterogeneous network in terms of quality SINR and throughput. As the distance of node increases from the reference, the corresponding value of SINR and the throughput associated with it also decreases. At the boundary region of a network, the RSS (Received Signal Strength) falls below the threshold value, requiring the deployment of pico cell or femto cell. A multi-tier stochastic environment amalgamated with deterministic network planning is the need of future. After deployment, the optimization of the performance parameters like antenna tilt, orientation, transmitted power etc. takes place using modern optimization techniques.

References

1. Elnashar, A., El-Saidny, M.A., Sherif, M.R.: Design, Deployment and Performance of 4G-LTE Networks—A Practical Approach. Wiley, Hoboken (2014)
2. Wang, Y.-C., Chuang, C.-A.: Efficient eNB deployment strategy for heterogeneous cells in 4G LTE systems. Elsevier J. Comput. Netw. **79**(14), 297–312 (2015)
3. Son, K., Eunsung, O., Krishnamachari, B.: Energy-efficient design of heterogeneous cellular networks from deployment to operation. Elsevier J. Comput. Netw. **78**(26), 95–106 (2015)
4. Zhang, Z., Sun, H., Hu, R.Q., Qian, Y.: Stochastic geometry based performance study on 5G non-orthogonal multiple access scheme. In: 2016 IEEE Global Communications Conference (GLOBECOM), Washington, DC, pp. 1–6 (2016)
5. Nguyen, H.D., Sun, S.: Closed-form performance bounds for stochastic geometry-based cellular networks. IEEE Trans. Wirel. Commun. **16**(2), 683–693 (2017)
6. El-Beaino, W., El-Hajj, A.M., Dawy, Z.: On radio network planning for next generation 5G networks: a case study. In: 2015 International Conference on Communications, Signal Processing, and Their Applications (ICCSPA), Sharjah, pp. 1–6 (2015)
7. Koutitas, G., Karousos, A., Tassiulas, L.: Deployment strategies and energy efficiency of cellular networks. IEEE Trans. Wirel. Commun. **11**(7), 2552–2563 (2012)
8. Boiardi, S., Capone, A., Sansó, B.: Radio planning of energy-aware cellular networks. Elsevier J. Comput. Netw. **57**(13), 2564–2577 (2013)
9. Xu, X., Li, Y., Gao, R., Tao, X.: Joint Voronoi diagram and game theory-based power control scheme for the HetNet small cell networks. EURASIP J. Wirel. Commun. Netw. (1), 213 (2014)
10. Moysen, J., Giupponi, L., Mangues-Bafalluy, J.: On the potential of ensemble regression techniques for future mobile network planning. In: 2016 IEEE Symposium on Computers and Communication (ISCC), Messina, pp. 477–483 (2016)
11. http://www.3gpp.org/specifications/specifications
12. Akyildiz, I.F., Gutierrez-Estevez, D.M., Balakrishnan, R., Chavarria-Reyes, E.: LTE-advanced and the evolution to beyond 4G (B4G) systems. J. Phys. Commun. **10**, 31–60 (2014)

Robust Features for Emotion Recognition from Speech by Using Gaussian Mixture Model Classification

M. Navyasri[1(✉)], R. RajeswarRao[2], A. DaveeduRaju[3],
and M. Ramakrishnamurthy[1]

[1] Department of Information Technology,
Anil Neerukonda Institute of Technology, Sangivalasa, Visakhapatnam, India
navyasrimullapudi@gmail.com,
ramakrishna.malla@gmail.com
[2] Department of Computer Science and Engineering,
JNTUK-UCEV, Vizianagaram, India
raob4u@yahoo.com
[3] Department of Computer Science and Engineering,
RamaChandra College of Engineering, Eluru, India
123.davidjoy@gmail.com

Abstract. Identification of emotions from speech is a system which recognizes the particular emotion automatically without basing on any particular text or a particular speaker. An essential step in emotion recognition from speech is to select significant features which carry large emotional information about the speech signal, speech signal has an important features. The features extracted from the shape of speech signal are used such as MFCC, spectral centroid, spectral skewness, spectral pitch chroma. These features have been modeled by Gaussian mixture model and optimal number of Gaussians is identified. IITKGP-Simulated Emotion Speech corpus is used as database and four basic emotions such as anger, fear, neutral and happy are considered. The different mixture of spectral features is extracted and experiments were conducted.

Keywords: Pattern recognition · Mel frequency · Gaussian · Centroid

1 Introduction

Emotion recognition based on a speech signal is one of the current research areas in the study of how people connect with computer systems. The objective of this exploration is to create the scope to which natives may naturally articulate feelings. The main application of identifying others feelings is, we can estimate other person's mental state. In the medical field psychiatric diagnosis analysis of patients is one of the major challenges. Therefore Identification of emotions is incredibly helpful to discover patient's mental disorders.

© Springer International Publishing AG 2018
S.C. Satapathy and A. Joshi (eds.), *Information and Communication
Technology for Intelligent Systems (ICTIS 2017) - Volume 2*, Smart Innovation,
Systems and Technologies 84, DOI 10.1007/978-3-319-63645-0_50

1.1   Emotion Recognition from Speech Issues

Speech characteristics are not enough to differentiate variety of expressions (such as happy, neutral, anger etc.). The main reason is, while recording the speech from the speakers in theater, there may be a chance of noise occur which leads to less clarity to understand voice. But human voice may demonstrate the various passions. Based on the environment and traditions the speaker's speaking style also alters. So it leads to intricate to identify the expression. Finding out the exact emotion from ambiguous responses of speaker is one of the complex chores to identify that speaker's feeling.

1.2   Speech Emotion Recognition System

Emotion Recognition from Speech mainly has three working steps which are Feature Extraction, Training and Testing (Fig. 1).

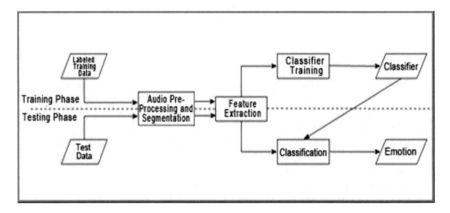

Fig. 1.   Pattern recognition task

Feature extraction is one of the important tasks in pattern recognition. The first objective is to make available an emotional speech data collections. Mel frequency cepstral coefficients are the preeminent characteristics of speech signal to identify emotions. The extracted features can go through the training phase; in this a classification algorithm is applied to train the features to classify the speech into emotional status. The best classification algorithms are artificial neural networks, Gaussian mixture model, Hidden markov model, support vector machines.

2   Related Work

In Shashidhar G. Koolagudi et al. [1], proposed that epoch parameters play a vital role in any of the speech tasks and explained the importance of epoch parameters of pitch period. In Iliou et al. [2] explained that an emotion recognition system based on echo processing might extensively improve human computer interaction. A large amount of speech features obtained from sound processing of speech be experimented in order to make a feature set adequate to make a distinction among seven emotions. In Iker Luengo et al. [4] described that, only pitch and energy associated features provide valuable information for identification of emotion. Intonation is extracted from the recordings, in linear scale. Initial and subsequent derivative curves are calculated, as the pitch might provide original valuable information for the recognition.

In the above approaches they have used the standard basic feature extraction techniques, the combination of MFCC and spectral features proposed in this paper which carries large information about the speech signal.

3 Feature Extraction

One of the important parts of emotion recognition from speech system is the required feature extraction process, selecting the exact features is essential for successful classification. Speech signal composed of huge number of features which specify emotion contents of it; changes in these features indicate vary in the emotions. Therefore appropriate selection of feature vectors is one of the most important task. The spectral features play a significant role in Speech emotion recognition. In anxious speech, the vocal tract spectrum is modulated resulting changes in whole spectrum. The analysis is performed using frame size 20 ms and window shift 10 ms.

3.1 MFCC

Mel Scale Frequency Cepstral Coefficients (MFCC) are the unique frequency components for every human. Thus it is a spectrum with unique frequency components. By using these features, a computer can able to identify the human feelings from their speech with the help of learning algorithms. This attribute play a vital role in language detections, gender identification also (Fig. 2).

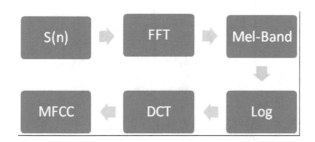

Fig. 2. Steps for extraction of MFCC features

In this, the continuous speech signal partitioned into minute frames of n samples. In this the frame size is 20 ms and the frame shift (next frame division) is 10 ms. To avoid the disorder windowing will be done at initial and at end of the each frame. The signal go through the several steps of MFCC extraction. The MFCC features will be extracted from each and every frame of that speech signal. The approximation about the existing energy at each spot is

$$Mel(f) = 2595 * \log_{10}(1 + f/700) \tag{1}$$

These frequency components are extracted for both gender voices.

3.2 Spectral Centroid

Spectral centroid is one of the best feature to find out the intensity of speaker voice. It is a mean calculation of fundamental frequency and formants. In the shape of speech spectrum, the pitch and next top frequencies which are called formants are used in recognizing the anger emotion of particular person. If the calculated spectral centroid of speech signal is high, then we can predict that the speakers intensity. It is calculated as mean of the fundamental frequency and formants of the signal determined by FFT with their magnitudes and weights (Fig. 3).

$$Spectral\ Centroid = \frac{\Sigma kF[k]\{k = 1\ to\ N\}}{\Sigma F[k]} \tag{2}$$

In practice, Centroid finds this frequency for a given frame, and then finds the nearest bin for that frequency.

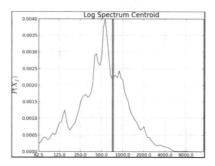

Fig. 3. Spectral centroid on a spectrum

3.3 Spectral Skewness and Pitch Chroma

If the shape of the spectrum of a speech signal is less than the center of gravity, then spectral skewness is a used to evaluate how much the shape is below or higher than the center of gravity (Fig. 4).

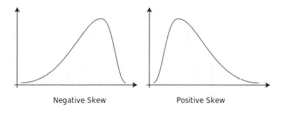

Fig. 4. Spectral skewness on a spectrum

The main advantage of skewness is, it shows the energy of a spectrum based on positive and negative skewness. For n samples, The skewness is

$$Skewness = \frac{\frac{1}{n}\sum_{i-1}^{n}(x_i - \bar{x})^3}{\left[\frac{1}{n-1}\sum_{i-1}^{n}(x_i - \bar{x})^2\right]^{3/2}} \tag{3}$$

Where, the mean is $\bar{x}$, standard deviation s, the third moment is represented by the numerator 3.

Chroma describes the position of pitch rotary motion as go across the spiral. In music applications, pitch chroma feature plays an important role. Two octave-related pitches will distribute the same angle in the chroma sphere, a relation that is not encapsulated by a linear pitch scale. It illustrates the angle of pitch rotation, chroma features are robust to noise and loudness.

4   Classification

After the suitable features are extracted from the acoustic signal, those features should be trained by one classification algorithm. Classification is a process of superintend learning approach. That means a class label is assigned for every model. In this the class labels are emotions (for instance neutral, anger, happy, sad). Hence the extracted attributes are applied as the input and after train those features, the generated output is model. Choosing the suitable training algorithm is one of the complex assignments. There are so many classification algorithms such as ANN, HMM, GMM etc. In this Gaussian mixture model classification algorithm is used. GMM classification method is more useful for speech spectrum compare to other classification methods.

The features which are MFCC, spectral centroid, spectral skewness, pitch chroma are extracted from male speaker and female speaker. The Gaussian probability density function, mean μ, $\sum$ is covariance matrix and variance σ^2. In the D-dimensional space it is defined in a matrix form as

$$N(x; \mu; \Sigma) = \frac{1 \exp^{-1}}{(2\pi)^{D/z|\Sigma|^{1/2}}}\left[\frac{(x - \mu)^T \Sigma^{-1}(x - \mu)}{2}\right] \tag{4}$$

The male and female acoustic signal training duration is 30 s. The testing duration is 2 s. The experiments are conducted as the GMM algorithm applied to train the male spoken signal of 30 s and male, female voice is given for testing. Similarly Female spoken signal is trained and male, female voice is given for testing. Various numerals of Gaussian components such as 2, 4, 8, 16, 32, 64, 128 and 256 are explored. The Classified models are anger, neutral, happy, sad.

5   Experimental Results

The experiments are conducted on IIT-KGP telugu speech corpus with a training period of 30 s and testing period of 2 s. The combination of MFCC and Spectral Skewness gives the better accuracy among all the combinations with the MFCC features. The classification error at optimal gaussian components is 17859.222405 (Tables 1, 2, 3 and 4).

Table 1. Emotion recognition with the extracted features MFCC and skewness

Emotions (%)	Training period: 30 s					
	M-Male		F-Female		M & F-Male and Female	
	Testing: 2 s					
	Male	Female	Male	Female	Male	Female
Anger	100	100	33.33	33.33	100	100
Fear	83.33	16.66	16.66	16.66	100	66.66
Happy	16.6	0	16.66	16.66	50	0
Neutral	66.66	33.33	33.33	100	33.33	100

Table 2. Emotion recognition with the extracted features MFCC, centroid and skewness

Emotions (%)	Training period: 30 s					
	M		F		M & F	
	Testing: 2 s					
	Male	Female	Male	Female	Male	Female
Anger	100	16.66	16.66	100	100	100
Fear	66.66	0	100	100	100	66.66
Happy	16.66	16.66	33.33	33.33	16.66	0
Neutral	33.33	33.33	100	100	83.33	100

Table 3. Emotion recognition with the extracted features MFCC, pitch chroma and skewness

Emotions (%)	Training period: 30 s					
	M		F		M & F	
	Testing: 2 s					
	Male	Female	Male	Female	Male	Female
Anger	100	0	16.66	100	100	100
Fear	66.66	66.66	50	100	100	100
Happy	33.33	0	0	50	50	50
Neutral	83.33	83.33	66.66	66.66	83.33	100

Table 4. Emotion recognition with the extracted features MFCC, centroid and pitch chroma

Emotions (%)	Training period: 30 s					
	M		F		M & F	
	Testing: 2 s					
	Male	Female	Male	Female	Male	Female
Anger	100	16.66	16.66	100	100	100
Fear	50	16.66	0	100	100	83.33
Happy	16.66	16.66	50	50	0	16.66
Neutral	66.66	66.66	50	100	50	100

6 Conclusion and Future Scope

The main passion of emotion recognition from speech is to detect feelings from the person speech without a text restraint. In this the shape of the acoustic spectrum is focused to extract the various attributes of a voice signal. Gaussian Mixture Model is used to train or classify the extracted features of male and female speech signals and generated the class models such as anger, neutral, happy, sad expressions. A Telugu language data base is collected with both gender voices for Analysis. It was observed that Emotion Recognition performance depends on speaker, Text. Emotion Recognition performance also varies from different features which are extracted from speech signal. The Training was conducted with Male database, Female database and combination of both male and female with training period 30 s and Testing was conducted with a period of 2 s. It was observed that Emotion Recognition performance increased for combination of MFCC and skewness features.

In future, a noise reduction method will be used for more clarity in speaker's voice. Feature dimensionality reduction using rough sets are used to select the best features, and train extracted features with different classification algorithms.

References

1. Koolagudi, S.G, Reddy, R., Rao, K.S.: Emotion recognition from speech signal using epoch parameters. In: 2010 International Conference on IEEE Signal Processing and Communications (SPCOM), pp. 1–5, July 2010. ISBN 978-1-4244-7137-9
2. Iliou, T., Anagnostopoulos, C.N.: Statistical evaluation of speech features for emotion recognition. In: Fourth International Conference on IEEE Digital Telecommunications, pp. 121–126. ICDT 2009, July 2009. ISBN 978-0-7695-3695-8
3. Bitouka, D., Vermaa, R., Nenkovab, A.: Class-level spectral features for emotion recognition. Speech Commun. **52**(7–8), 613–625 (2010)
4. Luengo, I., Navas, E., Hernáez, I., Sánchez, J.: Automatic emotion recognition using prosodic parameters
5. Koolagudi, S.G., Rao, K.S., Ramu, V.R.: Emotion recognition from speech using source, system and prosodic features. Int. J. Speech Technol. **16**, 143–160 (2013)
6. Theodoridis, S., Koutroumbas, K.: Pattern Recognition. Academic Press, Cambridge (1999). ISBN 0-12-686140-4
7. Duda, R.O., Hart, P.E., Stork, D.G.: Pattern Classification, 2nd edn. Wiley, Hoboken (2001)
8. Kämäräinen, J.-K., Kyrki, V., Hamouz, M., Kittler, J., Kälviäinen, H.: Invariant Gabor features for face evidence extraction. In: Proceedings of the IAPR Workshop on Machine Vision Applications, pp. 228–231. Nara, Japan (2002)
9. Bilmes, J.: A gentle tutorial on the EM algorithm and its application to parameter estimation for Gaussian mixture and hidden Markov models (1997)
10. Shete, D.S.: Zero crossing rate and energy of the speech signal of Devanagari script. IOSR J. VLSI Sig. Process. (IOSR-JVSP) **4**(1), 01–05 (2014)
11. Jiang, D., Cai, L.: Speech emotion classification with the combination of statistic features and temporal features. In: Proceedings on IEEE International Conference on Multimedia, Taipei, Taiwan, China, pp. 1967–1970 (2004)
12. Schuller, B., Reiter, S., Muller, R., Al-Hames, M., Lang, M., Rigoll, G.: Speaker independent speech emotion recognition by ensemble classification. In: Proceedings on IEEE International Conference on Multimedia and Expo, Amsterdam, The Netherlands, pp. 864–867 (2005)

Genetic Algorithm Based Peak Load Management for Low Voltage Consumers in Smart Grid – A Case Study

B. Priya Esther[1], K. Sathish Kumar[1(✉)], S. Venkatesh[1],
G. Gokulakrishnan[1], and M.S. Asha Rani[2]

[1] School of Electrical Engineering, VIT University, Vellore 632014,
Tamil Nadu, India
priyarafela@gmail.com, kansathh21@yahoo.co.in,
venkatesh.srinivasan@vit.ac.in
[2] Kaynes Interconnection Systems, Bengaluru, Karnataka, India

Abstract. To withstand the quick development with the demand for energy and the overall demand cost, enhanced effectiveness, dependability and adaptability, smart strategies should be carried forth in energy sector for our earth and vitality protection. In the electrical domain DSM can be a part of smart grid where the consumers can participate themselves to decrease the peak load and eventually the load profile can be reshaped. A portion of the DSM method is peak clipping, load shifting, valley filling and energy conservation. The paper involves the concept of load shifting to low voltage consumers using several types of appliances and in large numbers. Load shifting with respect to day ahead forecast is formulated as a minimization problem and are solved using learning based evolutionary algorithm. Simulations were carried out with a specific test case using Mat Lab and the results show a substantial peak reduction and cost savings for the future smart grid.

Keywords: Demand Side Management (DSM) · Heuristic algorithm · Load forecasting · Day ahead load shifting · Load scheduling · Peak load management (PLM) · Time of Day (ToD)

1 Introduction

With the increase in power demand and the rise of environmental concerns rather than dealing with the power generation, nowadays the focus is towards the control and management of power in the consumer side. The advent of smart grid technology further permits the usage of electricity in a smart and customer friendly way. One of the mechanism to control the power in the distribution side is Demand Side Management. DSM can reduce carbon emission levels and save the environment for future generation. In future residential consumption may be more than 35% of generation. Therefore, the latest focus is towards the distribution side rather than generation and transmission.

In most of the reviewed papers related to peak load management, direct load control [1–5] is used, but the drawback is that user's comfort is compromised in all these cases. Appliances can be scheduled to reduce the peak based on different tariff structures like

© Springer International Publishing AG 2018
S.C. Satapathy and A. Joshi (eds.), *Information and Communication Technology for Intelligent Systems (ICTIS 2017) - Volume 2*, Smart Innovation, Systems and Technologies 84, DOI 10.1007/978-3-319-63645-0_51

real time pricing as in [6] or it can also be a dynamic pricing as in [7]. Flat rate and ToD tariff structures are used so that the consumer can reduce their usage. Conventional methods like those that integer linear programming [8–10] also used in the scheduling and for the further optimization to reduce peak and cost. As these optimization methods can handle a less number of appliances, heuristic based [11–15] methods are used in replacement of these conventional methods. This paper discusses the peak load reduction based on load shifting and the day ahead forecasting techniques.

The paper is structured like that Sect. 1 deals with DSM introduction and the literature review whereas Sect. 2 deals with DSM concepts explaining DSM techniques, methodologies, classification and optimization approaches. Section 3 discusses the proposed PLM strategy with the architecture, algorithm and the problem formulation. Section 4 presents a case study with a test smart grid along with the observations results and discussions of PLM. Section 5 concludes with the PLM strategy and future work.

2   Demand Side Management Preview

2.1   DSM Techniques

The load curves depicted by residential, industrial and commercial users during peak and off peak hours can modified by various techniques. They are (a) Peak clipping, (b) Valley filling, (c) Strategic Conservation, (d) Strategic growth, (e) Flexible load and (f) Load shifting.

(a) Peak Clipping: It means during peak load demand when either there is no extra generation limit, with or without the concern of the consumer, the utility decides the curtailment. This can delay the requirement for an extra generation limit. However, its cause inconvenience to the consumer the net impact is a reduction in both top request and aggregate management utilization. Peak clipping can be accomplished by direct control of consumer apparatus.

(b) Valley Filling: It means decreasing the difference between the higher and lower load demand to eliminate the disparity in load consumption along with peak load reduction to maintain the security of the smart network. It mainly uses direct load control method.

(c) Strategic Conservation: It means the overall reduction in the load consumption and it is a non-traditional approach in DSM.

(d) Strategic Growth: It is the increase in overall utility load, which can change the complete load curve.

(e) Flexible Load: It can be an agreement between a utility and the consumer as on a needed basis may be in return for different motivations.

(f) Load Shifting: Load shifting includes moving peak loads to off peak hours. Well-known applications like storage water heating, space heating can be listed in the process.

These six techniques are independent of each other and in practical situations, it is better to adapt to any specific technique mainly for the peak load and cost reduction. Load shifting is a commonly used and the best load management method in the current scenario with the latest available techniques and consequently that is the background technique utilized in this paper.

2.2 DSM Methodologies

With the use of smart meters and smart instruments, DSM open up the achievability of applying such projects on a much bigger scale. Different DSM methodologies [16] have been proposed, some of them are

(i) Direct control method: Private appliance of an individual customer are grouped by their centrality. At the point when a given state is coming to, a central management grid as indicated by the predetermined demand can separate single appliance.

(ii) Control of apparatus: Rather than separating appliances, their energy necessities are reduced or their operations is suspended, It has been accounted that peak load decreases up to 40% if the control is done for cooking and washing apparatuses alone (shiftable load).

(iii) Differential tariff: By utilizing the different tariff, structure during on peak and off-peak hours, the peak load management can be achieved.

(iv) Conservation voltage decrease: Given that the voltage affects the power needed by some local loads, it is conceivable to adjusting to bring down the demand the set purpose of LV transformers.

2.3 Classification and Optimization Approaches of DSM

Optimization is the mathematical procedure of maximizing or minimizing a certain parameter in a finite dimensional Euclidean space determined by its functional inequalities.

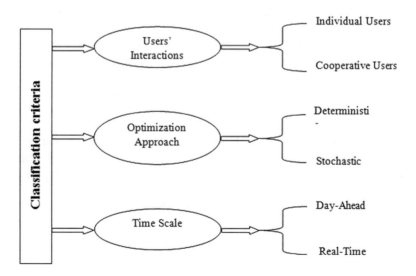

Fig. 1. Classification of DSM approaches based on optimization methods

Optimization includes branches, like linear optimization, convex optimization, discrete optimization, global optimization, etc. This paper presents a non-convex optimization technique. Figure 1 depicts the various classifications [16] based on optimization methods.

3   Proposed PLM Strategy

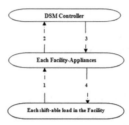

Fig. 2. Proposed architecture for PLM smart grid manager. 1. Seek connection permission with an ID. 2. Seek for a connection permission with ID, facility ID and time slot. 3. Provide connectivity details. 4. Provide details and order

This paper discusses day ahead load shifting technique of smart grid to minimize the peak load consumption. This section first discusses the architecture for PLM, then the problem is mathematically formulated and solved using [13] heuristic based algorithm.

The architecture is shown below (Fig. 2).

3.1   Proposed Algorithm

Step 1: Data to be taken from utility for specific periods with ToD tariff.
Step 2: Objective load curve is drawn inversely proportional to tariff.
Step 3: Integer Genetic Algorithm inputs are to be taken from step 1 and step 2.
Step 4: Load shifting is done so to match with the desired load curve.
Step 5: Execution of genetic algorithm mutually benefits consumers and utilities.

3.2   Problem Formulation

A load shifting technique is utilized to plan shiftable appliance of consumers at different time of the day to bring the last load curve much closer to the target load utilization curve. The load shifting procedure [13] is defined as minimization problem.[1]

$$\text{Minimize} \sum\nolimits_{t=1}^{N} (\text{PLoad}(t) - \text{Objective}(t))^2 \qquad (1)$$

Where Objective(t) is the estimation of the target load utilization curve at time t, and Pload(t) is given by

$$PLoad(t) = Forecast(t) + Connect(t) - Disconnect(t) \qquad (2)$$

Where forecast(t) is the approximate day ahead prediction of usage at time t, and Connect(t) and Disconnect(t) are the measure of the loads associated and disengaged at time t individually during the load shifting. 'N' is the quantity of time steps.

4 Case Study of Peak Load Management

4.1 Test Grid

To verify the performance of the methodology, the proposed DSM strategy is tried on a local location of a Mulbagal grid. Mulbagal is an area in Kolar area, which is an interior part of Karnataka having 20 townships. The system works at 11 kV. The hourly load utilization information is monitored by a substation. The goal is to reduce the peak demand and cost in this area. An objective load curve, which is inversely proportional to the tariff, is calculated.

The maximum demand in this area is around 2.5 MW. On a regular day, the control period was taken from the 8th hour of the present day in the 8th hour of the next day. Table 2 gives an idea of shiftable devices with time slots. There are about 2149 shiftable devices, which is about 11 different types.

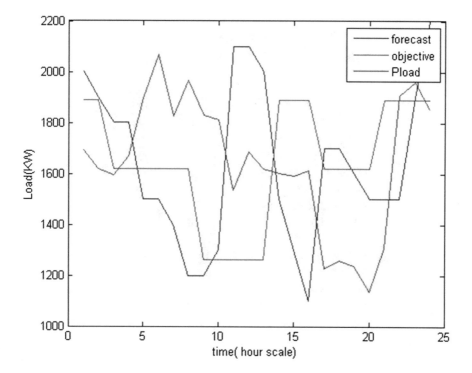

Fig. 3. DSM results of LV consumers on weekend

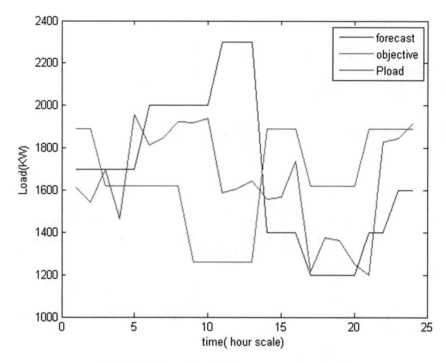

Fig. 4. DSM results of LV consumers on weekend

Table 1. Forecasted load with ToD tariff

Time	ToD tariff	Hourly forecasted load (Kwh)	
		Week day	Week end
8.00AM–9.00AM	2.8	2000	1700
9.00AM–10.00AM	2.8	1900	1700
10.00AM–11.00AM	3.6	1800	1700
11.00AM–12.00PM	3.6	1800	1700
12.00PM–1.00PM	3.6	1500	2000
1.00PM–2.00PM	3.6	1500	2000
2.00PM–3.00PM	3.6	1400	2000
3.00PM–4.00PM	3.6	1200	2000
4.00PM–5.00PM	4.2	1200	2000
5.00PM–6.00PM	4.2	1300	2000
6.00PM–7.00PM	4.2	2100	2300
7.00PM–8.00PM	4.2	2100	2300
8.00PM–9.00PM	4.2	2000	2300
9.00PM–10.00PM	4.2	1500	1400

(continued)

Table 1. (*continued*)

Time	ToD tariff	Hourly forecasted load (Kwh)	
		Week day	Week end
10.00PM–11.00PM	2.8	1300	1400
11.00PM–12.00AM	2.8	1100	1400
12.00AM–1.00AM	3.6	1700	1200
1.00PM–2.00PM	3.6	1700	1200
2.00AM–3.00AM	3.6	1600	1200
3.00AM–4.00AM	3.6	1500	1200
4.00AM–5.00AM	2.8	1500	1400
5.00AM–6.00AM	2.8	1500	1400
6.00AM–7.00AM	2.8	1900	1600
7.00AM–8.00AM	2.8	2200	1600

Table 2. Appliance scheduling

Device type	Hourly consumption (KW)			Number of devices
	1st hour	2nd hour	3rd hour	
Dryer	1.2	–	–	189
Washing machine	5	0.4	–	268
Microwave oven	1.3	–	–	279
Iron box	1	–	–	340
Vacuum cleaner	0.4	–	–	158
Fan	0.2	0.2	0.2	288
Kettle	2	–	–	406
Toaster	0.9	–	–	48
Rice cooker	0.85	–	–	59
Hair dryer	1.5	–	–	58
Coffee maker	0.8	–	–	56
Total	–	–	–	2149

4.2   Results and Discussions

Execution of algorithm was carried out for one-week day (Wednesday) and other weekend day (Sunday) The simulation results from load shifting DSM technique achieves reduction in utility operational costs and also reduce system peak load. The

Table 3. Percentage reduction of operational cost

Day	Operational cost without DSM (Rs)	Cost with DSM (Rs)	Reduction percentage (%)
Wednesday	1.3688 e+05	1.337 e+05	2.28
Sunday	1.4308 e+05	1.3607 e+05	4.88

Table 4. Percentage reduction in peak load

Day	Peak load without DSM (KW)	Peak load with DSM (KW)	Reduction percentage (%)
Wednesday	2200	1882.8	14.41
Sunday	2300	1980.2	13.39

simulation results obtained for residential consumers in Mulbagal region are shown in Figs. 3 and 4 taking the data from Tables 1 and 2.

4.3   Observations

The proposed load shifting hold good with genetic algorithm, which lead to substantial reduction in peak load and cost. Tables 3 and 4 depict the calculated values with and without DSM.

5   Conclusion and Future Work

Load shifting DSM method has been successfully simulated for a real distribution system of Mulbagal region, India. This technique reduces peak load and utility cost. ToD tariff implementation for residential consumers has been proposed, that may help utility to reduce their system peak and operational cost. The present work confirms good reductions in terms of peak load as well as cost to the end users and utilities. The simulation results are satisfactory and can be implemented with the future smart grid. The heuristic techniques used for optimization can be extended with other algorithms and the percentage of savings can be compared. In addition, the same strategy can be proposed and implemented to high voltage consumers.

References

1. Gomes, A., Antunes, C.H., Martins, A.G.: A multiple objective approach to direct load control using an interactive evolutionary algorithm. IEEE Trans. Power Syst. **22**(3), 1004–1011 (2007)

2. Ng, K.H., Sheble, G.B.: Direct load control—a profit-based load management using linear programming. IEEE Trans. Power Syst. **13**(2), 688–694 (1998)
3. Majid, M.S., Rahman, H.A., Hassan, M.Y., Ooi, C.A.: Demand side management using direct load control for residential. In: 2006 4th Student Conference on Research and Development, pp. 241–245, Selangor (2006)
4. Dong, X., Yi, P., Iwayemi, A., Zhou, C., Li, S.: Real-time opportunistic scheduling for residential demand response. IEEE Trans. Smart Grid. **4**(1), 227–234 (2013)
5. Teive, R.C.G., Vilvert, S.H.: Demand side management for residential consumers by using direct control on the loads. In: Fifth International Conference on Power System Management and Control, Conf. Publ. No. 488, pp. 233–237 (2002)
6. Joe-Wong, C., Sen, S., Ha, S., Chiang, M.: Optimized day-ahead pricing for smart grids with device-specific scheduling flexibility. IEEE J. Sel. Areas Commun. **30**(6), 1075–1085 (2012)
7. Zhao, H., Jia, H., Liu, G., Yang, Z., Fan, S.: Analysis of residential loads behaviours integrated with distributed generation under different pricing scenarios. In: IEEE Energytech, pp. 1–5, Cleveland, OH (2013)
8. Cohen, A.I., Wang, C.C.: An optimization method for load management scheduling. IEEE Trans. Power Syst. **3**(2), 612–661 (1988)
9. Zhu, Z., Tang, J., Lambotharan, S., Chin, W., Fan, Z.: An integer linear programming based optimization for home demand-side management in smart grid. In: IEEE PES Innovative Smart Grid Technologies (ISGT), pp. 1–5, Washington, DC (2012)
10. Babu, P.R., Kumar, K.A.: Application of novel DSM techniques for industrial peak load management. In: International Conference on Power, Energy and Control (ICPEC), pp. 415–419, Sri Rangalatchum, Dindigul (2013)
11. Graditi, G., Di Silvestre, M.L., Gallea, R., Riva, S.E.: Heuristic-based shiftable loads optimal management in smart micro-grids. IEEE Trans. Ind. Inform. **11**(1), 271–280 (2015)
12. Kunwar, N., Yash, K., Kumar, R.: Area-load based pricing in DSM through ANN and heuristic scheduling. IEEE Trans. Smart Grid **4**(3), 1275–1281 (2013)
13. Logenthiran, T., Srinivasan, D., Shun, T.Z.: Demand side management in smart grid using heuristic optimization. IEEE Trans. Smart Grid **3**(3), 1244–1252 (2012)
14. Kinhekar, N., Padhy, N.P., Gupta, H.O.: Demand side management for residential consumers. In: IEEE Power & Energy Society General Meeting, pp. 1–5, Vancouver, BC (2013)
15. Ravibabu, P., Praveen, A., Chandra, C.V., Reddy, P.R., Teja, M.K.R.: An approach of DSM techniques for domestic load management using fuzzy logic. In: IEEE International Conference on Fuzzy Systems, pp. 1303–1307, Jeju Island (2009)
16. Priya Esther, B., Kumar, K.S.: A survey on residential demand side management architecture, approaches, optimization models and methods. In: Renewable and Sustainable Energy Reviews (2016)

A Critical Analysis of Twitter Data for Movie Reviews Through 'Random Forest' Approach

Dubey Prasad Kamanksha and Agrawal Sanjay[✉]

DCEA, National Institute of Technical Teachers' Training and Research, Bhopal, India
kpdubeym.p@gmail.com, sagrawal@nitttrbpl.ac.in

Abstract. Using Sentiment analysis one can understand interaction of a user with the movies through their feedback. Here analysis is done based on the movie reviews that can be collected from many sources. Twitter is one among the foremost frequent on-line social media and micro blogging services. Due to the popularity of twitter it has become a useful resource for collecting sentiments through API or other data mining techniques. Our work here presents an examination on the evaluation of the machine learning algorithms (Random Forest, bagging, SVM and Naïve Bayes) in R together the public opinion for example opinion about 'Civil War' Movie. Here we have used 'Random Forest' to show its better performance in the analysis of movie reviews.

Keywords: Opinion mining · Sentiment analysis · Twitter · Sentiment classification · Natural language processing

1 Introduction

Sentiment analysis is a sub domain of the bigger domain natural language processing where one needs to mine text and use linguistic based processing to study the opinion sentiments and emotion associated with it. Associate opinion could also be viewed as an announcement within which the opinion holder makes a selected claim a couple of product, movie, article, post, etc. employing a sure sentiment. The opinions present on the social networks such as twitter and the facebook are valuable for sentiment analysis because of the expressive nature. We can make our mining process biased towards a particular topic. The intuitive nature of such opinion makes them an efficient tool by the bulk of researchers that build them the premise for creating selections concerning product review, research, exchange prediction, etc.

Sentiment analysis means the mining of an opinion's overall polarization and potency towards a specific subject. In this paper we propose an analysis of collective sentiments related to Movie 'Civil War'.

Our work here proposes the technique to get public opinion of the movie 'Civil War'. And this technique is based on supervised learning approach such as Random Forest, Bagging, Support Vector Machine and Naïve Bayes. Random Forest effectively classify the data as belonging to positive or negative sentiment.

© Springer International Publishing AG 2018
S.C. Satapathy and A. Joshi (eds.), *Information and Communication Technology for Intelligent Systems (ICTIS 2017) - Volume 2*, Smart Innovation, Systems and Technologies 84, DOI 10.1007/978-3-319-63645-0_52

2   Related Work

G. Vinodhini and R.M. Chandrasekaran [1], provide the brief idea about sentiment analysis. Since challenges are adhesive in nature for any kind of research work so sentiment analysis also cannot skip the challenges associated with it. The first one is the dualising associated with an opinion. Due to this kind of dualism it is very difficult to identify this kind of opinion as positive opinion or negative opinion. The second challenge is to track the intermediate opinion that neither say positive nor negative about a movie.

According to Vasu Jain [2] Social media is a big pool of content where the context are reach and information about people's preferences. Peoples express their expression about the movie by sharing their thoughts for example in the form of tweets. Due to the unstructuredness of the content as well as limited-length of texts, that can share. It is difficult to build one system which extract the opinion about a particular movie or any other object accurately.

In the work proposed by Neethu M.S. and Rajasree R. They conducted survey [3] which gives a brief explanation about to different levels of sentiment analysis. The first one is coarse level sentiment analysis that deals with the sentiment of complete document and Fine level deals with the sentiment analysis at attribute level.

Rishabh et al. [4] in their proposed work used an unsupervised learning technique. in the pattern of k-means clustering to cluster the tweets and combine and supervised learning methods (support vector machine and decision trees (CART)) to create an hybrid model. The model given by author is known as cluster-then-predict model. He showed the sentiment the users have towards the product iphone 6 s using R language.

Wanxiang Che, Yanyan Zhao et al. [5] have proposed a framework where they have included a sentiment sentence compression (Sent_Comp) step before the aspect-based sentiment analysis. The pre step i.e. the compression step here is needed to remove the unnecessary information from the sentence in consideration which in turn make a complicated sentence shorter in length.

K. Mouthami et al. [6] in their work proposed a sentiment analysis model based on natural language processing and text analytics. They have used a sentiment fuzzy classification algorithm for this purpose.

3   Proposed Methodology

This contains the steps that were taken to implement proposed model. The Fig. 1 contains the flowchart of implementation, showing the steps needed to be performed from collecting raw data to classification for movie 'Civil War', using the proposed approach.

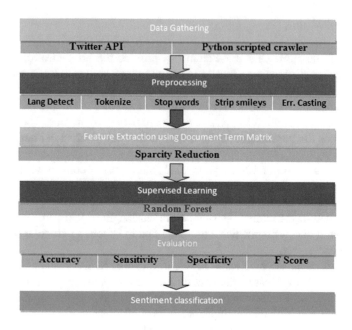

Fig. 1. Proposed methodology

Author proposes a supervised learning for predicting the opinions of public about Movie 'Civil War'. Because of the best trade-off in accuracy, precision, F-Score error rate random forest can be considered as one of the best supervised learning method for mining.

3.1 Data Collection

To verify the sentiment analysis model, twitter text as data is used for the recently released movie 'civil war'. Twitter data is publically available. The data can be collected from twitter through its streaming API service. The movie used for sentiment analysis in this research is 'Civil War', which was released on May 6, 2016 by the USA The tweets had been collected for a period of 15 days before and after the release date. Here we have used python's API name tweepy [7] to implement one streaming API of twitter. It is combination of libraries which intern extract dynamic tweets from the twitter. It stores the incoming tweets in CSV (comma separated values) file format in real time which uses python csv library functions. By this process we have marked 3006 tweets positive or negative manually.

3.2 Data (Tweets) Pre-processing

It is very difficult to make understanding by machine about raw twitter data due to there homonyms and metaphors as a first step towards finding a tweet's post is completed now we need to clean some noise and worthless symbols from the original text of tweets that do not contribute to a tweet's opinion. Fully understanding the text is very difficult but bag of words [8] provide it simple. As it counts the number of times each word appear. One part of cleaning the text is cleaning up irregularities, as text data often has many inconsistencies that will cause algorithm trouble e.g. @Civil War, '#civil war', civil war! will all count as just 'civil war'.

3.3 Feature Extraction Using Document-Term Matrix

For feature extraction from the twitter corpus, 'Bag-of Words' model was used. In 'Bag-of Words' techniques. The feature that is used for training a classifier is the frequency of each word in the tweets. Before this process data cleaning is much needed in raw data.

3.4 Machine Learning Algorithm

Machine learning is a branch of the big tree artificial intelligence which is used to make learned machines.

Naïve Bayes
The Naïve Bayes algorithm is commonly used for text classification in many opinion mining applications. The algorithm itself is derived from Bayes theorem:

$$P(A|B) - \frac{P(A)P(B|A)}{P(B)}$$

Random Forests
Ensemble learning focuses on techniques to associate the results of different trained models in order to produce a more accurate classifier. Ensemble models have considerably improved performance than that of a singular model. The random forest algorithm is an example of an ensemble method which was introduced by (Breiman 2001) [9], it is quite a simple algorithm but despite its simplicity it can produce state-of-art performance in terms of classification.

3.5 Evaluation

After collecting the tweeter data we have imposed 'Random Forest Model', for classification. Confusion matrix is used to compare the results of this experiment with the other existing method (Fig. 2).

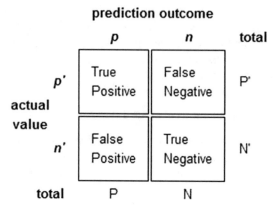

Fig. 2. Confusion matrix

For finding the accuracy, recall, f-measure of the classification, confusion matrix was used.

4   Experimental Setup

We used Python 2.7.11 for writing and executing python script to implement Twitter Streaming API, using Tweepy. We also used R GUI to writing and executing 'R' scripts for text mining. Apart from the above software and technologies, several 'R' packages like tm, caTools, rpart, etc. were used in implementation to fulfill the objectives.

5   Results and Discussion

After executing the Random Forest model for twitter sentiment analysis, the result was evaluated on several parameters, such as Accuracy, Precision, F-Score (Fig. 3).

$$Accuracy = \frac{TP + TN}{TP + FP + FN + TN}$$

$$Precision = \frac{TruePositive}{TruePositive + FalsePositive}$$

$$F\ Score = \frac{2 * TruePositive}{2 * TruePositive + FalsePositive + FalseNeagative}$$

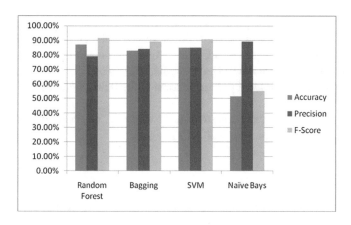

Fig. 3. Performance of several classifiers in twitter sentiment analysis

A review table (Table 1) is shown below which specifies various evaluation parameters and results of evaluation.

Table 1. Performance of proposed 'Random Forest' approach

Technique	Parameters		
	Accuracy (in %)	Precision (in %)	F score
Random forest (proposed)	87.03	79.06	91.55
Bagging	82.82	84.01	89.21
SVM	85.14	85.16	90.71
Naïve Bayes	51.55	89.37	55.18

Thus, it could be observed that the parameter which is most important, that is accuracy, F Score are higher as compared to other algorithms. Using the proposed 'Random Forest' approach, it was found that out of 785 tweets in the test data, 634 showed positive sentiments, and 151 showed negative sentiment towards movie 'Civil War'. So, using the proposed approach of sentiment analysis, it is known that public largely accepted and liked the Movie 'Civil War' and also gave good opinion about it on Twitter.

6   Conclusion

Bag of Words techniques has been used in our proposed work to collect data from twitter. After collecting the data and extracting features from the tweets using some various hybrid methods of classification they have been used to classify the sentiments or opinion of data into Positive and Negative class.

It is shown in this paper that proposed approach of 'Random Forest' has a major effect on overall accuracy of the analysis. This approach has an accuracy of around 87.03% for classification. Comparison between different algorithms and proposed approach shows that proposed approach is superior in critical evaluation parameters of accuracy and F Score. The solution obtained from the proposed approach is more interpretable as well. The simulation of the presented technique was provided in 'R'.

In future work, one major consideration would be to include a neutral class in the classification algorithm; this would have entailed clearly defining the positive, negative and neutral classes and collecting large amounts of neutral examples to train the algorithm.

In future, This works can be carried out to improve the accuracy of our proposed model using different algorithms with technologies.

References

1. Vinodhini, G., Chandrasekaran, R.M.: Sentiment analysis and opinion mining: a survey. Int. J. Adv. Res. Comput. Sci. Softw. Eng. **2**(6), 282–292 (2012)
2. Jain, V.: Prediction of movie success using sentiment analysis of tweet. Int. J. Soft Comput. Softw. Eng. **3**(3), 308–313 (2013)
3. Neethu, M.S., Rajasree, R.: Sentiment analysis in twitter using machine learning techniques. In: IEEE – 31661 (2013)
4. Soni, R., Mathai, J.: Improved twitter sentiment prediction through cluster-then-predict model. Int. J. Comput. Sci. Netw. **4**(4), 559–563 (2015)
5. Che, W., Zhao, Y., et al.: Sentence compression for aspect-based sentiment analysis. IEEE/ACM Trans. Audio Speech Lang. Process. **23**(12), 2111–2124 (2015)
6. Mouthami, K., et al: Sentiment analysis and classification based on textual reviews. In: IEEE (2013)
7. Yao, R., Chen, J.: Predicting movie sales revenue using online reviews. In: IEEE International Conference on Granular Computing (GrC) (2013)
8. Yu, X., Liu, Y., Huang, J.X., An, A.: Mining online reviews for predicting sales performance: a case study in the movie domain. IEEE Trans. Knowl. Data Eng. **24**(4), 720–734 (2012)
9. Breiman, L.: Random forests. Mach. Learn. **45**(1), 5–32 (2001)

An Approximation Algorithm for Shortest Path Based on the Hierarchical Networks

Mensah Dennis Nii Ayeh[(⊠)], Hui Gao, and Duanbing Chen

School of Computer Science, University of Electronic Science and Technology of China (UESTC), Chengdu, China
niiwise@live.com, ahuigao@163.com

Abstract. Social networks have become a "household name" for internet users. Identifying shortest paths between nodes in such networks is intrinsically important in reaching out to users on such networks. In this paper we propose an efficient algorithm that can scale up to large social networks. The algorithm iteratively constructs higher levels of hierarchical networks by condensing the central nodes and their neighbors into super nodes until a smaller network is realized. Shortest paths are approximated by corresponding super nodes of the newly constructed hierarchical network. Experimental results show an appreciable improvement over existing algorithms.

Keywords: Complex network · Hierarchical networks · Approximation algorithm · Shortest path

1 Introduction

The emergence of the internet and its technologies has attracted researchers' attention to design tools, models and applications in diverse fields in internet application technologies [11] such as communication [13], recommendations, social marketing, terrorist threats [12], and key nodes mining [10]. Efficiently computing the shortest path between any two nodes in a network is one of the most important concerns of researchers since it has a great potential for mobilizing people [5]. Exact algorithms such as Dijkstra's and Floyd-Warshall's algorithms have not performed so well on large scale networks due to their high computational complexity. We design an approximate algorithm to calculate the distances of the shortest paths in a modeled large scale network. Chow [4] presented a heuristic algorithm for searching the shortest paths on a connected, undirected network. Chow's algorithm relies on a heuristic function whose quality affects the efficiency and accuracy of estimation. Rattgan et al. [8] designed a network structure index (NSI) algorithm to estimate the shortest path in networks by storing data in a structure. Construction of the structure however consumes so much time and space. Tang et al. [6] presented an algorithm, CDZ, based on local centrality and existing paths through central nodes (10% of all nodes) and approximates their distances by means of the shortest paths between the central nodes computed by Dijkstra's algorithm. Although CDZ achieved high accuracy on some social networks within a reasonable time, it performed poorly on large scale networks due to the large number of central nodes. Tretyakov et al. [10] proposed two

© Springer International Publishing AG 2018
S.C. Satapathy and A. Joshi (eds.), *Information and Communication Technology for Intelligent Systems (ICTIS 2017) - Volume 2*, Smart Innovation, Systems and Technologies 84, DOI 10.1007/978-3-319-63645-0_53

algorithms, LCA and LBFS based on landmark selection and shortest-path trees (SPTs). Although these algorithms perform well in practice, they do not provide strong theoretical guarantees on the quality of approximation. LCA depends on the lowest common ancestors derived from SPTs and landmarks to compute the shortest path. LBFS adopts SPT's to collect all paths from nodes to landmarks by using best coverage approach and split the network into sub networks. Based on the usual BFS traversal in these sub networks, LBFS can approximate the shortest path.

Even though there exists a large variety of algorithms to calculate shortest paths, there are few approximate algorithms based on hierarchical networks. To deal with large scale hierarchical networks, we present a novel approximate algorithm based on the hierarchy of networks, which is able to efficiently and accurately scale up to large networks. To ensure high efficiency, we condense the central nodes and their neighbors into super nodes to construct higher level networks iteratively, until the scale of the network is reduced to a threshold scale. After which the distances of the shortest paths in the original network are calculated by means of their central nodes in the hierarchical network. The performance of our algorithm is tested on four different real networks. Experimental results show that the runtime per query is only a few milliseconds on large networks, while accuracy is still maintained.

2   Construction of Hierarchical Networks

Let $G = (V, E)$ be an undirected and unweighted network with $n = |V|$ nodes and $e = |E|$ edges. A path $P_{s,t}$ between two nodes $s, t \in V$ is represented as a sequence $(s, u_1, u_2, \ldots\ldots, u_{l-1}, t)$, where $\{s, u_1, u_2, \ldots\ldots, u_{l-1}, t\} \subseteq V$ and $\{(s, u_1), (u_1, u_2), \ldots\ldots, (u_{l-1}, t)\} \subseteq E$. $d(s, t)$ is defined as the length of the shortest path between s and t.

Based on G (parent network), we construct series of undirected and weighted networks with different scales. The original network G is taken as the bottom level or level 0 network. For each level of construction i.e. from bottom to top, we iteratively perform the following steps: A normal node with the largest degree (having lot of clusters) is selected as a central node and condensed with its normal neighbors (other than super nodes) into a super node; the edges between the normal nodes are redirected to their corresponding super nodes. Condensing of current level network is completed when all nodes are merged into super nodes. These nodes are regarded as normal nodes in the next level network. Edges between two super nodes in the previous hierarchy results in a single link for two normal nodes in the next level network. The weight of an edge between two adjacent nodes in the next level network represents the approximate distance between the nodes. The topmost network is obtained when the number of nodes in the next level network is below a given threshold t.

Figure 1 shows the process of constructing a hierarchical network. Figure 1(a) is the current level network whose black nodes represent central nodes. Central nodes condense with their neighbors (white nodes) into super nodes respectively, which results in the next level network as shown in Fig. 1(b). All super nodes in the previous level network (nodes in dash squares) are considered normal nodes in the next level network as shown in Fig. 1(b). The black node in Fig. 1(b) will be selected as the central node, and condense with its neighbors into a super node.

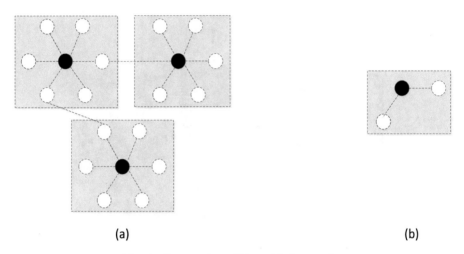

(a)   (b)

Fig. 1. Construction of hierarchical networks.

3   Algorithm Based on the Hierarchy of Networks

After transforming the original network into a number of hierarchical networks, the distance of the shortest path between any two nodes in the lower network can be estimated by considering their corresponding central nodes in the higher level network. Let $\hat{d}_i(s,t)$ be the approximate distance between nodes s and t in the level i network. The distance of shortest path between nodes s and t in the original network is approximated by $\hat{d}_0(s,t)$. In general, $\hat{d}_i(s,t)$ is iteratively computed by

$$\hat{d}_i(s,t) = \begin{cases} \hat{d}_i(s,c_s) + \hat{d}_i(t,c_t) & c_s = c_t \\ \hat{d}_i(s,c_s) + \hat{d}_{i+1}(c_s,c_t) + \hat{d}_i(c_t,t) & c_s \neq c_t \end{cases}, \quad i \geq 0, \tag{1}$$

where c_s and c_t are the central nodes of nodes s and t respectively. Figure 2 shows an example of shortest path approximation using Eq. 1. $\hat{d}_i(s,c_s)$ and $\hat{d}_i(t,c_t)$ are the

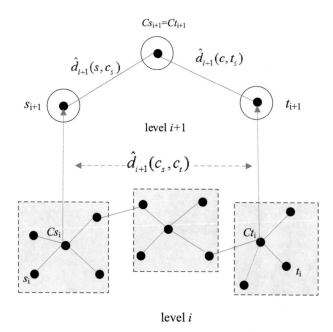

Fig. 2. Illustration of iterative approximation

approximate distances from nodes s_i and t_i to their respective central nodes c_{s_i} and c_{t_i} in the level i network. $\hat{d}_{i+1}(s, c_s)$ and $\hat{d}_{i+1}(c, t_s)$ are the distances from nodes s_{i+1} and t_{i+1} to their common central nodes in the level $i + 1$ network.

We define the longest distance from sub nodes to its corresponding central node as the radius of the super node. For example, assuming normal nodes in the level i network have the same radius r, then the radiuses of their corresponding super nodes will range from r to $3r + 1$. However computing for all radiuses of super nodes in the network hierarchy will consume so much memory and time. For this reason, we define an *appr radius* for all super nodes in the level i network by approximating the distance between two adjacent normal nodes in the level i network. As shown in Fig. 3, the *appr radiuses* of super nodes in the level $i + 1$ network are the approximate distances between the centers of two adjacent normal nodes (represented as hollow circles). Normal nodes in the level $i + 1$ network were derived from super nodes in the level i network whose *appr radiuses* were same. The *appr radius* of normal nodes in level i network is defined by

$$r_i = \begin{cases} 0, & i = 0 \\ 2r_{i-1} & 0 < i < k \end{cases}, \tag{2}$$

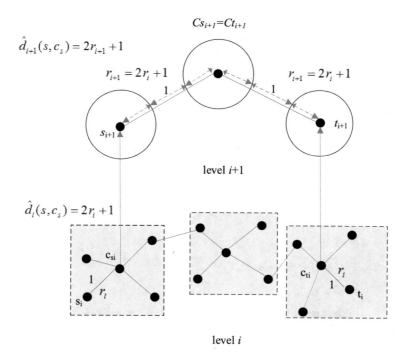

Fig. 3. Illustration of approximated distance r_i

where k is the number of hierarchical networks with different scales including the original network. Furthermore, Eq. (2) can be written as $r_i = 2^i - 1$ when $0 \leq i < k$. We approximate the distance from node s to its central node c in the level i network by

$$\hat{d}_i(s, c) = 2r_i + 1 \tag{3}$$

The approximate distance between adjacent nodes in level i network can also be written as $2r_i + 1$. Substituting Eqs. 2 and 3 into Eq. 1, the upper bound of $\hat{d}_i(s, t)$ can be calculated by

$$\hat{d}_i(s, t) = \begin{cases} 2 & i = 0, c_s = c_t \\ 2^{i+2} - 2 & 0 < i < k - 1, c_s = c_t \\ 2^{i+2} - 2 + \hat{d}_{i+1}(c_s, c_t) & 0 < i < k - 1, c_s \neq c_t \\ d_{k-1}(s, t) & i = k - 1 \end{cases} \tag{4}$$

The distance between any two adjacent nodes in the top level network is approximated as $2r_{k-1} + 1$. $d_{k-1}(s, t)$ in Eq. 4 is the length of shortest path between nodes s and t in the top level network computed by Dijsktra's algorithm.

The construction of each level of hierarchical network is described in Algorithm 1. Algorithm 1 consumes $O(n_i \log n_i)$ time to rank the n_i nodes in the level i network and $O(n_i + e_i)$ time to generate the level $i + 1$ network, where e_i is the number of edges in level i network. The space complexity of Algorithm 1 is $O(n + e)$.

Algorithm 1. SingleCluster

Notation. $old=(V, E)$ is current level network, $new=(V', E')$ is next level network, S is the set of super nodes constructed from old, S_c is a super node composed of a central node c and its neighboring normal nodes.

1: **function** SingleCluster(old)

2:   $C \leftarrow$ HighestDegree(old)

3:   **for** $c \in C$ **do**

4:     **if** $c \notin S$

5:       $S_c = \{c\} \cup \{c.neighbors \setminus S\}$

6:       $S \leftarrow S \cup \{S_c\}$

7:       Edges connected to the sub nodes inside S_c are redirected to S_c, also the multiple edges are merged into a single one.

8:     **end if**

9:   **end for**

9:   All super nodes in S are regarded as normal nodes in V' of new.

11: **return** new

12: **end function**

13: **function** HighestDegree(old)

14:   for each $v \in V$ let $d[v] \leftarrow$ degree(v)

15:     sort V by $d[v]$

16:   $v_{(i)}$ denotes the vertex with the i-th highest $d[v]$

17: **return sequence** $\{ v_{(1)}, v_{(2)}, \ldots\ldots, v_{(|V|)} \}$

18: **end function**

The construction of all other hierarchical networks is described in Algorithm 2.

Algorithm 2. HierarchyCluster

Require: Network *original=(V, E)*, *d[i][j]* records the shortest distance between node *i* and *j* in the top level network.

1: **function** HierarchyCluster

2:   Network *current = original*;

3:   **while** *current.size > threshold* **do**

4:     *next* = SingleCluster(*current*);// Algorithm 1

5:     *current = next*;

6:   **end while** // hierarchy networks

7:   Employ Dijsktra's algorithm to find shortest path between any pair of nodes on the top level network. Save the results to *d[i][j]*.

8:   **return** *d[i][j]*

9: **end function**

Algorithm 1 is repeatedly executed until the top level network is achieved, given a certain threshold value of nodes (e.g. 100). The time complexity for constructing hierarchical networks is given as $O\left(\sum_{i=0}^{k-1} (n_i \log n_i + n_i + e_i)\right)$ and space complexity as $O(kn+e)$. In the top level network, Dijsktra's algorithm require $O(m^2)$ time to compute the shortest paths between each pair of nodes and $O(m^2)$ space to store the distances, where m is the number of nodes in the top level network. Incorporating time and space complexity into Dijsktra's algorithm requires $O(kn+e+m^2)$ space to store hierarchical networks and corresponding distances in the top level network.

Finally, the approximate distances of the shortest paths in the original network can be calculated by Algorithm 3. Algorithm 3 requires at most $O(kn^2)$ time to compute the shortest paths for all pairs of nodes.

Algorithm 3. CalculateshortestPath

Require: *Network network=G(V, E)*, *d*[*i*][*j*] records shortest distances between nodes in the top level network calculated by Algorithm 2; *k* denotes the number of hierarchical networks constructed by Algorithm 2, including the original network; c_s and c_t are the central nodes of *s* and *t* respectively; supernode(c_s) is the super node which contains nodes c_s and *s* and its regarded as a normal node in the next level network.

1: **function** IterativeApproximation(*s,t,i*)

2: **if** $i < k$-1

3:    **if** $c_s = c_t$ in the level *i* network // hierarchy networks

4:      **return** $d(s, c_s)+d(t, c_t)$ // use Eq. 4 to calculate $d(s, c_s)$, $d(t, c_t)$

5:    **end if**

6:    **else if** $c_s \neq c_t$ in the level *i* network

7:      **return** $d(s, c_s)+$ IterativeApproximation(supernode(c_s), supernode(c_t), i+1)+ $d(t, c_t)$ // Eq. 4

8:    **end else**

9: **end if**

10: **else**

11:    **return** d[*s*][*t*] // use the distances between nodes in the top network computed by Dijsktra in Algorithm 2

12: **end else**

13: **end function**

14: **function** CalculateshortestPath

15:   **if** $s \in V$ is directly connected with $t \in V$ **then**

16:    *result* = 1;

17:   **end if**

18:   **else**

19:    *result* = IterativeApproximation (*s,t,*0) // iterative approximation for the $d_0(s,t)$

20:   **end else**

21:   **return** *result*

22: **end function**

Based on the above analysis, the time complexity for approximating the shortest distances in an undirected and unweighted network with n nodes and e edges can be calculated as $O\left(m^3 + \sum_{i=0}^{k-1}(n_i \log n_i + n_i + e_i) + kn^2\right)$, and the memory complexity as $O(kn + e + m^2)$.

We compared the complexity of CDZ [9] and LBFS [7] with our algorithm. CDZ algorithm selects c central areas in the network with n nodes and e edges and computes the distances between c and its central nodes by Dijkstra's algorithm. The time complexity of CDZ algorithm is given as $O(ed + n\log n + c^3)$. The number of central areas in CDZ algorithm is about 10% of the number of nodes. LBFS algorithm selects M pairs of nodes from the network to obtain i landmarks by "best coverage strategy" in a network with n node and e edges. The time complexity of LBFS is given as $O(M^3 + le + l^2D^2)$, where D is the average size of the sub network related to the landmarks. Sometimes the number of pairs of nodes in LBFS could be too large to make a best coverage and selection. The advantage of our algorithm over LBFS is that, the number of the nodes in the top network is relatively very small (m is below a certain threshold), which significantly reduces the complexity.

4 Experimental Results and Discussions

Our algorithm was implemented by java programming language, running on a PC with 2.93 GHZ CPU and 4 GB RAM. Performance of the algorithm was evaluated on four real undirected and weighted networks; Email-Enron [3], itdk0304_rlinks [1], DBLP [9] and roadNet. Email-Enron contains about half a million email communications among users whose nodes are the email addresses of senders or receivers and edges are the communication relationships. Tdk0304_rlinks contains the relationships among nodes that access the router where the nodes are users or websites and the edges are the relationships between them. DBLP network contains information on computer science publications, in which each node corresponds to an author; two authors are connected by an edge if they have co-authored at least one publication. The roadNet is a traffic network composed of roads and sites. The number of nodes V, edges E, diameter of network D, average degree <k>, and the largest degree k_{max} of these four networks are shown in Table 1.

Table 1. Basic information of four networks

	Email-Enron	Itdk0304_rlinks	DBLP	roadNet
V	36,692	190,914	511,163	1,405,790
E	367,662	1,215,220	3,742,140	23,442,590
D	12	24	25	30
<k>	10.020	6.365	7.320	9.433
k_{max}	1383	1 071	976	1244

We use *Path Ratio p* to assess the accuracy of the algorithm. P is defined as

$$p = \frac{\sum_{i=1}^{pr} P_{f_i}}{\sum_{i=1}^{pr} P_{O_i}}, \tag{5}$$

Where Pr is the total number of pairs of nodes, P_{f_i} is the distance between pairs of nodes computed by the approximation algorithm, and P_{O_i} is the accurate distances computed by Dijkstra's algorithm. The value of p is always greater than 1 since approximate distances are always longer than their corresponding accurate distances. Table 2 shows experimental results on preprocess time, average query time, and path ratio from our algorithm compared with that of CDZ. The threshold value is set at 100. T_{init} is the total time for preprocessing. T_q is the average runtime for 10,000 random queries. From Table 2, it can be construed that our algorithm performed relatively better on four different networks. It ran 10 times faster than CDZ especially on DBLP and roadNet networks. Moreover, it approximation for shortest path is more accurate on Email-Enron and itdk0304_rlinks, compared with CDZ on DBLP and roadNet.

Table 2. Runtime and accuracy of CDZ and our algorithm on four networks

Network	Our algorithm			CDZ		
	T_{init} (ms)	T_q (ms)	p	T_{init} (ms)	T_q (ms)	P
Email-Enron	12,845	1.38	1.022	112,448	11.38	1.026
itdk0304_rlinks	54,969	5.59	1.020	1,092,348	109.34	1.023
DBLP	103,126	10.41	1.020	8,623,459	862.44	1.020
roadNet	196,558	19.75	1.019	18,824,559	1882.55	1.019

Table 3 shows results from LCA, LBFS and our algorithm. It can be seen from Table 3 that our algorithm outperforms LBFS in terms of efficiency and accuracy. In DBLP and roadNet, our algorithm ran twice as fast as LBFS. Compared with LCA, our algorithm approximates more accurately but with a slightly higher runtime.

Table 3. Experimental results of LCA, LBFS and our algorithm

Network	Our algorithm		LCA		LBFS	
	T_q (ms)	p	T_q (ms)	p	T_q (ms)	p
Email-Enron	1.38	1.022	0.84	1.095	1.29	1.030
itdk0304_rlinks	5.59	1.020	3.57	1.083	5.60	1.028
DBLP	10.41	1.020	6.35	1.072	15.99	1.025
roadNet	19.75	1.019	10.70	1.067	37.89	1.022

From our experiments, we deduced that threshold t affects the efficiency and accuracy of our algorithm. Table 4 shows the influence of threshold t. From Table 4, it can be seen that runtime increases and path ratio decreases when threshold is increased. When there are more nodes at the top level network, approximation is more accurate

Table 4. Comparison of different thresholds

Network	Threshold				
	40	60	100	140	180
Email-Enron					
T_q (ms)	0.54	0.78	1.38	2.39	4.57
p	1.028	1.024	1.022	1.020	1.020
itdk0304_rlink					
T_q (ms)	2.73	4.31	5.59	9.63	18.32
p	1.030	1.026	1.020	1.019	1.018
DBLP					
T_q (ms)	6.54	8.86	10.41	20.29	43.32
p	1.026	1.022	1.020	1.018	1.018
roadNet					
T_q (ms)	12.93	15.42	19.75	39.34	72.58
p	1.026	1.020	1.019	1.018	1.018

but requires more time for Dijkstra's algorithm. When t increases from 40 to 100, Tq also increases a little, but p improves significantly. When threshold value increases from 100 to 180, Tq increases sharply, but p remains nearly constant. We therefore set the threshold value at 100 in order to obtain a good trade-off.

5   Conclusion

Based on hierarchical networks, we propose an approximate shortest path algorithm which is efficient and also with a high approximation accuracy on large scale networks. The algorithm condenses central nodes and their neighbors into super nodes to iteratively construct higher level networks until the scale of the top level meets a set threshold value. The algorithm approximates the distances of the shortest paths in the original network by means of super nodes in the higher level network. Performance of our algorithm was tested on four real networks. Results from our tests shows that our algorithm has runtime per query within few milliseconds and at the same time delivers high accuracy on large scale networks. Compared with other algorithms, our algorithm runs twice as fast as LBFS and 10 times faster than CDZ.

The proposed algorithm mainly focuses on undirected and unweighted networks. In the future, we seek to focus on directed and weighted networks by exploring the approximate distance between a node and its central node based on hierarchical networks. We will also consider an adaptive algorithm for different types of networks.

Acknowledgments. This work was partially supported by the National Natural Science Foundation of China under Grant No. 61433014, by the National High Technology Research and Development Program under Grant No. 2015AA7115089.

References

1. Broido, A.: Internet topology: connectivity of IP networks. In: SPIE Conference on Scalability and Traffic Control in IP Networks, Denver, pp. 172–187 (2001)
2. Slivkins, A.: Distance estimation and object location via rings of neighbors. Distrib. Comput. **19**, 313 (2007)
3. Klimmt, B., Yang, Y.: Introducing the Enron corpus. In: CEAS Conference (2004)
4. Chow, E.: A graph search heuristic for shortest distance paths. In: The Twentieth National Conference on Artificial Intelligence, Pittsburgh, PA, USA (2005)
5. Tang, J.T., Wang, T., Wang, J.: Shortest path approximate algorithm for complex network analysis. J. Softw. **22**, 2279 (2011). (in Chinese)
6. Freeman, L.C.: Centrality in complex networks: conceptual clarification. Soc. Netw. **1**, 215 (1979)
7. Potamias, M., Bonchi, F., Castillo, C., Gionis, A.: Fast shortest path distance estimation in large networks. In: Proceedings of the 18th ACM Conference on Information and Knowledge Management, pp. 867–876. ACM (2009)
8. Ley, M., Reuther, P.: Maintaining an online bibliographical database: the problem of data quality. In: EGC, pp. 5–10 (2006)
9. Rattigan, M., Maier, M.J., Jensen, D.: Using structure indices for efficient approximation of network properties. In: Proceedings of the 12th ACM SIGKDD International Conference on Knowledge Discovery and Data Mining, pp. 357–366. ACM (2006)
10. Qiao, S.J., Tang, C.J., Peng, J., Liu, W., Wen, F.L., Qiu, J.T.: Mining key members of crime networks based on personality trait simulation email analysis system. Chin. J. Comput. **31**, 1795 (2008). (in Chinese)
11. Stolfo, S.J., Hershkop, S., Wang, K., Nimeskern, O., Hu, C.W.: Behavior profiling of email. In: Proceedings of the 1st NSF/NIJ Conference on Intelligence and Security Informatics, pp. 74–90 (2003)
12. Yang, Y.B., Li, N., Zhang, Y.: Networked data mining based on complex network visualizations. J. Softw. **19**, 1980 (2008). (in Chinese)
13. Schwarzkopf, Y., Rákos, A., Mukamel, D.: Epidemic spreading in evolving networks. Phys. Rev. E **82**, 036112 (2010)

Intelligent Text Mining Model for English Language Using Deep Neural Network

Shashi Pal Singh[1](✉), Ajai Kumar[1], Hemant Darbari[1],
Balvinder Kaur[2], Kanchan Tiwari[2], and Nisheeth Joshi[2]

[1] AAIG, Center for Development of Advanced Computing, Pune, India
{shashis,ajai,darbari}@cdac.in
[2] Banasthali Vidyapith, Rajasthan, India
balvinderkaur818@gmail.com, kanchan040790@gmail.com,
jnisheeth@banasthali.in

Abstract. Today there exist various sources that provide information in very massive amount to serve the demand over the internet, which creates huge collection of heterogeneous data. Thus existing data can be categorized as unstructured and structured data.

In this paper we propose an idea of a tool which intelligently preprocesses the unstructured data by segmenting the whole document into number of sentences, using deep learning concepts with word2vec [11] and a Recurrent Neural Network [13]. At the beginning step we use word2vec which was introduced by Tomas Mikolov with his team at Google, to generate vectors of the inputted text content which will be further forwarded to Recurrent Neural Network. RNN takes this series of vectors as input and trained Data Cleaning Recurrent Neural Network model will perform preprocessing task (including cleaning of missing, grammatically incorrect, misspelled data) to produce structured results, which then passed into automatic summarization module to generate desired summary.

Keywords: Text processing · Sentence processing · Deep neural network (DNN) · Recurrent Neural Network (RNN) · Summarization

1 Introduction

Due to the fast growth of data usage by various firms, organizations, and institutions over the internet there is a condition of a huge amount of data generation on daily basis. This exponential growth of big data over web is a huge challenge for us to manage and store this big data in a structured format. According to IDC Big data is termed as "a new generation of information technologies and design architecture to economically extract values from a very large volume of wide variety of data, enabling high velocity of capture, discovered and/or analysis" [2]. Hence, extracting reliable information from large amount of available documents is the biggest problem in today's time.

On wide categorization the data is categorized into three categories: Unstructured data, Semi-structured data and structured data. On the basis of some survey work by researchers it is found that 95% of digital data is in unstructured format [1].

© Springer International Publishing AG 2018
S.C. Satapathy and A. Joshi (eds.), *Information and Communication Technology for Intelligent Systems (ICTIS 2017) - Volume 2*, Smart Innovation, Systems and Technologies 84, DOI 10.1007/978-3-319-63645-0_54

Figure 1 explains existing problem as now a days each user gathers data from the web which is in dispersed form, from various data sources. Hence while integrating data derived from various sources makes difficult for common user to extract essential information. Hence on this step data needed to be intelligently processed to make it understandable.

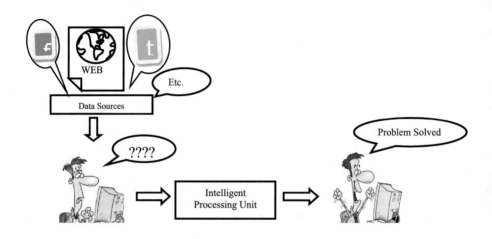

Fig. 1. Existing problem

2   Literature Review

Unstructured data is referred as the information which does not have any predefined relational database or data model. This unstructured data cannot be used for any analytical process because of improper data structure which leads to meaningless information for further operations or implementations. This unstructured data can be obtained from various sources, some of those sources are:

- Multimedia data—audio files and video clips
- Data retrieved from satellites
- Web logs
- Results of various kinds of surveys
- Data from social media facebook, twitter, etc.
- Electronic mails

Structured data is a data which is stored in a relational database system. Data of this type is highly organized and it can be managed by SQL and its supports many relational databases such as ODBC, RDBMS. In general terms it can be said that structured data is stored in rows and columns in a tabular.

2.1 Data Extraction

Data extraction is the process of retrieving information from unstructured or poorly structured document for further processing of extracted knowledgeable text. Before cleaning data is needed to be extracted from various sources in various formats (pdf, doc, mp3, xlsx, rtf, jpg, png etc.). For this purpose Apache tika is used.

2.1.1 Apache Tika
Apache Tika [22] is a library function used for document type detection and content extraction. Tika includes various existing parsers and java files to detect the type of file (pdf, txt, doc, mp3, mp4, rtf, jpg etc.) and then extract content from it. Tika can be easily integrated in any project and act as a universal file type detection and file content extraction. Tika architecture consists of following: [21]

- Language detection mechanism.
- MIME detection mechanism.
- Parser interface.
- Tika Facade class.

2.2 Data Preprocessing

After the extraction of content data preprocessing is an important aspect which will convert raw data having irregularities into consistent structured data. The cleaning operations were performed with the help of regular expressions and for some cases regular expressions does not work for those we train our recurrent neural network (DCRNN) by applying deep learning concept for fast cleaning.

2.2.1 Deep Learning
Deep learning is a sub part of machine learning field. It is a learning technique to train the neural network with respect to the set of input data provided to the input layer of neural network with intent to achieve a desired output. In deep learning the multi layer neural network is used for training purpose i.e. the network consist of input layer, hidden layers and an output layer. In case of deep learning there are more than one hidden layers between the input layer and output layer of neurons. The deep learning network can learn either by supervised learning (learning under the supervision of the input data set provided for training) or by unsupervised learning (self learning by experience over the time duration). Here in our model we will use supervised learning concepts [11].

2.2.2 Word2vec
Word2vec [11] is an open source neural network to generate vectors for the words, it consist of two layers- input layer and an output layer. It is not a learning network it is only used to generate vector set of the input words, the input for the word2vec is a text corpus and the output is set of vectors which are further used as an input for the deep networks. The main idea behind the usage of word2vec is to keep similar words in nearest domains. Word2vec perform normal for small input data but it performs best in

case of large data, if enough amount of data is given to word2vec then it performs accurate guess about the similarity between the words on the basis of its past experiences i.e. word2vec mathematically detects the similarity between words. The output generated by word2vec is a vocabulary which consists of words and their respective vectors which are used as an input for the deep learning neural network for further learning or it can be used to find the relationship between words.

Word2vec perform the vector generation by two ways either by using CBOW or by skip-gram algorithm. Word2vec for java code by deeplearning4j uses skip-gram method as it produces better accurate result for large data sets. After generation of word2vec vector space we can perform vector operations (dot product, addition and multiplication) on generated vectors. The dot product of two vectors divided by product of the length of those vectors is equal to the cosine of angle between vectors. When this cosine value is applied to semantic vector spaces then it is defined as cosine similarity score [12], which generally lies between −1.0 to 1.0, where −1.0 means that the terms are completely different and 1.0 means the words or terms are semantically same. This cosine similarity score is used for ranking of similar terms with reference to the desired target term.

2.2.3   RNN (Recurrent Neural Network)

A recurrent neural network is a type of artificial neural network in which the connections between neural units form a directed cycle, RNN have its internal memory to process sequence of inputs. [13] In case of traditional neural network it is assumed that all inputs and outputs are independent to each other but for some tasks where previous knowledge is required for processing like upcoming next word prediction in a sentence it's not applicable so in that case RNN is useful because it perform same task for each element of input sequence and the output depends on the previous computation results stored in its memory [13]. We can create a recursive neural network providing the same set of weights in a recursive manner on a graph like structure, and RNN is a special type of recursive neural network whose structure leads to a linear chain and this RNN uses a tensor-based composition function for each and every node in the tree structure [14].

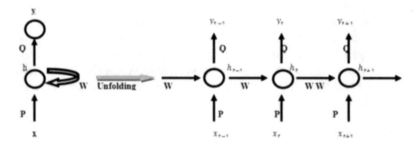

Fig. 2. Recurrent Neural Network

Figure 2 of recurrent neural network shows the unfolding of network into a full neural network. Unfolding means if a sequence is of sentence having 8 words in it then the network will be unfolded into eight-layer neural network i.e. one layer for each word.

Here:

- x_t Is the input to the network at time 't'.
- y_t Is the output at time 't'. If a user wants to predict next upcoming word in a sentence then it can be done by calculating the vector of probabilities across the generated vocabulary, $y_t = softmax(Qh_t)$.
- h_t is the hidden state of the network at time t and it is considered as the memory of the recurrent network. The value of h_t is calculates on the basis of network's previous hidden layers and the input to layer at current state: $h_t = f(Px_t + Wh_{t-1})$ where function f is nonlinearity e.g. tan h.

Hence it is clear that a RNN share the same parameter (P, Q, and W) across all the steps which show that in recurrent neural network at each step same task is performed with different inputs.

3   Related Work

Following are some already implemented works related to the text processing using deep learning:

1. Deep Text

Deep Text is a text understanding engine used by facebook which is based on the unsupervised deep learning of neural network [6]. Deep text can understand the text format posts on the facebook with a similar accuracy to a human understanding capacity i.e. it can understand the meanings of textual posts and also their sentiments. In aggregate deep text can understand some thousands posts per second which were written in more than twenty languages [6].

2. Open Refine

Open Refine [20] is a data cleaning tool which is formerly known as Google refine. It is a desktop application and with cleaning it also transform the files into other formats. Open refine working behavior is like a database i.e. it performs operation on rows of a table, the open refine project is created in the form of table and users have to manually provide conditions for cleaning.

3. Mead

Mead [7] is a framework for multi lingual summarization of text. This platform implements many summarization algorithms some of them are centroid based algorithm, position based algorithm, and query based approach [7]. On this framework the quality evaluation of the summarized content is done by both methods extrinsic and intrinsic. Mead also allows users to implement their own algorithms over the framework.

4   Proposed Work

Internet today becomes an ultimate source of data which clearly defines that we have entered an era of 'big data' [5]. The model defines in this paper, download data from web which may be in any format say pdf, txt, doc, rtf etc.

From Fig. 3, the first step is data extraction which returns the file type of the uploaded document and its text content.

The extracted raw data is forwarded to the cleaning module which includes cleaning of inconsistencies and misspelled data with the help of spell checker, this also includes data standardization. Then sentence segmentation is to be done on the basis of some delimiter and normalization is applied on these segmented sentences. Then these sentences were tokenized via parallel processing of multiple sentences at a time to check grammatical correctness of these sentences.

After preprocessing we have list of meaningful sentences and phrases which were further combined to obtain a structured data stored in a structured database.

Last module performs summarization with the help of natural language processing concepts and graph based algorithms.

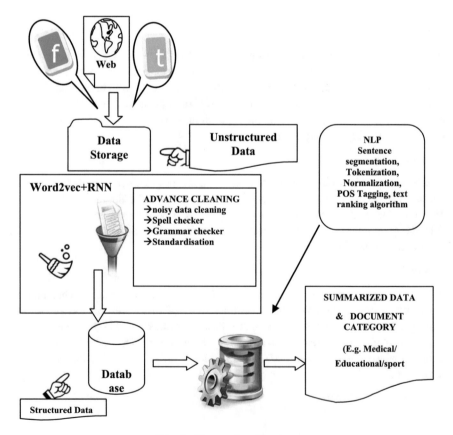

Fig. 3. Proposed architecture

A. **Sentence Processing Module**

Document downloaded from the internet needed to be processed including advance cleaning methods. This includes segmenting the documents into number of sentences and performs methods to clean data by removing inconsistencies, inaccurate content so that data must be used for further processing.

Inconsistent sentences include:

   i. Data incorrectly entered at the time of submission.
  ii. Data may consist of null values or incomplete information.
 iii. Misspelled data.
 iv. Redundant data.
  v. Data having symbols like $, @, %, ^, &,* etc.
 vi. Grammatically incorrect data.

Figure 4 explains whole sentence processing module which will remove inconsistent sentences from the document and then includes document summarization (which is explained in the further defined part C(Trained DCRNN Model). Below define Table 1 shows how Regex (regular expressions) and sql queries are used to make inconsistent sentences into consistent one.

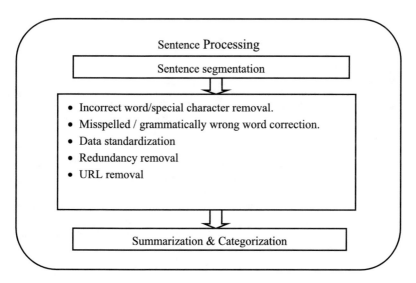

Fig. 4. Sentence processing

Table 1. Queries to correct inconsistent sentences

Incorrect sentences	Description	Query to be used for sentence processing	Processed sentences
After the extr@ction of content data$ processing is = important aspect	Extra symbols appearing In between of sentences	**Regex-** [^a-zA-Z0-9\s./]|(a-zA-Z =)	After the extraction of content data processing is important aspect
The following is the given @ email id for contact xyz@gmail.com	Email validation and removal of @ symbol appearing apart from email id	**Regex-** [_a-zA-Z0-9 + -\] + (\. [a-zA-Z0-9-] +)*@ [a-zA-Z0-9-] + (\.[a-zA-Z0-9] +) *(\.[a-zA-Z]{2,})	The following is the given email id for contact xyz@gmail.com
Minimum percentage % requires for % eligibility criteria is 50.00% and for direct interview percentage should be more than 85%	Numeric percentage validation and removal of % symbol appearing apart from percentage	**Regex-** (\d + (\.\d +)?\s%)|(\d + (\.\d +)? %)|([a-zA-Z]%)|([^a-zA-Z0-9]%)	Minimum percentage requires for eligibility criteria is 50.00% and for direct interview percentage should be more than 85%
Data incorrectly ^ % entered * at the time _ # of $ submission.	Removal of all special characters	**Regex-** [^a-zA-Z0-9\s./]|(a-zA-Z =)	Data incorrectly entered at the time of submission
http://zeenews. india.com Demonetization leads to digital India	Removal of url from the text	**Regex-** (?:(?:https?|file|ftp):\/\/|www\.|ftp \.)(?:[-A-Z0-9\/@_# ~=%|&! $?,:. +])*	Demonetization leads to digital India
You can't use any existing tools for the development of new work It's been a grt day The total expenditure must be under 1 lac Hey, r u dere? Data claening performs methods to claen data by removing inconsistencies I'm very hungyyyyyyyyyyy	Apostrophe words are replaced with correct word extracted from the database Misspelled words are replaced with correct word extracted from the database	**SQL Query-** ResultSet rs = Statement. executeQuery("select correct_word from Apostrophe_table where incorrect_word = '" + word + "'"); if (rs.next()) { replace(word,correct_word); }	You can not use any existing tools for the development of new work It is been a great day The total expenditure must be under 1 lakh Hey, are you there? Data cleaning performs methods to clean data by removing inconsistencies I am very hungry

B. Training the Model

For training the model the desired task is performed using natural language processing tools and regular expressions. Regular expressions (Regex) are used for defining text patterns that are matched along with given input to validate the text content. Different regular expressions are used for defining different string patterns. Figure 5 shows network training module for preprocessing.

Regex can be used for following purposes:

Fig. 5. Network training

- Matching or finding text content within a given input data
- For validation of data values
- For conversion of data into different forms
- For case sensitive and case insensitive matching
- Parsing of data
- Data standardization

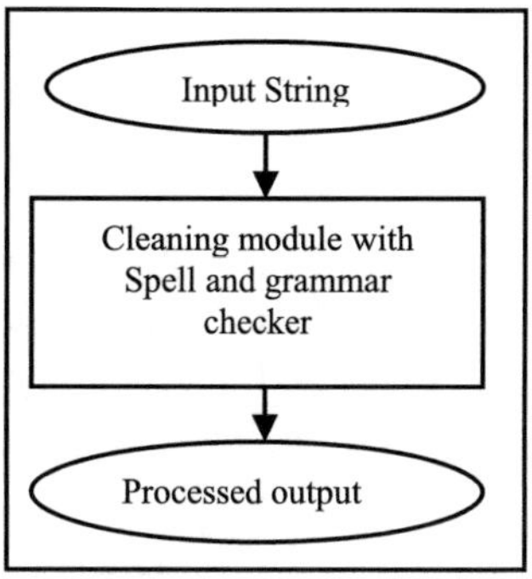

C. Trained DCRNN Model:

Trained DCRNN (Data Cleaning Recurrent Neural Network) Model is trained in such a way that it takes text as input and performs preprocessing on text using word2vec and Recurrent Neural Network automatically with supervision.

I. **Encoding:** Inputted text is passed to word2vec which will convert bags of strings into their corresponding resultant vectors.

II. **DCRNN:** Data Cleaning Recurrent Neural Network takes vectors of strings produced by word2vec as input to perform desired data preprocessing functionality as defined at the training step.

III. **Decoding:** Processed vectors generated by DCRNN are needed to be decoded into words for further processing.

IV. **Automatic Summarization:** Data summarization is a necessity of today's world due to shortage of time. We all aware of the fact how much amount of data produces daily from various sources like social media sites, news bulletin, it industries, educational sites, research areas etc. ence it is required to filter the data daily to figure out which data is of use and which data should be destroyed immediately. Although it is not possible for us to read all of the available data thoroughly therefore need of automatic summarization is required (Fig. 6).

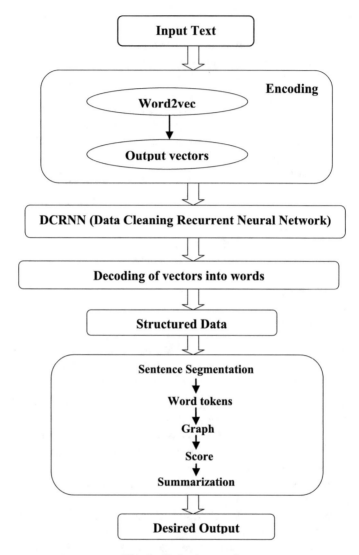

Fig. 6. Trained network

Following are the steps to summarize the document:

- Tokenize the text into sentences
- Tokenize each sentence into a collection of words
- Convert the sentences into graphs
- Score the sentences via page rank
- Arrange the sentences from highest score to lowest
- Summarize the document using top 4 sentences (Fig. 7)

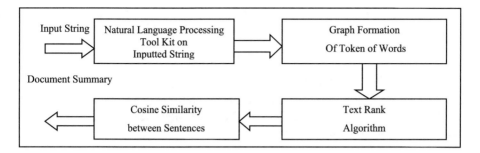

Fig. 7. Text summarization

a. **Sentence Segmentation:** The text given as input must be broken into sentences. This is required as the whole text consist of several sentences. Hence they need to be separated on the basis of the delimiters such as ",", ".", "!" etc.

b. **Word Tokenization:** Segmented sentences are tokenized in order to identify key terms.

c. **Stop Word Removal:** Stop words are the words which are removed after or before processing of data i.e. words such as-are, is, a, to, have etc. which doesn't have much effect on the meaning of the sentence and these words are needed to be removed before further processing.

d. **Graph Formation:** After preprocessing using deep learning and natural language processing modules, the uploaded document is left with tokens of words. These tokens are used to form the undirected weighted graph with V set of vertices connected with E set of edges. In this proposed model tokens of words are used to represent vertex and their lexical relationship is used to represent edge between them [8].

e. **Sentence Score Calculation:** Sentence score is used to find out the importance of the sentence using graph based algorithm [9]. Sentence score is used to find out the summary of the whole text. Most appropriate and highest scored sentences are grouped together to form the summary.

   Text Rank [4] is a graph-based, surface-level algorithm. In this algorithm, the similarity values of the edges are used to weight the vertices. The basic premise of a graph-based ranking model is similar to voting or recommendation. Consider a graph G with V vertices and E edges, where vertex $V = \{v_i : 1 < i < n\}$, and edge eij is the edge between vertex v_i and vj [8] the score of each vertex can be calculated by following given equation.

$$\text{Score}(V_i) = (1 - d) + d * \sum_{i=1}^{n} \frac{1}{Out(V_j)} S(V_j) \tag{1}$$

f. **Text Summarization:** Text summarization [10] is an important application of information extraction. Text rank algorithms [4] are applied on the graph to calculate the similarity between two words. Cosine similarity is used to make a similarity matrix using vectors of sentences [8]. This similarity matrix is used to make graph of scored sentences and the top scored sentences are to be considered to calculate the summary of the document (Fig. 7).

D. Back Propagating the Error

In this paper we define a neural network with 'n' number of inputs, 'n' hidden neurons, and 'n' output neurons. Additionally, the hidden and output neurons will also include a bias. Figure 8 shows basic structure.

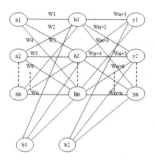

Fig. 8. Basic network structure

Here $x_1, x_2, x_3 \dots x_n$ denotes input vectors $y_1, y_2, y_3 \dots y_n$ denotes output vectors and $w_1, w_2, w_3 \dots w_n$ denotes weight vectors. The neural network maps x input to y outputs using w weights. The main aim of backpropagation is to make neural network learn to correctly maps inputs to outputs by adjusting weights.

This algorithm performs using forward pass and a backward pass. When an input is passed to inputted layers to form a desired output. The output produced is compared to required output and error value calculated is backpropagated to adjust weights.

(1) Forward pass

Neural network in the forward pass randomly chooses weights and biases. Following equation defines how to calculate total input for hidden layer h_1.

$$h_1 = w_1 * x_1 + w_2 * x_2 + b_1 \tag{2}$$

Following equation defines net output produced using logistic function by taking output produced by hidden layer as input.

$$\text{out}_{h_1} = \frac{1}{1 + e^{h1}} \tag{3}$$

We repeat the above procedure for all the neurons in order to produce net output by using hidden layer output as its input.

Here's the output for output y_1:

$$net_{y_1} = w_5 * out_{y_1} + w_6 * out_{h_2} + b_1 \tag{4}$$

$$out_{h_1} = \frac{1}{1 + e^{net_{y_1}}} \tag{5}$$

Calculating the Total Error
Now error function is used to calculate the error of each neuron. Following defined is the total error calculation.

$$Error(total) = \sum \frac{1}{2} \left(required_{output} - actual_{output}\right)^2 \tag{6}$$

(2) **Backward pass**

Backward pass in backpropagation is used to minimize the error by propagating from output layer to input layer. The main of backward pass is weight updation from output layer to hidden layer and back to the input layer.

5   Conclusion and Future Scope

In this work we present an intelligent preprocessing of text data which performs data cleaning of uploaded document and then its automatic summarization. The model we are proposing produce an efficient and accurate results as this is based on most demanding technologies of todays like deep learning, word2vec, Regex, natural language processing. The study proposed "Intelligent pre-processing unit" that performs all the operations simultaneously presently which is performed by various individual machines. In this intelligent preprocessing unit, unstructured documents (txt, pdf, doc, xlsx) having text contents are preprocessed and then converted into structured data files and its automatic summarization.

In future there will be a very good scope of this approach as most of the data over the web is dispersed and is in unstructured format.

References

1. Gandomi, A., Haider, M.: Beyond the hype: big data concepts, methods, and analytics. Int. J. Inf. Manage. **35**(2), 137–144 (2015)
2. Kanimozhi, K.V., Venkatesan, M.: Unstructured data analysis-a survey. Int. J. Adv. Res. Comput. Commun. Eng. **4**(3) (2015)
3. Chakraborty, G., Pagolu, M.K.: Analysis of Unstructured Data: Applications of Text Analytics and Sentiment Mining. Paper 1288-2014
4. Mihalcea, R., Tarau, P.: Text Rank: Bringing Order Into Texts

5. Brown, B., Chui, M., Manyika, J.: Are you ready for the era of Big Data? McKinsey Q. **4**, 24–35 (2011)
6. Network World. http://www.networkworld.com/article/3077998/internet/understanding-deep-text-facebooks-text-understanding-engine.html
7. MEAD. http://www.summarization.com/mead
8. Pawar, D.D., Bewoor, M.S., Patil, S.H.: Text rank: a novel concept for extraction based text summarization. Int. J. Comput. Sci. Inf. Technol. **5**(3) (2014)
9. Mihalcea, R.: Graph-Based Ranking Algorithms for Sentence Extraction, Applied to Text Summarization. Department of Computer Science University of North Tex. 1974–July 2004)
10. Gupta, V., Lehal, G.S.: A survey of text summarization extractive techniques. J. Emerg. Technol. Web Intell. **2**(3), 258–268 (2010)
11. DeepLearning4j. https://deeplearning4j.org/word2vec
12. Chanen, A.: Deep learning for extracting word-level meaning from safety report narratives. IEEE (2016). 978-1-5090-2149-9
13. AI, Deep Learning, NLP. http://www.wildml.com/2015/09/recurrent-neural-networks-tutorial-part-1-introduction-to-rnns/
14. Socher, R., Perelygin, A., Wu, J.Y., Chuang, J., Manning, C.D., Ng, A.Y., Potts, C.: Recursive deep models for semantic compositionality over a sentiment treebank. In: EMNLP (2013)
15. Hemalatha, I., Varma, G.P.S., Govardhan, A.: Preprocessing the informal text for efficient sentiment analysis. IJETTCS **1**(2) (2012)
16. Collobert, R., Weston, J.: A unified architecture for natural language processing: deep neural networks with multitask learning. In: 25th International Conference on Machine Learning. Helsinki, Finland (2008)
17. You, L., Li, Y., Wang, Y., Zhang, J., Yang, Y: A deep learning-based RNNs model for automatic security audit of short messages. In: IEEE 16th International Symposium on Communications and Information Technologies (ISCIT) (2016)
18. Ouyang, X., Zhou, P., Li, C.H., Liu, L.: Sentiment analysis using convolutional neural network. In: IEEE International Conference on Computer and Information Technology; Ubiquitous Computing and Communications; Dependable, Autonomic and Secure Computing, Pervasive Intelligence and Computing (2015)
19. Chen, K.-Y., Liu, S.-H., Chen, B., Wang, H.-M., Jan, E.-E., Hsu, W.-L., Chen, H.-H.: Extractive broadcast news summarization leveraging recurrent neural network language modeling techniques. IEEE Trans. Audio Speech Lang. Process. **23**(8), 1322–1334 (2015)
20. Open Refine. http://openrefine.org/
21. Tutorials Point. http://www.tutorialspoint.com/tika
22. Apache Tika. https://tika.apache.org

Intelligent English to Hindi Language Model Using Translation Memory

Shashi Pal Singh[1](✉), Ajai Kumar[1], Hemant Darbari[1], Neha Tailor[2], Saya Rathi[2], and Nisheeth Joshi[2]

[1] AAIG, Center for Development of Advanced Computing, Pune, India
{shashis,ajai,darbari}@cdac.in
[2] Banasthali Vidyapith, Banasthali, Vanasthali, Rajasthan, India
nehatailor2424@gmail.com, sayarathi22ll@gmail.com,
jnisheeth@banasthali.in

Abstract. English to Hindi Translator using Translation Memory is a translation tool that transforms an English sentence to its proper Hindi meaning. This translator works on both exact match and fuzzy match. When the input source in English is divided into segments and is completely matched with the database then the appropriate translation for that input is directly fetched from the database. The matching process between the input and the source file is done with the help of Edit Distance Matching algorithm. The case when the input is not 100% matched with the database that is considered as the case of fuzzy match so; in that case N-gram modeling is performed in order to give an appropriate translation to the user. The case in which the input is not completely matched with database (fuzzy match) score is calculated to get the match percentage. The case when the input in the English language is not matched with the database at all, then an appropriate algorithm is used that gives word to word translation of that particular input.

Keywords: Fuzzy matching · Edit distance · N-gram · Match score · Translation memory

1 Introduction

Translation memory is a tool that is used to monitor and help in the translation process from a language to the other one. Translation memory (TM) a database that contains "segments", of any sentence, paragraph, heading or element in list, that have already been translated in order to help human translators. Translation memory contains the source text and its corresponding required translation in language pair called "translation unit".

Programs that use translation memories sometimes called *translation memory managers.* Although it is well known to us that there is the requirement of English to Hindi translator, and various translators are already there available to us, but there is not proper and enough tools for the translators to use, because there are many difficulties and complexities in it, and the main reason behind these difficulties are the semantic level, word level and phrase level translation difficulty and this is because of the use of different languages and different cultures and lack of a good translation memory [3].

© Springer International Publishing AG 2018
S.C. Satapathy and A. Joshi (eds.), *Information and Communication Technology for Intelligent Systems (ICTIS 2017) - Volume 2*, Smart Innovation, Systems and Technologies 84, DOI 10.1007/978-3-319-63645-0_55

As in India there are 29 states and each state has its own language and various other regional languages known as dialects so it creates a problem to understand each and every language.

As English and Hindi languages are considered as the most common national or international language which helps to communicate with the people from different regions, that is why this translator is gaining too much importance in day to day life.

Translation memory system is having a very wide scope in today's life as it is used in various fields in daily life, as this project is going to handle the condition of fuzzy match so it is going to be advancement in the field of translation because most of the translators work only on the condition of exact match.

2   Methodologies

2.1   N-grams Algorithm

N-grams [10] approach can also be used for efficient approximate matching (fuzzy matching). In this approach a single sentence or a segment gets broken down in a set of *n*-grams, thus allowing the set to be compared to other set in sequences in an efficient manner.

1.   Tokenize the given segment.
2.   Count no. of tokens
3.   if tokens<5
- Then make bigrams of given sentence by combining two tokens together
- Else
- goto 4
4.   if tokens<10
- Then make trigrams of given sentence by combining three tokens together.
- Else
- goto 5
5.   if tokenA10
- Then make four grams of given sentence by combining three tokens together.

2.2   Edit-Distance Matching Algorithm

3-operations (Levenshtein Distance) Edit Distance approach [9] is used as a string matching function. It is used for efficient retrieval from translation memory.

Levenshtein Distance is a character based matching approach that computes, the minimum number of edit operations. Edit operations may be insertion, deletion and replacement required for transforming one N-gram to another and will act as a source for matching of current input and the data in TM and retrieve the translation in target language which help human translator to either accept it, replace it with fresh translation or modify the translation.

This is Dynamic Programming based algorithm which is a tabular computation of D (n, m), which helps to solve problems by combining solutions to sub-problems.

This is a based on bottom-up approach here:

- We compute D(i, j) for i, j
- And compute larger D(i, j) on the basis of previously computed smaller values that is compute D (i, j) for all i($0 < i < n$) and j ($0 < j < m$).

Algorithm.
Steps:
Input- Two strings A1 and A2.
Output- Score of similarity.
1.   int m[i,j] = 0
2.   for i <- 1 to |A1|
3.   do m[I,0] = i
4.   for j <-1 to |A2|
5.   do m[0,j] = j
6.   for i <- 1 to |A1|
7.   do for j <- 1 to |A2|
     do m[i,j] = min{m[i-1,j-1]+if(A1[i] = A2[j]
then 0
else 1,
          m[i-1, j]+1, m[i,j-1]+1}
8.   Return m[|A1|,|A2|]

3   Literature Survey

Related to Machine Translation we have read various research papers and gone through various translators approaches and tools, here are some of the Machine Translation system.

3.1   Google Translator [11]

Till now there are a number of translators available to us such as Google translator developed by Google, but Google translators do not show its main focus on Hindi language, it performs translation for various languages that is why it also has various drawbacks in English to Hindi conversion as per our survey.

3.2   MANTRA-Rajbhasha [12]

Mantra-Rajbhasha MAchiNe assisted TRAnslation Tool is developed for Department of Official Language (DOL), Ministry of Home Affairs, Government of India to facilitate the translation of documents pertaining to Personnel Administration, Finance, Agriculture, Small Scale Industries, Information Technology, Healthcare Education and Banking domains from English to Hindi.

3.3   MANTRA-Rajya Sabha [13]

Mantra-Rajya Sabha system, which translates from English to Hindi language pertaining to Papers to be laid on the Table [PLOT], List of Business [LOB], Bulletin Part-I, Bulletin Part-II with an accuracy of 90–95% is working since 2007 at Rajya Sabha Secretariat.

3.4   ANUVADAKSH [14]

English to Indian Languages Machine Translation System [EILMT] allows translating the text from English to 08 Indian languages such as Tamil, Oriya, Hindi, Bengali, Marathi, Urdu, Gujarati and Bodo in Tourism, Health care and Agriculture Domains.

3.5   MATRA MT System

MATRA is the human supported transfer based translation system that converts English to Hindi. The system has been applied mainly in the area of news, technical phrases and annual reports. The work is supported under the TDIL Project, currently CDAC Mumbai is working on MATRA [3].

3.6   SHAKTI MT System

This system works on three languages as Hindi, Marathi, and Telugu. If the user is not satisfied with the translation then it ask for other meanings of input sentences components as word, phrases and even sentence level from user and retranslate it [3].

4   System Architecture

The system architecture of the translator contains various components that perform various different tasks that are required in order to translate source file (English) to the target (hindi) file.

In order to achieve the target various operations has to be performed and they are as follows (Fig. 1):

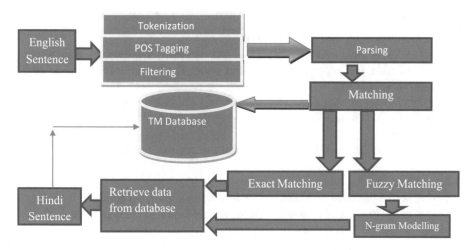

Fig. 1. Architecture diagram of translation memory

4.1 Implementation

The database of our translation memory contains a source text and a corresponding target text. While using translation memory for translation the very first task is to break the source text (to be translated) into segments, after the segmentation process tokenization process is applied in order to get tokens. As tokens are obtained then *POS Tagging* has to be performed. In POS tagging process the data entered for the translation is assigned with its category, whichever category it belongs to that is noun, pronoun, verb, adjective etc. On tagged data then parsing is applied in order to produce a parse, so that reordering rules can be applied [1].

After reordering matching of the input data is performed to get the desired output. There are various translators that works only on 100% matched data that is the output will be received only if the source text (to be translated) gets exactly matched with the database and this matching process can be done by using any algorithm that can calculate the distance between the words,

In order to apply matching algorithms Bi-gram or Tri-gram approach has to be applied depending upon the length and complexity of the sentence, but there may be the case when matching is not 100% that means the condition *of fuzzy match* then N-gram approach is used, with a different type of algorithm that can handle the fuzzy match in order to obtain similar segments which are presented to the translator with differences flagged.

The condition in which the source text (to be translated) will not match at all then translation will done by translator manually or it returns the word to word translation of the particular source text. Every time the newly translated document gets stored in database so that it can be further used for future translations and also in the case when repetitions of the particular segment takes place in the current text.

4.2   Pre Processing

4.2.1   Segmentation and Tokenization

The text in TM database is stored in units called segments. Segments are considered as the most suitable and appropriate translation units. These segments can be sentence, headings, paragraph etc.

Segmentation is the process of breaking the big document into small segments such as sentences, phrases, headings etc. The text in the TM database is stored in the units called segments.

Segments are the most appropriate, useful and suitable translation units (TUs).

4.2.2   Tagging

We have built a database of 50,000 words in which the category to which it belongs is also mentioned in front of it thereby making it easy to assign the tags to the individual word/token such as noun, verb, adjective, etc.

4.3   Parsing

Parser is used for generating the parse tree. Parse tree [6] is obtained from syntactic analysis that represents the syntactic structure of sentence. In a parse tree, an interior node is called a non-terminal/non-leaf node of the grammar, while the leaf node is a word called a terminal of the grammar.

For Parsing *Stanford Parser* is used to generate parse tree.

English grammar rules for generating parse tree [3]:

- S->NP VP
- NP->PRON NOUN
- VP->AUX Y
- Y->ADV ADJ

Input: This book is very good (Fig. 2).

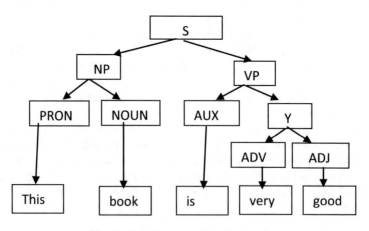

Fig. 2. Result generated by the parsing

4.4 Matching

Matching is the process of calculating or determining the similarity and dissimilarity between two strings in order to conclude whether they are same or not.

In matching process there are three possibilities:

- Exact Match
- Fuzzy Match
- Not Matched

4.4.1 Exact Match

In our project we first compare the generated tokens with the TM database if the tokens gets exactly matched (100%) then the output is retrieved from the database directly otherwise the condition of fuzzy match may appears.

For matching edit-distance algorithm is used:

Edit-distance algorithm can be explained by the following example:

String 1: CAT

String 2: MAT

As C is not matched with M

So distance between String 1 and String 2 is 1.

Therefore by using Edit-distance algorithm one can find similarity between the given two strings, and hence makes us able to calculate the *matching score* for that particular string.

Matching Score

The whole consideration in matching approach for match score [8] calculation is to get a match score which ranges between 0 and 1, where 0 represents a "no match", means not matched at all, 1 represents a "precise match (exactly matched)" and a value in-between represents a "partial match".

$$\text{Match}\% = \frac{\text{Number of tokens matched} * 100}{\text{Total number of tokens}}$$

4.4.2 Fuzzy Match

Fuzzy match [7] is the condition when the input is not 100% matched with the database i.e. the matching percentage lies between 0–100.

To handle this fuzzy match situation n-gram modelling has to be performed. When on applying matching algorithm the input source is not completely matched with database then n-gram modelling is performed depending upon the size of the segment.

As in fuzzy match the input is not exact match that is, it is approximate match, so with the help of generated grams from the segments the matched percentage is calculated.

Working procedure for n-gram algorithm for fuzzy matching is explained by the following example:

A1: The drought caused distress in a large number of chronically deficit areas.

The bi-gram sequences are:

{"The drought$_1$", "drought caused$_2$", "caused distress$_3$", "distress in$_4$", "in a$_5$", "a large$_6$", "large number$_7$", "number of$_8$", "of chronically$_9$", "chronically deficit$_{10}$", "deficit areaA1$_1$"}.

A2: The drought affecting large number of parts of the country especially rural areas.

The bi-gram sequences are:

Table 1. Matching of A1 and A2

The drought$_1$	The drought$_1$
drought caused$_2$	drought affecting$_2$
caused distress$_3$	affecting large$_3$
distress in$_4$	large number$_4$
in a$_5$	number of$_5$
a large$_6$	of the$_6$
large number$_7$	the country$_7$
number of$_8$	country especially$_8$
of chronically$_9$	especially rural$_9$
chronically deficit$_{10}$	rural areaA1$_0$
deficit areaA1$_1$	

{"The drought$_1$", "drought affecting$_2$", "affecting large$_3$", "large number$_4$", "number of$_5$", "of the$_6$", "the country$_7$", "country especially$_8$", "especially rural$_9$", "rural areaA1$_0$"} (Table 1).

Here out of 10 sequence of tokens only 3 bigrams are matched so the matched percentage is 30%

$$\text{Match percentage} = (3/10) * 100 = 30\%$$

4.4.3 Not Match

The case when the input string is 0% matched with the database or we can say that string is not matched at all, in that case one to one/word translation of the text is performed and for that one to one translation we have built a database of 3, 60,000 words and phrases which is bilingual lexicon of English word and a Hindi meaning corresponding to that particular word [7].

5 Retrieval and Updation

In retrieval process the translator makes a choice which one to accept or modify. Now the modified or newly entered translation are learned and updated in TM database so as to enrich the TM database with the new translation so that they can be used in future to produce quality translation.

Example:
Sentences in Translation Memory:
A1: I play cricket. - मैं क्रिकेट खेलता हू
A2: Ram plays cricket. – राम क्रिकेट खेलता है
S3: Ram plays football. – राम फुटबॉल खेलता है
Input Sentence:
I play football.
Output:
मै फुटबॉल खेलता हू

6 Conclusion and Future Scope

This approach has been tested and developed on 30,000 sentences aligned corpus with a database of 50,000 words and phrases. This approach has helped us to achieve fast and accurate translation with minimum efforts, this approach can applied for other Indian languages and also on foreign languages and can made more efficient by enhancing the Parallel corpus it automatically have provision of machine learning in it.

References

1. Naskar, S., Bandyopadhyay, S.: Use of Machine Translation in India: Current Status. Jadavpur University, Kolkata (2005)
2. David Peter S., Nair, L.R.: Machine translation system for Indian languages. Int. J. Comput. Appl. (0975–8887) **39**(1), 24–31 (2012)
3. Gehlot, A., Sharma, V., Vidyapith, B., Singh, S., Kumar, A.: Int. J. Adv. Comput. Res. **5** (19). 4AAIG, Centre for Development of Advanced Computing, Pune, India (2015). ISSN (Print) 2249-7277; ISSN(Online) 2277-7970
4. Dhomne, P.A., Gajbhiye, S.R., Warambhe, T.S., Bhaga, V.B.: Accessing database using NLP. IJRET Int. J. Res. Eng. Technol. eISSN 2319-1163; pISSN 2321-7308
5. Dungarwal, P., Chatterjee, R., Mishra, A., Kunchukuttan, A., Shah, R., Bhattacharyya, P.: The IIT Bombay Hindi, English Translation System at WMT. Department of Computer Science and Engineering, Indian Institute of Technology, Bombay (2014)
6. Roy, M., Walivadekar, M., Kadam, P.: Sentence validation using natural language processing. In: Proceedings of IRF International Conference, 23rd February 2014, Pune, India. ISBN 978-93-82702-61-0
7. Ashrafi, S.S., Kabir, Md. H., Anwar, Md. M., Noman, A.K.M.: English to Bangla machine translation system using context—free grammars. IJCSI Int. J. Comput. Sci. **10**(3) (2013)
8. Ludwig, S.A.: Fuzzy Match Score of Semantic Service Match. IEEE
9. https://en.wikibooks.org/wiki/Algorithm_Implementation/Strings/Levenshtein_distance
10. http://www.decontextualize.com/teaching/dwwp/topics-n-grams-and-markov-chains/
11. https://translate.google.com/
12. https://mantra-rajbhasha.rb-aai.in/
13. https://www.cdac.in/index.aspx?id=mc_mat_mantra_rajyasabha
14. https://cdac.in/index.aspx?id=mc_mat_anuvadaksha

Building Machine Learning System with Deep Neural Network for Text Processing

Shashi Pal Singh[1(✉)], Ajai Kumar[1], Hemant Darbari[1], Anshika Rastogi[2], Shikha Jain[2], and Nisheeth Joshi[2]

[1] AAIG, Center for Development of Advanced Computing, Pune, India
{shashis,ajai,darbari}@cdac.in
[2] Banasthali Vidyapith, Banasthali, Rajasthan, India
anshikarastogi1992@gmail.com, shikhaj959@gmail.com,
jnisheeth@banasthali.in

Abstract. This paper provides the method and process to build machine learning system using Deep Neural Network (DNN) for lexicon analysis of text. Parts of Speech (POS) tagging of word is important in Natural language processing either it is speech technology or machine translation. The recent advancement of Deep Neural Network would help us to achieve better result in POS tagging of words and phrases. Word2vec tool of Dl4j library is very popular to represent the words in continuous vector space and these vectors capture the syntactic and semantic meaning of corresponding words. If we have a database of sample words with their POS category, it is possible to assign POS tag to the words but it fails when the word is not present in database. Cosine similarity concept plays an important role to find the POS Tags of the words and phrases which are not previously trained or POS Tagged. With the help of Cosine similarity, system assign the appropriate POS tags to the words by finding their nearest similar words using the vectors which we have trained from Word2vec database. Deep neural network like RNN outperforms as compare to traditional state of the art as it deals with the issue of word sense disambiguation. Semi-supervised learning is used to train the network. This approach can be applicable for Indian languages as well as for foreign languages. In this paper, RNN is implemented to build a machine learning system for POS-tagging of the words in English language sentences.

Keywords: Natural Language Processing (NLP) · Machine learning · Recurrent Neural Network (RNN) · Cosine similarity · Word2vec

1 Introduction

Machine learning is one of the most popular research topic in NLP application area. Different techniques and strategies are available to train the system. Deep learning is very hot topic and this approach have proven its importance to build a better machine learning system as compare to state-of-the-art [6]. It is a science that provide us tools and techniques that make a system capable to sense the input from the environment and perform like a human brain [11]. Our main focus exclusively on text processing that

© Springer International Publishing AG 2018
S.C. Satapathy and A. Joshi (eds.), *Information and Communication Technology for Intelligent Systems (ICTIS 2017) - Volume 2*, Smart Innovation, Systems and Technologies 84, DOI 10.1007/978-3-319-63645-0_56

how learning methods can be implemented using deep neural networks so that the system produce more accurate results.

1.1 Deep Neural Networks

Deep learning uses neural networks but with two or more hidden layers. These networks are called deep neural networks (DNNs). DNNs have the capability to learn features so they are very useful in machine learning processes. They are very good in learning semantic and syntactic representations of text (words, sentences, structures etc.). These networks go through the training phase followed by inference phase. Structure and training process of DNN depend on the performing task [4].

In Training Phase, Feedback loop is present here, in order to improve the system on the basis of error (difference between actual output and target). Weights are modified with each iteration to minimize the error and to build well efficient and robust system. On the other hand, In Inference Phase, a well-trained system produces output according to the given input. There are only feed-forward network connections (Fig. 1).

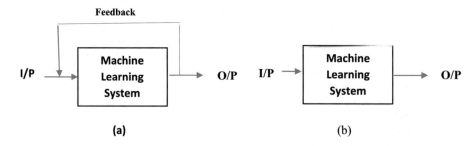

Fig. 1. (a) Training phase and (b) interface phase

1.2 Machine Learning

Machine learning is an important part of artificial intelligence. Machine learning methods are used in training of the computer system and make it capable to perform specific task. Learning can be based on rules, algorithms, features etc. A well-trained system sense the environment and with the help of its knowledge, it will produce the required output.

2   Vector Representation of Words

First task is to find out vector value of a word. Word embedding concept is used here. It is the vector representation in a dense and low dimensional space. A large corpus wiki [14] (3 GB) is go through the training which is necessary for learning purpose. The word2vec is neural network, required to produce the vectors that will be used as input for deep networks. The idea use here is to find the numeric value of a word that is based on the context words. Different models and learning algorithms are available for this task. The whole training is based on unsupervised learning [7, 8].

There have been a number of models, algorithms to address the general word embedding problem with neural network. Here, the approach defined is closely related to word2vec that first map input to the continuous space vector representation and then word vectors works as a lookup table.

Skip-gram and CBOW Models are two ways of creating the "task" for the neural network, where we create "labels" for the given input. Both architecture describe how the neural network "learns" the underlying word representation for each word. Since learning word representation is essentially unsupervised, learning algorithms perform well to "create" labels to train the model [8].

2.1 Learning Algorithms

Two popular Learning algorithms are hierarchical softmax and Negative sampling. In hierarchical softmax sampling use Huffman tree concept and assign short codes to frequent words. But it is not useful as number of epochs increases. In this neighbouring word is pulled closer or away from a word subset. This word subset is chosen from tree structure and may be different for each word. Negative sampling method is a good algorithm that close the neighbour words and some words are pushed away. This complete working is depending on the approach, maximization problem and the idea use here is minimization of the log-likelihood of sampled negative instances [9].

2.2 Dimensionality Reduction

Dimensionality Reduction is also a part of word2vec processing. Initially, one dimension per word is allocated. To convert a high-dimensional input into a fixed-size vector representation, dimension reduction technique is used [12] (Fig. 2).

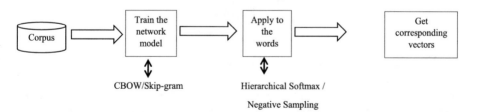

Fig. 2. Block diagram of word2vec processing

Word2vec places the words in continuous space where each word is represented by vectors of fixed dimensions (usually 100–500). Original features in text processing is too sparse so we need distribute representation that enables more flexibility when it uses in language modelling step. Vectors of a word can be find out if the vectors of other words in same dimension are known [8, 9]. For example

$$\text{Vector}[\textbf{girl}] = \text{Vector}[\textbf{man}] + \text{Vector}[\textbf{woman}] - \text{Vector}[\textbf{boy}] \qquad (1)$$

Following figures visualize this more clearly (Fig. 3);

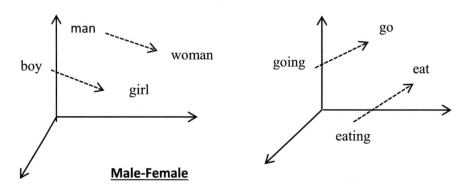

Fig. 3. Continuous space representation of words

Now, these vectors can be used in different machine learning applications according to need. For example, to find the similarity between the words or to find the vector of a new word, in sentiment analyses, text summarization etc. [12, 15].

2.3 Cosine Similarity

It is a way to find the similarity among the words by finding the dot product of their vector representation. The cosine similarity between two vectors can be given by the equation: [13]

$$cos\theta = \frac{\vec{a}.\vec{b}}{\|\vec{a}\|\|\vec{b}\|} \qquad (2)$$

3   Word2vec Implementation in NLP

We can use vector representations of words in various NLP applications to train the neural network systems. Word-vectors capture syntactic and semantic information which is important in text processing tasks like POS-tagging, semantic analyses, machine translation etc. [10]. To assign the POS-tags to the token in the input sentence, we have created a table in database which contains some sample of words with their POS-category. When tokens generate, next step is to find its cosine similarity with sample words and assign the POS-tag of the word to the token that is most similar to it. Here we are using concept of cosine similarity because it will find the nearest word in the vector space correspond to the each given token [1, 12].

4   Machine Learning Using DNN

One issue with the above method is that it can't handle the words which belong to more than one category according to their use in sentence. For example, book can be noun or verb. This will be depended on neighbouring words because tags of neighbouring words are dependent. Some tags are very unlikely to be followed by other tags (e.g. verbs never follow the determinant) and some tags represents the word chunks (e.g. proper noun can consist more than one word) (Fig. 4).

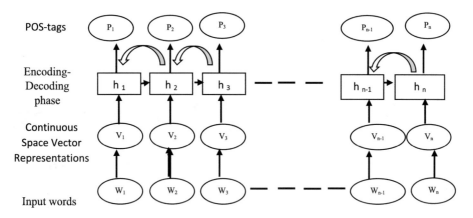

Fig. 4. POS-tagging through RNN

To resolve these problems, we require a well-trained system that can provide the most suitable tag to a word by using the information of its neighbour. RNN has the capability to maintain the history of the words in a given sentence and then it uses this information in decoding phase [2, 3]. That is the reason, it is used in building a machine learning system for POS-tagging [5].

5   Methodology

We have trained a 3 GB wiki file with word2vec. These word vectors required in neural network processing. In our experiment, semi-supervised learning method train a machine for POS-tagging of words.

First task is applying pre-processing on input that includes sentence separation, contraction removal, and tokenization.

After the pre-processing, next step is to find vector values of words using trained corpus. Suppose we have an input sentence S_1 that consists words $w_1, w_2 \ldots w_n$. Trained wiki file is used to find the corresponding vectors of the words $v_1, v_2 \ldots v_n$, respectively. To learn the system for POS-tagging, a small database is required that have some sample words with their POS-category.

Let us take an example 1, I have a book (Table 1).

Table 1. Sample words in database table

Words	POS-category
The	Determinant
Bad	Adjective
He	Noun
Was	Verb

We can represent this whole process with the help of Fig. 5.

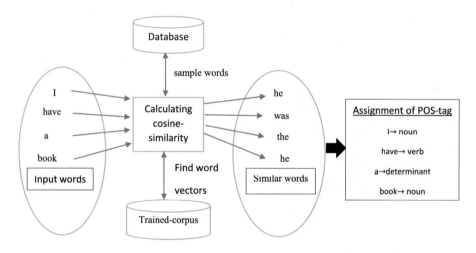

Fig. 5. POS-tagging of words with the help of word vectors

If $<S_1, V_1>$ is the pair of set, where S_1 is the set words of input sentence w_1, w_2, ... w_n and V_1 contains their corresponding vector values v_1, v_2, ... v_n. Similarly, $<S_2, V_2>$ is the pair of set, where S_2 is the set of sample words.

$w'_1, w'_2, \ldots w'_m$ that exist in the database and V_2 contains their corresponding vector values $v'_1, v'_2, \ldots v'_m$.

$$S_1 = \{w_1, w_2, \ldots w_n\}$$
$$V_1 = \{v_1, v_2, \ldots v_n\}$$
$$S_2 = \{w'_1, w'_2, \ldots w'_m\}$$
$$V_2 = \{v'_1, v'_2, \ldots v'_m\}$$

Where v_i, $v'_i \,\varepsilon\, R$ for all i, j ε N.

Find the similarity of each input word vectors with all sample word vectors and find out the most similar sample word for each input word using following equation

$$C = \sum_{k=1}^{m} f\left(v_i, v'_k\right) \text{ where } C \in R \tag{3}$$

Here the function

$$f\left(v_i, v_k'\right) = \frac{k_1}{k_2}$$

$K_1 = V_i \cdot V_k'$ and $k_2 = \|v_i\| \cdot \|v_k'\|$.

We have $<P_1, P_2>$, pair of set that contains category values of sample words and input words, respectively.

Initially,

$P_1 = \{p_1', p_2', \dots p_m'\}$ and $P_2 = \{p_1, p_2, \dots p_n\}$

Suppose v_j' is the most similar to v_i then assign the POS tag value of w_j' to the w_i POS category as shown in the following equation

$$p_i = p_j' \tag{4}$$

When we use recurrent neural network to deal with the disambiguity issue then we find out the POS tag using the given equation

$$p_{i+1} = f(p_i, v_{i+1}) \tag{5}$$

Here $p_i \in P_2$ and $v_i \in V_1$.

When we apply only cosine similarity concept to find out the POS tags of the words then wrong values of the words can be obtained because some words come under more than one POS category. For e.g., book can be used as noun (as shown in above example 1) or as a verb (as shown in following example 2) Book my tickets for the evening show.

Hence, we are using RNN to assign correct tag to the word. In example 1, book follows the word "a" which is a determinant. According to English grammar rule, a verb follows a determinant in more unlikely cases. RNN assign the tag "noun" to the book because it has the information about previous word i.e. "a".

6   Testing

We can calculate the accuracy of the POS tag system with the help of following formula,

$$A = \frac{N_C}{N_T} * 100 \tag{6}$$

where N_C is the number of words to which the system has assigned correct POS tags and N_T is the total number of words in the given input sentence.

7   Conclusion

The recent advancement of Deep Neural Network would help us to achieve better result in POS tagging of words and phrases so we have opted the Machine leaning using DNN,

RNN and word2vec. For the testing and analysis purpose we have used 3 GB wiki data/ file to do machine learning for identifying the POS tagging of the words and phrases. RNN has the capability of feature learning and sense the accurate meaning based on previous history. In this task RNN provide better results in POS tagging as it can handle the issues related to word sense disambiguity.

8   Future Work

Result obtained from the above approach can be used in various NLP applications like speech reorganization, semantic analyses, machine translation etc. We can implement this idea on Indian languages as well as on foreign languages. It is difficult to train the large data file on CPU and deep neural network also need a powerful hardware support for training. GPUs can be used to improve the system performance by reducing training time period.

References

1. Nougueira dos Santos, C., Zadrozny, B.: Learning character-level representation for part-of-speech tagging. In: Processing of the 31st International Conference on Machine Learning, Beijing, China, vol. 32 (2014). JMLR: W&CP
2. Larochelle, H., Bengio, Y., Louradour, J., Lamblin, P.: Exploring strategies for training deep neural networks. J. Mach. Learn. Res. **1** (2009)
3. Sutskever, I., Vinyals, O., Le, Q.V.: Sequence to Sequence Learning with Neural Networks. Google
4. Zhang, J., Zong, C.: Deep Neural Network in Machine Translation. Institute of Automation, Chinese Academy of Sciences
5. Perez-Ortiz, J.A., Forcada, M.L.: Parts-of-speech tagging with recurrent neural network. IEEE (2001)
6. Deng, L., Yu, D.: Deep Learning: Methods and Applications, vol. 7. Microsoft Research, Redmond (2013)
7. Chen, M.: Efficient vector representation for documents through corruption. In: ICLR. Criteo Research, Palo Alto (2017)
8. Mikolov, T., Sutskever, I., Chen, K.: Distributed representations of words and phrases and their compositionality. Google Inc.
9. Mikolov, T., Chen, K., Corrado, G., Dean, J.: Efficient estimation of word representation in vector space. arXiv:1301.3781 v3, 7 September 2013
10. Zheng, X. Chen, H., Xu, T.: Deep learning for Chinese word segmentation and POS tagging. In: Proceedings of the 2013 Conference on Empirical Methods in Natural Language Processing, Seattle, WA, USA, 18–21 October 2013
11. http://Www.Deeplearningbook.Org/
12. http://Nlp.Cs.Tamu.Edu/Resources/Wordvectors.Ppt
13. http://Www.Minerazzi.Com/Tutorials/Cosine-Similarity-Tutorial.Pdf
14. http://www.cs.upc.edu/~nlp/wikicorpus/
15. https://deeplearning4j.org/word2vec

Exploration of Small Scale Wood Industries in Nanded District, Maharashtra, India Using Statistical Technique

Aniket Avinash Muley[✉]

School of Mathematical Sciences, Swami Ramanand Teerth Marathwada University,
Nanded 431606, Maharashtra, India
aniket.muley@gmail.com

Abstract. The present study is an attempt for exploring parameters which represents the overall performance of Small Scale Wood Industries (SSWI) in Nanded district using statistical technique. This study specially focuses on some important parameters viz. availability of human resources, financial, production, transportation and marketing management aspect. To study the various in sights of the SSWI based on the focused parameters. The importance of this study on both the academic and the application levels is attributed to SSWIs, despite their contributions to the economy, have not been given due attention as the research of performance has been biased towards large enterprises.

Keywords: Small and medium scale enterprises · Nanded · Wood industries · Statistical analysis · Data mining

1 Introduction

Now days in the development of Indian economy and increase of trade, the small scale industrial sector has plays an important role [25]. To survive in the global market entrepreneurs should have to overcome certain factors [26]. The small business sector has played a vital role in the process of labour absorption [15]. The South African government has put into place programmes to encourage growth of the sector interventions such as creation of an enabling legal framework, access to markets, finance, training, infrastructure, capacity building, taxation and financial incentives among others [26]. Bennet noted about successful strategies about economical growth, democracy, self-reliance and independence [2]. Baumbark highlighted that, the government had implemented laws aimed at preventing large scale businesses from competing unfair with small business [1]. The increasing trend of the employment Germany's wood-based industries mainly is attributed to regional factors such as comparatively higher subsidies for new investments, lower labour costs, lower land values or infrastructural peculiarities [13]. The divergent employment trends in wood-based industries of the forest cluster have been studied for the first time for Germany as a whole as well as its individual federal states. [4, 11, 12, 14, 17, 20–24].

The problems and restrictions of forestay development in about: lack of awareness; lack of linkage; low level of technology; inadequate research and extension; weak

© Springer International Publishing AG 2018
S.C. Satapathy and A. Joshi (eds.), *Information and Communication
Technology for Intelligent Systems (ICTIS 2017) - Volume 2,* Smart Innovation,
Systems and Technologies 84, DOI 10.1007/978-3-319-63645-0_57

planning capability; government involvement and control; low level of people's participation and involvement by NGOs; lack of private sector participation; unwanted restrictions on felling, transport and marketing of produce from non-industrial forests; lack of inter-sectoral coordination; and weakness and conflicting roles of public forest administration [3, 9, 10, 18, 19, 25].

The objective is to explore and identify the factors affecting on the performance on SSWI in Nanded District. The objectives of the study are: Collection of information based on various parameters and identify its effects. To establish the effects of capital input and access to finance/credit, to measure the effects of technology on the growth, try to evaluate the association of various problems occurs in SSWI through factor analysis technique by SPSS 22 version software [5–7]. In the next subsequent sections detailed methodology, result and discussion have been given.

2   Study Area

The Nanded district is in nature placed at $18''15'$ N and $77''7'$ to $78''15'$ E. Since 1960, the 1272 number of small scale industries in the Nanded District Industrial Center (DIC) was recorded [25].

3   Methodology

In this study, primary as well as secondary data is collected. First of all, secondary data is obtained from DIC and Government Forest office, Nanded. This research adopted to perform exploratory study of entrepreneurs of SSWI in Nanded district. Population study is more envoys as one and all have equal opportunity to be incorporated [3, 16].

The target population size of SSWIs is 117 in Nanded district. The population of 117 respondent a sample of 100% from each group in proportions that each group allowed to the population as a whole was taken since the study was a census survey. The structured questionnaire methodology is preferred for the primary data collection purpose. The questionnaire divided into sections representing the various variables adopted and for each section included closed structured, open ended questions which collected the views, opinion, and attitude from the respondent. The questions were designed to collect qualitative and quantitative data. The open ended questionnaires gave unrestricted freedom of answer to respondents.

4   Analysis and Interpretation

To achieve the objective of the study, data is gathered and tabulated in SPSS data file and further it has been utilized for calculating principal component analysis. To represent the collected data graphically MS-Excel were used.

4.1 Exploratory Analysis

Preliminary, data is explored through the unsupervised learning's and graphical techniques. Various graphs (Figs. 1, 2, 3, 4, 5, 6, 7, 8 and 9) represent the information about SSWI in Nanded district. In this study it is observed that, there were 110 male and 7 female of proprietors. It simply shows the domination of males in this field.

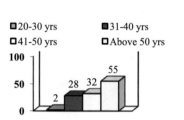

Fig. 1. Age-wise frequency distribution table

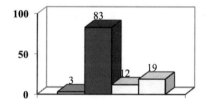

Fig. 2. Education-wise frequency distribution

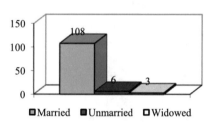

Fig. 3. Marital status of proprietors

Fig. 4. Various kinds of training taken by proprietors

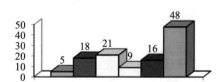

Fig. 5. Proprietors experience in SSWI

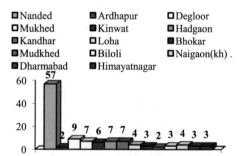

Fig. 6. Taluka-wise distribution of SSWI in Nanded district

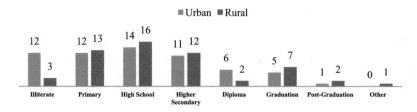

Fig. 7. Location and education-wise distribution of SSWI

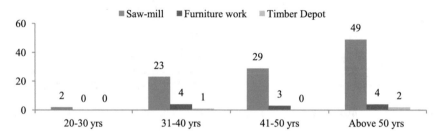

Fig. 8. Agewise distribution types of SSWI

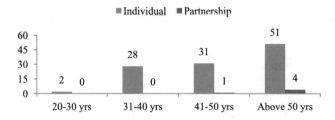

Fig. 9. Agewise classification of proprietorship

Figure 1 represents the information about different age group proprietors, most of them i.e. 56 of them are above age of 50 years old, 32 of the proprietors are age group of 41–50 years old. 28 of them are 31–40 age groups and only 2 of them are of 21–30 age group proprietors. Figure 2 explores education-wise distribution and it is observed that, most of the people's qualification is high school i.e. 30, primary 25 and 23 were higher secondary proprietors.

Figure 3 shows proprietor's marital status, 108 are married and 6 are unmarried and 3 of them are widowed. Figure 4 explores that, out of 117 proprietors 83 of the peoples are family learnt; 12 of them had taken vocational training, 10 of them other way and 3 of them apprenticeship.

Figure 5 describes that, 48 of the proprietors are having more than 25 years work experience, 21 are having 10–15 years experience. 18 proprietors are having 5–10 years, 16 of the proprietors are having 20–25 years and 5 proprietors are having less than five years work experience. The data explains that, family background of 54 proprietors is agriculture, 35 of them business, 16 are from government service. 10 and 2 of the peoples are belongs to other sources and private service respectively. Among 117 proprietors,

47 due to their friends they entered in this field, 45 of themselves are interested, 15 are due to family pressure and 10 do not have any other option. Figure 6 shows that, the distribution of SSWI. There are 16 Taluka places in Nanded district. Nanded tehsil is having 57; Degloor is having 9; Mukhed, Hadgaon and Kandhar are having 7; Kinwat are having 6; Loha tehsil are having 4; Bhokar, Biloli, Dharmabad and Himayatnagar are having 3; Ardhapur tehsil and Mudkhed are having SSWI. There are no SSWI in Mahur and Umri techsil of Nanded district.

Among 117 SSWI's, 61 of them are located at urban region and 56 are situated at rural region. Figure 7 illustrates the detailed classification of proprietors according to their education as well as residential location wise.

Among 117 it is observed that, 103 are saw mills, 11 are of furniture works and 3 are of timber depot types of SSWI. Figure 8 represents the detailed age-group wise distribution SSWI.

Overall, there are 112 of individuals and 5 are only having partnership proprietorship. Figure 9 represents the age-wise and proprietorships wise classification. The data revealed that, 112 of the proprietors are having own land and 5 of them are having land on rent. Nature of operation performed in SSWI is found that, 69 are full time operating, 47 are seasonal operating and 1 is occasional working SSWI industries.

4.2 Principal Component Analysis (PCA)

The PCA analysis is performed through SPSS22v software. PCA is powerful tool for pattern recognition that attempt to explain the variance of large set of intercorrelated variables. PCA extracts the Eigen values and Eigen vectors from the covariance matrix of original variables. The principal components are uncorrelated variables with Eigen vectors and Eigen values of the principal components are the measure of their associated variances. PCA provides six components with Eigen values > 1 explaining 62.85% of the total variance of a dataset. The scree plot (Fig. 10) is the way of identifying a number of useful factors, wherein a pointed break in sizes of Eigen values results in a change in the slope of the plot could be observed and is given below:

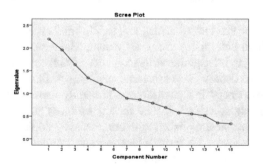

Fig. 10. Scree plot

Factor analysis is an extensively used multivariate statistical method to reorganize original variables into fewer underlying factors to retain as much information contained in the original variables as possible. Factors are produced according to an Eigen value analysis of the correlation matrix and factor loadings and factor scores are the main measurements. The first step of factor analysis is to standardize the raw data.

The second step is to estimate the factor loadings that express the degree of nearness between the factor and variables. The last step linearly transforms factors associated with the initial set of loadings by factor rotation to maximize variable variances and to obtain a better interpretable loading pattern. Factor scores are work out for each individual case to represent the contribution of each factor in each case. This study performed factor analysis to determine the factors affecting on the performance of SSWI in Nanded district. Factor extraction was carried out by principal components, where only eigenvalues greater than one were retained [6, 7] and is graphically represented by Scree plot (Fig. 10). The factor loading matrix was rotated to obtain uncorrelated factors by varimax rotation. In factor analysis [6–8], rotated varimax method with Kaiser Mayer normalization has been applied and six components show significant variation among the study. The component reveals that, in the first component problems slight addition of one variable of material with marketing, infrastructure, transportation and power shows key role on their performance. Rest of the component shows the same variables in the subsequent components affecting on the performance of the SSWI. Factor analysis is an exploratory tool used to make decisions. In the six components, various factors are affecting the performance of the SSWIs. It is observed that, in the first components, transportation (0.834), infrastructure (0.736) and marketing (0.70) problems are found to be more significant. The second component explores that, age (0.861) and working experience (0.800) of the entrepreneur's factors shows the significance. In the Third component, education (0.575) and type of industry (0.543) observed significant factors. In the fourth to sixth components, the material problem (0.733), types of industry (0.492) and special training taken by them (0.877) factors are found to be significant. It is observed that, in SSWIs, 94.02% of entrepreneur are males and 5.98% of females in this sector. The interesting fact is found that, majority of the entrepreneurs educated up to higher secondary level. Also, married peoples are working hours are observed more than unmarried people.

The study revealed that lack of managerial training and experience affect the growth of small and medium enterprises to a great extent. Mainstream of the SSWI owners applied the following to managerial skills to greater extent, controlling activities, coordination, directing and deployment of material, planning, staffing, design of organization structure and deployment of finance resources. The study also established that, management and management experience is vital for any business growth for it enhances the process of getting tasks accomplished with and through people by guiding and motivating their efforts, those with more education and training are more likely to be successful in the SSWI sector. Education and skills are needed to run micro and small enterprises; every enterprise concerned with the deployment of material, human and finance resources with the design of organization structure.

5   Conclusions

In this study, the collection, tabulation and exploration of data is one of the important tasks of the study. Tabulation and exploratory analysis is performed through the SPSS and MS-Excel software. It is observed that, in SSWI's, 94.02% of entrepreneurs are males and 5.98% of females in this sector. The interesting fact is found that the majority of the entrepreneurs educated up to higher secondary level. Also, married peoples are working hours are observed more than unmarried people. Further, PCA analysis established that six extracted components showing 62.85% of the cumulative percentage of total variance. This study reveals that managerial training and experience in SSWI sector have a positive effect on the growth. The study established that in most cases, SSWI's had limited access to collaterals and that financial institution lacked appropriate structure to deal with SSWI's on matters of finance which would ensure that the SSWI's acquire the relevant finances to carry on with their businesses. This study recommends that, the management of the financial institutions should consider reviewing their policies regarding access to finance which brings ease in obtaining of credit from the financial institutions to encourage entrepreneurship. The entrepreneurs should consider advancing their technological expertise by even training on use of modern technology in order to acquire information that is of essential to their businesses and also investing in technology in order to benefit from modern technology. The government should consider organizing managerial training forums for SSWI's owners; this time focusing on challenges facing the growth of SSWI in Nanded district.

Acknowledgements. Author would like to thank University Grant Commission for providing financial support under Major Research Project (F. No. 42-44/2013(SR), 20 Dec. 2013, Statistics).

References

1. Baumbark, C.M.: How to Organize and Operate a Small Business. Prentice Hall, Upper Saddle River (1979)
2. Bennett, M., James, P., Klinkers, L. (eds.): Sustainable Measures: Evaluation and Reporting of Environmental and Social Performance. Greenleaf Publishing, Sheffield (1999)
3. Cooper, D.R., Schindler, P.S.: Business Research Methods. McGrawHill, Irwin (2003)
4. Dubey, P.: Investment in small-scale forestry enterprises: a strategic perspective for India. Small-Scale For. **7**(2), 117–138 (2008)
5. Field, A.: Discovering Statistics Using SPSS, 3rd edn. Sage Publications, London (2010)
6. George, D., Mallery, P.: SPSS for Windows Step by Step: Simple Guide and Reference 110 Update, 4th edn. Allyn and Bacon, New York (2003)
7. Hair Jr., J.F., Anderson, R.E., Tatham, R.L., Black, W.C.: Multivariate Data Analysis, 3rd edn. Macmillan Publishing Company, New York (1995)
8. Hair Jr., J.F., Black, W.C., Babin, B.J., Anderson, R.E., Tatham, R.L.: Multivariate Data Analysis, 6th edn. Pearson Prentice Hall, Upper Saddle River (2006)
9. Hallberg, K.: A Market-Oriented Strategy for Small and Medium Scale Enterprises. International Finance Corporation, Discussion paper, IFD 40

10. Hieu, P.S., Thuy, V.H., Thuan, P.D.: Main characteristics of statistical data and the statistical system for wood and wood-processing products in Vietnam. Small-Scale For. **10**(2), 185–198 (2011)

11. Jaensch, K., Harsche, J.: Der Cluster Forst und Holz in Hessen. Bestandsanalyse und Entwicklungschancen. HA Hessen Agentur GmbH **7**(12), 109 (2007)

12. Kies, U., Mrosek, T., Schulte, A.: A statistics-based method for cluster analysis of the forest sector at the national and subnational level in Germany. Scand. J. For. Res. **23**, 445–457 (2008)

13. Klein, D., Kies, U., Schulte, A.: Regional employment trends of wood-based industries in Germany's forest cluster: a comparative shift-share analysis of post-reunification development. Eur. J. For. Res. **128**(3), 205–219 (2009)

14. Kramer, M., Möller, L.: Struktur- und Marktanalyse des Clusters Forst und Holz im Freistaat Sachsen und in ausgewählten Regionen des niederschlesischen und nordböhmischen Grenzraums unter den Bedingungen der EU-Osterweiterung (Cluster-Studie). Internationales Hochschulinstitut Zittau, p. 200 (2006)

15. Lambin, J.Jean: Market-Driven Management, Strategic and Operation Marketing. Palgrave Macmillan, London (2000)

16. Mrosek, T., Kies, U., Schulte, A.: Clusterstudie Forst und Holz Deutschland 2005. Forst- und Holzwirtschaft hat sehr große volkswirtschaftliche und arbeitsmarktpolitische Bedeutung. Holz-Zentralblatt **84**, 1113–1117 (2005)

17. Mugenda, O.M., Mugenda, A.G.: Research Methods, Quantitative and Qualitative Approaches. ACTS, Nairobi (2003)

18. Nanded District Industrial Center (NDIC). www.nandeddic.in

19. Prasad, R.: The present forest scenario in India. Compilation of papers for preparation of national status report on forests and forestry in India (Survey and Utilization Division). Ministry of Environment & Forests, Government of India, New Delhi, pp. 1–20 (2006)

20. Rongo, L.M., Msamanga, G.I., Burstyn, I., Barten, F., Dolmans, W.M., Heederik, D.: Exposure to wood dust and endotoxin in small-scale wood industries in Tanzania. J. Eposure Sci. Environ. Epidemiol. **14**(7), 544–550 (2004)

21. Schulte, A.: Nordrhein-Westfalen zieht Bilanz für Forst und Holz. Cluster-Studie weist unerwartete volkswirtschaftliche Größe der Forst- und Holzwirtschaft aus. Holz-Zentralblatt, vol. 74, 1018–1019 (2003)

22. Schulte, A. (ed.): Clusterstudie Forst und Holz Nordrhein-Westfalen. Gesamtbericht. Schriftenreihe der Landesforstverwaltung NRW 17. MUNLV Ministerium für Umwelt, Naturschutz, Landwirtschaft und Verbraucherschutz, Düsseldorf, p. 138 (2002). http://www.forst.nrw.de/nutzung/cluster/cluster.htm. Accessed Nov 2007

23. Schulte, A., Mrosek, T.: Analysis and assessment of the forestry and wood-processing industry cluster in the State of North-Rhine Westphalia, Germany. Forstarchiv **4**, 136–141 (2006)

24. Seegmüller, S.: Die Forst-, Holz- und Papierwirtschaft in Rheinland-Pfalz. Clusterstudie. FAWF Forschungsanstalt für Waldökologie und Forstwirtschaft Rheinland-Pfalz, Trippstadt, p. 63 (2005)

25. SME: Annual Report, 2009–2010, Ministry of Micro, Small and Medium Enterprises (MSME) (2010)

26. Spieler, R., van Eeden, M: Drawings (in collaboration with the exhibition with Marcel van Eeden. Drawings, Museum Franz Gertsch, Burgdorf, Germany 2004), Burgdorf, Germany (2004)

Remote Controlled Solar Agro Sprayer Robot

Hareeta Malani, Maneesh Nanda[(✉)], Jitender Yadav,
Pulkit Purohit, and Kanhaiya Lal Didwania

M.L.V. Textile and Engineering College, Bhilwara, Rajasthan, India
hmalani03@gmail.com, maneeshnnd@gmail.com,
raosahabjitender420@gmail.com, pulkitpurohitprince@gmail.com,
kldidwania111@gmail.com

Abstract. This paper provides with the exposition of how robotics can offer to various places of agriculture. As we know that agriculture is the backbone of our country. We can make improvement in agriculture by replacing man's power by using latest technologies like robots. The main motive of our prototype is to help farmers in various agricultural operations like spraying of pesticides and fertilizers by using the solar energy. As we know that pesticides and fertilizers contain harmful chemicals which affects human body. This prototype will provide safety to farmers, precision agriculture and high speed. The cost of our prototype is effective as it works on the solar energy (Renewable energy) and it is based on the wireless technology (RF communication) by using Embedded system.

Keywords: Renewable energy · Embedded system · RF communication

1 Introduction

There are both short and long-term outcomes that happen from way to destroy unwanted pests exposures. In the short- term, exposures can lead to rashes, blisters, stinging eyes, loss of vision, nausea, faintness, problems, coma, and loss of life. Some outcomes are delayed, however, and may not be instantly obvious. Such long-term outcomes include sterility, beginning problems, hormone disruption, sensing problems, and melanoma. Our Aim of this paper is to prevent the farmers from these short term and long term diseases and to reduce the farmer work. In our robot, solar energy is used to charge the battery. The non-renewable types of energy are losing, probably may last within the years to come or formerly and to reduce the working cost of the looking system, we are in an attempt to incorporate the above mentioned features in our farming robot. This prototype will not depend on non renewable sources like petrol, Diesel, Coal. Due to this reason it is economical and Eco friendly. Productivity enhances to the greater extent.

Figure 1 represents the conversion of farmer work into machine work.

© Springer International Publishing AG 2018
S.C. Satapathy and A. Joshi (eds.), *Information and Communication*
Technology for Intelligent Systems (ICTIS 2017) - Volume 2, Smart Innovation,
Systems and Technologies 84, DOI 10.1007/978-3-319-63645-0_58

Fig. 1. Aim of solar agro sprayer robot.

This idea of applying robotics in agriculture based on solar energy is tremendous. In agriculture, the opportunities for software enhanced productivity are remarkable. The Current venture is made in creating an amazing robotic vehicle which can be handled quickly through RF communication. The main aim for our venture has been to build a solar operated looking system, which is a solar energy panel technology handled. In this prototype, we used a solar panel and turn solar energy into electrical energy which often is used to charge two 12 V battery energy pack, Which then gives the necessary capability to DC motors, DC pumps.

A Special remote contains several buttons that are designed as a transmitter to which signal is transmitted to the receiver which is connected to the controller and controller manage the movement of the wheels. The direction is decided by wireless remote. This technology of RF communication makes the operation of the robot simpler and easier. Maximum speed achieved by robot is approximately 5 km per hour. Speed totally depends on the irregularities on the land. Our aim is to create a Model Multi Purpose Agricultural Application which can perform the following functions like spraying pesticides and solar energy use for domestic purposes. The venture's objective is to create a program vehicle which can apply water or pesticides and look after the crop, these whole functions of the robot synchronize with battery energy pack and the solar energy panel technology.

1.1 To reduce human effort in the farm field with the use of small software.
1.2 To carry out this feature at a single time, hence increases production and helps you to not put things off.
1.3 The speed and precision of the venture is more than the human being.
1.4 The use of solar energy panel technology can be utilized for Battery charging. As the venture works in the region where the rays of the sun can be used for solar energy panel technology.
1.5 To increase the efficiency, the solar energy panel technology is used and the Power result can be enhanced.

2 Methodology

2.1 Architectural Block Diagram

See (Fig. 2).

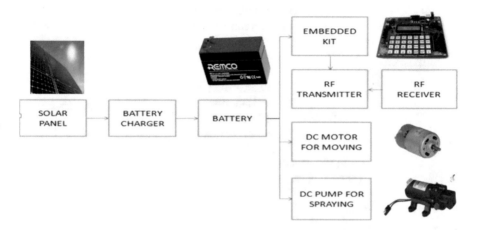

Fig. 2. Architectural block diagram of solar agro sprayer robot.

2.2 Working Model

See (Fig. 3).

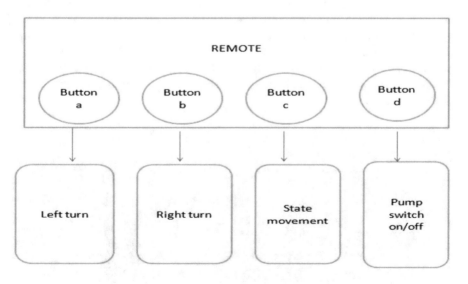

Fig. 3. Working model of robot.

2.3   Calculation

POWER REQUIRED FOR MOTOR: -
Power= (friction force)*(velocity)
Friction force= (friction coefficient)*(Reaction Force)
Friction coefficient u=0.4
Friction force=R*0.4
=mg*0.4
=20*9.81*0.4
=78.48 N
Velocity=5km/hr
=1.38.m/s.
Power= ff*v
= 78.48*1.38
=109W
POWER REQUIRED FOR PUMP= 24W
TOTAL POWER = 109W+24W=133W

3   Implementation

3.1   Hardware Implementation

The Solar agro sprayer robot is made up of the metal sheet 1.5 cm * 1.5 cm in which this could hold weight upto 30 kg. The dimension of the vehicle is 45 cm in length and 30 cm in breadth and 30 cm height. We have designed this robot in a way that the robot can balance its weight equally over it. The wheel drive system used in robot is 4 wheel drive system in which each wheel of the vehicle is connected to a DC motor of torque about 70 RPM which could carry 30 kg weight over the vehicle. The entire wheel has a radius of 11 cm which is connected to the axle of DC motor. Sprayers are used to perform the operation of spraying pesticides or water. Embedded system is very useful in this structure (Fig. 4).

Fig. 4. Prototype model

3.2   Software Implementation

The software used for the implementation of the robot is as follows:

3.2.1   AVR studio - It is used to write code.
3.2.2   Proteus professional - It is used for simulation on a computer.

4   Result

The main motive of our robot is to reduce human efforts that are present in agricultural fields in today's Scenario to a greater extent. We have made trial of the robot to nearby farm for functional testing. The operation of this robot is suitable for crops like wheat, mustard and also on some vegetables. This type of technique will prove to be very effective in farming techniques.

5   Conclusion

In agriculture, the chances for robot-enhanced productivity are fantastic and the robots are displaying on farms in different types and in enhancing numbers. The other problems associated with personal town equipment can probably be get over with technology. Vegetation manufacturing may be done better and cheaper with a help of little gadgets than with a few large ones. One of the advantages of little gadgets is that they may be more appropriate to the non-farming team. The jobs in agriculture are in a shift, dangerous, need intelligence and quick, though incredibly options, hence robots can be successfully customized with personal proprietor. The highest quality products can be seen in gadgets (color, solidity, weight, stability, ripeness, sizing, shape) definitely. Robots can enhance the outstanding top excellent high quality of our way of farming, but there are some limitations that it will not work for some crops like sugarcane, rice etc. The exclusive conditions in our country are that all the agricultural machines are operating on petrol engine or tractor is expensive. To implement one kind of discovering, improving program system within the limited available source and the economy. This venture may bring the revolution in the field of agriculture.

5.1   It can be experienced with further development using awesome techniques.
5.2   It may become an achievement if our venture can be implemented throughout our country.
5.3   We could improve accuracy by bringing out pesticides on crop and save farmer's time as the speed is high.

References

1. Fernando, A.A.C., Ricardo, C.: Agricultural robotics, unmanned robotic service units in agricultural tasks. IEEE Ind. Electr. Mag. **7**, 48–58 (2013)
2. Slaughter, D.C., Giles, D.K., Downey, D.: Autonomous robotic weed control systems: a review. Comput. Electron. Agric. **61**(1), 63–78 (2008)

3. Rosell, J.R., Sanz, R.: A review of methods and applications of the geometric charactersization of tree crops in agricultural activities. Comput. Electron. Agric. **81**, 124–141 (2012)
4. Astrand, B., Baerdveldt, A.: A vision based row-following system for agricultural field machinery. Mechatronics **15**(2), 251–269 (2005)
5. Zhang, C., Geimer, M., Patrick, O.N., Grandl, L.: Development of an intelligent master-slave system between agricultural vehicles. In: IEEE Intelligent Vehicles Symposium, San Diego, CA, pp. 250–255 (2010)
6. Uto, K., Seki, H., Saito, G., Kosugi, Y.: Charaterization of rice paddies by a UAV-mounted miniature hyperspectral sensor system. IEEE J. Appl. Earth Observ. Remote Sens. **6**(2), 851–860 (2013)
7. Freitas, G., Hamner, B., Bergerman, M., Singh, S.: A practical obstacle detection system for autonomous orchard vehicles. In: IEEE International Conference on Intelligent Robots and Systems, (IROS), Vilamoura, Portugal (2012)
8. Guillet, A., Lenain, R., Thuilot, B.: Off-road path racking of a fleet of WMR with adaptive and predictive control. In: IEEE International Conference on Intelligent Robots and Systems, (IROS), Tokyo (2013)
9. Costa, F.G., Ueyama, J., Braun, T., Pessin, G., Osorio, F.S., Vargas, P.A.: The use of unmanned aerial vehicles and wireless sensor network in agricultural applications. In: IEEE International Geoscience and Remote Sensing Symposium (IGARSS), Munich (2012)
10. Godse, A.P., Godse, D.A.: Microprocessor and Microcontrollers, 1st edn. Technical Publications, Pune (2009)

Intelligent System for Automatic Transfer Grammar Creation Using Parallel Corpus

Shashi Pal Singh[1(✉)], Ajai Kumar[1], Hemant Darbari[1], Lenali Singh[1], Nisheeth Joshi[2], Priya Gupta[2], and Sneha Singh[2]

[1] AAIG, Centre for Development of Advanced Computing, Pune, India
{shashis,ajai,darbari,lenali}@cdac.in
[2] Banasthali Vidyapith, Vanasthali, Rajasthan, India
jnisheeth@banasthali.in,
priyagupta.banasthali@gmail.com,
snehasingh1001@gmail.com

Abstract. In this paper we describe an Intelligent System for Automatic Transfer Grammar creation using parallel corpus of source and target language. As we know about English and Hindi language, the structure of Hindi is Subject-Object-Verb (SOV) while in English the structure is Subject-Verb-Object (SVO). Now the system has to decide in which order to translate the given source language (English) to the given target language (Hindi). The grammatically parsing source sentence has to generate target language on basis of grammar rule which we have created based on parallel corpus. These unique rules are always applicable when the same input structure is found and regenerate the output on basis of the grammatical rules. Thus it gives better accuracy in terms of quality of translation. Reordering [1, 2] is important part of Transfer Grammar which is very helpful in language pairs of distant origin. We focus on designing a system that gives correct reordering for English-Hindi Machine Translation System including simple, compound as well as complex sentences. Reorder sentences gets generated based on probability. The system is evaluated on the basis of precision, recall and f-measure.

Keywords: Tagging · Parsing · Transfer rules · Transliteration · Context free grammar · Part of speech

1 Introduction

The growth and evolution of Internet has led to huge amount of data including written, audio and visual data online. These data need to be consumed but however the problem lies in its consumption. The language barrier is one of the major hindrances for this information to be shared among all. The data or the Information available is in the language that the user cannot understand. Talking about India, only 10% of the total population understands English and on the contrary, maximum population understands Hindi. So here the main challenge of availing information or data to the user can be

© Springer International Publishing AG 2018
S.C. Satapathy and A. Joshi (eds.), *Information and Communication Technology for Intelligent Systems (ICTIS 2017) - Volume 2*, Smart Innovation, Systems and Technologies 84, DOI 10.1007/978-3-319-63645-0_59

helped by designing a system that converts the English language to Hindi language. Such a system could be designed by the help of Machine translation. Machine Translation is the process of converting a piece of text from one source language like English to another target language like Hindi. It is also known as automated translation. There are various approaches of machine translations like Rule Based and Corpus based. Rule Based Machine Translation deals with morphological, syntactic and semantic information of source language and target language. It uses bilingual dictionaries and grammar rules. It has various approaches as Direct Approach, Transfer-Based, Knowledge-Based. In our project we are using Corpus-Based approach. Corpus-Based approach makes use of bilingual aligned corpora as a basis to the translation process. It is also known as Data-Driven Machine Translation. They learn to translate by analyzing large amounts of data for each language pair. The two main Corpus-Based Machine Translation approaches are Statistical Machine Translation and Example-Based Machine Translation. Statistical Machine Translation deals with automatically mapping sentences from one human language to another human language.

Structural information plays a great role in the process of transfer, like how does the ordering of words or group of words change when translated from one source language like English to another target language like Hindi. The structure of Hindi is Subject-Object-Verb (SOV) while in English the structure is Subject-Verb-Object (SVO). This structural information is added to system, then the composition and hierarchical structure of a sentence is analyzed, and manual rules are given for how the hierarchy transfers into another language. The manual rules are used because they are more reliable and produce better results as they handle various kinds of rules ranging from specific to generic.

Example:

> *Input:* This city is very good.
> *Output:* यह शहर बहुत अच्छा है।

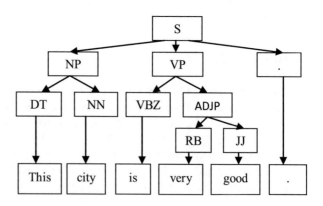

Transfer or reorder rules are applied to the above parsed sentence in order to produce the reordered sentence.

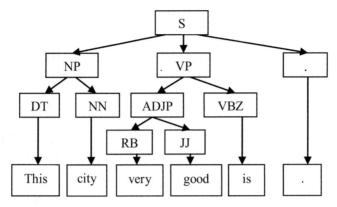

The reordered sentence is translated into the target sentence by following the Hindi model.

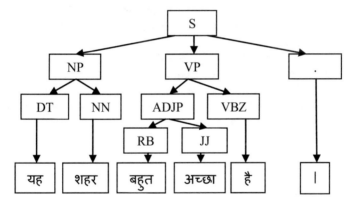

2 Literature Review

Natural language processing (NLP) is a path for computers to analyze, understand and extract meaning from human language in a better and efficient way. It is a part of artificial intelligence. In NLP, there exists hierarchical structure of language like several words make a phrase, several phrases make a sentence and ultimately, sentences convey some meaning. It is used in various real-world applications like sentence segmentation, parts-of-speech tagging, parsing and more. Instead of hand-coding large sets of rules, NLP can depend on machine learning to automatically learn these rules by analyzing a set of examples like a large corpus and making a statistical inference. In other way, the more data analyzed, the more accurate the model will be.

Machine translation is the process of converting a piece of text from one source language like English to another target language like Hindi. It is also known as automated translation. The various types of machine translation are Rule-based Machine Translation, Corpus-based Machine Translation, Neural Machine Translation and Hybrid Machine Translation.

Rule-Based Machine Translations also knew as knowledge Driven Machine Translation. It is widely driven by linguistics and rules, making use of manually created rules and resources as a basis to the translation process. It uses a combination of language and grammar rules plus dictionaries for common words. The two main Rule-Based Machine Translation approaches are Transfer-Based Machine Translation and Interlingua-Based Machine Translation. Now it is slowly overtaken by Corpus-Based approaches.

Corpus-Based Machine Translations also known as Data-Driven Machine Translation. Corpus-Based makes use of bilingual aligned corpora as a basis to the translation process. It has no knowledge of language rules. Instead they learn to translate by analysing large amounts of data for each language pair. The two main Corpus-Based Machine Translation approaches are Statistical Machine Translation and Example-Based Machine Translation. Statistical Machine Translation deals with automatically mapping sentences from one human language to another human language. This process can be considered as stochastic process. Statistical Machine Translation has various approaches such as string-to-string mapping, trees-to-strings and tree-to-tree models. The main idea behind these approaches is that translation is automatic with models driven from parallel corpora.

Neural Machine Translation is a new approach for machines that make system learn to translate through one large neural network (multiple processing devices modelled on the brain). It shows better translation performance in many language pairs.

Hybrid Machine Translation is a combination of Rule-based Machine Translation and Statistical Machine Translation.

3   Methodology

3.1   System Architecture

The analysis of the source language is done on the basis of linguistic information such as morphology, Part of speech, syntax, semantic, etc. Stanford Parser [3] is used to parse the source language and derive the structure of the text to be translated. Manually created reordering rules are used by which source language syntax structure is translated into target language syntax structure by help of the Transfer System. The working procedure could be well understood by the following steps:

Step 1: Source Language is given in the form of English as the input to the system which is further passed on to the Pre-processing phase.

Step 2: Pre-processing text is called tokenization or text normalization. It consist number of operations which are applied to input data to make it executable by translation system. It involves handling of punctuation and special characters. Tokenizer is used to segment a sentence into meaningful unit called as tokens. POS tagging allocates one of the parts of speech to the given word. It includes nouns, verbs, adverbs, conjunction and their sub-categories.

Step 3: The output of the pre-processing phase is passed on to the Parsing as the input for the further processing. Parsing involves breaking up of the sentence into its component parts and describing its grammatical structure. This type of analysis is vital because significant part of the meaning of a sentence is encoded in its grammatical structure. It also resolves the problem of structural ambiguity.

Step 4: After the task of parsing is completed the output is fed to the transfer system. The transfer system in itself consists of three subparts: transfer rules, Hindi model, and transliteration. Transfer rules are created and applied onto the output of the parsing phase in order to obtain the Hindi model. Hindi Model obtained is in the form of Subject-Object-Verb (SOV). Transliteration is used in case of proper noun instead of translation. Transliterated are those words or tokens which are not found in the database.

Step 5: Target language in the form of Hindi is obtained after the processing of the above phases (Fig. 1).

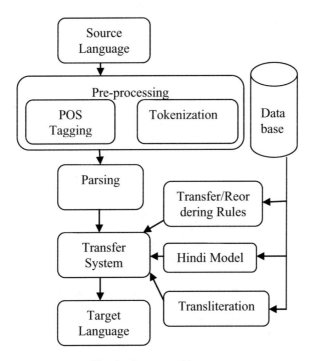

Fig. 1. System architecture

3.2   POS Tagging and Parsing

POS Tagging and Parsing of source sentences are done by using Stanford Tagger [4, 5] and Stanford Parser respectively.

3.3   Shift Reduce Parsing

Shift reduce parser [6, 7] is a simple kind of bottom-up parser. A shift-reduce parser finds out sequences of words and phrases that correspond to the right hand side of a grammar production, and replace them with the left-hand side, until the complete sentence is reduced to anode as its root. It requires input string and an LR parsing table for grammar with functions ACTION and GOTO. The Parse Table maintains the state of each row, an ACTION column for each terminal symbol and a GOTO column for each symbol which is the left-hand side of a production in grammar. The action table consists of rows and columns. The rows are indexed by states and columns are indexed by terminal symbols. When the parser points to some state s and the current look-ahead terminal is l, the action initiated by the parser depends on the contents of action[s][l], which has four different kinds of entries: shift, reduce, accept and error. The goto table consists of rows and columns. The rows are indexed by states and columns indexed by non-terminal symbols. When the parser is in state s immediately after reducing by rule M, then the next coming state to enter is given by $goto[s][M]$.

Algorithm:

Name: parseLR

Input: action table, goto table, input string

```
push start state onto the stack
let p be the first input symbol
repeat forever
switch Action [stack . top ( ) , p]
case shift[ s ]
push s onto the stack
p is now the next input symbol
case Reduce [A⟶ α]
pop |α| items from stack
push Goto [stack . top ( ) , A] onto the stack
case Accept
halt the parse with successful result
case Error
halt the parse with error result
```

3.3.1 Breaking CFG Rules up to Two Level

After getting the parsed source sentence, we need to break the parsed source sentence upto two level.

Example:

> *Input:* (ROOT (S (NP (NNP Jaipur)) (VP (VBZ is) (NP (NP (NN capital))(PP (IN of) (NP (NNP Rajasthan)))))(. .)))
> *Output:*(ROOT(S)
>   (S (NP) (VP) (. .))
>   (NP (NNP))
>   (NNP Jaipur)
>   (VP (VBZ) (NP))
>   (VBZ is)
>   (NP (NP) (PP))
>   (PP (IN) (NP))
>   (NP (NN capital))
>   (IN of)
>   (NP (NNP Rajasthan)))

3.4 Adding POS to Reordered Sentence with Respect to English Sentence

Reordered sentence also get POS tagged with respect to English Sentence.

> *Example:*
> *Input:* 1. Jaipur/NNP is/VBZ capital/NN of/IN Rajasthan/NNP ./.
>   2. Jaipur Rajasthan of capital is.
> *Output:* Jaipur/NNP Rajasthan/NNP of/IN capital/NN is/VBZ ./.

3.5 Creation of Transfer/Reordering Rules

For training the system, we need to create the transfer rules. Transfer/Reordering rules have been created manually as per Subject-Verb-Object (SVO) to Subject-Object-Verb (SOV) form (Table 1).

> *Example:*
>
> *Input Sentence:*   The deficiency of blood means deficiency of iron.
> *Parsed Input Sentence:*(ROOT
>   (S
>   (NP
>   (NP (DT The) (NN deficiency))
>   (PP (IN of)
>   (NP (NN blood))))
>   (VP (VBZ means)
>   (NP
>   (NP (NN deficiency))
>   (PP (IN of)
>   (NP (NN iron)))))
>   (. .)))
> *Reordered Sentence :*blood of The deficiency iron of deficiency means.

Table 1. English and reordered rules

English rules	Reordered rules
S NP VP	S NP VP
NP NP PP	NP PP NP
NP DT NN	NP DT NN
PP IN NP	PP NP IN
NP NN	NP NN
VP VBZ NP	VP NP VBZ
NP NP PP	NP PP NP
NP NN	NP NN
PP IN NP	PP NP IN
NP NN	NP NN

3.6 Breaking Reordered Sentence with Respect to English Parsed Sentence

Reordered sentence is further reduced with respect to the English parsed sentence (Table 2).

Example:

Input: 1. Output of breaking of CFG rule.
2. English parsed sentence.
3. POS tagged reordered sentence.

Output Break Reordered Sentence with respect to English Parsed Sentence

Table 2. Break reordered sentence with respect to English parsed sentence

Break English parsed sentence	Break reordered sentence
(ROOT(S)	(ROOT(S)
(S(NP)(VP))	(S(NP)(VP))
(NP(NNP Jaipur))	(NP(NNP Jaipur))
(VP(VBZ is)(NP))	(VP(NP)(VBZ is))
(NP(NP)(PP))	(NP(PP)(NP))
(NP(NN capital))	(NP(NN capital))
(PP(IN)(NP))	(PP(NP)(IN))
(IN of)	(IN of)
(NP (NNP Rajasthan))	(NP (NNP Rajasthan))

3.7 Probability Calculation of Transfer Rules

Probability Calculation [8, 9] of Parallel Transfer rules based on formula

$$P(H|E) = \frac{\text{Count}(E, H)}{\text{Count}(E)} \tag{1}$$

Where H is Hindi rules, E is English rules and P (H | E) is Probability of Hindi rules with respect to English rules which is calculated by frequency of both English and Hindi rules appearing together divided by frequency of English rule.

4   Evaluation Based on Precision, Recall and F-Measure

Evaluation can be done on the basis of precision [10, 11], recall [10, 11] and f-measure [10, 11] (Table 3).

Table 3. Data used for training and testing the system

Training of system	6000 sentence Parallel corpus
Testing of system	1000 English sentence

4.1   Precision

The range of precision lies between 0 and 1. More correct will be the system, if precision is higher. It can be measured as

$$\text{Precision} = \frac{\text{Real errors found}}{\text{Sentence tagged as erroneous}} \tag{2}$$

4.2   Recall

Recall counts for the number of errors in the text. This factor checks for the completeness of the system. The range of recall also lies between 0 and 1. It can be measured as

$$\text{Recall} = \frac{\text{Real errors found}}{\text{Errors in the text}} \tag{3}$$

4.3   F-Measure

F-measure is a measure of a test's accuracy. It is also known as F_1 score or F-score. The traditional F-measure or balanced F-score (**F_1 score**) is the harmonic mean of precision and recall multiplying the constant of 2 scales the score to 1 when both recall and precision are 1.

$$F_1 = 2 * \frac{\text{Precision, recall}}{\text{Precision} + \text{recall}} \tag{4}$$

5   Conclusion and Future Work

In this paper, we have discussed about Automatic Transfer Grammar Creation using Parallel corpus and we have created almost all possible reordering rules for simple, compound and complex type English structures. It reorders all given sources sentences into target language which lie in same construct frame structures. The problem lies when the input sentences does not qualify the grammar what we have created then it becomes complex to parse such sentences to get better accuracy. As a future work this approach can be extended to other Indian and Foreign languages to get better accuracy.

References

1. Avinesh, P.V.S.: Transfer Grammar Engine and Automatic Learning of Reorder Rules in Machine Translation Language Technologies. Research Centre International Institute of Information Technology Hyderabad (2010)
2. Herrmann, T., Weiner, J., Niehues, J., Waibel, A.: Analyzing the potential of source sentence reordering in statistical machine translation. In: Proceedings of the International Workshop on Spoken Language Translation (IWSLT) (2013)
3. Klein, D., Manning, C.D.: Accurate unlexicalized parsing. In: Proceedings of the 41st Meeting of the Association for Computational Linguistics, pp. 423–430 (2003)
4. Tagging, Automatic POS: Corpus Linguistics (L615). Corpus, vol. 2, p. 31 (2013)
5. http://www.nlp.stanford.edu/nlp/javadoc/javanlp/edu/stanford/nlp/tagger/maxent/MaxentTagger.html
6. Shieber, S.M.: Sentence disambiguation by a shift-reduce parsing technique. In: Proceedings of the 21st Annual Meeting on Association for Computational Linguistics. Association for Computational Linguistics (1983)
7. Sagae, K., Lavie, A.: A best-first probabilistic shift-reduce parser. In: Proceedings of the COLING/ACL on Main Conference Poster Sessions. Association for Computational Linguistics (2006)
8. Junker, M., Hoch, R., Dengel, A.: On the evaluation of document analysis components by recall, precision, and accuracy. In: Proceedings of the Fifth International Conference on Document Analysis and Recognition, ICDAR 1999. IEEE (1999)
9. Fruzangohar, M., Kroeger, T.A., Adelson, D.L.: Improved part-of-speech prediction in suffix analysis. PLoS One **8**(10), e76042 (2013). doi:10.1371/journal.pone.0076042
10. Powers, D.M.: Evaluation: from precision, recall and F-measure to ROC, informedness, markedness and correlation (2011)
11. Melamed, I.D., Green, R., Turian, J.P.: Precision and recall of machine translation. In: Proceedings of the 2003 Conference of the North American Chapter of the Association for Computational Linguistics on Human Language Technology: Companion Volume of the Proceedings of HLT-NAACL 2003—Short Papers, vol. 2. Association for Computational Linguistics (2003)

Erudition of Transcendence of Service and Load Scrutinizing of Cloud Services Through Nodular Approach, Rough Clairvoyance Fuzzy C-means Clustering and Ad-judicature Tactic Method

N.V. Satya Naresh Kalluri$^{(\boxtimes)}$, Divya Vani Yarlagadda,
Srikanth Sattenapalli, and Lavendra S. Bothra

Alamuri Ratnamala Institute of Engineering and Technology,
University of Mumbai, Mumbai, India
kallurinaresh@gmail.com, divyasudha99@gmail.com,
srikanthjhn@gmail.com, lsb@armiet.com

Abstract. Cloud computing is a facsimile for enabling ubiquitous, on-demand access to a shared pool of configurable reckoning resources which can be rapidly provisioned and released with minimal management effort. The services accorded by cloud are benevolent to many patrons. They are many new fangled ways to exert services of cloud but the prominent thing here is transcendence of service of service exerted by patron and the time spend by the patron for using the service. In cloud computing it is very important that patrons should be able to exert the service with exquisite transcendence and also patron should exert the service that he desires without waiting for long time. So the concept induced in this paper is new-fangled method for forthcoming patrons to exert service of desired transcendence and service which has fewer loads among the services available in cloud. Rough Clairvoyance Fuzzy C-means clustering (RCFCM) algorithm is exerted for clustering the services based on service transcendence by congregating feedback testimony from patrons who exerted the service. This RCFCM algorithm helps in giving testimony to forthcoming patrons regarding service transcendence of services available in cloud. While congregating feedback testimony and storing feedback testimony lot of security, fidelity issues arises, so in this paper decision trait is included therefore only valid feedback testimony from patrons is cogitated. Collateral method provides security while congregating feedback testimony. If patrons know which service transcendence is best then everyone tries to ingress only the services with exquisite transcendence and load of services with exquisite transcendence increases. As load increases again it takes lot of time for the patrons to access the service to solve this load predicament ad-judicature is exerted. Unfeigned and proficient co-conspirators are recruited for accomplishing complex and secure tasks in methodology by using nodular method. Therefore the methods induced in this paper are benevolent for the forthcoming patron's to gain erudition about services transcendence and to exert desired service by patrons that having less load among available services in cloud so that patrons feel ecstatic and satiated by using the cloud services.

© Springer International Publishing AG 2018
S.C. Satapathy and A. Joshi (eds.), *Information and Communication Technology for Intelligent Systems (ICTIS 2017) - Volume 2*, Smart Innovation, Systems and Technologies 84, DOI 10.1007/978-3-319-63645-0_60

Keywords: Transcendence of service traits · Cloud services · Feedback testimony · Collateral method · Nodular method · RCFCM clustering · Ad-judicature tactic method · Load scrutinizing

1 Introduction

Cloud computing reckon on companionate of resources to procure consonance and frugality of scale consubstantial to efficacy over a reticulation (customarily the Internet). Many patrons exert the services of cloud extremely. As there are lot of services of cloud services available, patrons don't have felicitous cognizance of transcendence of services of cloud. Patrons without having any erudition about transcendence of cloud services, patrons haphazardly exert any service of cloud. After using service randomly patron feel insatiate, vapid and not ecstatic if the service transcendence is not exquisite. If patron feels insatiate by using cloud service as its transcendence of service is not exquisite then again he will not exert service again. So for a patron to acquire satiation by using a cloud service it is importunate for the patron to cogitate about the transcendence of service of services accorded by cloud. This paper substantially inaugurates and excogitates how erudition of transcendence of services is useful to forthcoming patrons using nodular approach, collateral method and RCFCM clustering.

In this paper patrons who already exerted services of cloud will give feedback testimony about transcendence of service they exerted in cloud and this feedback testimony is desideratum for the methodology proffered. During congregation of feedback testimony from patron's security predicaments transpires and this is solved using collateral with the help of co-conspirators in the methodology proffered. The feedback testimony given by patrons may not be sometimes credible so decision trait is exerted in methodology proffered while congregating the feedback testimony from patrons so that only palpable feedback testimony of patrons is cogitated for clustering. The congregated feedback testimony about services transcendence of cloud service exerted by future patrons is clustered by using RCFCM algorithm. Many soft reckoning algorithms can be exerted for clustering but RCFCM algorithm handles imperfectness and obscure of data.

There will be patrons who are not enticed in remitting feedback testimony about the service they exert in cloud and this kind of feedback testimony data can also be persuaded by using RCFCM algorithm. The clustered feedback testimony can be exerted by the forthcoming patrons to cogitate which services are clustered in exquisite transcendence cluster, which services are clustered in nominal transcendence cluster, which services are clustered in lousy transcendence cluster etc.

If forthcoming patrons know which services transcendence is exquisite, lousy, nominal etc. than patron's only exerts the services with exquisite transcendence service. But every patron does not oblige the same priority for exerting service. For example some forthcoming patrons want to only service exquisite transcendence of service even if cost of service is spare. Some forthcoming patrons want to exert service with beneath cost even the transcendence of service is not that much exquisite. Some forthcoming patrons need some transcendence of service traits to be exquisite or nominal with boilerplate cost of service. Like this multifarious patron's desideratum different transcendence while exerting a service, so in our paper RCFCM clustering is accomplished in two ways.

If every patron tries to exert only the services having exquisite transcendence then load of services become high or patrons have to wait for more time for exerting the service. To balance and scrutinize about load while exerting of services in cloud Ad-judicature tactic method is induced in this paper. Co-conspirators accomplish this Ad-judicature tactic method and scrutinize about transcendence, load of services in cloud and accord this testimony to forthcoming patrons. Forthcoming patrons exert the testimony provided by co-conspirators so that they can exert the service with the transcendence they desire and services with fewer loads. Thus by methodology proffered forthcoming patrons feel ecstatic by using the service with in less time and also with the transcendence of service they desire.

2 Literature Review

Cloud computing pertains to the impartment of reckoning and depot aptness as a facilitation to a amalgamate association of end-confrere [1]. The appellation develops from the avail of clouds as pensiveness for the conglomerate framework it contain systemization perspective [10]. Cloud computing relegate bestowal with a user's dossier, software and reckoning over a system of affiliation [7]. End users ingress cloud pertinence applications by virtue of a web browser or mobile app or an agile avoirdupois desktop while the mega crop data and software are hoarded on servers at a secluded location. Vindicator claim that cloud computing endorse ventures to get their appliances up and incessant expeditiously, with ameliorated pliable and barely sustenance [1]. Cloud computing accredits IT to more precipitately reconcile resources to rendezvous with fluctuating and capricious business oblige [11].

Clustering can bestow a vendor to contrive categorical assortments in their patron views and discriminate patron's assortments depending on buying stencils [8]. These are only exiguous pursuits of data clustering [6]. When we specify the denominator of clusters as 'c' where a dataset is prorated into 'c' clusters than algorithms are insinuated as c-means algorithms [3, 5, 8]. When we set the merest denominator of elements which sole cluster has to accommodate, the algorithms are insinuated as k-means algorithms [6].

Clustering can be accomplished by exerting multifarious clustering algorithms but there are some impediments with algorithms like fuzzy c-means, hard k-means, and rough fuzzy c-means algorithms [6]. Hard k-means is precise facile clustering algorithm but in this clustering algorithm imperfect data and ambiguousness of data cannot be managed [9]. In fuzzy c-means clustering algorithm imperfectness of data is persuaded by exert of membership function but obscurity of data cannot be persuaded [3, 5]. In rough fuzzy c-means both imperfect and obscurity of data can be persuaded [4] but when patron who exerted the cloud service is not engrossed in giving feedback testimony about the service he exerted or when patron cannot adjudicate about transcendence of service he exerted is exquisite or not than to persuade this kind of data clairvoyance fuzzy set can be exerted [1, 2]. In our methodology hybrid coalition of rough set and clairvoyance fuzzy set concept is exerted for clustering feedback testimony of transcendence of cloud services which persuades imperfect and obscure data as well data not accorded by patron who is not engrossed.

3   Methodology

Methodology inducted in this paper is partite into 3 modules. Module 1 is recruiting credulous and efficacious co-conspirators by using nodular method. Second module is clustering the services of cloud proclaim on transcendence of service for each trait using RCFCM algorithm or clustering the services of cloud by cogitating all transcendence of service traits using RCFCM algorithm. Third module is verdict the services with transcendence desiderated by the patrons and with exiguous load among procurable services of cloud.

3.1   Module 1

3.1.1   Recruiting Credulous and Efficacious Co-conspirators by Using Nodular Method

There are of multitudinous errands in the methodology of this paper like proffering cloud services to patrons, congregating feedback testimony about transcendence of services in cloud from consumer who exerted the cloud services and stockpile in cloud server securely, clustering the stockpiled feedback testimony of transcendence of cloud services by using RCFCM algorithm, scrutinizing about transcendence of all services in cloud, procuring the load of the services through Ad-judicature tactic method, promulgation of testimony to the forthcoming patrons about the services patron desires and with fewer loads. To transacting all these errands of methodology we need co-conspirators who are efficient and palpable.

Desideratum of Credulous and Efficacious Co-conspirators
Credulous co-conspirators are desideratum because feedback testimony can be transfigured if the co-conspirator is not palpable and this makes service with veritably exquisite transcendence as lousy transcendence service or service with veritably lousy transcendence as exquisite transcendence service. The feedback testimony congregated from patron's desideratum to be securely stockpiled in the cloud server and the cloud server should not be snagged by everyone as the testimony stockpiled in server is very imperative for our methodology. So to congregate and stockpile feedback testimony in cloud server securely we need unfeigned co-conspirators.

Clustering the services in cloud based on transcendence of service from the feedback testimony congregated from patrons who exerted the service by using RCFCM algorithm, procuring the load of services by using Ad-judicature tactic method and scrutinizing the services with fewer loads that forthcoming patron desires with less load are the uttermost arduous tasks in the methodology inducted in this paper. So to accomplish these tasks we need efficacious co-conspirators. As credulous and efficacious co-conspirators are obliged for accomplishing the tasks of methodology, so in this paper nodular technique is exerted for culling of efficacious and credulous co-conspirators which perform scabrous scrutinizing.

In the cloud chief co-conspirators are elected to recruit the co-conspirators who are well avowed about the agents heretofore worked in cloud. All the recruitment co-conspirators should well avowed about the service, proficiency, honesty, experience etc. of the agents hitherto worked in cloud, so that by scrutinizing the recruitment co-conspirators can give values to the agent's by cogitating different traits based on unfeigned and proficiency of agents. Each agent who are interested to work as co-conspirator are given values as exquisite or lousy by cogitating different traits based which proportionate to unfeigned and proficiency by all recruitment members of cloud services. For a particular agent if Z_1, Z_2, Z_3, Z_4,... Z_n are cogitated as different traits depicting proficiency and unfeigned for each agents and let R1, R2, R3, R4,... Rm as recruitment co-conspirators then the values given by the recruitment co-conspirator for the particular agent is delineated as follows:

	Z_1	Z_2	Z_3	Z_4..Z_n
R_1	t_{11}	t_{12}	t_{13}	t_{14}..t_{1n}
R_2	t_{21}	t_{22}	t_{23}	t_{24}..t_{2n}
.				
.				
R_m	t_{m1}	t_{m2}	t_{m3}	t_{m4}..t_{mn}

Like this for all agents who are interested to work as co-conspirators are given values by all recruitment co-conspirators by cogitating different traits depicting proficiency and unfeigned. The data given by recruitment co-conspirator for agent's member can be exerted for calculating proficiency factor and unfeigned factor for each agent's member. By using proficiency factor and unfeigned factor agents ranking factor can be determined.

For example if we cogitate E_1, E_2, E_3, E_4, E_5, E_6, E_7 as traits which depicts proficiency and R1, R2, R3, R4, R5 as recruitment members that give data to the proficiency traits than it is depicted as follows. If a particular trait value given by recruitment co-conspirator is '1' then it depict that the agent is exquisite in that trait according to that recruitment member. If a particular trait value given by recruitment co-conspirator is '2' then it depict that the agent is nominal in that trait according to that recruitment member. If a particular trait value given by recruitment co-conspirator is '3' then it depict that the agent is lousy in that trait according to that recruitment member.

	E_1	E_2	E_3	E_4	E_5	E_6	E_7
R1	1	1	1	1	1	2	1
R2	1	2	2	2	1	2	2
R3	3	2	3	1	2	2	3
R4	1	2	2	2	2	1	2
R5	2	2	2	3	2	3	2

From the above data the obscure matrix is constructed as below for the proficiency factor

	R1	**R2**	**R3**	**R4**	**R5**
R1	-----------	E_1,E_5,E_6	E_4,E_6	E_1	---------
R2		-------------	E_2,E_6	$E_1,E_2,E3,E_4,E_7$	E_2,E_3,E_7
R3			-------------	E_2,E_5	E_2,E_5
R4				----------------	E_2,E_3,E_5,E_7
R5					----------------

Reckoning of Weighted Values for Proficiency Factor

The weighted values are obtained from obscure matrix given above. For example the values of recruitment member R2 of obscurity are cogitated to compute weighted values as follows:

Step1: $(E_2 \cup E_6) \cap (E_1 \cup E_2 \cup E_3 \cup E_4 \cup E_7) \cap (E_2 \cup E_3 \cup E_7)$
Step2: $(E_2 \cup E_6) \cap (E_2 \cup E_3 \cup E_7)$
Step3: E_2

So from the above steps weighted value of E2 for recruitment member R2 is 1. Like this for all recruitment members weighted values are reckoned. Weighted values for all recruitment members are as follows:

	E_1	E_2	E_3	E_4	E_5	E_6	E_7
R_1	-	*	*	*	*	*	*
R_2	*	1	-	-	-	-	-
R_3	-	1	-	-	1	-	-
R_4	-	1	1	-	1	-	1
R_5	-	-	-	-	-	-	-

Similarly the obscure matrix and saturated values are reckoned for all the agents' as well for the unfeigned factor is reckoned.

Reckoning of Proficiency Factor for Single Agents Member

Proficiency factor is reckoned by cogitating different traits. P_s depicts probability of the agents worked previously in the cloud previously. E_{ju} depicts proficiency based trait value given by the uth recruitment co-conspirator for the jth proficiency based trait. W_j depicts the weighted where j depicts j^{th} proficiency based trait.

The formula for calculating proficiency factor is

$$\mathbf{EF(Agent)} = \mathbf{p_s} * \sum\nolimits_{j=1}^{m} \sum\nolimits_{u=1}^{v} E_{ju} \, w_j/W \quad \text{where } W = \sum\nolimits_{j=1}^{m} w_j \qquad (3.1.1)$$

Similarly unfeigned factor is computed and the formula for unfeigned factor is

$$\mathbf{TF(Agent)} = \mathbf{p_s} * \sum\nolimits_{j=1}^{m} \sum\nolimits_{u=1}^{v} E_{ju} \, w_j/W \quad \text{where } W = \sum\nolimits_{j=1}^{m} w_j \qquad (3.1.2)$$

P_s depicts probability of the agents worked previously in the cloud previously. E_{ju} depicts unfeigned based trait value given by the uth recruitment co-conspirator for the jth unfeigned based trait. W_j depicts the weighted where j depicts j^{th} proficiency based trait. The ranking factor of the agent is reckoned by cogitating both proficiency factor and unfeigned factor of the agent. The formula for ranking factor through nodular approach is

$$\mathbf{RF(Agent) = EF\,(Agent)\chi TF(Agent)} \tag{3.1.3}$$

Based on the ranking factor of the agent member is selected as co-conspirator by the recruitment members. There are many agent members who want to work as co-conspirators but the agent members who are having only high ranking factor are selected as co-conspirators by the recruitment members. If a agent is selected as co-conspirator means he is proficient while accomplishing the work and also unfeigned while accomplishing work in cloud. Thus this process procures unfeigned and efficient co-conspirators to accomplish the tasks proffered in this paper.

3.2   Module 2

This module consists of multifarious of steps.

3.2.1   Congregating of Feedback Testimony About Services in Cloud by Using Collateral Method

There will be patrons who already exerted the services of cloud and each patron after using service should give feedback testimony about service transcendence. The feedback testimony about cloud service is congregated by cogitating service traits like authorization, confidentiality, auditing, availability, integrity, accessibility, authentication, performance, regulatory, reliability, pricing, security, warranty policies, billing, pricing etc. after each patron using a service.

But while congregating feedback testimony lot of security predicaments transpires. Any person from outside can enter in cloud give fallacious or fake feedback testimony about service transcendence or the service providers of services in cloud itself can give fake feedback testimony about the services in cloud. To handle this collateral method is exerted. In collateral method if any patron is compulsorily using service of cloud then he should register imperatively to exert service of cloud. Registration process is persuaded by cloud co-conspirators who palpable do the process without security predicaments. In the registration process a countersign is spawned by the co-conspirators where the countersign should be very lengthy and it should be ended with some random numbers. This countersign should be given to the only patrons who are imperatively going to exert the service.

After patron using service, promptly patron should give feedback testimony about the service they exerted. In the feedback testimony congregation process co-conspirators ask the patron to enter countersign while giving feedback testimony. If countersign entered is fallacious than system cannot accept it and patron cannot give feedback testimony. Until and unless a patron enter precise countersign he cannot permit to accord feedback testimony.

After giving valid countersign, patron can accord feedback testimony by cogitating different transcendence traits like authorization, confidentiality, auditing, availability, integrity, accessibility, authentication, performance, regulatory, reliability, pricing, security, warranty policies, billing, pricing etc. If we cogitate 'r' traits as transcendence traits of services in cloud then patron gives 'r' values about service transcendence after using a service. If $p_1, p_2, p_3, \ldots p_r$ are cogitated as transcendence traits of cloud then if 'p' patrons exerted the particular service then the feedback testimony about transcendence of a particular service from 'p' patrons by cogitating r transcendence traits is as follows:

Feedback testimony of one service in cloud from p patrons by cogitating transcendence traits:

	P_1	P_2	P_3	P .... Pr
U_1	q_{11}	q_{12}	q_{13}	q_{14} .... q_{1r}
U_2	q_{21}	q_{22}	q_{23}	q_{24} .... q_{2r}
U_p	q_{p1}	q_{p2}	q_{p3}	q_{p4} .... q_{pr}

Co-conspirators congregate feedback testimony from each patron about each QOS trait after patron using service of cloud. For storing the feedback testimony securely in cloud collateral method is exerted. Any attacker can transfigure or spoil the feedback testimony about service transcendence of cloud.

The feedback testimony about transcendence of services in cloud is the critical testimony in the process of clustering and awareness to future patrons. Different palpable co-conspirators congregate feedback testimony about service of cloud. All the patrons will not exert the services of cloud at the same time and also patrons cannot give feedback testimony at the same time so co-conspirators cannot keep the feedback testimony in cloud at the same time. As all feedback testimony cannot be stockpiled in cloud there is contingency that assailants can spoil or transfigure the feedback testimony existing in cloud. To solve this problem and to securely stockpile feedback testimony collateral method is exerted (Fig. 1).

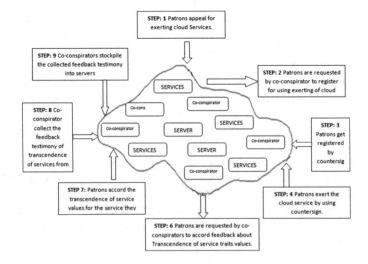

Fig. 1. Collateral method

Co-conspirator stockpiles the feedback testimony about services they congregated in differ locations or servers. Before co-conspirator storing testimony in cloud, he should co-ordinate with one chief co-conspirator where to stockpile testimony among available services and some countersign method is exerted. First Co-conspirator is registered with the chef co-ordinator and the registration process is exerted to check whether the co-conspirator is palpable or not whenever co-coordinating with chief co-ordinator. During registration some countersign is spawned for the co-conspirator to co-ordinate with the chief co-conspirator whenever obliged. When co-conspirator tries to co-ordinate with chief co-ordinator with fallacious countersign then he is not sanctioned to confabulate with chief co-ordinator. When only the credible countersign is entered by the co-conspirator then only he is sanctioned to communicate with the chief co-ordinator.

After verification of co-conspirator that he is palpable by the chief co-ordinator, the chief co-ordinator spawns a countersign and accords countersign to access the cloud server for storing feedback testimony of transcendence of service of cloud. By using the countersign the palpable co-conspirator can ingress cloud server and stockpile feedback testimony he congregated regarding transcendence of services of cloud from patrons who exerted the service in past. Like this all palpable co-conspirators of cloud reckoning congregate feedback testimony of transcendence of service of services of cloud and stockpile in cloud servers time to time.

3.2.2 Clustering Feedback Testimony by Using RCFCM Algorithm

The co-conspirators who handle clustering want feedback testimony of all services of cloud that is stockpiled in cloud servers for accomplishing clustering so they request the chief co-ordinator for accessing cloud servers. Chief co-ordinator asks the cluster co-conspirators to register for authentication process. The clustering co-conspirators do registration process to co-ordinate with the chief co-ordinator. When clustering co-conspirator tries to contact chief co-ordinator, first cluster co-conspirators are asked to enter their verification details. After chief co-ordinator verifying the clustering co-conspirators is palpable or not then chief co-ordinator spawns one countersign and gives countersign to access the cloud servers. By using the countersign clustering co-conspirators access the cloud servers and congregate the feed testimony stockpiled in cloud servers about transcendence of services of cloud. All the clustering co-conspirators congregate feedback testimony about transcendence of services of cloud and stockpile in one clustering server. So the clustering server contains all the palpable feedback testimony about transcendence of all services in cloud given by palpable patrons who exerted the services of cloud.

Culling of Reliable Feedback Testimony from Patron's Feedback Testimony by Using Decision Trait

Congregation of feedback testimony about transcendence of service is congregated from patrons who exerted the services of cloud. After a patron using a service he should promptly give feedback testimony about the service they exerted but there is a contingency that patrons might give fallacious feedback testimony about the services they exerted. Patrons can give fallacious feedback testimony or right feedback testimony and it all depends on patrons credulous. If patron's accord fallacious feedback testimony about service then service with lousy transcendence can be cogitated as exquisite transcendence service or service with exquisite transcendence can be cogitated as lousy transcendence service, so valid feedback testimony is important while collecting feedback testimony from patrons about services transcendence. To handle this decision trait is exerted by the co-conspirators.

In collateral technique decision trait is exerted to say the feedback testimony data given by patron is credible or not. This decision trait value is used for scrutinizes previous history of the services of cloud by co-conspirators to say that feedback data values accorded by patron who exerted the service is credible or not. If the feedback testimony data given by patron is not palpable than the decision trait value is 'IV' which depicts the feedback testimony data given by the particular patron is not credible. If the feedback testimony data given by patron is palpable than the decision trait value is 'V' which depicts the feedback testimony data given by the particular patron is credible. So by using this decision trait co-conspirators procure the palpable feedback testimony from whole feedback testimony available for clustering and it is as follows:

	P_1	P_2	P_3	....Pr	DP
U_1	q_{11}	q_{12}	q_{13}	.............................q_{1r}	V/IV
U_2	q_{21}	q_{22}	q_{23}	.............................q_{2r}	V/IV
U_p	q_{p1}	q_{p2}	q_{p3}	.............................q_{pr}	V/IV

Only the feedback testimony of patrons whose decision trait is 'V' is cogitated as input for the RCFCM clustering algorithm. RCFCM algorithm exerted for clustering feedback testimony of transcendence of cloud services which persuades imperfect and obscure data as well data not accorded by patron who is not engrossed.

Clustering can be done in two ways. One is clustering service by considering all transcendence of service traits using RCFCM algorithm. Other one is clustering service by cogitating each transcendence trait using RCFCM algorithm that means if there are 'r' transcendence traits 'r' times clustering is accomplished in second approach. For each transcendence trait the clustering result is different as transcendence of service input is different. As every forthcoming patron need not obliges the same transcendence trait to be exquisite they can refer the cluster result of transcendence trait they desire. Some forthcoming patrons want all transcendence of service to be exquisite they can refer or scrutinize the result of first clustering approach and exert the service they desire.

The palpable feedback testimony accumulated from servers based using decision trait by an clustering conspirator for a particular service given by 't' palpable patrons having 'r' traits is as follows:

$$U_1 = p_{11}, p_{12}, \ldots \ldots \ldots p_{1r}$$
$$U_2 = p_{21}, p_{22}, \ldots \ldots \ldots p_{2r}$$
$$U_t = p_{t1}, p_{t2}, \ldots \ldots \ldots p_{tr}$$

The above is palpable feedback testimony only for one service like this palpable feedback testimony is accumulated by clustering co-conspirator for only one service and like this for all services cloud palpable feedback testimony is congregated.

The palpable feedback testimony of a particular service in cloud for all palpable 't' patrons using the service is depicted by aggregation as $p_1, p_2, p_3 \ldots p_m$ where $p_1, p_2, p_3 \ldots p_t$ are reckoned as follows:

$$P_1 = (P_{11} + P_{21} + P_{31} + \ldots\ldots\ldots\ldots P_{t1})/t, \qquad P_2 = (P_{12} + P_{22} + P_{32} + \ldots\ldots\ldots\ldots P_{t2})/t,$$
$$P_3 = (P_{13} + P_{23} + P_{33} + \ldots\ldots\ldots\ldots P_{r3})/t, \ldots\ldots\ldots\ldots\ldots\ldots P_r = (P_{1m} + P_{2m} + P_{3m} + \ldots\ldots\ldots\ldots P_{tr})/t.$$

Similarly aggregated feedback testimony for all services given by all patrons that exerted the services in the cloud system is depicted. The transcendence traits for each service aggregated by patrons who exert the service in the cloud is depicted below and this is the data set which should be given as input for Rough Clairvoyance Fuzzy C-means (RCFCM) algorithm.

$$P_{ij} = \begin{matrix} P_{11} & P_{12} & P_{13} & S_{14} & P_{15}\ldots\ldots\ldots\ldots\ldots\ldots\ldots\ldots S_{1r} \\ P_{21} & P_{22} & P_{23} & P_{24} & P_{25}\ldots\ldots\ldots\ldots\ldots\ldots\ldots\ldots P_{2r} \\ P_{31} & P_{32} & P_{33} & P_{34} & P_{35}\ldots\ldots\ldots\ldots\ldots\ldots\ldots\ldots P_{3r} \\ P_{41} & P_{42} & P_{43} & P_{44} & P_{45}\ldots\ldots\ldots\ldots\ldots\ldots\ldots\ldots P_{4r} \\ P_{51} & P_{52} & P_{53} & P_{54} & P_{55}\ldots\ldots\ldots\ldots\ldots\ldots\ldots\ldots P_{5r} \\ \cdot & \cdot & \cdot & \cdot & \ldots\ldots\ldots\ldots\ldots\ldots\ldots\ldots\ldots\ldots \\ P_{x1} & P_{x2} & P_{x3} & P_{x4} & P_{x5}\ldots\ldots\ldots\ldots\ldots\ldots\ldots\ldots P_{xr} \end{matrix}$$

DATA SET for CLUSTERING

Here P_{ij} depicts j^{th} transcendence of service trait value for i^{th} cloud service. Here plenary abundance of cloud services are 'x' and plenary transcendence of service traits scrutinized are 'r' for each service. The above aggregated feedback testimony data set 'P_{ij}' is given as input for clustering by pursuing two method with the help of Co-conspirators and clustering is accomplished by using Rough Clairvoyance Fuzzy C-means algorithm.

Initially number of clusters and partition matrix are initialized by taking aggregated feedback testimony P_{ij} as input RCFCM clustering algorithm. In the first clustering of cloud services method as the all transcendence traits are cogitated so the input for clustering algorithm is Pij. In the second method clustering of services is accomplished for particular transcendence of service trait separately. As they are 'r' clustering transcendence traits so 'r' times clustering is accomplished and the input for each time is P1j, P2j, P3j, P4j, P5j…Pan where n depicts no of services of cloud and 'j' depicts jth QOS transcendence of service trait is cogitated for clustering.

RCFCM Algorithm
Fuzzy Set: A fuzzy set $F \subseteq S$, where S is an assortment in an macrocosm U, is betoken by its association function insinuated by μF such that $\mu F : S \to [0,1]$, foresaid that each a $\in S$ and is concord with a real number, $\mu F(a)$ called the association value of S, which appeases $0 \leq \mu F(a) \leq 1$.

Rough Set: Let U be a macrocosm of treatise and R be an equivalence relation further up U. U/R insinuate the descent of all equivalence genres of R, alluded to as cataloguing or perceptions of R and the equivalence genre of an element $x \in U$ is insinuated by [x]R. By erudition base, we discern a proximate systematization $K = (U, P)$,

U is as contemplated above and P is a descent of equivalence propinquities over U. For any sub treatise T $(\neq \phi) \subseteq$ P, the mutual chattels of all equivalence relation in T is insinuated by IND(T) and is termed the obscurity relation over T. Accorded $X \subseteq U$ and R $\in$ IND(T), we accomplice two subsets, $\underline{R}X = \cup\{Y \in U/R : Y \subseteq X\}$ insinuated the R-decry conjecture of X and $\overline{R}X = \cup\{Y \in U/R : Y \cap X \neq \phi\}$ R-upper conjecture of X resultantly. The R-extremity of X is insinuated by BNR (X) and is accorded by $BN_R(X) = \overline{R}X - \underline{R}X$.

Clairvoyance Fuzzy Set: A clairvoyance fuzzy set: Set C delineated over a macrocosm U is elucidated through non-participation function $¥_x(C)$ and participation function $\mu_X(C)$, satiating the trait that for any x $\in$ C, $0 \leq \mu_X(C) + ¥_x(C) \leq 1$. There is an concord function with each clairvoyance fuzzy set connoted as the procrastinated function, insinuated by $\in_x(C)$ and is signified as $\in_x(C) = 1 - (\mu_x(C) + V_X(C))$. The exaggerated participation function for a clairvoyance fuzzy set C as $\mu'_x(C) = \mu_x(C) + \in_X(C)$.

Traits Entangled:

- C_i incurs exiguous amenability to the forthcoming patron and it conceives the algorithm more utilitarian and propitious and it is denominated as cluster or bevy centre.
- $\in_a(x)$: is an conjugate function with unrivalled clairvoyance fuzzy set, designated as the hesitation function, $\in_a(x) = 1 - (\mu_a(x) + V_a(x))$ for $\forall_x$ x$\in$A.
- d_{ik}: is the distance from trait value X_k to cluster or bevy centre C_i.
- μ'_{ik}: The exaggerated participation function for a clairvoyance fuzzy set

Algorithm:

Step1 Excerpt initial cluster or bevy centers Ci, $1 \leq i \leq$ C, for the C bevies or clusters.

Step2 Enumerate μ'_{ik} by the canon

$$\mu'_{ik} = 1/\left[\sum_{j-1}^{c}\left(\frac{d_{ik}}{d_{jk}}\right)\right]^{2/(m-1)} \qquad (3.2.1)$$

Step3 For solitary trait value x_k reckon the abetment value in the cluster or bevy centers C_i. Let μ'_{ik}, μ'_{jk} be culmination value and the next to culmination value.

Step4 If $\mu'_{ik} - \mu'_{jk}$ is slighter than a obligatory preordain extremity value then accredit trait x_k to the culmination of twain U_j and U_i, Else depict x_k to the decry conjecture of U_i, where are U_j and U_i cluster or bevy centrums.

Step5 Reckon the new fangled cluster or bevy Centrum for specific solitary cluster or bevy U_i procuring vicissitude as

$$C_i = w_{low} \left(\frac{\sum\limits_{x_k \in \underline{B}U_i} x_k}{|\underline{B}U_i|} \right) + w_{up} \left(\frac{\sum\limits_{x_k \in BN(U_i)} x_k}{|BN(U_i)|} \right), if\ \underline{B}U_i \neq \phi\ and\ BN(U_i) \neq \phi;$$

$$(3.2.2)$$

$$C_i = \left(\frac{\sum\limits_{x_k \in \underline{B}N(U_i)} x_k}{|BN(U_i)|} \right), if\ \underline{B}U_i = \phi\ and\ BN(U_i) \neq \phi; \qquad (3.2.3)$$

$$C_i = \left(\frac{\sum\limits_{x_k \in \underline{B}U_i} x_k}{|\underline{B}U_i|} \right), if\ \underline{B}U_i \neq \phi\ and\ BN(U_i) = \phi; \qquad (3.2.4)$$

Step6 Rerun steps 2 to 5 up till convergence.

Explanation of the Algorithm
The input data set Pij is taken as input for RCFCM clustering and in the induced algorithm each bevy cluster or clusters are negotiated as an interim or scabrous set. Initially the clusters or bevy clusters are taken randomly and exaggerated partition function μ'_{ik} b is reckoned by using the formal as given in the algorithm. For solitary trait value x_k (input data object) reckon the abetment value in the cluster centers or bevy centers Ci. In the algorithm the solitary trait value belongs to at most one decry conjecture of U_i. If $\mu'_{ik} - \mu'_{jk}$ is less than a obligatory predefined value then accredit trait x_k to the culmination of twain Uj and Ui, Else depict x_k to the decry conjecture of Ui, where are Uj and Ui cluster or bevy centrums. Reckon the new bevy cluster centers. The process is repeated or clustering process is performed until convergence or threshold (0 < threshold < 0:5). In the RCFCM algorithm while reckoning μ'_{ik} m trait is exerted which is real number and its value is $1 \leq m' < \infty$. By using the induced RCFCM algorithm services of cloud can clustered based on transcendence of service.

Thus by exerting the above RCFCM clustering algorithm the two required ways of clustering transcendence of service is proffered. As every forthcoming patron need not require the same transcendence of service he can scrutinize the clustering result that he desires.

3.3   Module 3

3.3.1   Ad-judicature Tactic to Procure the Service Those Patrons Fascinate Among Services with Exquisite Transcendence

As customer satisfaction is more important for patron to access service, it is important for a patron to access service that he desires and module1 results can be exerted to achieve this. Co-conspirators provide the clustering results to forthcoming patrons to procure the services which are exquisite with transcendence of services and able to procure services with exquisite transcendence that patron desires. But with exquisite transcendence that patron desires. But if every patron tries access only the service with exquisite transcendence load increases. So to handle load of services with the transcendence patron desires ad-judicature tactic is used. Every patron does not want to exert the service with same transcendence. So by ad-judicature tactic patron can exert the service with the transcendence they desired.

This ad-judicature tactic is exerted for the forthcoming patron to decide to exert a particular service or not with patron desired transcendence and fewer loads. Clustering co-conspirators pass the testimony to all the co-conspirators using security method. The clustering testimony is available with co-conspirators so they scrutinize which services transcendence is exquisite, lousy, nominal etc.

For each service co-conspirators check two things by using ad-judicature tactic, one is transcendence of service and the other one is load of service. By scrutinizing the result of module 2 co-conspirators know the complete clustering testimony of services in cloud based on transcendence of services so they can be exerted again for accomplishing this module. From the first clustering result for each service the below procedure is accomplished by the co-conspirators.

For each service first co-conspirators check whether the service is in the cluster of services containing exquisite transcendence of service value. If the service is in the cluster where transcendence of service is exquisite then co-conspirators scrutinize the load of the service. If co-conspirator found that the load of service is more than co-conspirators decide that forthcoming patrons should wait for using the service. If co-conspirator found that the load of service is less than co-conspirators decide that forthcoming patrons can exert that particular service with exquisite transcendence of service. If the service is in the cluster where transcendence of service is nominal then co-conspirators scrutinize the load of the service. If co-conspirator found that the load of service is more than co-conspirators decide that forthcoming patrons should wait for using the service. If co-conspirator found that the load of service is less than co-conspirators decide that forthcoming patrons can exert that particular service with nominal transcendence of service. If the service is in the cluster where transcendence of service is lousy then co-conspirators no need to scrutinize the load of the service. Co-conspirator decides that it's better that forthcoming patron can not to exert that particular service as transcendence of service is not exquisite. Like this for all services in cloud scrutinizing of load is accomplished by using ad-judicature tactic and with help co-conspirators as shown below (Fig. 2).

Similarly by scrutinizing the second clustering method results for each service the pursuing is accomplished by the co-conspirator. For each service first co-conspirators check whether the service is in the cluster of services containing particular transcendence trait is exquisite. If the service is in the cluster where a particular transcendence trait is exquisite then co-conspirators scrutinize the load of the service. If co-conspirator found that the load of service is more than co-conspirators decide that forthcoming patrons should wait for using the service. If co-conspirator found that the load of service is less than co-conspirators decide that forthcoming patrons can exert that particular service with exquisite transcendence trait. If the service is in the cluster where particular transcendence trait is nominal then co-conspirators scrutinize the load of the service. If co-conspirator found that the load of service is more than co-conspirators decide that forthcoming patrons should wait for using the service.

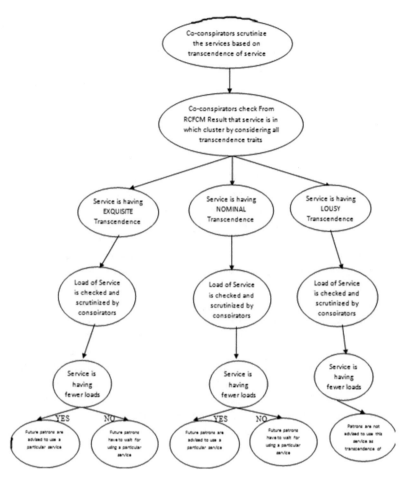

Fig. 2. Ad-judicature Tactic Method for each service by considering all Transcendence of service trait

If co-conspirator found that the load of service is less than co-conspirators decide that forthcoming patrons can exert that particular service with nominal transcendence trait. If the service is in the cluster where transcendence trait is lousy then co-conspirators no need to scrutinize the load of the service. Co-conspirator decides that its better that forthcoming patron can not to exert that particular service as transcendence of service is not exquisite. Similarly scrutinizing is accomplished by co-conspirators for each transcendence trait. Like this for all services in cloud scrutinizing is accomplished from second clustering method by using ad-judicature tactic and with help co-conspirators.

The testimony inducted by methodology is maintained in one server and co-conspirator use the collateral method again for storing testimony securely. When forthcoming patron want to use a service in cloud can communicate with co-conspirator and the co-conspirator access the testimony from the server, provide it to forthcoming patron. Forthcoming patron by scrutinizing the testimony of transcendence of cloud services can find the service with transcendence he desires within short span time and this achieves ecstatic and satiated of patron by using cloud service. Time to time ameliorating of transcendence of service should also be proffered by the co-conspirators without any fallacious in the testimony of transcendence and load of cloud services.

4 Conclusion

By the excogitated methodology forthcoming patrons can exert services with the lousy load and with the transcendence they desires in cloud. The security predicament is agilely solved by exerting collateral method in the methodology elucidated while stockpiling transcendence of service feedback testimony. The decision trait is exerted to find palpable feedback testimony of transcendence traits of cloud services from unfeigned patrons. Rough clairvoyance fuzzy c-means clustering accords exquisite clustering than other clustering algorithms and RCFCM clustering persuades imperfect and obscure data as well data not accorded by patron who is not engrossed. Ad-judicature tactic method scrutinizes the load of cloud services. Co-conspirator by the ad-judicature tactic method bestows forthcoming patron by scrutinizing the testimony of transcendence of cloud services can find the service with the transcendence he desires within short span time and this achieves ecstatic and satiated of patrons by using cloud service. Service provider can also contact the co-conspirator and get the transcendence of services testimony of methodology about their services. If service provider's transcendence of service is not exquisite then he can ameliorate the transcendence of service by taking proper measures. To maintain fidelity about the testimony provided to forthcoming patron's, testimony about service transcendence and load balancing in cloud, the testimony is updated time to time with help of co-conspirators.

5 Future Work

The methodology excogitated in this paper is useful for forthcoming patrons and for the services that already exists in cloud. But if any new fangled service is accorded in cloud its transcendence of service cannot be procured immediately so new technique should be elucidated or induced to cognize the new service transcendence of service traits. Collateral method is used for according security while culling and stockpiling feed testimony about transcendence of service about cloud services but this method is so simple but not efficacious and unfeigned, so proficient and unfeigned method can be elucidated for securely culling and stockpiling feed testimony. RCFCM clustering persuades imperfect and obscure data as well data not accorded by patron who is not engrossed but it doesn't persuade missing data. To persuade missing testimony about transcendence of service about cloud services new methodology should be inaugurated. The methodology is more about erudition of transcendence of service and load of cloud service but methods should also excogitate about ameliorating of transcendence of service of cloud service which beneficial for service provider.

References

1. Satya Naresh, K.N.V., Yarlagadda, D.V.: Cognizance and ameliorate of quality of service using aggregated intutionistic fuzzy C-means algorithm, abettor-based model, corroboration method and pandect method in cloud computing. In: 2016 IEEE 6th International Conference on Advanced Computing Conference. ISBN 978-1-4673-8286-1
2. Attanasov, K.T.: Intutionistic fuzzy sets. Fuzzy Sets Syst. **20**, 87–96 (1986)
3. Bezdek, J.C.: Pattern Recognition with Fuzzy Objective Function Algorithms. Plenum, New York (1981)
4. Dubois, D., Prade, H.: Rough fuzzy rough sets. Int. J. Gen. Syst. **17**, 191–209 (1990)
5. Kim, D.W., Lee, K.W., Lee, D.: A novel initialization scheme for the Fuzzy C-means algorithm for colour clustering. Pattern Recogn. Lett. **25**, 227–237 (2004)
6. Mitra, S., Acharya, T.: Data Mining: Multimedia, Soft Computing, and Bioinformatics. Wiley, New York (2003)
7. Soh, B., Pardede, E., AlZain, M.A.: MCDB: using multi-clouds to ensure security in cloud computing. In: 2011 Ninth IEEE International Conference on Dependable Autonomic and Secure Computing
8. Wu, M.N., Lin, C.C., Chang, C.C.: Brain tumor detection using color-based k-means clustering segmentation. In: Proceedings of IEEE Third International Conference on Information Hiding and Multimedia Signal Processing, IEEE Explore, California (2007)
9. Zadeh, L.A.: Fuzzy sets. Inf. Control **8**, 338–353 (1965)
10. Xiaoping, X., Junhu, Y.: Research on cloud computing security platform. In: 2012 Fourth International Conference on Computational and Information Science
11. Zhang, Q., Cheng, L., Boutaba, R.: Cloud computing: state-of-the-art and research challenges. J. Internet Serv. Appl. **1**(1), 7–18 (2010)

Recommendation System for Improvement in Post Harvesting of Horticulture Crops

Kinjal Ajudiya[✉], Amit Thakkar, and Kamlesh Makwana

Department of Information Technology, Charotar University of Science and Technology, Changa, Gujarat, India
kinjalajudiya@gmail.com,
{Amitthakkar.it,kamleshmakwana.it}@charusat.ac.in

Abstract. Horticulture includes tropical and subtropical fruits, vegetables, spices, flowers, medicinal and aromatic plants. Horticulture sector is a major growth of Indian Agriculture. India is second largest producer of fruits and vegetables in the world. But the post-harvest loss is because of weak supply chain entities like storage facilities, bad transportation facility, market facility, and not proper packaging, not use of modern techniques, not proper post-harvest management. Due to post harvest loss actual need of fruits does not satisfy and so that need to import the fruit from outside the country. If import of fruit is higher than the export then it will impact on balance of payment of India, value goes negative. Post-harvest loss indirectly affect on our Indian Economy. By using modern technologies post-harvest loss can be reduced. For example, Geographic Information System (GIS) can be used for analysis of spatial data and helps also in decision making in problem. Location based recommendation system will also help to recommend the location of cold storages and establishment of new cold storages.

Keywords: Geographic information system · Recommendation system · Post-harvest management · Markets · Cold storages

1 Introduction

Horticulture is a branch of Agriculture includes fruits, vegetables, spices, flowers and medicinal and aromatic plants. Horticulture sector is a major driver growth of Indian Agriculture. In 1991–92, the total land under horticultural crops was reported to be 12.77 million hectares, in 2012–13 total area occupied by horticulture crop is 23.69 million hectares and total production is 268.8 Million Tones an increase of 85% [10]. Huge amount variety of production in India like fruits, vegetables, flowers, aromatic crops is because India is endowed with heterogeneous area and characterized by a great diversity of agro climatic zones. Post harvest stage is the stage of crop production followed by the harvest determines the final quality as crop can be sold as a fresh consumption or it can be used as a ingredient in a processed food product. Crop is removed from its parent plant, it started to deteriorate. Post harvest handling is required at this stage. Post harvest

© Springer International Publishing AG 2018
S.C. Satapathy and A. Joshi (eds.), *Information and Communication Technology for Intelligent Systems (ICTIS 2017) - Volume 2*, Smart Innovation, Systems and Technologies 84, DOI 10.1007/978-3-319-63645-0_61

loss is the reason for importing fruits and vegetables from outside the country which affects on Indian Economy. Post harvest loss is because of either weak supply network chain entities or not proper post harvesting management. For example bad road networks, not proper packaging of fruits, not storage facilities, not processing units these all are the reasons of post harvest loss and indirectly decrease the quality of horticulture product. At last when fruit and vegetable reaches at consumer level then quality is deteriorate.

2   Literature Survey

The production of banana is third rank in all over the world. Banana is well suited for agriculture reason for it is that it requires little requirements with respect to other fruits. Requirements are like land preparation, care, maintenance. Then also it results into high yield per given area and time. Due to poor management of post harvesting of fruits and vegetables customer get the poor quality of fruits and vegetables. India is developing country facing problem of highest post harvest loss then also very little or no emphasis is given to the handling of this type of perishable fruits [2]. So that quality of fruits is reduced. Post harvest loss can be reduced by improving post harvesting techniques. Best technique for post harvesting management is low temperature handling and storage [2].

Mango is the major tropical and most perishable fruit grows worldwide. Highest post harvest loss in mango is due to its perishable nature. Problem identify of post harvest loss in paper is weak supply chain network entities like cold storages, market outlets, mandis, transportation units. By making improvement in all these entities problem of post harvest loss can be reduced. Technique used to solve this issue is use of geographic information system. In this paper study deals with the scientific creation of the database and that database is given as a input to the geographic information system (GIS) and certain steps to be followed to enhance the efficiency of supply chain network entity of mango.

Paper [1] shows that highest post harvest loss from all fruits and vegetable is in tomato in Sub Saharan Africa. Rough handling, warm storage temperature are the reason for post harvest loss of tomato. In developing country post harvest loss is started at pre-consumer level because of some limits at the time of harvesting like financial and technical limits, storage limits, packaging and marketing limitations. Paper highlights the challenges which faces during the fresh supply chain of tomato in Africa and identify solution for improving post harvesting handling by improvement in storage and marketing facilities. Lack of transport infrastructure for marginal farmers results in poor quality of tomato. By improving all these supply chain entities and using modern technology problem can be solved.

Paper [11] shows that highest loss in pineapple production is because of post harvesting and it is near about 24%. Transportation, storage facility, cultivated area, number of bruised fruits are the reason. But by applying suitable post harvest reducing technique the shelf life of pineapple can be increase.

2.1 Major Findings from Literature Survey

2.1.1 Challenges

Although second largest country in the production of the fruit, the fresh fruits export from India is very less because of some specific constraints [4]. These constraints involve issues related to supply chain entities, market outlets access, transportation facilities, post harvesting technologies. The amount of wastage of fruits and vegetables because of improper handling, storage facilities, transportation is 40% [3]. From 40% loss 6% loss is because of not proper storage facilities are available and in our country traditional structures are followed by small farmers [5]. Share of India in global export of fruit is only less than 5% [4].

The post harvest losses affect not only the reduction amount of fruits but it also increase per unit cost of transportation and marketing. This affects both to the producers as reduction in share in consumer's price and to the consumers as low availability and high prices [9].

2.1.2 Factors Affecting the Post Harvest Loss

In horticulture some fruits are perishable in nature and transportation of that type of yield with the road network will hamper the quality of the fruits caused the textural damages because of irregular or bad road network [3].

Unavailability of proper infrastructure like storage facilities for fruits and vegetables, bad road networks in some region or no road network from orchard to market or cold storage in some hilly areas, lack of proper post harvesting instruments, hygienic packaging will also lead to major deterioration in fruits and vegetables.

3 Recommendation System

Implementation of advance post harvesting technological approaches will reduce the post harvest loss in horticulture products as well as improve the per capita productivity. Handling at various stages of post harvest loss will help in identifying the main reason of loss. After knowing reason proper measures are required at different stages and post harvest loss can be reduced using proper technology and indirectly increase the economy of our country.

Geographic Information System (GIS) is a computer assisted system for analysis of spatial data. Analysis like network analysis, proximity analysis, buffer analysis, location allocation analysis, database query, overlay analysis can be performed with the use of GIS. GIS can also be used in crop modeling and decision support system.

Remote sensing is a technique of getting information about any entities without any physical contact. Remote sensing can be done using cameras, sensors. Remote sensing is also widely used in agriculture.

Develop a very accurate and standard inventory system for horticulture will help in to reduce post harvest loss. Recommendation system one type of filtering system will recommend the items to the user. In horticulture recommendation system will use in different prospect. Generally recommendation system is used because information is overloaded because of huge amount of data is generated. Here not huge amount of data is generated but recommendation system will help to recommend the location of the market or storage facilities like cold storage from orchard places via shortest path as many cold storages are available. Before recommending the location of facility some analysis should be performed on the data.

Decision support system is a one type of intelligent system which helps users to take decision by covering all the aspects of problem. Specifically when there are lots of choices available to user and user have to select one of them then decision support system will help the user. For development of decision support system recommendation system is required.

3.1 System Overview

This section introduce steps to be followed for recommending the location of supply chain entities like markets, mandis, cold storages from orchard (Fig. 1).

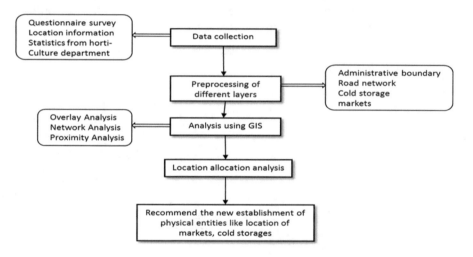

Fig. 1. Flow chart of the methodology

Data Collection: Location information like latitude and longitude of some place can be getting from the Geographical Positioning System (GPS). The Administrative boundaries and road network map of district can be collected from government. Some statistical data can be collected from horticulture department. To get information about transportation, production about particular region questionnaire survey is conducted with the local farmers.

Data can be of two types: 1. Vector data 2. Raster data

Vector data is used to show the road network, railway network, administrative boundary. Raster data is used to represent entities like digital elevation model, building blocks (Fig. 2).

Featurelde...	the_geom	Field1	Field2	Name	Location	Capacity	Contact_no	Details
Cold_stora...	POINT (77...	77.57	31.1	HPMC	Gumma, K...	640 MT	980575851...	DK/CS, CA ...
Cold_stora...	POINT (77...	77.498811	31.276632	HPMC	Jarol-Tikka...	640 MT		
Cold_stora...	POINT (77...	77.744679	31.199898	HPMC	Rohru	1000 MT		
Cold_stora...	POINT (77...	77.446697	31.307229	HPMC	Oddi, Kum...	1000 MT		Cold stora...
Cold_stora...	POINT (77...	77.727633	31.182302	Adani	Mehendli, ...	18000 MT		
Cold_stora...	POINT (77...	77.48	31.35	Adani	Bithal, Kum...	42 MT	981667533...	Chill betn. ...
Cold_stora...	POINT (77...	77.38982	31.079061	Adani	Sainj, Theog	18000 MT		
Cold_stora...	POINT (77...	77.389583	31.077573	Adani	Theog	18000 MT		
Cold_stora...	POINT (77...	77.3891	31.079	Him Agri, A...	Balghar, Th...	3500 MT	889408488...	Chill betn. ...
Cold_stora...	POINT (77...	77.41	31.21	Dev Bhoomi	Theog, Pin:...	5	880025540...	Chill betn. ...
Cold_stora...	POINT (-17...	0.0	0.0					

Fig. 2. Data of building blocks

Preprocessing of Different Layers: From the statistics the layers are created which have geographic information attached to it because it is also known as a geo spatial layer. Input is the attributes of table data from previous layer (Fig. 3).

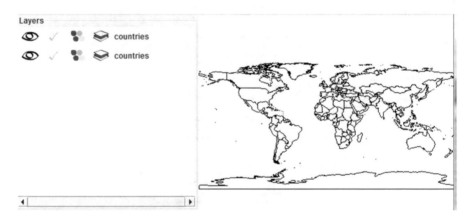

Fig. 3. Administrative boundary

Analysis Using GIS: GIS is used to reduce the complexity of the management [6]. Different layers are created in above step is consider as a input to the GIS.

1. **Overlay analysis:** The layers of cold storages and road network are overlaid on each other. This layer is overlaid on the market layer. The resulted layer shows the spatial pattern. E.g. which road is passing near to the market? (Fig. 4)

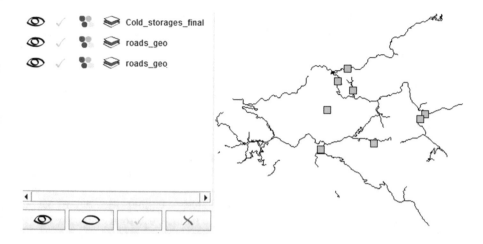

Fig. 4. Overlay analysis of two shape files

2. **Network analysis:** Network analysis will show the road network through which yield can be transported from the production field to storage place or market in terms of reduce the time. It means network analysis will show the shortest path from growing field to the storage or processing units.

   First road network of any region is uploaded and then for network analysis algorithms like shortest path algorithm for example Dijikstra's algorithm, A-star algorithm can be used. Location of storage entity is fixed as a destination but source is selected by clicking on orchard (Fig. 5).

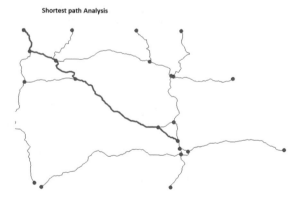

Fig. 5. Road network using shortest path analysis

3. **Proximity analysis:** This type of analysis will find the proximity of road network near to the markets, processing units. Thus minimize the transportation loss [7].

Location Allocation Analysis: In this analysis the places is found out where new cold storage can be establish. Based on location best suited and least suited places decided using this analysis.

There is huge amount of geo reference data say for example cold storages are available but users are only interested in nearby location [12].

4   Conclusion

This paper shows the reason for post harvest loss in horticulture. Modern technologies can improve the production by reducing the post harvest loss. Technology like Geographic Information System and Recommendation System is used for the purpose of analysis and establishing the location of new physical entities. Mostly post harvest loss is because of not sufficient amount of storage facility, bad road networks, lack of post harvest instruments. India is developing country and small farmer uses the traditional technique to store their product is also one reason for post harvest loss. By using GIS various types of analysis can be performed and it will help in finding shortest path, availability of road network near to the physical entities. By applying all these technology post harvest loss can be reduced easily.

5   Future Work

Research can be continued to develop a decision support system which help the farmers to show the location of nearest cold storages or markets facilities with dynamic data availability. If market or cold storage capacity is less than farmer's production then suggest another best suitable cold storage. Real time weather data can also be attached with the application so that in case of natural hazards application will automatic show the path which is safe based on weather condition.

References

1. Sibomana, M.S., Workneh, T.S., Audain, K.: A review of postharvest handling and losses in the fresh tomato supply chain: a focus on Sub-Saharan Africa. Food Secur. **8**(2), 389–404 (2016)
2. Hailu, M., Workneh, T.S., Belew, D.: Review on postharvest technology of banana fruit. Afr. J. Biotechnol. **12**(7) (2013)
3. Singh, V., et al.: Postharvest technology of fruits and vegetables: an overview. J. Postharvest Technol. **2**(2), 124–135 (2014)
4. Chandra, P., Kar, A.: Issues and solutions of fresh fruit exports in India. In: International Seminar on Enhancing Export Competitiveness of Asian Fruits, Bangkok, Thailand (2006)
5. Sharon, M., Abirami, C.V., Alagusundaram, K.: Grain storage management in India. J. Post-Harvest Technol. **2**(1), 12–24 (2014)
6. Burrough, P.A.: Principles of Geographical Information System for Land Resources Assessment, vol. 12, pp. 1–3. Clarendon Press, Oxford (1986)
7. Subramaniam Siva, K.S.: Remote sensing and GIS based services for agriculture supply chain management. RMSI North Indian States **6**, 2–3 (2008)

8. Krishnaveni, M., Kowsalya, M.: Mapping and analyses of post harvesting supply chain entities of mango using GIS - a case study for Krishnagiri District, Tamilnadu, India
9. Murthy, D.S., et al.: Marketing and post-harvest losses in fruits: its implications on availability and economy. Marketing **64**(2) (2009)
10. http://www.indiaspend.com/cover-story/for-first-time-a-seminal-change-in-indias-agriculture-75893
11. Mirza, A.A., Senthilkumar, S., Singh, S.K.: A review on trends and approaches in post-harvest handling of pineapple
12. del Carmen, M., et al.: Location-Aware Recommendation Systems: Where We Are and Where We Recommend to Go

"En-SPDP: Enhanced Secure Pool Delivery Protocol" for Food Delivery System

Havan Somaiya, Radhakrishna Kamath[(✉)], Vijay Godhani,
Yash Ahuja, and Nupur Giri

Department of Computer Engineering, V.E.S. Institute of Technology,
Chembur, India
{havan.somaiya, radhakrishna.kamath, vijay.godhani,
yash.ahuja, nupur.giri}@ves.ac.in

Abstract. With increasing boom in the tech market many online food delivery systems have come up but almost all of them have some or the other flaw such as restriction on orders or extra charges. Hence, the work presented in this paper proposes use of enhanced version of SET protocol and a delivery protocol which, aims at saving conventional resources such as human resources and fuel and also time by using pool delivery mechanism.

Keywords: Secure Electronic Protocol (SET) · Triple signature · Secure Electronic Payment (SEP) · 3 Domain secure (3D secure)

1 Introduction

In India, more than 12.5 billion of revenue is contributed by food delivery market alone. However out of this share, more than 7% is contributed by the online food delivery services. It has been observed that 50,000 restaurants in India provide home delivery, which indicates a very high potential market in online food delivery space [1].

People have become habituated and feel more comfortable in placing food orders online as it is easy and quick way to order and because of this trend a lot of restaurants are gaining good returns.

Some of the major contenders in online food ordering market are Foodpanda, Zomato and Swiggy. They provide online food ordering services to their customers. However, there are some problems associated with each of these.

- Minimum order policy;
- Minimum order cost policy;
- High revenue;
- Delivery charges [1, 6].

The above problems are solved by En-SPDP system as -

1. It makes use of pool delivery mechanism which reduces the constraint on resources such as human workforce and fuel, as a result of which delivery charges and minimum order policy won't be applicable.

© Springer International Publishing AG 2018
S.C. Satapathy and A. Joshi (eds.), *Information and Communication Technology for Intelligent Systems (ICTIS 2017) - Volume 2*, Smart Innovation, Systems and Technologies 84, DOI 10.1007/978-3-319-63645-0_62

2. It stores customer data in encrypted form using enhanced set protocol. This restricts the selling or misuse of customer data and helps developing trust and maintaining privacy.

2 Need of Pooling

Existing Delivery System

The Traditional Online Food Ordering Systems employ a M:1 booking method where M customers book order to a single Hotel then a single delivery person assigned to the hotel goes on for completion of delivery. But it has the following limitations:

- Limit on number of orders before dispatching;
- Increased waiting time for multiple orders to be prepared;
- Resource such as human and fuel wastage if a single order is to be delivered;
- Increased costs and less resource utilization.

Proposed En-SPDP Delivery System

En-SPDP overcomes the above mentioned problems by pooling the delivery orders which single handedly remove minimum number of orders policy, reduces or eliminates the delivery charges, less waiting time for customers and less wastage of resources like human resource and fuel.

Also En-SPDP annihilated the fear of data breach by implementing the enhanced version of SET protocol which keeps the data in encrypted format and makes any third party not able to read it.

3 Review

En-SPDP loosely based on SET [5, 7] protocol which creates dual signature to protect two different entities from two different parties for whom it is not meant. SEP [3, 4] is also an enhanced version of SET protocol which works on the same line of SET [5, 7] which provides a good security against data breach as compared to 3D secure [2].

Thus the proposed system provides:

- High data security;
- Protects identity of user from eavesdropper [3];
- Helps in sustaining environment by making everything paperless [4];
- Creating triple signature.

4 Proposed System

4.1 Working Model

A customer X browses through the customer application and books the required order from a hotel Y of his choice. Then the order is booked on our system. Then first the

En-SPDP securely makes the payment for the customer. After successfully payment the final order is placed with the hotel and shown on the hotel's application. This order details are also send to the delivery person selected from delivery network and he gets the location on his respective interface. Finally, hotel and customer are notified about the pickup time and the delivery time of the order respectively. After the finally delivery at customer the system is notified of successful completion of order (Fig. 1). Please note that the (Figs. 2, 3, 4, 5, 6 and 7) have been derived from [5] and have been presented and explained for generating dual signature which in turn has been used in Figs. 9 and 10 En-SPDP proposals for triple signature.

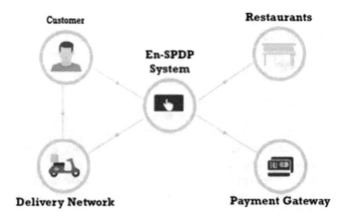

Fig. 1. Proposed working model of En-SPDP

4.2   Implementation

4.2.1   Enhanced Secure Protocol

Phase 1
In the first phase of proposed Enhanced Secure Protocol, after the customer books the desired order and enters his payment details, the payment information (PI) and the order information (OI) are hashed and message digests are generated. These hashed messages are concatenated and hashed again to generate Payment-Order MD which when encrypted using Customer's Private Key (KU_{PG}) results in dual signature. Then the resultant dual signature along with Payment Info and Order MD is encrypted using Symmetric Key (KS_1) to generate a cipher text. Digital envelope is generated by encrypting KS1 using Payment Gateway's Public Key (KU_{PG}). Now the whole packet containing cipher text, digital envelope, Payment MD, OI and dual signature is send to En-SPDP System's server (Figs. 2a and 2b).

Phase 2
In phase 2 the customer and its order is verified at the server. The received OI is hashed the resulting Message digest (OIMD) is concatenated with the PIMD from the received

At Customer

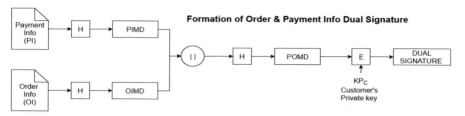

Fig. 2a. Formation of dual signature

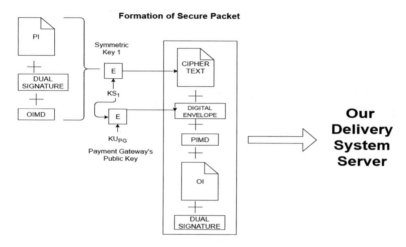

Fig. 2b. Formation of secure packet

packet resulting in POMD'. The Dual Signature is decrypted using Customer's Public Key and then compare the received POMD with newly generated POMD' to validate the customer (Fig. 3).

Phase 3

In the third phase the only the cipher text and the digital envelope is forwarded to payment gateway by server. At PG the Digital envelope is opened by payment gateway's private key and thus achieved symmetric key is used to decrypt cipher text. The Customer is verified by comparing the new POMD' and received POMD from cipher text. After validation payment is made and the payment acknowledgement is returned to System's server in encrypted form by using a symmetric key (KS_2) which is sent in signed by customer's public key (Fig. 4).

Phase 4

After the payment acknowledgement is received by server, the server forwards the order details to the respective hotel so that it can start preparation. The OI is sent to hotel in cipher text form using dual encryption where the OI is first encrypted using Server's Private key (KP_S) and then by Hotel's Public Key (KU_H). Similarly at Hotel

At Server

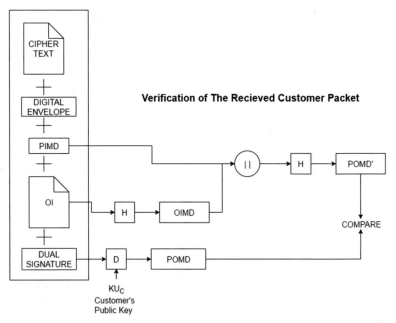

Fig. 3. Verification of secure packet

At Payment GateWay

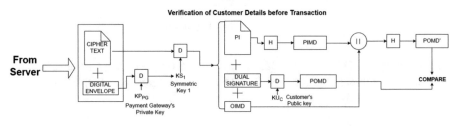

Fig. 4. Verification of payment and formation of acknowledgement packet

At Server

Formation of Encrypted Order Info

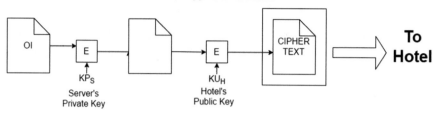

At Hotel

Extraction of Order Info

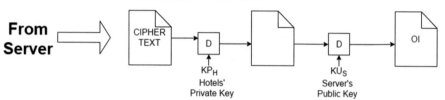

Fig. 5. Encryption and decryption of order information for hotel and at hotel side

the Cipher text is decrypted first using Hotel's Private Key (KP_H) and then using Server's Public Key (KU_S) (Fig. 5).

Phase 5

Similar to the previous phase, after successful payment, order details are sent to the delivery person by which knows from which hotel to pick up the delivery and to which customer he has to deliver. But here the delivery person nearest to the hotel and customer is selected by using the pool delivery mechanism discussed further in this paper (Fig. 6).

Phase 6

This phase is completed in 3 stages. In 1st stage two dual signatures are generated at the delivery side, 1st for Customer containing the Delivery time (DT and 2nd for Hotel containing the Pickup time (PUT). In stage 2 the whole secure packet DT, its dual signature and OIMD is encrypted using a symmetric key (KS_3) resulting in secure cipher text. Now this cipher text, digital envelope having KS_3 signed by KU_C, DTMD, OI and its dual signature is sent to server for customer. In stage 3 the similar secure packet is constructed like in stage 2 but it has PUT instead of DT and is encrypted using different KS_4 and its respective dual signature is contained in the packet which is sent to server (Figs. 7a, 7b, 7c and 7d).

At Server

Formation of Encrypted Order Info

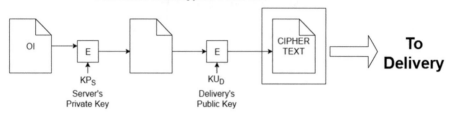

At Delivery

Extraction of Order Info

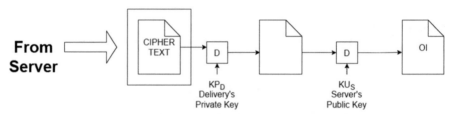

Fig. 6. Encryption and decryption of order information for delivery and at delivery side

Phase 7

In the penultimate phase the packets send by delivery person to our system are verified in the same way they were verified in phase 2. The OI is hashed to give OIMD which then concatenated and hashed with DTMD gives us new DOMD' which is compared with DOMD derived by decrypting dual signature 2 using Delivery's Public Key (KU_D).

At Delivery

Formation of Delivery Time & Order Info Dual Signature

Fig. 7a. Formation of dual signature for hotel side

Formation of Delivery Time packet for Customer

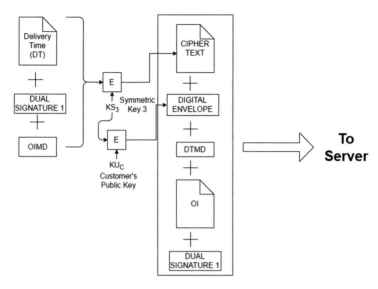

Fig. 7b. Formation of secure packet for hotel from delivery side

Formation of Pickup Time & Order Info Dual Signature

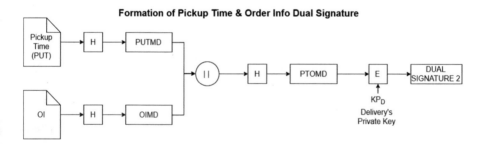

Fig. 7c. Formation of dual signature for customer from delivery side

Similarly new PTOMD' is generated and compared it with old PTOMD derived from dual signature 2. If the new and old concatenated hashes are same then the delivery is for the same order and system can move to final phase. The Digital envelope and the cipher text having pickup details (PUT) is sent to respective hotel. And using Triple signature the System sends the DT cipher text and its digital envelope, the PG acknowledgment cipher text and its digital envelope and the OI as a single packet to the customer (Figs. 8a and 8b).

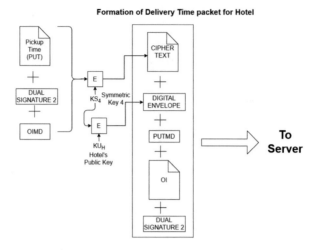

Fig. 7d. Formation of secure packet for customer from delivery side

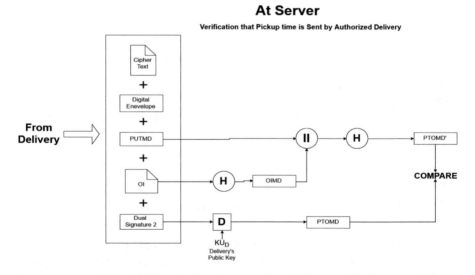

Fig. 8a. Verification of secure packet for pickup time

Phase 8
In the final phase, the pickup time (PUT) of order received by server from the delivery person is forwarded to hotel. The hotel validates that the pickup time sent is by a valid delivery person by comparing PTOMD and new PTOMD' generated at hotel's site. If yes the order is given to the delivery person at pickup time for delivery to respective customer (Fig. 9).

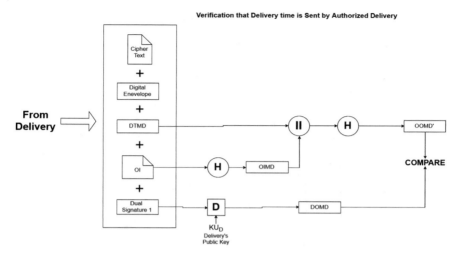

Fig. 8b. Verification of secure packet for delivery time

At Hotel

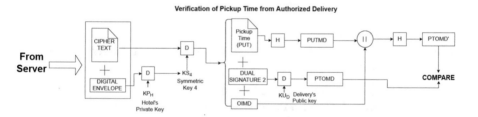

Fig. 9. Decryption at hotel side

Phase 9

In the final stage, the customer receives a triple signature packet from the server. Customer then gets the payment acknowledgement by decrypting the P_{Ack} cipher text using KS_2 got by opening the digital envelope using Customer's Private Key (KP_C). Also the Customer gets the delivery time (DT) for its order by decrypting DT cipher text using KS_3. It validates the delivery details by creating a new DOMD' and comparing it with the received DOMD. And if everything is validated then the customer is satisfied that the order received is from proper hotel by proper delivery person (Fig. 10).

Thus this paper proposes on creating an overall transparent system for the customer which will secure all these transactions using En-SPDP System.

At Customer

**Extraction of Triple Signature Packet Recieved from Server
to verify the Delivery Details & to Get Payment Acknowledgement**

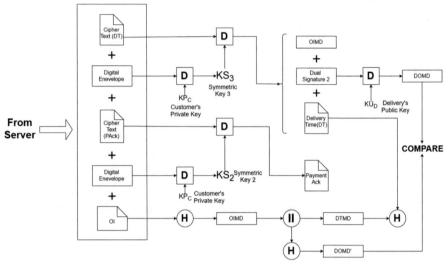

Fig. 10. Decryption at customer side

Pool Delivery Mechanism

Delivery system mainly focuses on pooling of food orders. The delivery system uses a formula proposed below:

X = Decision Factor for giving delivery order
T(L$_{curr}$, L$_{hotel}$) = time taken to reach Hotel from current location
S = Speed of the vehicle
L$_{curr}$ = current location (a set of longitude and latitude)
L$_{hotel}$ = hotel location (a set of longitude and latitude)
L$_{cust}$ = customer location (a set of longitude and latitude)
D = delay due to traffic
pd = preparation delay
T(pd) = a minimum window period required to make food
T(L$_{hotel}$, L$_{cust}$) = time taken to reach customer location from Hotel
P(t; μ) = probability distribution function of getting order between hotel to customer location

$$P(t; \mu) = (e^{-\mu})(\mu^t)/t! \qquad (1)$$

for t $\leq$ 0 and μ is average number of order recorder per unit time

$$dlon = \text{longitude2} - \text{longitude1} \qquad (2)$$

$$dlat = \text{latitude2} - \text{latitude1} \qquad (3)$$

$$a = \left(\sin\left(\frac{dlat}{2}\right)\right)^2 + \cos(\text{lat1}) * \cos(\text{lat2}) * \left(\sin\left(\frac{dlon}{2}\right)\right)^2 \quad (4)$$

$$c = 2 * \text{atan2}(\text{sqrt}(a), \text{sqrt}(1-a)) \quad (5)$$

$$d = R * c(\text{where R is the radius of the Earth}) \quad (6)$$

$$T = d/S; (\text{time required to reach from one location to another}) \quad (7)$$

$$X = \text{Max}(T(L_{\text{curr}}, L_{\text{hotel}}) + D, T(\text{pd})) + T(L_{\text{hotel}}, L_{\text{cust}}) + D + \int_{t=L_{hotel}}^{L_{curr}} P(t; \mu)dt \quad (8)$$

If that X is less than a threshold value then he will be given orders. That value can depend upon locations and traffic conditions.

Equation (8) will give nice chances of pooling and will ensure that the delivery person will not get new order delivery if his current delivery gets late because of this new order deliver.

5 Advantages

- Over the current system proposed system saves up resources like petrol, paper and time;
- Secondly no information is stored at server and complete secrecy is kept;
- More efficient delivery network is implemented which has a upper hand on existing system;
- All the principles of security like confidentiality, integrity, non-repudiation, and authentication is achieved using proposed Enhance Security protocol.

6 Conclusion

The proposed system has a safe and secure transaction protocol which is a customized version of SET protocol. The proposed paper improves data secrecy, privacy and ensures safe transactions. The end user order experience is improved as well as the delivery mechanism is changed by using pooling methods saving resources giving more efficiency and better output.

References

1. Bhotvawala, M.A., Balihallimath, H., Bidichandani, N., Khond, M.P.: Growth of food tech: a comparative study of aggregator food delivery services in India. In: Proceedings of the 2016 International Conference on Industrial Engineering and Operations Management Detroit, Michigan, USA, 23–25 September 2016

2. Murdoch, S.J., Anderson, R.: Verified by visa and mastercard securecode: or, how not to design authentication. In: Murdoch, S.J., Anderson, R. (eds.) Financial Cryptography and Data Security 2010, 25–28 January 2010
3. Ismaili, H.E., Houmani, H., Madroumi, H.: A secure electronic transaction payment protocol design and implementation. Int. J. Adv. Comput. Sci. Appl. (IJACSA) **5**(5), 172–180 (2014)
4. Ahamad, S.S., Sastry, V.N., Udgata, S.B.: Enhanced mobile SET protocol with formal verification. In: 2012 Third International Conference on Computer and Communication Technology (ICCCT), 23–25 November 2012. doi:10.1109/ICCCT.2012.65, http://ieeexplore.ieee.org/document/6394714
5. Stallings, W.: Cryptography and Network Security, 3rd edn.
6. Deshpande, A.: Zomato - market and consumer analysis. Proc. Int. J. Adv. Sci. Res. Eng. Trends **1**, Issue Sept 6, 146–152 (2016). ISSN (Online) 2456-0774
7. Cheng, H.: Privacy protection based on secure electronic transaction protocol in e-commerce. In: Advanced Research on Computer Science and Information Engineering. Communications in Computer and Information Science, CCIS, vol. 153, pp. 449–453 (2011)

Internet of Emotions: Emotion Management Using Affective Computing

Vinayak Pachalag$^{(\boxtimes)}$ and Akshay Malhotra

Symbiosis Institute of Technology,
Symbiosis International University, Pune, India
info@pvinayak.com, akshay9864@yahoo.com

Abstract. The many advantages of increase in Human Machine Interaction are obvious but it has also led to issues such as emotional imbalance, depression, reduction in interpersonal communication etc. Internet of Emotions can be broadly categorized as internet based technologies which aim to mitigate these problems and facilitate better Human to Human interaction in real world. IoE can be defined as an ecosystem where emotion packets travel via internet to manage user's real time experience. We propose a system which will detect emotional state of the user, categorize it and actuate outer net elements to manage the emotion of the user. Detailed algorithm is given which includes use of the passive sensors, smartphone, big data analytics and machine learning. The framework is further explained with example of stress management. The proposed system based on affective computing will play a vital role in development of products and platforms which emphasises user involvement.

Keywords: Emotion · Internet of Things · Algorithm · Affective computing · System · Design · Sensors · Stress

1 Introduction

WHO fact sheet on depression published in April 2016 states that almost 350 million people are suffering from depression. It emphasises over usage of technology as one of the major reasons for this phenomenon [1]. Tremendous speed of development and usefulness of technology has created a situation where we spend a large amount of time with various computational devices. This situation is contradictory to the basic psychological need of humans to communicate with another humans. This has resulted in emotional imbalance as well as depression [2]. Emergence of Augmented Reality, Virtual Reality, and Artificial Intelligence will make the situation much worse by increasing Human Machine Interaction exponentially and hereby, significantly reducing the need for Human Interaction.

It is the need of the hour to develop technologies which will help a person manage his/her emotions while he is interacting with these machines in a way that enhances his experience in the real world. Rosalind Picard initiated basic work in this field by introducing the term affective computing which measures physical response to quantify emotions [3]. Affective computing is the study and development of systems and devices that can recognize, interpret, process, and simulate human affects. It is an

© Springer International Publishing AG 2018
S.C. Satapathy and A. Joshi (eds.), *Information and Communication
Technology for Intelligent Systems (ICTIS 2017) - Volume 2*, Smart Innovation,
Systems and Technologies 84, DOI 10.1007/978-3-319-63645-0_63

interdisciplinary field spanning computer science, psychology, and cognitive science [4]. Many other researchers mainly from MIT Media Lab and Cambridge University have devised methodologies to detect human emotions. Kaliouby and Robinson made a significant contribution in detecting emotions via facial recognition by using head gestures and captured facial expressions in frontal view [5] whereas Guan et al. present methodologies to detect emotion using speech analysis using techniques like maximum likelihood, K nearest neighbour for audio analysis. Decision level modalities and feature level integration is used to obtain superior results [6, 7].

Mobile, especially smartphones, have become such a necessary and ubiquitous part of human life, that their presence and use is taken for granted and they do not intrude in or effect the personal behaviour of the user. Rachury et al. use this characteristic in their project 'Emotion Sense' to detect emotions using mobile as a passive sensor and Gaussian Mixture models. Their system has an accuracy level above 80% when tested on eighteen samples [8]. Increased respiration, elevated screen temperatures, cortisol level in the body are few physiological parameters which suggests change in emotion and can be measured using wearable sensors [9].

A lot of emphasis is currently being given to the area of giving emotions to technology i.e. making machines more humane by giving emotions to robots, AR devices which react as per emotional input etc. [10–12]. Science and Technology has made tremendous progress in building neural abilities in machine. We can use same building block of technology to manipulate human emotions in real world using emotion tracking and actuating real world situations remotely to manage emotions. Currently, there is very less focus in this direction of the research.

We propose a system which advances the work currently done in the field by Emotion detection by collaborating all methodologies of detection which will lead towards better results. The system has a unique proposition of using technology for management of human emotion in real world using Big Data Analysis.

Section 2 describes the detailed system which can detect, analyze and modify emotions with the help of individual's digital data. Section 3 provides the prototype of the system for the particular case of stress management whereas Sect. 4 comprises of the algorithm for stress management with minute details. Section 5 elaborates various applications of the framework and Sect. 6 concludes the concept explained in the paper.

2 Methodology

The general system can be categorized in three stages namely Emotion Detection alias Quantified Self, programmed decision and implementation. Work of each stage can be summarized in the following diagram (Fig. 1).

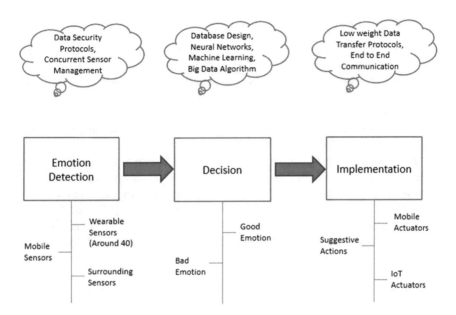

Fig. 1. Architecture of the system which elaborates three major stages of the system as well as technologies which are proposed to be utilized in each stage

Stage wise functioning of the system is as follows.

2.1 Detection Stage

This stage will include all the sensors which are available in the user's smartphone and can provide indicators regarding emotion. All other sensors which the user possesses as a wearable sensor in various formats i.e. fitness band etc. will also be enlisted and enumerated. Data from non-intrusive passive sensors data will be accepted if the user is surrounded by such sensors at any point of time. User location, call and messaging details, people he/she is surrounded with will also be stored after user's prior permission which will help in giving better results.

All the sensors will be identified with a unique id attached to the user and data will be continuously stored in the cloud storage database as well as mobile database. Mobile sensors will provide data in interrupt mode whereas wearable and surrounding sensors will provide data in polling mode. The data will travel in encrypted format and then be decrypted at receiver side to maintain privacy and secrecy of the user. Sensors in surrounding area may get data of multiple users. Classification of data will take place before storage of the data. Data which is not used in decision will be dumped after certain pre-determined intervals.

Use of multiple sensors will reduce the possibility of error and detection accuracy level will be high as the result of one type of sensor will be confirmed by another type of sensor in the analysis stage as explained in the next section. This methodology is proposed to avoid false positive results.

2.2   Analysis Stage

Algorithms will run at the database to categorize the incoming data in to six universal emotions namely happiness, sadness, surprise, fear, disgust, and anger. These emotions were selected on the basis of work done by Hung and Kim [13]. The next stage will depend on the number of inputs available at the given point of time i.e. whether mobile data is available, mobile as well as wearable data is available and/or surrounding sensors are available. The algorithm also takes into account the number of sensors available under each classification mentioned above.

A decision number will be calculated based on the data received from various sensors in decision stage. Permutation and combination will be used as number of sensors available may change frequently. This number will be compared with prede-fined threshold which are calculated from the dataset of WHO [14]. These thresholds will act as a supervised learning dataset for the system. An emotion will be classified as good or bad emotion based on its position with respect to threshold values. If the current state does not fall into any category, then it will be termed as the neutral state of the user.

These results will be periodically informed to the user and feedback will be obtained. Threshold will be adjusted as per the user feedback. This correction will mature the system. Neural Networks are proposed to be used for this update. Self-Organized Incremental Neural Network (SOINN) [15] will be the best fit Neural Network architecture for the database due to their ability to adapt, quick response time and robustness Weights will be assigned to each input parameter as per the impact they are generating on the result which will generate multivariable equations.

Data about user outer net i.e. surroundings will also be mapped to each of the emotion and cross correlation is found for each surrounding data with user emotion to decide which activity makes the most impact on user emotion. We will be able to quantify the impact of various surroundings to the emotional state of the user.

2.3   Implementation Stage

This will only trigger only when the detection number will go beyond action threshold

$$\text{Action threshold} > \text{Decision threshold} \tag{1}$$

Implementation will follow two basic rules

a. Generate Preventive action if the user is in a state of negative emotion.
b. Activate possible actuators which may shift users mood to Good condition, these actuators are decided from the correlation of user emotional state and user activity which is available in the database.

User Actuators are already predefined in the system. Actuators are the devices, software programs and IoT receivers to which the user has given the permission to change the state automatically. User may allow full access for few devices as well as suggestive access to few. Suggestive actuators will be the ones where system cannot change this state autonomously but can only do so after the user's permission every time. For example, user may allow system to automatically switch off TV or native mobile application but will not allow blocking messaging function without prior permission.

Implementation will also depend upon the number of actuators which are actually available with the user at a given point of time as all the devices may not be up and running all the time. In the scenario, the system will take decision as per the priority of the each actuator which is decided in advance using machine learning.

3   Stress Management Using IoE

As per the WHO Report, 'Work Organization and Stress' published in 2003, stress is the prime factor for reduction in productivity which in turn generates financial losses [16]. We are proposing a stress management system based on above framework and algorithm which can cater to the said problem. This prototype will illustrate detailed executive information about the IoE Algorithm (Fig. 2).

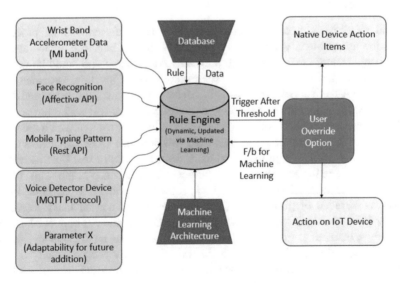

Fig. 2. Stress management system architecture with flow of data and details of execution with exact elements used in each stage and protocols for communication

We have chosen four elements as sensors in detection phase namely Wrist Band, Face Recognition, Key stroke data and voice level detection based on the work of Mozos et al. for stress detection [17]. Those four elements are explained in detail below.

3.1   Wrist Band

Wrist band contains a heart rate meter. For a person in stress, the heart rate, arterial tension and testosterone production increase, cortisol (the stress hormone) decreases, and the left hemisphere of the brain becomes more stimulated [18]. Abnormal heart rate beyond certain threshold is indication of a person being in stress. We propose use of wrist band for heart rate monitoring as its heart rate accuracy ranges between 71 and 84% [19].

Heart rate monitoring is based on optical sensing technology. These optical sensors throw light on the body, particularly the wrist, and measure artillery pumping rate by illuminating user capillaries with a LED. A sensor adjacent to the light measures the frequency at which the user's blood pumps (aka user heart rate).and the user gets a BPM (beats per minute) reading. Stress and heart rate have a linear relationship [20] which we propose to utilize to correlate change in heart rate with change in stress.

3.2   Face Detection

To avoid repetitive collection of database which is a time consuming process, we propose the use of already available Application Program Interface, which are generally available free of cost for research purposes. An Android app could be developed using available API to detect facial recognition. Various institutes provides training data and API. Affectiva, developed by MIT Media Lab is one of the most used API call for face detection [21]. Facial recognition of emotion is done using 22 reference points.

We can use expression invariant points efficiently in the system which are nose tip, root, nostrils, inner and outer eye corner, head yaw and roll. These parameters are scale invariant [22]. A tracker is proposed to track these markers in real time as well as in offline mode, where the basic measuring parameter is face bunch graph. This method is selected as it shows satisfactory results irrespective of the ethnicity of the subject, the presence or absence of glasses etc. [22]. Dimensions of normal head positions are generally 55° upward pitch and 55 roll, generally a combination of both occurs. Anchor points are generated using 2D projection of approximate model and distances are calculated with reference to that. Anchor point is generally chosen exactly below the middle of the lower lip and fixed in the initial frame. Polar distance between each of the two mouth corners is also calculated at the beginning (Fig. 3).

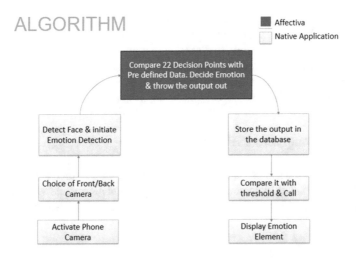

Fig. 3. Facial detection algorithm indicating the functioning of application and coordination with API

3.3 Key Stroke Data

It is expected that a person will make more mistakes in typing when in stress. Hence, Key stroke is proposed as a mobile sensor to detect stress level of the person. This data should include typing speed, number of mistake per interval of time, amount of spell check usage, frequency of using backspace button etc. It has been observed by Po-Ming Lee [23] that variables like keystroke duration and latency forms a pattern with the emotional state of the user. These patterns will be calculated at the database.

Sentiment Analysis of the sentence is generally used to determine the polarity of the sentence i.e. the extreme characteristics. Sentiment analysis can also be termed as Opinion Mining Sentiment Analysis of the text can also be included as a part of the key stroke data which will help to distinguish between stress and joy as patterns mentioned above are almost similar in both emotions [23]. Exact analysis of the each word typed by the user can suggest his/her emotional state. Sentences are parsed in the system word wise and polarity of the each word is calculated and stored. Predefined wordlists are used for Supervised Machine learning and weight of each word is calculated after comparison. We propose to use AFINN word list due to its comprehensiveness [24]. Scaling and distribution of the sentence depends on the weighing mechanism used. Mechanism which may be used are Support Vector Method or Naive Bayes Method [25]. Final result of the sentence will provide exact emotion of the user which should be used in deciding further action.

3.4 Voice Level Detection

Voice Level Detection can be categorized as the surrounding sensor. Anger is one of the indicators of the stress. An individual's voice level is generally higher than normal in a state of anger. Hence, detecting changes in voice level can be one of the measures to detect stress level. Weight of this input should be kept low as anger and excitement both will have change in sound level which might not be possible to distinguish. The input i.e. sound intensity is prone to error due to number of factors such that distance of user from the sensor, noise level, sensitivity of the sensor used etc. We recommend that VLD at best be used as a confirmatory input signal along with other parameters and should not be used as a standalone input in the system.

4   Stress Management Algorithm

Data processing and decision making will take place at the central database. Database is in synchronization with detection as well as implementation stage via two way communication. The algorithm which will run in the database of the system is as described in the Fig. 4.

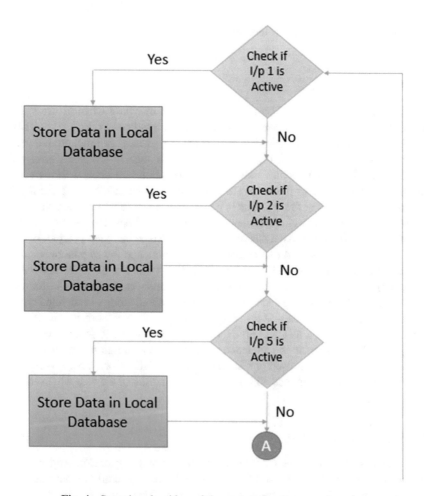

Fig. 4. Stepwise algorithm of the system for stress management

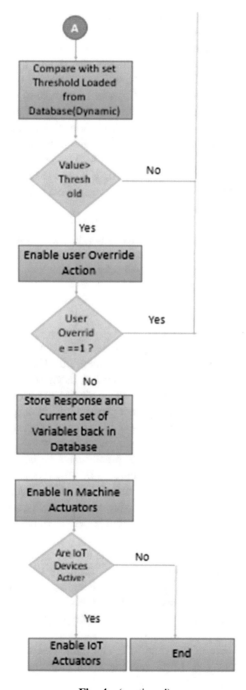

Fig. 4. (continued)

The system will continuously check if the first input is active. The above algorithm is meant for sequential processing of data. In parallel processing, we can keep system polling for all 5 inputs coming from various devices at a time. In the above system, the inputs are checked one by one and the data is stored into the local database. This process will be followed till all the inputs are sensed and the data will be stored in a single table. Once all the data is received, it will be compared with the set thresholds which are pre-determined. Threshold can be determined separately for each user after a short survey or after usage for first few interactions. Alternatively, we may add threshold as per our need for various permutation and combination of logics/elements. The system will go ahead only when the output is above certain threshold, else it will rotate in the same loop. System will ask the user if he/she is agreeing with the output or not i.e. User Override Option for the user to give consent to system i.e. it will be in a suggestive mode before performing any task.

Action points can be within device or via other peripheral device though the final control remains in the mobile phone only. Finally, the actions will be saved as a feedback for the improvement of the results. Unsupervised Machine Learning algorithm will be continuously running in the background.

5   Applications

The proposed system has applications in various domains namely healthcare, business process management, human resource management etc. Quantitative analysis of the data will help governing bodies such as corporation, business management to prioritize their spending on mental health as well as to devise employee based tailor made solutions for each employee to increase their productivity. Workplace happiness index, city happiness index can be devised in line with national happiness index which could make impact on formulation and implementation of policy.

Regular data analysis will help medical practitioners to predict a candidate's stress at a very early stage which can then be curated by generation/simulation of interpersonal activities which can switch his/her mood than medication. Correlation between user's emotional state and his/her physiological activities will give detail insights about the cause of certain diseases and their possible remedies which will be tailor made for the said user.

It can be determined that what process in the certain organizations generates stress to the employee from the emotion analysis of majority of the workers working in different processes of the businesses. Necessary changes and modifications can be done and the impact can be tested immediately from the system. Quick pilot projects with immediate feedback will reduce cost and time to implement a positive change in the organization.

An emotional bio data of the candidate, complimented with his/her regular portfolio, will help recruiters' select right candidates for the job and provide personalized training based on the traits of the candidate.

The framework can act as an open source architecture on which various platforms for each vertical in the industry can be developed.

Privacy issues and data security are two major issues which will decide the usage of the proposed system. Dedicated algorithms for ethical practices as well as data security can be one of the verticals of development.

6 Conclusion

We have proposed a system which advances the current methodologies used for emotion detection and takes into consideration affective, cognitive and contextual factors. These detection methodologies are part of a framework proposed in the paper which will manage and manipulate emotions of the user for his/her betterment. It has unique characteristics of mapping user behavior and interaction in real world as well as computing devices with his/her current state of emotion. This correlation will be used to manage various devices surrounded by the user in a way to attain better state of mind of the user.

The methodology as well as technologies which are proposed to be used in this solution are mentioned in detail. Big Data Analysis as well as low weight transmission protocols play a key role in the solution.

Practical implementation of the framework is elaborated with the case of stress management. This example elaborates the intelligent use of certain sensors complimented with exhaustive algorithms to help user mitigate impact of certain emotion. Similar frameworks can be built for other emotions, in this case, anger.

The last section of the paper gives a comprehensive update on the possible applications of the architecture in various interdisciplinary fields which reiterates the importance of the proposed architecture and timeliness of the system.

References

1. World Health Organization. http://www.who.int/en/
2. Marcus, M., Yasamy, M.T., Ommeren, M.V.V., Chisholm, D., Saxena, S.: Depression: a global public health concern. PsycEXTRA Dataset
3. Picard, R.W.: Affective computing for future agents. In: Cooperative Information Agents IV: The Future of Information Agents in Cyberspace. Lecture Notes in Computer Science, p. 14 (2000)
4. Calvo, R., D'mello, S., Gratch, J., Kappas, A., Picard, R.W.: The promise of affective computing. In: The Oxford Handbook of Affective Computing (2015)
5. Kaliouby, R.E., Robinson, P.: Generalization of a vision-based computational model of mind-reading. In: Affective Computing and Intelligent Interaction. Lecture Notes in Computer Science, pp. 582–589 (2005)
6. Bhatti, M., Wang, Y., Guan, L.: A neural network approach for human emotion recognition in speech. In: Proceedings of the 2004 IEEE International Symposium on Circuits and Systems (IEEE Cat. No. 04CH37512)
7. Sebe, N., Cohen, I., Gevers, T., Huang, T.: Emotion recognition based on joint visual and audio cues. In: Proceedings of the 18th International Conference on Pattern Recognition (ICPR) (2006)

8. Rachuri, K.K., Musolesi, M., Mascolo, C., Rentfrow, P.J., Longworth, C., Aucinas, A.: EmotionSense. In: Proceedings of the 12th ACM International Conference on Ubiquitous Computing, Ubicomp (2010)

9. Hernandez, J., Mcduff, D., Infante, C., Maes, P., Quigley, K., Picard, R.: Wearable ESM. In: Proceedings of the 18th International Conference on Human–Computer Interaction with Mobile Devices and Services, MobileHCI (2016)

10. Breazeal, C.: Emotion and sociable humanoid robots. Int. J. Hum Comput Stud. **59**, 119–155 (2003)

11. Hollinger, G., Georgiev, Y., Manfredi, A., Maxwell, B., Pezzementi, Z., Mitchell, B.: Design of a social mobile robot using emotion-based decision mechanisms. In: Proceedings of the 2006 IEEE/RSJ International Conference on Intelligent Robots and Systems (2006)

12. Shah, J., Wiken, J., Williams, B., Breazeal, C.: Improved human–robot team performance using chaski, a human-inspired plan execution system. In: Proceedings of the 6th International Conference on Human–Robot Interaction, HRI (2011)

13. Project Deidre (II). https://people.ece.cornell.edu/land/OldStudentProjects/cs490-95to96/HJKIM/deidreII.html

14. Depression. http://www.who.int/mediacentre/factsheets/fs369/en/

15. Furao, S., Ogura, T., Hasegawa, O.: An enhanced self-organizing incremental neural network for online unsupervised learning. Neural Netw. **20**, 893–903 (2007)

16. Publications A-M. http://www.who.int/occupational_health/publications/en/

17. Mozos, O.M., Sandulescu, V., Andrews, S., Ellis, D., Bellotto, N., Dobrescu, R., Ferrandez, J.M.: Stress detection using wearable physiological and sociometric sensors. Int. J. Neural Syst. **27**, 1650041 (2017)

18. Hjortskov, N., Riss, N.D., Blangsted, A.K., Fallentin, N., Lundberg, U., Sogaard, K.: The effect of mental stress on heart rate variability and blood pressure during computer work. Eur. J. Appl. Physiol. **92**, 84–89 (2004)

19. Rettner, R.: How accurate are fitness tracker heart rate monitors? http://www.livescience.com/56459-fitness-tracker-heart-rate-monitors-accuracy.html

20. Stress and Heart Health. http://www.heart.org/HEARTORG/HealthyLiving/StressManagement/HowDesStressAffectYou/Stress-and-HeartHealth_UCM_437370_Article.jsp#.WHxHdP197IU

21. Emotion Recognition Software and Analysis. http://www.affectiva.com/

22. Kaliouby, R.E., Robinson, P.: Mind reading machines: automated inference of cognitive mental states from video. In: Proceedings of the 2004 IEEE International Conference on Systems, Man and Cybernetics (IEEE Cat. No. 04CH37583)

23. Lee, P.-M., Tsui, W.-H., Hsiao, T.-C.: The influence of emotion on keyboard typing: an experimental study using auditory stimuli. PLoS One **10**(1), 81 (2015)

24. Hansen, L.K., Arvidsson, A., Nielsen, F.A., Colleoni, E., Etter, M.: Good friends, bad news: affect and virality in twitter. In: Future Information Technology. Communications in Computer and Information Science, pp. 34–43 (2011)

25. Joachims, T.: Text classification. In: Learning to Classify Text Using Support Vector Machines, pp. 7–33 (2002)

Monitoring of QoS in MANET Based Real Time Applications

Mamata Rath[1(✉)] and Binod Kumar Pattanayak[2]

[1] Deptarment of Information Technology, C.V. Raman College of Engineering,
Bhubaneswar, India
mamata.rath200@gmail.com
[2] Department of Computer Science and Engineering, Siksha 'O' Anusandhan University,
Bhubaneswar, Odisha, India
binodpattanayak@soauniversity.ac.in

Abstract. Specialty of Mobile Adhoc Networks (MANET) is that in current mobile technology it is the most promising and highly developed elucidation due to its remarkable performance in offering network connectivity even in very radical situations of adversity where there is maximum chance of link failure and more necessity of quick set up of networks. The elemental routing procedure in a MANET involves facilitating unremitting communication in the network system between two mobile stations during required period of time and the basic factor being selection of the most suitable forwarding node to proceed the real time packets from source towards destination in such a way that the optimization of the network can be achieved by maximum utilization of available resources. Considering that real time applications are one of the most challenging issue in MANET, due to transportation of high volume of data including audio, video, images, animation and graphics, this paper presents a monitoring approach for checking the Quality of Service (QoS) conditions during competent routing using the concept of Mobile Agents. An intelligent mobile agent is designed in the proposed QoS platform that monitors and controls the QoS processing tasks using longest critical path method at the forwarding node to select it as the best option out of all neighbor nodes. Simulation results shows higher packet delivery ratio and comparatively reduced bandwidth consumption overhead when it is compared with other similar approaches.

Keywords: MANET · QoS · Real time applications · Mobile agent · AODV protocol

1 Introduction

In the discussed research paper, a Quality-of-Service (QoS) based Real time Platform in Mobile Adhoc Network has been designed that uses the Mobile Agents [20] for implementing the QoS parameters in the channel on demand and it checks the availability of network resources properly at every intermediate node to the concerned QoS path. Mobile Agent Based QoS platform [28] has been designed which imposes priority

© Springer International Publishing AG 2018
S.C. Satapathy and A. Joshi (eds.), *Information and Communication
Technology for Intelligent Systems (ICTIS 2017) - Volume 2,* Smart Innovation,
Systems and Technologies 84, DOI 10.1007/978-3-319-63645-0_64

on real time applications such as multimedia transmission with high loads of audio, video, graphics and animated features [10]. During the mobility of nodes, mobile agents [13] of the registered applications also move from one station to the next forwarding station in order to check consistent availability of the required resources and to continue that QoS flow irrespective of the constraints like frequent link failure, topology reconfiguration, faster mobility of mobile stations, limited residual battery power etc. Task scheduling according to their order of precedence during Quality of Service [21] maintenance in a specialized network called Mobile Adhoc Network is a challenging task, because there are many decision taking operations which are exceptional, non-repetitive and very much transparent in view of their range, purpose and time constraints. The interdependent tasks performed by the mobile agent in mobile adhoc network for satisfaction of quality of service at every intermediate node can be scheduled properly using Critical Path Method (CPM) for network optimization.

2   Literature Review

A systematic survey has been carried out in [1] about the protocols supporting real time applications in Mobile Adhoc Network. To implement security measures in a MANET, paper [2] proposes a well developed mobile agent based architecture for Intrusion Detection System with unique features for handling risks. Limited residual battery power has always been a constraint for mobile nodes during transmission. An improved and optimized energy efficient algorithm has been designed for real time applications in MANET [3]. There are many challenging issues regarding the design of protocols which are to be considered for better network performance. A detailed survey has been carried out based on multiple issues in [4]. To handles sensitive issues of MANET communication between different layers of TCP/IP Protocols is very much essential. A Cross-layer based MANET protocol has been designed and implemented in [5] for military applications. Real time data transmission [10] involves loads of data to be transmitted securely over the network without any interruption. Behavior of Routing protocols is necessary for further research in this area to handle video streaming and jitter related problem in MANET [6]. Adhoc On demand Distance Vector (AODV) Protocol is one of the leading reactive protocols of Mobile Adhoc Network (MANET) described in RFC 3561 [16].

3   The Proposed System

This proposed model is an improvement of our previous research work [9] where an energy efficient MANET routing protocol has been designed. In the proposed model additional Quality of Service (QoS) provision has been incorporated with regular monitoring system in the designed Real Time platform [9] using cross layer communication system between well developed Power and delay optimized AODV Routing Protocol [5] and Improved Channel Access Scheme in MAC Layer [10]. We have implemented Mobile Agent for Quality of Service [22] checking at every intermediate node during packet transmission in order to check if the QoS criteria are satisfied in the channel during routing. To achieve this, the Mobile Agent visits from node to node and every time it validates the activities presented in Table 2. All the activities are part of QoS checking system and they are prioritized with their probable time of processing depending on many network parameters which are stored in the mini database [3] of the concerned node's routing engine module that gets updated by the static agent of the concerned station. Using the heuristic method [18] probable processing time for every activity is computed for each intermediate node by the mobile agent. Then the total possible delay calculation is carried out using the longest Critical Path Method(CPM) [19], if the total delay estimated is below the constraint range of QoS criteria, then that node will be selected for the next forwarding node. This processing logic has been implemented with incorporation of Mobile agent in our proposed system. Using Critical Path Method [19] we have calculated the minor difference in delay during Quality Of Service criteria checking by the Mobile Agent at the intermediate node. A Critical Path Method helps to schedule the tasks related to QoS management and minor delay on execution of any task will cause delay in entire real time transmission resulting failure of deadline. For our proposed QoS Checking System, the critical path is the minimum longest path that covers all the activities throughout the process. Critical Path can be determined using 'Earliest Time' forward pass followed by a 'Latest Time'. The 'Earliest Starting Time' of every task is attached with every activity. It refers to the longest time of any path from a former point. The 'Latest Starting Time' of every activity is also attached with the events. It denotes the longest path from any succeeding event. The Critical Path is the path along which the earliest time and latest time are the same for all events, and the early start time plus activity time for any activity equals the early start time of the next activity. Table 2 describes the activities performed by our proposed MA Based QoS Delay Checking system. For Delay calculation at various stages of QoS Checking, in larger Mobile Adhoc Networks, CPM methodology [19] offers an efficient solution provided that the probable duration of tasks are possibly known. Stages in the said approach include calculating the latest and earliest schedules to completely avoid delay in the discussed QoS Checking system. Therefore for every activity related to QoS Checking, the following values are to be calculated (Table 1).

Table 1. Time durations in CPM

Time	Description
EST	Earliest time in which a task if no delay occurs.
EFT	Earliest Finish time in which a task can be over if there is no delay in the system
LST	Latest Start Time in which the task can start when there is no delay in the completion of any task
LFT	Latest Finish Time in which a task can finish if there is no delay in the system

From starting node to the end node a forward pass value is used to calculate EST and EFT. A backward pass value is used to calculate the LST and LFT. To calculate the critical path and the total delay in QoS Calculation, this improved statistics involves the stages for estimation of start time and end time for each sub-activity under QoS system. Fundamentally the basic approach for forward pass value denotes that all immediate predecessors has to be completed before another task may start.

As shown above, as per the Quality of Service Policy, a mobile agent executes the tasks mentioned in Table 2 at every intermediate node before forwarding packets along the next suitable station towards destination. For simulation purpose in Table 2 task completion time using heuristic method [18] for a single scenario has been considered as an example. Selection of best suitable station for next hop is determined according to the optimized route selection algorithm [5]. In our previous work, in [9, 10], the QoS mechanism of the proposed real time platform has been described where Activity 1 refers to Real time flow detection which is executed by a special module of PDO AODV [9] that filters the real time packets from the channel.

Table 2. Activities in the MA based QoS system

Activity no.	Activity	Completion time (ms)
1	Real time flow detection	6
2	Analysis of QoS criteria	2
3	Calculate available bandwidth	3
4	Calculate energy consumption	2
5	Bandwidth reservation	4
6	Energy reservation	1
7	Prioritized real time policy	1
8	Delay constraint control	6
9	Uniform jitter variation	3
10	Total delay estimation	1
11	Selection of next node	1

4   Simulation Parameters

Table 3 shows the simulation parameters used in the proposed system. Bandwidth can be distinctly denoted as the transmission of number of bits per second in a channel [21]. For example when the capacity of a network is said to be 100 Mbps, then it can send 100 Mega bits per second.

Figure 1 illustrates Bandwidth consumption in our approach of CPM-MA Qos Routing which increases with increase in node mobility. As the number of mobile nodes increases, there is more processing task at the intermediate nodes to forward, send and receive packets along with handling the challenges of route failure, reduced battery power and decreasing transmission capacity. When compared with other similar Mobile Agent based improved strategy to handle QoS, our proposed strategy consumes comparatively less bandwidth and efficiently transmits the real time data packets.

Table 3. Simulation and results

Parameters	Values
Area	1000 × 1000
Mac	802.11 e
Radio range	250 m
Simulation time	50 s
Routing protocol	CPM-MA routing
Traffic source	CBR & VBR
Packet size	512 byte
Mobility model	Random way point
Speed	5–10 m/s
Pause time	5 s
Interface queue type	Drop tail
Network interface type	Wireless Phy
Simulator	Ns2.35

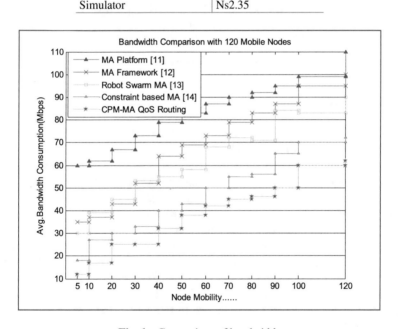

Fig. 1. Comparison of bandwidth

In CPM-MA based QoS Routing, the Packet Delivery Ratio (PDR) increases with increase of node mobility in MANET because in random mobility model, the nodes can move randomly and their speed may vary. Due to enhanced mechanism employed for route failure, load balance path and improved channel access method in our approach. From Fig. 2 we can observe that the proposed framework imposes better PDR in comparison to other similar approaches.

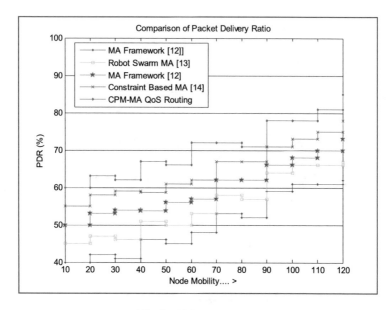

Fig. 2. PDR comparison

5   Conclusion

This research piece presents an intelligence based Mobile Agent architecture that controls the Quality of Service activities in a real time platform that has been designed in our previous work. Here, the basic function of mobile agent is to calculate optimized delay estimation of QoS activities at the neighboring nodes before forwarding the real time packets specifically multimedia transmission towards the load balanced path so that the challenging issues of real time transmission such as video streaming problem, variation of jitter and deadline mission etc. can be handled effectively. The Minimum but longest possible critical path method has been employed at the next hop station of a concerned node by the mobile agent with assistance of the static agent. Advantage of using mobile agent in MANET include modular development of mobile agent with maximum cohesive functionality. Simulation evaluation and results shows better performance of the proposed QoS platform in terms of better throughput, optimized delay and improved network lifetime.

References

1. Rath, M., Pattanayak, B.: A methodical survey on real time applications in MANETS: focussing on key issues. In: 2014 International Conference on High Performance Computing and Applications (ICHPCA), 22–24 December 2014, pp. 1–5 (2014)
2. Pattanayak, B., Rath, M.A.: Mobile agent based intrusion detection system architecture for mobile ad hoc networks. J. Comput. Sci. **10**, 970–975 (2014)

3. Rath, M., Pattanayak, B.K.: Energy competent routing protocol design in MANET with real time application provision. Int. J. Bus. Data Commun. Netw. **11**(1), 50–60 (2015)
4. Rath, M., Pattanayak, B., Rout, U.: Study of challenges and survey on protocols based on multiple issues in mobile adhoc network. Int. J. Appl. Eng. Res. **2015**(10), 36042–36045 (2015)
5. Rath, M., Pattanayak, B.K., Pati, B.: Energy efficient MANET protocol using cross layer design for military applications. Def Sci. J. **66**(2) (2016)
6. Rath, M., Pattanayak, B.K., Pati, B.: MANET routing protocols on network layer in realtime scenario. Int. J. Cybern. Inf. **5**(1) (2016)
7. Rath, M., Pattanayak, B., Pati, B.: Comparative analysis of AODV routing protocols based on network performance parameters in mobile Adhoc Networks. In: Foundations and Frontiers in Computer, Communication and Electrical Engineering, pp. 461–466. CRC Press, Taylor & Francis Group (2016). ISBN 978-1-138-02877-7
8. Rath, M., Pattanayak, B., Pati, B.: A contemporary survey and analysis of delay and power based routing protocols in MANET. ARPN J. Eng. Appl. Sci. **11**(1) (2016)
9. Rath, M., Pattanayak, B.K., Pati, B.: Inter-layer communication based QoS platform for real time multimedia applications in MANET. In: IEEE WiSPNET, Chennai, pp. 613–617 (2016)
10. Rath, M. Pati, B., Pattanayak, B.K.: Cross layer based QoS platform for multimedia transmission in MANET. In: 3rd International Conference on Electronics and Communication Systems, ICECS 2016, Coimbatore, pp. 3089–3093 (2016)
11. Park, J., Youn, H., Lee, E.: A mobile agent platform for supporting ad-hoc network environment. Int. J. Grid Distrib. Comput. **1**(1), 9–16(8) (2008)
12. Kakkasageri, M.S., Manvi, S.S., Goudar, B.M.: An agent based framework to find stable routes in mobile ad hoc networks (MANETs). In: 2008 IEEE Region 10 Conference, TENCON 2008, Hyderabad, pp. 1–6 (2008). doi:10.1109/TENCON.2008.4766704
13. Li, Y., Du, S., Kim, Y.: Robot swarm MANET cooperation based on mobile agent. In: 2009 IEEE International Conference on Robotics and Biomimetics (ROBIO), Guilin, pp. 1416–1420 (2009). doi:10.1109/ROBIO.2009.5420763
14. Bindhu, R.: Mobile agent based routing protocol with security for MANET. Int. J. Appl. Eng. Res. **1**(1), 92–101 (2010)
15. Sharma, V., Bhadauria, S.: Mobile agent based congestion control using AODV routing protocol technique for mobile adhoc network. Int. J. Wirel. Mobile Netw. **4**(2) (2012)
16. Perkins, C.E., Royer, E.M.: The ad hoc on-demand distance-vector protocol (AODV). In: Perkins, C.E. (ed.) Ad Hoc Networking, pp. 173–219. Addison-Wesley, Reading (2001)
17. Network Simulator. http://www.isi.edu/nsnam/ns. Accessed 20 Apr 2016
18. http://www2.cs.uni-paderborn.de/cs/ag-monien/PERSONAL/SENSEN/Scheduling/icpp/node4.html. Accessed 20 Apr 2016
19. Hendrickson, C., Tung, A.: 11. Advanced Scheduling Techniques. Project Management for Construction, 2.2 edn. Prentice Hall (2008). cmu.edu. ISBN 0-13-731266-0
20. Bhati, P., Chauhan, R., Rathy, R.K., Khurana, R.: An efficient agent-based AODV routing protocol in MANET. Int. J. Comput. Sci. Eng. **3**(7), 2668–2673 (2011)
21. Forouzan, B.A.: Data Communications and Networking. McGraw-Hill Companies Inc., New York (2007)
22. Lablod, H.: Wireless Adhoc and Sensor Networks. Wiley, New York (2008)
23. Wang, X., Li, J.: Improving the network lifetime of MANETs through cooperative MAC protocol design. IEEE Trans. Parallel Distrib. Syst. **26**(4) (2015)
24. Dougherty, D., Robbins, A.: Sed & Awk, 2nd edn. O'Reilly Media (1997)

25. Biradar, R.C., Manvi, S.S.: Agent-driven backbone ring-based reliable multicast routing in mobile ad hoc networks. IET Commun. **5**(2), 172–189 (2011). doi:10.1049/iet-com. 2010.0002
26. Karia, D.C., Godbole, V.V.: New approach for routing in mobile ad-hoc networks based on ant colony optimisation with global positioning system. IET Netw. **2**(3), 171–180 (2013). doi: 10.1049/iet-net.2012.0087
27. Oh, H.: A tree-based approach for the Internet connectivity of mobile ad hoc networks. J. Commun. Netw. **11**(3), 261–270 (2009). doi:10.1109/JCN.2009.6391330
28. Derr, K., Manic, M.: Adaptive control parameters for dispersal of multi-agent mobile ad hoc network (MANET) swarms. IEEE Trans. Ind. Inf. **9**(4), 1900–1911 (2013)
29. Bridges, C.P., Vladimirova, T.: Towards an agent computing platform for distributed computing on satellites. IEEE Trans. Aerosp. Electron. Syst. **49**(3), 1824–1838 (2013)
30. Umamaheswari, S., Radhamani, G.: Enhanced ANTSEC framework with cluster based cooperative caching in mobile ad hoc networks. J. Commun. Netw. **17**(1), 40–46 (2015)

Use of an Adaptive Agent in Virtual Collaborative Learning Environment

Nilay Vaidya[1(✉)] and Priti Sajja[2]

[1] Uka Tarsadia University, Bardoli, Gujarat, India
vaidyanilay@gmail.com
[2] Sardar Patel University, Vallabh Vidyanagar, Gujarat, India
priti@pritisajja.info

Abstract. Personalized education in an ICT enabled environment is a contemporary matter today. The adaptation of education to diverse types of student is becoming a big challenge. Proposing personalized learning in the digital era, we obtained a new dimension called an agent based adaptive learning. This has led to adopt knowledge management practices that provide innovation in knowledge clustering for active learning. In the ICT enabled learning, the learners are geographically scattered. Agent-Oriented System simulates the teaching-learning pedagogy by sensing the environment, listing several traits, observing the user behavior, finding pattern and learning pace of the learner. This leads towards imparting intelligence to an agent that helps both learner and the tutor to build a smart teaching-learning environment. This paper proposes an agent which works as a middleware which uses semi supervised learning mechanism with both forward and backward chaining for the inference to impart intelligence in learning environment.

Keywords: Agent-oriented system · Backward chaining · Forward chaining · Knowledge management · Semi-supervised learning

1 Introduction and Current Status

An e-Learning may be truly effective when the pedagogy, offering and measurement technique aligns with the learner's ability and proficiency skill. The talent, skills and pedagogical view must be treated in an effective way to make the e-learning environment an interactive and informative environment. The Virtual Collaborative Learning Environment (VCLE), which is an agent-based system, wherein the system infers the knowledge from the available knowledge. The overlap in learner, domain and technological knowledge in a collaborative learning environment encourages the agent based learning environment to connect and capture the knowledge which is implicitly available either in the domain database or with the learner (Fig. 1).

© Springer International Publishing AG 2018
S.C. Satapathy and A. Joshi (eds.), *Information and Communication
Technology for Intelligent Systems (ICTIS 2017) - Volume 2*, Smart Innovation,
Systems and Technologies 84, DOI 10.1007/978-3-319-63645-0_65

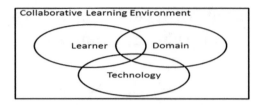

Fig. 1. Knowledge overlapping in VCLE

In this Help and Learn environment, a centered view of knowledge is not just residing in learners mid but in the interaction between leaner and between learner and content stored. In this open context, other Internet resources are being utilized which makes a rich learning environment.

An agent is an encapsulated computer application designed for the specific domain which is capable of performing actions that helps in making decisions in the prescribed environment in order to meet the design objective. In the progressive advancement in the field of computer science together with the multiple disciplines, universal environment, no barrier bars in physical location, and diversity of community with no or more learning obsession, makes researcher interested in proposing such a module that helps in filling the gap between the learner as well as the tutor. Agent which is a middleware that works as a bridge between an interface and database or knowledge base that allows and offers the personalized interface to the user in different environment. In the area of an artificial intelligence, an agent stands for in an environment which makes its own conclusions and acts in the environment through actuators. Here the approach proposed is an "Agent-Oriented" design environment in the problem domain in the teaching learning paradigm.

Agent-Oriented Computing environment is a mixture of Computer Science, Artificial Intelligence and Object Oriented paradigm which imparts intelligence to the system. In the given domain area the agent oriented computing environment exploited as a mechanism to solve the complex problems and for developing a system which is called an intelligent system (Fig. 2).

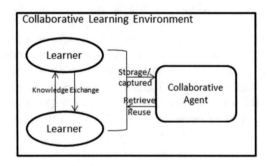

Fig. 2. Knowledge management in VCLE

2   Basic Need for an Adaptive Agent

Individuals learn differently. An adaptive learning is used to cope with individual differences in the environment. More and more researches are focusing on applying the recent technologies to personalize the pedagogy and content. Aptitude Treatment Interaction (ATI) theory states that is the choice of the instruction method matches with the aptitude of learner, the outcome will be higher. Learning Oriented Assessment proposes the iterative and recursive process till the competency scale is achieved. Learning oriented assessment focuses not only on assessment but also on how instruction and feedback can be tailored. Learning systems provides feature that identify the need of learner, tailor the content and provides the personalized approach to the learner. As per the literature regarding varying requirements of the learner as well as the tutor, we put forward the adaptive agent functionalities for teaching-learning environment that not only helps both learners as well as tutors. Model of learning that suits the individual needs of students is based on the new paradigm of personalization of education environment by considering the learning ability, current knowledge, and learning pattern

3   Characteristics of Agent-Oriented Design

Agent-mediated knowledge management comes as a solution a dynamic collaborative learning environment. It exhibits flexibility in behavior, providing the knowledge with "reactive" and "pro-active" manner. It serves as a tailored assistant, which maintains the user's log and behavior. General tendency and acceptance of smart electronic gadgets such as laptops, handheld PDAs, smart phones etc. are becoming so much popular in daily routine that not only offers fundamental platforms, but also raises issues on how to take benefits of these gadgets to enjoy this global computing environment. With increasing popularity and wide acceptance of such environment, the demands are also increasing that satisfies the users in the universal environment. This changes the access method and interaction pattern of the users in the environment and demands a pattern that recognizes the need, and pattern of the user and accordingly fetches the facilities and offers it to the users. As a solution to this, an agent-oriented approach provides these requirements with key advantage of autonomy, collaborative environment, and also an intelligent approach to the user as well as the admin of the domain which is teaching learning environment.

Various components are being identified in the teaching learning environment that promotes use of the agent based system. This components needs to interact with each other as data may be distributed and vary. System may need to interact with different other external agents for the effective services. System may need to monitor the user and user activities in the system; based on this, system may need to update the user details or other content available. Typical features that help the framework and works as a middleware which implements artificial intelligent model in the given domain.

Some of the silent characteristics of an Agent are:

(1) Robust degree of independence
(2) Reactivity i.e. responding in a timely manner to the change in environment and deciding when and what to act
(3) Proactivity i.e. the agent should respond in best possible action that are anticipated to happen
(4) Communicability i.e. should support facility to communicate with other agents
(5) Elastic behavior
(6) Multi-threaded control
(7) Supports concurrent and distributed approach and many more.

Some of the silent properties of an Agent for the teaching learning environment are:

(1) Flexibility
(2) Mobility
(3) Adaptivity
(4) Rationality
(5) Collaborative ability and many more.

4   Agent and Teaching Learning Environment

In the era of digitization, users are moving from manual to digital world, classroom teaching is more or less being supported by online resources. This increases the competition amongst the service providers in providing more and more advanced feature that makes the environment more user friendly and user centric. Here the environment encapsulates the agent which is not a part of it. Teaching learning environment are equivalents that have some properties which are then being used by an agent that generates an output which would be more powerful and useful for the users of the environment. Here users include both the learner as well as the admin.

Virtual environment, which provides platform to the learner that enhances learning skill and also allows the admin/tutor a platform that can identify learners, and learners competitive skill.

Major topological division of agent in a learning environment that should provide cooperative, autonomous and learning environment which provides user friendly interface which makes an environment a "smart environment".

Role of an agent in VCLE is to encourage the flow of knowledge of the learner in which learner feel free to share insights, experiences and know-how. The environment allows learner to perform this by using the blog as well as discussion forums. This gives the learner to be more innovative, thoughtful and competitive.

5   Conceptual Design Architecture

The layered design of an environment:
Mainly the architecture is divided into four layers (Fig. 3):

(a)   An interface layer
(b)   The database layer
(c)   The Internet Resource Layer
(d)   The agent layer (Hidden).

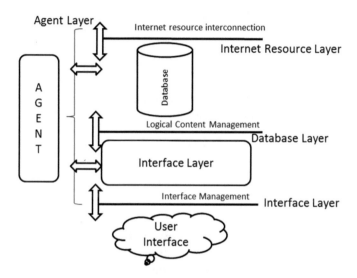

Fig. 3. Layered design of VCLE

5.1   Agent Layer

The agent identifies mainly three parameters: (i) Pedagogical view (ii) peer-learning view (iii) demonstrative view. The agent layer which works side by side in the working portal which has the functionalities which continuously monitors the activities being performed by the user. It monitors every task being performed using an interface layer by the user and accordingly sets the weights and at glance gives a competency scale to the user in the environment. Every offered features by an interface layer; agent works independently and captures the details of the user with its weights and on the action as well as the reaction of the agent allows the database to bifurcate the content.

Agent layer also helps admin to prepare and arrange content, prepare and arrange the question pool, helps managing the evaluation pattern based on average competency skills.

This agent uses combination of supervised as well as unsupervised learning to make it more user-friendly and flexible.

The feedback and the responses of the learners in the environment are tracked and being utilized after classifying learners and then being applied the regression method on it. So one can say the supervised learning is used by the agent in identifying the learner behavior in the system. Here the all types of responses and feedbacks are captured and then on the basis of the previous responses by other learners or by same learner the agent applies the regression mechanism to read the behavioural aspect of the learner in the environment.

The semi supervised approach is also taken into consideration by the system in which the situations and cases are identified where the resources are utilized and feedback or returns are not filled or ignored. The system uses the active learning approach in which the responses are filled and then utilized in a way that the decision theory can be applied on the data and can be used in the best manner.

As the system uses all the parts of Expert System architecture to make the system a "Smart System".

(a)   Knowledge Base (consists of IF-THEN statements i.e. rules)
(b)   Database (stores current situations)
(c)   Inference engine (which is Knowledge Base+Database)
    (i)   Forward chaining (data driven)
    (ii)   Backward chaining
(d)   Explanation and reasoning mechanism
(e)   User Interface.

6   Conclusion

On the basis of the preferences, personal characteristic and qualities, learner will be presented with learning content which suits them the most. One of the basic building blocks of adaptive learning is the storage of study materials. In order to provide tailored learning to each type of learner, content must be prepared in many different variants, in various forms. For different virtual learner, different learning content are proposed. The system also evaluates the learner's effectiveness and efficiency continuously. The simulation of teaching – learning cycle via an agent records the whole learning process. It identifies the correction of pedagogical aspect of the learner in real process and then makes necessary corrections in offerings and evaluation parameters. Several factor triggered interest in the agent for Virtual Collaborative Learning Environment (VCLE) which are innovation, globalization, easier navigation, quicker interaction between learners and between learner and tutor. The major functioning of the agent here is to trigger out the learner behaviour and pattern that infers the competency level. This helps tutor in identifying the group competency and accordingly the portal sets the evaluation pattern for the learner.

References

Vaidya, N., Sajja, P.: Ubiquitous computing agent to determine effective content and recommend curriculum in collaborative learning environment. Recent Trends Comput. Commun. Technol. 1(1), 96–99 (2016a)

Vaidya, N., Sajja, P., Gor, D.: Evaluating learning effectiveness in collaborative learning environment by using multi-objective grey situation decision making theory. Int. J. Sci. Eng. Res. **6**(8), 41–45 (2015)

Vaidya, N., Sajja, P.: Learner ontological model for intelligent virtual collaborative learning environment. Int. J. Comput. Eng. Res. **6**(2), 20–23 (2016b)

Vaidya, N., Sajja, P.: Feasibility study for assessing readiness to a collaborative e-learning environment. Int. J. Res. Eng. IT Soc. Sci. **6**(6), 10–16 (2016c)

Ashabi, A., Khalil, S.: Agent-oriented software engineering characteristics and paradigm. J. Multidiscip. Eng. Sci. Technol. **1**(4) (2014). ISSN 3159-0040

Jennings, N., Wooldridge, M.: Agent-Oriented Software Engineering

Chugh, R.: Knowledge sharing with enhanced learning and development opportunities. In: IEEE International Conference on Information Retrieval and Knowledge Management 2012, Kuala Lumpur, Malaysia, pp. 100–104 (2012)

Vafaee, P., Suzuki, Y., Pelzl, E.: How Aptitude-Treatment-interaction Studies Can Benefit Learning-Oriented Assessment, University of Maryland

Reed, Z.: Collaborative Learning in the Classroom. Center for Faculty Excellence, United States Military Academy, West Point (2014)

Kostolányová, K., Šarmanová, J.: Use of adaptive study material in education in elearning environment. Electron. J. e-Learn. **12**(2), 172–263 (2014)

Ahmad, N., Tasir, Z. et al.: Automatic detection of learning styles in learning management systems by using literature-based method. In: 13th International Educational Technology Conference, vol. 103, pp. 181–189 (2013)

The Recent Trends, Techniques and Methods of Cloud Security

Ravinder Yadav[(✉)], Aravind Kilaru, and Shambhavi Kumari

Manipal University Jaipur, Jaipur, India
ravinder11@gmail.com, kilaru.arvind@gmail.com,
shambhavikumari@muj.manipal.edu

Abstract. Security is one of the most important challenge that users face in migrating to cloud services. Users will lose the direct control over their data and they need to trust the cloud service provider (CSP) for security and access control. Therefore, this raises concern about security and privacy of data. Due to vulnerability of data being stored at these super data center, leads to more introspection about security in cloud computing. Thus, the security is the more prominent concern that need to be addressed to provide safe services. Many researches are going on for improving the security of the cloud storage systems. This survey paper focus on the trusted storage cloud computing proposed by different researchers.

Keywords: Cloud computing · RDS · TDS · ACPM · SNM

1 Introduction

In the recent history, Cloud computing seems to be most emerging technology, where users stores their data at remote location; super data centers and based on their needs they can utilize the services. Cloud computing is a shared/multi-tenant environment with enormous pools of various resources; these resources can be requested/auto provisioned as per need and can be made available via the Internet. List of advantages of the cloud computing are: need base auto provisioned servicing, abundant network resources access, geographic boundaries independent resource pooling, rapid provisioning of resources, pricing as per utilization [1]. Majorly cloud offers three categories of services i.e. (1) SaaS, (2) PaaS (3) IaaS. SaaS; a business model, hosted and managed by providers. PaaS provides a platform. The user can develop their application using any programming language; these platforms have been initially build to support the software. Among researcher, IaaS is more popular where an organization outsources the equipment/hardware/networking resources required to support operations. The provider is the owner of equipment/hardware/networking and responsible to manage these resources. The main concern of the providers is to ensure the secure storage for the sensitive data stored within cloud. Provider has to provide the trusted database to users.

© Springer International Publishing AG 2018
S.C. Satapathy and A. Joshi (eds.), *Information and Communication Technology for Intelligent Systems (ICTIS 2017) - Volume 2*, Smart Innovation, Systems and Technologies 84, DOI 10.1007/978-3-319-63645-0_66

2   Trusted Storage Cloud Computing

Users store their data on the cloud. These data centers should be trustworthy. Client should not face the loss of data while processing and migrating to data centers. Trusted storage refers as a highly secured database where Meta data can be stored. Only authorized or trusted users can access theses data. Security is the major issues in public cloud. To avoid security problems one can use hybrid cloud where sensitive data can store in private cloud and rest in public. Drawback of the hybrid cloud is to more expensive and also managing the database increases the complexity. To overcome these issues some researchers has proposed various methods, which can enhance the security of the storage computing.

3   Towards a Secure Distributed Storage System

A secured network storage system [2] architecture by attaching a new layer between the application and storage nodes which interacts with users and management layers. Three layer architecture improves the trustworthiness of storage computing using cryptographic method and also manage the access control policies. Three layered architecture basically has three layers: application layer, storage management layer and storage resource layer. Application layer, which lies on top of architecture, maintains the client's information i.e. private key and cryptographic information. Storage layer, the lies on the bottom of architecture. Storage layer is the collection of databases and responsible for storing the data. Management layer; the middle layer in the architecture and perform the core functionality. It receives the data access request from the user application layer and process the request. It is also listing the availability and effectiveness of the storage nodes. Two main component of the management layer namely: a) the storage management component (SMC) b) the secure database system. SMC is composed of various components:

(1) Interface to Virtual storage system: it provide an interface to the user to access like a traditional file system.
(2) ACPM (access-control and policy-management): it help in administration and mandate the policies related to security.
(3) TDS (threshold-distribution and scheduling): it manages the distribution of cipher data and perform fragmentation as per the given threshold.
(4) SNM (storage-node management): it determine the assembling and estimation capabilities of storage nodes.
(5) DC (Data integrity-checking module): it ensure authentication and integrity of data when returned from cloud storage.

4   Trustful Cloud Based Sealed Storage

Trustful cloud based sealed storage environment [3] grant the ability for cloud users to remotely authenticate the cloud environment. Remote delegation services (RDS) prevent the directly revelation of the cloud resources. Functionality of Trustful Cloud Based Sealed Storage are built within the storage management and storage resource layer. Remote attestation delegation services (RDS) is added to the management layers. Each node have a virtual machine monitor (VMM), trusted boot ability, remote authentication module (RAM) and sealed storage module (SSM) [1]. VMM prevents the administrator or third party admin to access cloud users' private data. RDS uses the services of resources management software, RAM & SSM to perform the following tasks [1–5]:

(1)   On the launch of VM, it authenticate the integrity and return the result to users.
(2)   Keep an eye on integrity changes of the node and ensure the sealed storage as per the integrity constraints of users.

Cloud provider can control the RDS, which leads to the access of the cloud private information, to avoid this problem RDS has its confidentiality or secrecy so that provider cannot inspect inside the space and has a control features so that provider cannot modify what resides in the space. As per the research [8–10], secrecy the primary concern for the business computing. The cloud service provider implement the security with the help of two concepts i.e. RAM and SSM in the VMM. RAM take care the integrity and report to RDS and SSM perform the functionality of data encryption and decryption. TSSC perform the remote attestation on the user behalf remotely sure it is un-tampered and trustworthy. As shown in Fig. 1, TSSC also seal the user data with hash function.

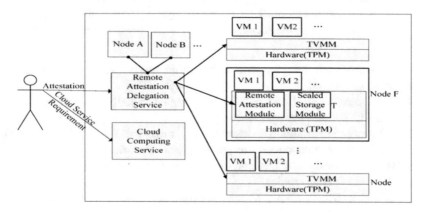

Fig. 1. TSSC architecture

VMM registration: RAM is responsible for recording the integrity information to the RDS, before the availability of cloud backend node. The platform configuration Register (PCM) record the integrity information for the node, which is accessed during RAM operation.

RDS remote authentication: the cloud users certify Integrity of RDS. It generate a pair of asymmetric key for the user, and accept designation key as user's the private key. The cloud users can perform the verification of the integrity by using backend nodes, which is stored in RDS, if integrity information comply the security requirement. They sign the nodes integrity information and specify the location of data to seal, and return them to RDS. The user also attach the hash value of his VM image.

Nodes Attestation: after the registration of node, RDS acquires the node's PCR value signed by AIK and compare with the value signed by cloud user to ensure that node has not been compromised, which prevent it from unsafe state. If the node is in unsafe state then will not acquire the resources. If the node's integrity complies with cloud user's expectations, the RDS will generate a sealed storage symmetric key for the user and send it to SSM in that node.

Sealed storage: during the deployment of the virtual machine, SSM encrypt and decrypt the sensitive data using sealed storage symmetric key. On the expected level of integrity of virtual machine deployed in the cloud, SSM encrypt the specified data dynamically on store operation and decrypted on read operation.

5   Trust Modeling in Cloud Computing

In this research paper, the trust value of four different layer (physical layer, PaaS, IaaS, and SaaS) is evaluated to compare the trustworthiness of the model [4]. Overall trust value will be show the trustworthiness of the model, which is evaluated on the cloud platform using fuzzy logic. Trust value for IaaS can be computed as multiplicative sum of the trust levels of individual physical hosts and resources.

$$T(\text{IaaS layer}) = T(R1) * T(R2) * \ldots * T(Rn)$$

The zones and firewalls secure PaaS layer. To obtain the trust of the PaaS layer, researcher uses defense capability i.e.: Intrusion Detection System (IDS) engine. IDS engine is runs separately within a VM whose trust may be evaluated by:

$$T(\text{IDS}, \text{app1}) = \max\{T(\text{IDS}, \text{app1}), T(\text{IDS}, \text{app2}), \ldots, T(\text{IDS}, \text{appn})\}$$

where $T(\text{IDS}, \text{appn})$ is the acronyms of trust measure given by the IDS engine for any specific application. IaaS and PaaS trust relationship may be computed as:

$$T(\text{IaaS}, \text{PaaS1}) = \max\{T(\text{IaaS}, \text{PaaS1}), T(\text{IaaS}, \text{PaaS2}), \ldots, T(\text{IaaS}, \text{PaaSn})\}$$

For each PaaS it can be modelled as

$$T(\text{PaaSx}) = T(\text{IaaS}, \text{PaaSx}) + T(\text{IDSx}, \text{appx})$$

SSL or SSH is used to secure the SaaS layer in cloud implementation. SSH, connections represent direct-trust between two parties. While SSL requires a third party certification authority.

$$T(SSH) = \{0 \text{ if identity - trusts exists but is not matched}\}$$
$$\{1 \text{ if identity - trust is matched}\}$$

$$T(SSL) = \{0 \text{ if identity - trust is NOT verified}\}$$
$$\{0 \text{ if identity - trust is revoked}\}$$
$$\{1 \text{ if identity - trust is verified}\}$$

The SaaS layer trust may be modelled as

$$T(SaaS) = (T(conn) + T(DP))/2$$

PaaS and a SaaS application trust can be computed as:

$$T(SaaS) = T(PaaS, SaaS)*((T(conn) + T(DP))/2)$$

6   Towards Publicly Auditable Secure Cloud Data Storage Services

Keeping the data secure and ensuring the privacy is the biggest challenge for the cloud service providers. Cloud service providers (CSP) has authority to access the customers data and for some time CSP can be unfaithful towards cloud customers such as discarding data that has not been or is rarely accessed [5], hiding the fact of data loss incidents to maintain a reputation [6]. Traditional cryptographic method does not guarantee for data integrity as well as for data availability. To ensure the security of data integrity, architecture has been suggested which enable overtly auditable storage services [4], where cloud users can recourse to an external agency for audit (TPA) to validate.

Third party auditing provides method that ensure the trust relationship between the data owner and service provider. In this, the owner delegates TPA to audit their data, which is on cloud servers. The TPA has to edit the cloud data without being able to store the local copy of data.

7   TruXy: Trusted Storage Cloud for Scientific Workflows

Cloud computing has evolved with business computing needs but it is also accepted by scientific communities [8]. In the bioinformatics domain, Taverna [9] and Galaxy [10, 11]. are two most popular workflow systems. With the passage of time, Galaxy has gained the popularity and lead to development of more of packages to integrated with cloud services such as BioBlend [12], CloudMap [13] and Galaxy Cloud [14]. TruXy [15] is a unique app which does not only provides the security to the data also ensure availability, confidentiality, integrity, and data sharing on public Clouds. TruXy uses algorithm proposed in TrustStore [16], where raw data divided into subparts and assigned to different service entities: CSP and KeyS. Raw data can restored only if all the separate parts are assembled by a single entity. TruXy architecture is shown in the Fig. 2.

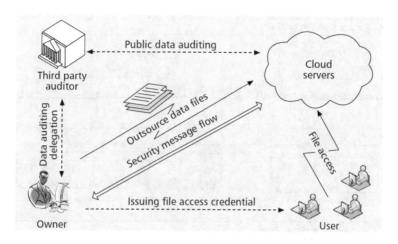

Fig. 2. Cloud Data Storage Service Architecture

TruXy provides security based on the following assumptions:

(1) Secure environment for client computer so that sensitive data can be secured.
(2) User perform encryption and decryption method to protect their data right after being created.
(3) There is a mutual trust between (CSP) and the TruXy Middleware Layer (TML) and no complication between them.

They protects data from both internal and external attacks. In TruXy, KeyS generate and distribute strong encryption key for cloud customer while all the operation of encryption and decryption are performed by TruXy applications deployed at the client side. TruXy utilities the cryptographic method to secure the sensitive data stored in public cloud. To protect the Data integrity, TruXy adopted the concept of DIaaS from TrustStore that defines a dedicated service – the Integrity Management Service (IntS) - to deal with the problem of integrity violations of data stored on the Cloud [13]. TruXy opt the principles of RAID (Redundant Array of Independent Disks) [14] and RAIN (Reliable Array of Independent Nodes) [15] to address the availability of the data. TruXy also provides data access control mechanism for confidential data sharing at store level. To ensure the sharing is authorized, each users must takes the permission to access the data from the KeyS (Fig. 3).

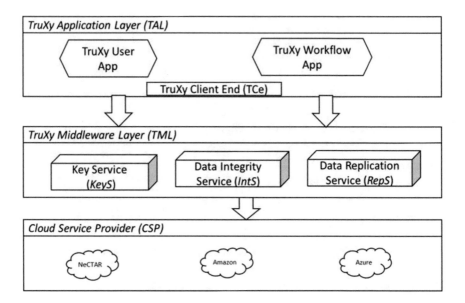

Fig. 3. Architectural Design of TruXy

8   Conclusion

Cloud computing has broken the geographical boundaries and created a new trend of providing the services through internet. It introduces a platform where user can be charged only based on their use. For a short span of utility, it gives freedom to user from setting up entire environment by investing huge amount of cost. It also opens the gateway of using shared pool of services with dynamic provisioning as per demand. This introduces the risk of security. Since threat can be injected from internal or external agencies, so it difficult to ensure the security. Cloud service provider uses different mechanism to ensure the security like TPA etc. but not a single method guarantee for complete security. Security is an ongoing work. Some of the securing techniques have been presented in the paper while much more is expected from the scientist and researcher.

References

1. Mell, P., Grance, T.: Draft NIST Working Definition of Cloud Computing (2009). http://csrc.nist.gov/groups/SNS/cloud-computing/index.html
2. Zhang, M., Zhang, D., Xian, H., Chen, C., Feng D.: Towards a secure distributed storage system
3. Juels, J., Kaliski, B.S.: PORs: proofs of retrievability for large files. In: Proceedings of ACM CCS 2007, October 2007, pp. 584–597 (2007)
4. Ateniese, G. et al.: Provable data possession at untrusted stores. In: Proceedings of ACM CCS 2007, October 2007, pp. 598–609 (2007)

5. Hoffa, C., Mehta, G., Freeman, T., Deelman, E., Keahey, K., Berriman, B. et al.: On the use of cloud computing for scientific workflows. In: IEEE Fourth International Conference on eScience, eScience 2008, pp. 640–645 (2008)
6. Oinn, T., Addis, M., Ferris, J., Marvin, D., Senger, M., Greenwood, M., et al.: Taverna: a tool for the composition and enactment of bioinformatics workflows. Bioinformatics **20**, 3045–3054 (2004)
7. Goecks, J., Nekrutenko, A., Taylor, J.: Galaxy: a comprehensive approach for supporting accessible, reproducible, and transparent computational research in the life sciences. Genome Biol. **11**, R86 (2010)
8. Afgan, E., Goecks, J., Baker, D., Coraor, N., Nekrutenko, A., Taylor, J.: Galaxy: a gateway to tools in e-science. In: Guide to e-Science, pp. 145–177. Springer, Heidelberg (2011)
9. Sloggett, C., Goonasekera, N., Afgan, E.: BioBlend: automating pipeline analyses within Galaxy and CloudMan. Bioinformatics **29**, 1685–1686 (2013)
10. Minevich, G., Park, D.S., Blankenberg, D., Poole, R.J., Hobert, O.: CloudMap: a cloud-based pipeline for analysis of mutant genome sequences. Genetics **192**, 1249–1269 (2012)
11. Afgan, E., Baker, D., Coraor, N., Goto, H., Paul, I.M., Makova, K.D., et al.: Harnessing cloud computing with Galaxy Cloud. Nat. Biotechnol. **29**, 972–974 (2011)
12. Yao, J., Chen, S., Nepal, S., Levy, D., Zic, J.: Truststore: making amazon s3 trustworthy with services composition. In: Proceedings of the 2010 10th IEEE/ACM International Conference on Cluster, Cloud and Grid Computing, pp. 600–605 (2010)
13. Nepal, S., Chen, S., Yao, J., Thilakanathan, D.: DIaaS: data integrity as a service in the cloud. In: IEEE International Conference on Cloud Computing (CLOUD), pp. 308–315 (2011)
14. Long, D.D., Montague, B.R., Cabrera, L.-F.: Swift/RAID: a distributed RAID system. In: Computing Systems (1994)
15. Bohossian, V., Fan, C.C., LeMahieu, P.S., Riedel, M.D., Xu, L., Bruck, J.: Computing in the rain: a reliable array of independent nodes. IEEE Trans. Parallel Distrib. Syst. **12**, 99–114 (2001)

A Description of Software Reusable Component Based on the Behavior

Swathy Vodithala[1(✉)] and Suresh Pabboju[2]

[1] KITS, Warangal, India
swathyvodithala@gmail.com
[2] CBIT, Hyderabad, India
plpsuresh@gmail.com

Abstract. Component Based Software Engineering (CBSE) is one of the specialized methodologies in the process of developing the software. The motivation behind the CBSE is software reuse that is using off the shelf components. The software component in reuse may be a design, document or a piece of code. The components considered in this paper are source codes, in particular functions. In order to have an efficient reuse of the components, they are to be described effectively and clustered so as to retrieve the components with a minimum effort. This paper shows the description of a software component based on the facets. An important facet in the description of software component is the behavior. The behavior is extracted from the comment lines present in the code, later these comments are converted to first order predicate.

Keywords: CBSE · Code reuse · Behavior · First order predicate · Parser · Translator

1 Introduction

Software engineering (SE) is the application of a systematic and disciplined approach i.e., engineering to the design, development, operation, and maintenance of the software. Component-based software engineering (CBSE) (also known as component-based development (CBD)) focuses on the development of applications based on software components, so that the applications are easy to maintain and extend. A software component is a software element that conforms to a component model and can be independently deployed and composed without modification according to a composition standard. Software element contains a sequence of abstract program statements that describe the computation to be performed by a machine [2, 8].

The software engineering has been widely used area nowadays, since it accepts even the other technical areas into it which strengthens this stream. The major area in software engineering is the software reuse. Software reusability is the use of existing software or software knowledge to construct new software. Reuse may be on design pattern, program elements or tools [5, 6]. The software repository must be developed in an efficient manner in order to reduce searching time to reuse the component. There are three major areas in software engineering which has to be focused when considering the components for software reuse [1]. These are described as

© Springer International Publishing AG 2018
S.C. Satapathy and A. Joshi (eds.), *Information and Communication Technology for Intelligent Systems (ICTIS 2017) - Volume 2*, Smart Innovation, Systems and Technologies 84, DOI 10.1007/978-3-319-63645-0_67

(a) Describing the components wanted.
(b) Classifying the components needed.
(c) Finding the appropriate components.

This paper is structured in the following manner. The next section in the paper describes about the related work. Section 3 describes about the proposed work. Section 4 describes the implementation results and the further sections explains conclusion, future scope and references.

2   Related Work

The related work explains the faceted description and translating a sentence to first order predicate in order to form the basis for the understanding of the proposed work.

2.1   Faceted Description

Faceted scheme was first proposed by Prieto-Diaz and Freeman in 1987 [3, 4] that depends on facets which are extracted by the experts to describe the features about the components. Ruben Prieto-Diaz has proposed a faceted scheme that uses six facets.
The functional facets are: Function, Objects and Medium.
The environmental facets are: System type, Functional area, setting [1].
Similarly there is an approach named attribute value classification scheme that uses a set of attributes to classify a component [6]. There is no limit on the choice of having multiple number of attributes. The difference between attribute valued and faceted is that facet considers only limited terms or facets where each facet is described by some facet values [10, 11]. The faceted classification is used in many real world software companies like IBM when compared to the attribute value scheme [7].

2.2   Conversion from Statement to Predicate Calculus

The conversion of English statement to first order predicate can be done by following two stages

1. English statement to parse tree
2. Translate parse tree to first order predicate

The parse tree checks syntax, grammar etc. for the sentences while parsing and removes the unnecessary/stop words. Then the translator with the specified rules will infer semantics and from these will create predicate calculus (Fig. 1).
Clocksin parse tree is an example considered to explain the translation of a sentence. When parsing is done, it takes the structure as a tree. The interior nodes define the different parts of speech and leaf nodes define the words of the language [9].

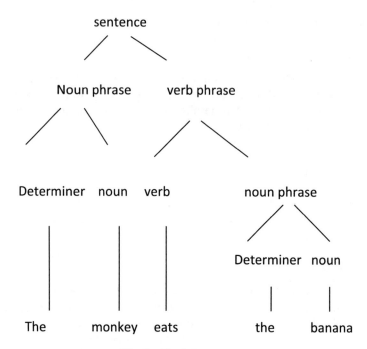

Fig. 1. Clocksin parse tree

3 Proposed Work

A software component is an independent part of the software system that is having complete functionalities. The CBSE focuses on e four levels of component reuse.

- Design level products
- Analysis level products
- Code level components (libraries, modules, procedures, subroutines etc.)
- Entire applications

The proposed work considers the level of component which is to be reused is the code level component (libraries, modules, procedures, subroutines) i.e., in particular subroutines or functions. There are many techniques which describes the software component based on the informal description but in the proposed work the component is described in a formal method by using the behavior.

The major issue for retrieving the components in software reuse mainly relies on the designing of software repository. The prerequisite for the designing of repository i.e., clustering of the components is the description of the component. The description of software component plays a vital role in measuring the efficiency of the software reuse.

There are many code reuse repositories available in the real world which itself cannot be sufficient for the efficient and effective reuse. These source codes or components must be described in a such way that they can be further processed for a specific application.

The proposed works explains the description of software component based on facets. Among the facets which are considered for component description the important facet is behavior of the component. Generally, each source file consists of the comments as per the standards prescribed by the companies. The behavior of the component is extracted from the comments of the source code file and later these comments are converted to first order predicate logic i.e. describing a code in a formal method. An important point to be noted while considering the comments is that not all the comments are converted to first order predicate but the comments that includes information about the input, output and some other important operations in the code are only converted. Since each line of code has a precondition and post condition it is not possible to consider all pre and post conditions, so we take into the consideration of only few pre and post conditions like input of the function, output of the function and some other important operations in the code of the function.

Example: Consider the component (binary search subroutine)
The binary search function works as follows: it takes an array or list as an input and a key value which is to be found as the output from the list. The prerequisite of the binary search is that the list should be in an sorted order. The list is further divided into two halves such that the merging of two lists gives the original list i.e., no loss of elements must happen. The function code searches the element in both the halves of the list so as to minimize the time.

The code for binary search is taken as a sample to explain the process of the proposed work

```
//alist is the list of  integer elements
//item is the integer element to be found
//alist should be sorted form

def binarySearch(alist, item):
//assigning low and high indices
            first = 0
        last = len(alist)-1
        found = False
    // divides the list into two halves
            // searching in either of the divided arrays
        while first<=last and not found:

                    midpoint = (first + last)//2
            if alist[midpoint] == item:
                found = True
            else:
                if item < alist[midpoint]:
                    last = midpoint-1
                else:
                    first = midpoint+1

            return found
//returns the position and item value found
```

Step 1:

The above source code file, we have the documentation i.e., comments regarding the code. We consider the comments related to the input, output and other important functions or operations performed. They are

- alist is the list of integer elements
- item is the integer element to be found
- alist should be sorted form
- divides the list into two halves
- returns the position and item value found.

Step 2:

The next step is to convert these English statements to first order predicate.

1. English sentence: alist is the list of integer elements
   First order predicate: is(alist,list)
2. English sentence: item is the integer element to be found
   First order predicate: is(item,integer)
3. English sentence: alist should be in sorted form
   First order predicate: sort(alist)
4. English sentence: divides the list into two halves
   First order predicate: equals((alist1.alist2),alist)
5. English sentence: returns the element is found or not
   First order predicate: is_in(item,alist1)
   is_in(item,alist2)

The behavior of the software component of binary search is described as follows:

- is(alist,list)
- is(item,integer)
- sort(alist)
- equals((alist1.alist2),alist)
- is_in(item,alist1)
- is_in(item,alist2)

The component is described by many facets and as described the behavior facet is taking a crucial role for describing the components. The other facets for the software component considered in the proposed work are space complexity, time complexity, operating system and programming language.

These facets for binary search are described as follows:

- Space complexity: $O(n)$
- Time complexity: $O((\log(n)))$
- Programming language: python
- Operating system: Unix

The overall description of the component (binary search) can be done as follows

- Space complexity: O(n)
- Time complexity: O((log(n))
- Programming language: python
- Operating system: Unix
- Behaviour: is(alist, list)
  is(item,integer)
  sort(alist)
  equals((alist1.alist2),alist)
  is_in(item,alist1)
  is_in(item,alist2)

4   Implementation Results

The implementation of the proposed work is done by considering the code reuse repositories. Figure 2 explains about the uploading of a particular component i.e., a code of a function such as addition of two numbers, binary search etc., from the repository and also displays the uploading status.

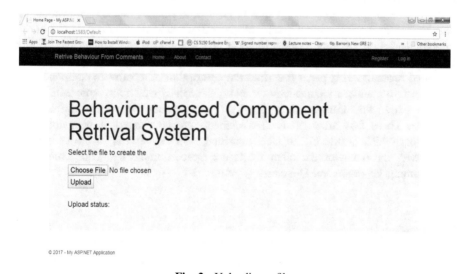

Fig. 2. Uploading a file

Figure 3 extracts the comment lines from the source code file that was uploaded and converts the English sentences into first order predicate.

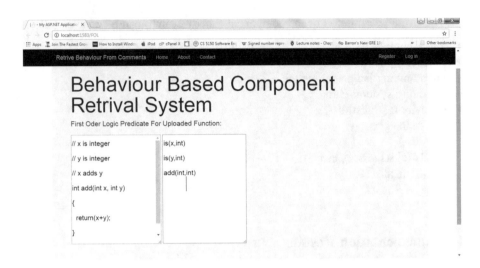

Fig. 3. Conversion to first order predicate

5 Conclusion

There are many techniques in the literature survey to describe the components. The components can be described by using keywords, attributes, facets, enumerated, full text, hyper text etc. The above described methods have its own pros and cons but yet are used based on the applications. Many of the techniques come under the informal description of the software component but the component in the proposed work is described formally. This paper describes the description of a software component in a very effective manner by considering the comment lines which are represented in the English statements. Further, these English sentences are converted to first order predicate. These first order predicate statements are considered as behavior of the component which is one of the facet considered for description of the component. Along with the behavior the other facets are Space complexity, Time complexity, Programming language and Operating system.

References

1. Vodithala, S., Niranjan Reddy, P.: A resolved retrieval technique for software components. IJARCET **1**(4), 653–656 (2012)
2. Kaur, V., Goel, S.: Facets of software component repository. IJCSE. **3**(6), 2473–2476 (2011)
3. Jones, G., Prieto-Diaz, R.: Building and managing software libraries. IEEE 228–236 (1998)
4. Prieto-Diaz, R.: Implementing faceted classification for software reuse. IEEE **34**, 300–304 (1990)
5. Kamalraj, Khannan, R.: Stability based component clustering for designing software reuse repository. Int. J. Comput. Appl. **27**(3), 33–36 (2011)

6. Gupta, A.K., Yadav, V.K., Kumar, S.: Vishal: A robust retrieval scheme for software component reuse. Int. J. Eng. Innov. Technol. **2**(1) (2012). ISSN 2277-3754. ISO 9001:2008
7. Poulin, J.S., Yglyesias, K.P.: Experiences with a faceted classification scheme in a large reusable software library (RSL). IEEE, 1–5 (1993). doi:10.1109/CMPSAC.1993.404220
8. Councill, B., Heineman, G.T.: Definition of a software component and its elements (chapter 1)
9. Harwitz, J.E.: Natural language to predicate calculus, logic module, CSC-173
10. Guru Rao, C.V., Niranjan, P.: An integrated classification scheme for efficient retrieval of components. J. Comput. Sci. **4**(10), 821–825 (2008). ISSN 1549-3636. Science Publications
11. Guru Rao, C.V., Niranjan, P.: A mock-up tool for software component reuses repository. Int. J. Softw. Eng. Appl. **1**(20) (2010)

A Novel Anti-phishing Effectiveness Evaluator Model

Shweta Sankhwar[(⌧)], Dhirendra Pandey, and R.A. Khan

Department of Information Technology,
Babasaheb Bhimrao Ambedkar University, Lucknow, Uttar Pradesh, India
shweta.sank@gmail.com, prof.dhiren@gmail.com,
khanraees@yahoo.com

Abstract. Phishing is a fraudulent way that is used to entice innocent users to a scheming website by employing legitimate looking email and messages for illicit purposes. In this paper, a brainstorming study on vulnerability causing email phishing is determined. These vulnerabilities are spared in three categories on the basis of email structure i.e. Page-content vulnerability, Domain vulnerability and Code-scripting vulnerability. Here, an Anti-Phishing Effectiveness Evaluator Model (APEE Model) is proposed to examine the effectiveness of existing Anti-Phishing Mechanism. The implementation of this Anti-Phishing Effectiveness Evaluator Model (APEE Model) is done with on existing Anti-phishing mechanism. Their effectiveness is evaluated and a result metric is listed. In this research paper, Major finding are reported which could lead the recent researcher to deliver effective Anti-Phishing Solution.

Keywords: Phishing · Email phishing · Web vulnerability · Internet security · Information security · Cyber security

1 Introduction

In the last few years, naïve users are increasingly targeted by phishing attacks as they permit to access information ubiquitously. E-commerce and its convenience are embraced by consumers as well as criminals and thus phishing has made online commerce dubious for internet users and consumers. Phishing is a way of enticing unwary internet users to a fraudulent website by using email and messages that look authentic and execute fraudulent purposes. Phishers employ various attack methods to lure victims and get sensitive information like passwords, usernames or credit card details. They mostly prefer to entice users by sending an email that seems to be sent by a trustworthy organization, banks or ISPs. It becomes difficult for the user to understand whether they are visiting a genuine website or a malicious website. In this research paper, an Anti-Phishing Effectiveness Evaluator Model (APEE Model) is proposed to examine the effectiveness of existing Anti-Phishing Mechanism. The implementation of this APEE Model is done on existing Anti-phishing Solution or mechanisms. Their effectiveness is evaluated and a result metric is listed. Exhaustive literature of the various ways adopted by phishers to attack online users are discussed. Various prominent authors have described different vulnerability used in phishing

© Springer International Publishing AG 2018
S.C. Satapathy and A. Joshi (eds.), *Information and Communication
Technology for Intelligent Systems (ICTIS 2017) - Volume 2*, Smart Innovation,
Systems and Technologies 84, DOI 10.1007/978-3-319-63645-0_68

attack. These vulnerability are categorized on basis of email structure i.e. Page-content vulnerability, Domain vulnerability and Cross-site scripting vulnerability. These vulnerability are exploited by attackers to enter email security and acquire user's personal information. After the exhaustive literature review and critical analysis some major finding also been identified.

This research paper is organized as follows: Sect. 2. Anti-Phishing Effectiveness Evaluator Model (APEE Model) is proposed; Sect. 4 exhaustive literature review is done to identified vulnerabilities causing email phishing separated three categories i.e. Page-content, Domain, Cross-cite scripting; Sect. 5 highlights the findings of the critical study and analysis to examines the effectiveness of existing security mechanism or approaches. Furthermore, at last Sect. 6 concludes the paper.

2 Anti-phishing Effectiveness Evaluator Model (APEE)

Phishing has turned into a substantial menace for the users over the internet and has become a significant root of financial losses. In such attacks, attackers try to obtain credential information of users so as to make them fall victim. Several anti-phishing techniques developed to combat such phishing attacks. These techniques resist the cybercriminals from exploiting the various vulnerabilities that can lead to a phishing attack. Here, **Anti-Phishing Effectiveness Evaluator Model (APEE Model)** is being proposed to evaluate the effectiveness of existing Anti-Phishing Solutions Tank (APST) with the help of identified vulnerability tank shown in Fig. 1. This Identified vulnerability tank contains all those vulnerabilities that are likely to be exploited by phishers to victimize the users and launch phishing attack.

APEE Model contains Identified vulnerability tank and Anti-Phishing Solutions Tank (APST). The identified vulnerability tank is classified into three categories i.e. Page-content Vulnerabilities (PV), Domain vulnerabilities (DV) and Code-Scripting Vulnerability (CV) as shown Fig. 1 and APEE Model runs existing APS individually and evaluates the efficiency of mechanism in covering (eliminating or detecting) the vulnerabilities that are categorized into various categories. Each category checks the number of vulnerabilities that are being eliminated by the Anti-Phishing solution and provides a numerical value of its effectiveness as shown Fig. 1.

Effectiveness Evaluator of Anti-Phishing techniques are denoted by EE_n and can be represented as,

$$EE_n = \frac{\textit{Number of vulnerabilities covered by Anti} - \textit{phishing mechanism}}{\textit{Total number of identified vulnerabilities of catetory n}} \quad (1)$$

where, for n is the number of categories of identified vulnerabilities,

Therefore, through Eq. (1), **Total Effectiveness Evaluator (TEE)** of Anti-Phishing Solution (APS) will be denoted as

$$\mathbf{TEE} = \sum_{i=1}^{n} \frac{EEi}{n} * \mathbf{100}$$

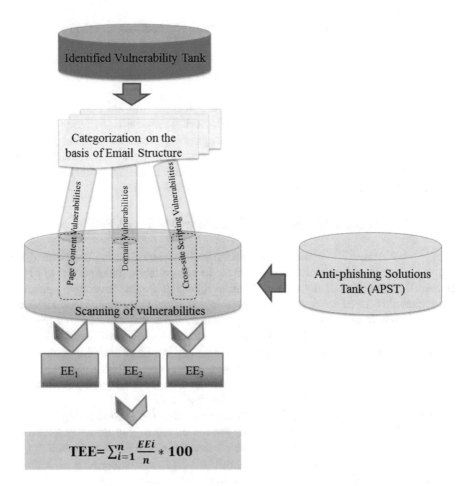

Fig. 1. Anti-Phishing Effectiveness Evaluator Model (APEE Model)

3 Implementation of Anti-phishing Effectiveness Evaluator Model (APEE Model)

In APEE Model, the vulnerabilities are gathered in 'identified phishing tank' and then classified into three categories on the basis of email structure i.e. Page-content Vulnerabilities (PV), Domain Name Vulnerabilities (DV) and Code-Scripting Vulnerabilities (CV) as shown Fig. 1. On the other side, Anti-Phishing Solutions Tank (APST) contains number of existing anti-phishing security mechanism. In order to calculate effectiveness of existing Anti-Phishing Solution (APS) it is required to pass all categorized vulnerabilities through the mechanism one by one i.e. PV, DV and CV. Therefore in the evaluation Model (APEE) existing Anti-Phishing Solution (APS) is picked one by one from the APST and its capability is measured on the basis of covered vulnerability from identified vulnerabilities i.e. how much it is capable to patch among the identified vulnerabilities.

We can see that the first category is Page-content Vulnerabilities. It will calculate the existing Anti-Phishing Solution (APS) effectiveness by checking the number of vulnerabilities that are covered by the APST.

1. *Effectiveness Evaluation of developed or existing Anti-Phishing Solution in covering Page-content Vulnerabilities (PV) will be denoted as,*

$$EE_1 = \frac{PVc}{PVi} \tag{1}$$

Here, PV_C is number of covered Page-content vulnerabilities (PV) by the existing Anti-Phishing Solution (APS) and the PVi is the total number of identified vulnerabilities of PV.

Similarly, the mechanism's effectiveness is evaluated under other categories viz. Domain vulnerabilities and Code-Scripting vulnerabilities and evaluates its effectiveness, denoted as EE_2 and EE_3 respectively and are represented as follows:

2. *Effectiveness Evaluation of developed or existing Anti-Phishing Solution in covering Domain Vulnerabilities(DV) will be denoted as,*

$$EE_2 = \frac{DVc}{DVi} \tag{2}$$

Here, DV_C is number of covered Domain vulnerabilities (DV) by existing Anti-Phishing Solution (APS) and the DV_i is the total number of identified vulnerabilities of DV.

3. *Effectiveness Evaluation of developed or existing Anti-Phishing Solution in covering Code-Scripting vulnerabilities (CV) will be denoted as,*

$$EE_3 = \frac{CVc}{CVi} \tag{3}$$

Here, CV_C is number of covered Code-Scripting vulnerabilities (CV) by the existing Anti-Phishing Solution (APS) and the CV_i is the total number of identified vulnerabilities of CV.

After evaluating the mechanism's effectiveness into defined categories Eqs. (1), (2) and (3) the total effectiveness of existing *Anti-Phishing Solution* will be calculated as,

$$\text{TEE} = \frac{EE_1 + EE_2 + EE_3}{3} * 100$$

Total Effectiveness Evaluator (TEE) of existing Anti-Phishing Solution will be denoted as

$$TEE = \sum\nolimits_{i=1}^{n} \frac{EEi}{n} * 100$$

where, n is the number of categories into which various vulnerabilities are classified and Ef_n is nth category (Table 1).

Table 1. Effectiveness metric of existing Anti-Phishing Solution (APS) from APST

SN.	Existing anti-phishing mechanism	Vulnerabilities				TEE
		Page-content$_{(EE1)}$	Domain$_{(EE2)}$	Cross-site scripting$_{(EE3)}$	Total	$\sum_{i=1}^{n} \frac{(EEi)}{n} * 100$
APS1	Linkguard	0.176	.030	0	7	26.33%
APS2	An integrated approach to detect phishing mail attacks: a case study 2	0.23	0.384	0.363	13	32.56%
APS3	Learning to detect phishing e-mails (Pilfer)	0.23	0.384	0.272	12	29.53%
APS4	PHONEY: mimicking user response to detect phishing attacks	0.05	0.076	0.0909	3	7.23%
APS5	Identification and detection of phishing emails using natural language processing techniques	0.23	0	0	4	7.66%
APS6	PhishGillNet-phishing detection methodology using probabilistic latent semantic analysis, AdaBoost, and co-training	0.117	0	0	2	3.33%
APS7	Tool for prevention and detection of phishing e-mail attacks	0.058	0	0	1	1.93%
APS8	Phishing e-mail detection based on structural properties	0.058	0.076	0.0909	3	7.46%

4   A Slew of E-mail Phishing Vulnerabilities

The number of ways adopted and innovated by phishers to trick the users. The phishing emails are crafted in such a way that seems to be sent by banks, online organizations or ISPs. Mostly it is very difficult for users to understand whether the email is authentic or not and turns them into a potential victim of phishing. An exhaustive literature review is done here. A slew of phishing vulnerability is identified as well as defined and described how malicious users exploit these vulnerabilities and trick innocent users. These vulnerabilities are spared in three categories on the basis of email structure i.e. PV, DV and CV. PV includes the sub criteria used in phishing sites, DV includes characteristics of hyperlinks and URL and CV includes the use of JavaScript to launch a sophisticated phishing attack. All Vulnerability are explained to understand how malicious users exploit these vulnerabilities and trick email or online users in Table 2.

Table 2. Anti-Phishing Solution (APS) from APST

Anti-phishing security mechanism from APST		Vulnerabilities		
		Page content vulnerabilities	Domain vulnerabilities	Cross-site scripting
APS1	Linkguard [1]	1. The hyperlink provides DNS domain names in the anchor text. But there is a mismatch between the destination DNS name in the visible link and that in the actual link 2. The link by encoding alphabets into their corresponding ASCII codes 3. By using special characters	1. Instead of DNS name, Dotted decimal format, IP address is used 2. The destination information in its anchor text is not provided by hyperlinks instead of that DNS names are used in its URI. The DNS name in the URI usually is similar with a famous company or organization 3. Blacklist or Whitelist DNS names 4. Pattern matching	
APS2	An integrated approach to detect phishing mail attacks a case study2 [1]	1. Using pop-ups 2. Disabling Right click 3. Using @ symbol to confuse fake address bar 4. Spelling error in domain name	1. Long URL Address 2. Replacing similar characters of URL 3. Adding prefix or suffix 4. Using hexadecimal character codes 5. Copying website	1. URL redirection 2. Double redirect tricks 3. Using on Mouse Over to hide the link 4. Server from handler
APS3	Learning to detect phishing e-mails (Pilfer) [2]	1. Non Matching URL 2. "here" links, "click" links to non-modal domain name 3. Number of links 4. Number of dots in links	1. Age to linked-to-domains names 2. IP based URLs 3. Number of domains 4. Number of dots in URL 5. Spam filter Output	1. Contains JavaScript to create pop-up window, change status bar of web browser, email client 2. Contains Java Script to hide information 3. Contains Java script to embed in something like link

(continued)

Table 2. (*continued*)

Anti-phishing security mechanism from APST		Vulnerabilities		
		Page content vulnerabilities	Domain vulnerabilities	Cross-site scripting
APS4	PHONEY: mimicking user response to detect phishing attacks [3]	1. Embedded links and HTML from email message body	1. Spoof URL act as legitimate URL using browser (IE) vulnerability	1. Attacker launches denial-of-service attacks by sending emails with URLs of real domain
APS5	Identification and detection of phishing emails using natural language processing techniques [4]	1. Absence of recipient's name 2. Mention of money 3. Reply Inducing 4. Sense of Urgency		
APS6	PhishGillNet-phishing detection methodology using probabilistic latent semantic analysis, AdaBoost, and co-training [5]	1. URL redirection to phishing sites 2. Collective term and document ambiguity		
APS7	Tool ffor prevention and detection of phishing E-mail Attacks [6]	1. Collective term and document ambiguity		
APS8	Phishing e-mail detection based on structural properties [7]	1. Embedded e-mail attachment	1. Using Browser vulnerability	1. Session Hijacking

5   Findings

During this exhaustive literature review and experimental analysis some finding are identified and are listed below:

1. The various vulnerabilities described by some of the authors' shows greater number of Page-content Vulnerabilities and are more susceptible to phishing as shown in Fig. 2. Phishers using Page-content vulnerabilities are most likely to trick naïve users who remain unsuspicious and can easily be fooled as shown in Fig. 2.
2. Domain vulnerabilities and Code-Scripting vulnerabilities are prone to bigger scams and uses advanced technology to trick naïve as well as proficient users as shown in Fig. 2.
3. Implementation of (APEE Model) on existing Anti-Phishing Solution (APS) is done and Effectiveness Metric of existing APS is reported in Table 1. Through this

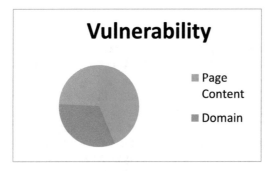

Fig. 2. A slew of vulnerabilities

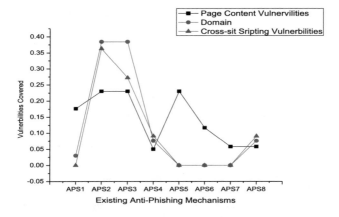

Graph 1. Effectiveness of existing Anti-Phishing Solution (APS)

developed metric it observed that-as number of covered vulnerability increases the effectiveness of anti-phishing mechanism also increases.

4. **Effectiveness metric of existing Anti-Phishing Solution (APS):** Effectiveness of existing Anti-Phishing Solution (APS) is calculated by the APEE Model as shown in shown in Table 1. Effectiveness metric of existing Anti-Phishing Solution (APS) is shows effectiveness evaluation of existing Anti-Phishing Solution (APS) of PV, DV, CV and Total Effectiveness of Anti-Phishing Solution. It is observed by the Graph 1 that none of existing Anti-Phishing Solution (APS) is effective as effectiveness is counted very less. None of existing Anti-Phishing Solution (APS) covered i.e. eradicated, reduced, detected all above defined categories of Vulnerability resulting phishing attack. It is easily observed by the Graph 1 that Effectiveness is very less and existing Anti-Phishing Solution (APS) does not cover all the vulnerabilities of Page-content Vulnerability (PV), Domain vulnerability (DV), Cross-site Scripting vulnerability (CV).

5. Through the critical analysis done in this research paper, it is recommended to cover more number of vulnerability to deliver robust Anti-Phish Solutions.

6 Conclusion

In this paper, brainstorming assessment is done to identify almost all email based vulnerabilities which are exploited by phishers to craft phishing email. Thus, it is extracted that, phishing is a continual threat and is commonly used by phishers to acquire personal and confidential information that can cause security damage that ranges from refusal of access of e-mails to considerable financial loss. Several anti-phishing solutions are developed to combat phishing attack. These anti-phishing solutions prevent phishing email by fixing the vulnerabilities at email- level. Here, an Anti-Phishing Effectiveness Evaluator Model (APEE Model) is proposed. Further, finding on identified vulnerabilities and its effect is made which reflects an immediate need to improve web & email security and introduce better techniques in order to avoid phishing attacks and provide secured access to internet users. An immediate need to revamp internet security is required in order to tackle such phishing attacks.

References

1. Suriya, R., Saravanan, K., Thangavelu, A.: An integrated approach to detect phishing mail attacks: a case study. In: SIN 2009 Proceedings of the 2nd International Conference on Security of Information and Network, pp. 193–199
2. Fette, I., Sadeh, N., Tomasic, A.: Learning to Detect Phishing Emails
3. Chandrasekaran, M., Chinchani, R., Upadhyay, S.: Phoney: mimicking user response to detect phishing attacks. In: Proceeding of 2006 International Symposium on a World of Wireless, Mobile and Multimedia Network, WoWMoM 2006 (2006)
4. Aggarwal, S., Kumar, V., Sudarsan S.D.: Identification and detection of phishing emails using natural language processing techniques. In: SIN 2014 Proceedings of the 7th International Conference on Security of Information and Network
5. Ramanathan, V., Wechsler, H.: phishGILLNET—phishing detection methodology using probabilistic latent semantic analysis, AdaBoost, and co-training. EURASIP J. Inf. Secur. (2012)
6. Firake, S.M., Soni, P., Meshram, B.B.: Tool fort prevention and detection of phishing e-mail attacks. In: Advances in Network Security and Applications. Series Communications in Computer and Information Science, vol. 196, pp. 78–88
7. Chandrasekaran, M., Narayanan, K., Upadhyay, S.: Phishing E-mail Detection Based on Structural Properties

Biometric Identification Using the Periocular Region

K. Kishore Kumar[1(✉)] and P. Trinatha Rao[2]

[1] Department of ECE, IcfaiTech School, IFHE University, Hyderabad, India
kkishore@ifheindia.org
[2] Department of ECE, GITAM School of Technology, GITAM University,
Hyderabad, India
trinath@gitam.in

Abstract. The face is the fundamental basis for the identification/recognition of a person. Recognition of a individual by the face data and methods to extract the unique facial feature points from the facial image have been significantly increased during the last decade. The periocular region has become the powerful alternative for unconstrained biometrics with better robustness and high discrimination ability. In this paper, multiple descriptors are used for deriving the discriminative features of the periocular region and city block distance is used to compute the similarity between the feature vectors. Feature extraction techniques employed in recognition using the periocular region are Local Binary Patterns (LBP), Local Phase Quantization (LPQ) and Speeded Up Robust Feature (SURF). Results of Periocular modality are then compared with the results of face patterns. Periocular region showed significant accuracies compared to face with only using 25% of the full face. Experimentations are carried on FRGC database, and accuracies of both the periocular and face regions are compared.

Keywords: Periocular region · Feature extraction · Local Appearance-based approaches · Keypoint based approaches · Equal Error Rate · Rank-1 Recognition Rates

1 Introduction

Face and Iris Biometric systems [1] present high performance with almost reaching 100% accuracies as these systems work under well-controlled constraint circumstances. These systems have been extensively studied for the last decade, with their approaches reaching the state of maturity. The limits of these modalities are clearly defined. With the success of Face and Iris Biometric systems, now the researchers are concentrating on the modalities with better accuracies under the non-ideal conditions. Iris biometric systems achieves 100% accuracies, which requires high-resolution data which had taken at the closest distance from a cooperative subject. Under the unconstrained scenarios, effectiveness of the iris systems will limit when the data is captured from a distance or on the move, or from a non-cooperative subject. Occlusion and dilated pupil will have an enormous impact on the recognition accuracies of the face and iris biometric systems

© Springer International Publishing AG 2018
S.C. Satapathy and A. Joshi (eds.), *Information and Communication
Technology for Intelligent Systems (ICTIS 2017) - Volume 2*, Smart Innovation,
Systems and Technologies 84, DOI 10.1007/978-3-319-63645-0_69

as these obstructive elements (beard, mask, sunglasses, high illumination) prevent the full regions from being useful.

The periocular region is the sub-region of the face in the locality of an eye with eyebrows, eyelids and eyefolds. The region of interest around the eye for feature extraction (Fig. 1).

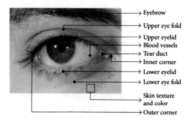

Fig. 1. Periocular region

Periocular region [2] is the most discriminative area of the face, gaining importance as an alternative or a complement to face and iris systems under non-ideal conditions. Periocular-based recognition is still a relatively new field, and most of their methods are adaptations from other biometric modalities. A comprehensive study of the periocular region is desirable to design algorithms specific to the periocular region. Designing the age invariant [16] face recognition systems are complex and challenging as face changes texture and shape with age. This region can offer superior recognition accuracies than the face, as the periocular region suffers less distortion than the face with changes in the age of the subject. The periocular region can be easily acquired from the face or eye as it is an integral region requiring no additional storage and acquisition cost [5, 6]. For obtaining the better recognition accuracies, the periocular region can be used for the fusion with face/Iris modalities. Large face templates [10] used in face biometric systems makes them slower in handling large database in the real-time scenarios. Recognition through periocular region [8] is faster compared to the face, as it is approximately 25% of the whole face image. Iris/Face recognition [11] fails for the following examples, where periocular biometrics can be favourable (Fig. 2).

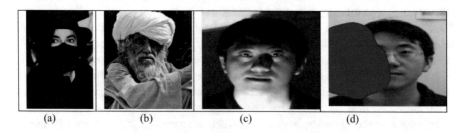

Fig. 2.   (a) Mask covering the face, (b) beard, (c) high illumination, (d) occlusion

Sato et al. [22] proposed the partial face recognition using the sub-regions of face: the nose, eyes, ear and eye region achieved better recognition rates compared to others suggesting us the periocular region could be a major region of the face with better discriminative features.

Savvides et al. [23] proposed partial face recognition using the sub-regions of face the nose, eyes, mouth on the Facial Recognition Grand Challenge (FRGC) dataset and the results showed eye region achieved better recognition rates compared to others. Teo et al. [24] compared the performance of full face recognition with the eye based partial face recognition. Park et al. [2] presented the periocular based recognition system using the Local Appearance-based approaches for the feature extraction of the periocular region and the performance is compared with the face. Miller et al. [25] proposed similar analysis with a different dataset and different feature extraction methodology. Lyle et al. and Merkow et al. presented the use of the periocular biometric for the gender classification [4]. Woodard et al., Santos and Hoyle and Tan et al. showed the fusion of iris and periocular modalities [25, 26]. Miller et al. explored the effects of data quality in a periocular biometric system [3].

2 Proposed Methodology

Periocular [8] Biometric is composed of the following steps: Data preprocessing, testing/training sets partitioning, feature extraction and comparison methods (Fig. 3).

2.1 Face Recognition Grand Challenge (FRGC)

FRGC database is used to extract the periocular regions from the face images. Face Recognition Grand Challenge (FRGC) database is prominently used for the periocular biometric. It consists of 16,029 still frontal face images, high-quality resolution face images of size 1200 × 1400 with different sessions and variable expressions. It was chosen for this work because the large face images will lead to relatively large periocular region images.

2.2 Preprocessing of Face/Periocular Imges

Preprocessing of Face/Periocular Images involves converting a raw colour facial image into the preprocessed periocular images, geometric normalization and histogram equalization.

2.3 Periocular Region Extraction

The next preprocessing step is to extract periocular region images from normalized and equalized facial images, which is accomplished by placing a square bounding box around each eye, centred on the post geometric normalization eye centre locations (Fig. 4).

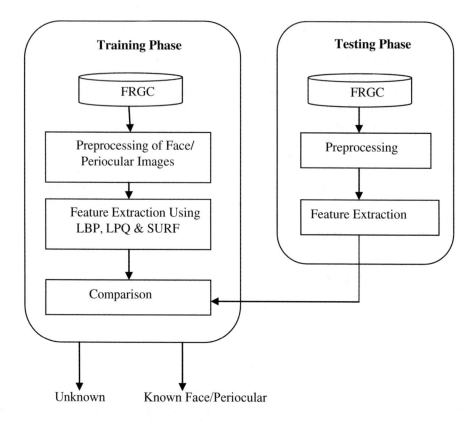

Fig. 3. Block diagram of proposed approach

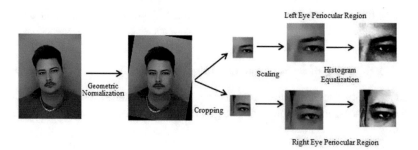

Fig. 4. Process flow for the periocular region extraction

2.4   Feature Extraction

Each feature extraction technique transforms a two-dimensional image into a one-dimensional feature vector through its unique process. Multiple descriptors are used for deriving the discriminative features of the periocular region, and the similarity between the feature vectors are computed by the city block distance method. Feature extraction techniques employed on the periocular region are Local Binary Patterns (LBP), Local Phase Quantization (LPQ) and Speeded Up Robust Feature (SURF) [9].

2.4.1   Local Binary Pattern (LBP)

Local Appearance-based approaches [3] are the class of feature extraction techniques which collect statistics within local neighbourhoods around each pixel of an image providing the information related to the occurrence of certain textures, patterns. The outcome of these approaches is one-dimensional feature vectors. Local Binary Patterns (LBP) is a texture classification method that was developed by Ojala [13]. LBP collects the texture information from an image into a feature vector by labelling pixels with a binary number by placing a threshold on the neighbourhood around each pixel. A histogram of these values forms the output feature vector. LBP is used extensively for both facial recognition [14] and periocular recognition [15].

The LBP value of pixel of concern P_k, is a function of intensity changes in the neighbourhood of M sampling points on a circle of radius r, then the LBP operator is given by,

$$LBP_{M,r} = \sum_{n=0}^{M-1} s(g_n - g_c)2^n \tag{1}$$

where

$$S(p) = 1 \text{ if } p > 0$$
$$= 0 \text{ if } p < 0$$

g_c intensity of the pixel of concern at the centre with the pixels on the circumference of a circle with values of g_n, where $n = 0, ..., M - 1$. In the proposed work, all LBP calculations are made from a circle of radius 1 pixel with 8 pixels along the circumference of the circle.

2.4.2   Local Phase Quantization

LPQ proposed by Ojansivu et al. [17] is a texture descriptor which quantizes the phase information of a discrete Fourier transform (DFT) in patch-sized neighbourhoods of an image. LPQ is robust to image blurring which has been used for face recognition [3]. Like LBP, the resulting LPQ codes are formed into a histogram.

In LPQ the local spectra at a pixel p is calculated from a short-term Fourier transform and is given by

$$F(u, p) = \sum_{l \in p_x} f(p - l)e^{-j2\pi u Tl} \tag{2}$$

where Pu is a pixel in a $M \times M$ neighbourhood around u. At frequency points $u_1 = [a, 0]^T$, $u_2 = [0, a]^T$, $u_3 = [a, a]^T$, $u_4 = [a, -a]^T$, local Fourier coefficients are computed where a is $1/M$. The phase portion of the Fourier coefficients is defined as the sign of the real and imaginary components of $F(u, x)$ given by

$$q_j(\mathrm{u}) = 1, \quad \text{if } q_j(\mathrm{u}) > 0$$
$$= 0, \quad \text{otherwise} \tag{3}$$

The LPQ score is the binary coding of the eight binary coefficients $qj(u)$ and all LPQ calculations were made on a 9×9 pixel window.

2.4.3   Speeded Up Robust Features

Speeded Up Robust Feature (SURF) descriptors are the class of keypoint-based approaches presented by Bay et al. [12]. It is faster and more robust to different transforms of an image than SIFT. SURF keypoints are found from maxima in the determinant of the Hessian matrix of images; a matrix made up of the convolution of the Gaussian second order derivatives with the image. SURF was originally used for object recognition and has also been used in facial recognition contexts over the years.

2.5   Classification Using City Block Distance Metric

City Block Distance Metric is used to determine the closeness between two feature vectors obtained from the feature extraction methods.City Block Distance Metric is given by

$$d(k, l) = \sum_{j=0}^{n} |k_i - l_i| \tag{4}$$

where k and l are feature vectors extracted from two images that have a feature dimensionality of n.

3   Performance Measures

Following performance measures defines the robustness of the designed system using face/Iris modalities:

Rank 1 recognition rates illustrate the successfulness of the scheme in identifying the best match for a subject.

The rate at which the False Rejection Ratio (FRR) and False Acceptance Ratio (FAR) are equal is defined as Equal Error Rate (EER).

FRGC protocol advocates the verification rate at 0.1% false accept rate be used to compare the performance between two methods.

D shows the separability between the similarity score distributions of the set of true matches and false matches.

4   Results

In the proposed paper, three feature extraction methods are applied to three image regions i.e., left eye, right eye and full face images of the FRGC Experiment 1 dataset. Tables 1, 2 and 3 shows the performance statistics [18] for the LBP [19–21], LPQ and

SURF feature extraction methods in terms of Rank-1 Accuracy, Equal Error Rate (EER), VR at 0.1% FAR, and D. In most of the cases, face region performs better when compared with the periocular regions as it has more information in terms of nodal points. Periocular region showed significant accuracies compared to face with only using 25% of the full face information. LPQ scheme produces the best performance results when extracted from face images (Figs. 5, 6, 7 and 8).

Table 1. Results obtained from Local Binary Pattern

Region	Local Binary Pattern (LBP)			
	Rank-1	EER	VR @ 0.1% FAR	D
Left eye	99.7058	8.8312	64.8837	2.7330
Right eye	99.7003	8.2005	69.7148	2.8245
Face	99.9178	7.0837	69.9876	2.9423

Table 2. Results obtained from Local Phase Quantization

Region	Local Phase Quantization (LPQ)			
	Rank-1	EER	VR @ 0.1% FAR	D
Left eye	99.7681	7.1179	75.9171	2.8653
Right eye	99.7805	6.7217	76.6564	2.9434
Face	99.9427	5.4145	79.9201	2.9911

Table 3. Results obtained from Speeded Up Robust Feature

Region	Speeded Up Robust Feature (SURF)			
	Rank-1	EER	VR @ 0.1% FAR	D
Left eye	99.6501	9.6850	60.4445	2.2050
Right eye	99.5373	9.4563	59.8141	2.2185
Face	99.9303	7.6782	64.6383	2.5890

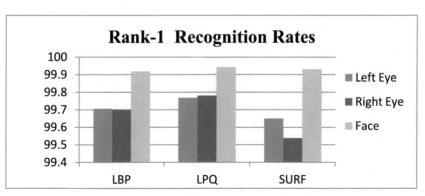

Fig. 5. Rank-1 recognition rates for the LBP, LPQ and SURF methods

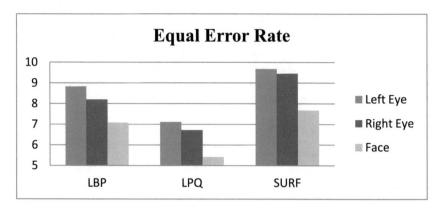

Fig. 6. Equal Error Rates (EER) for LBP, LPQ and SURF methods

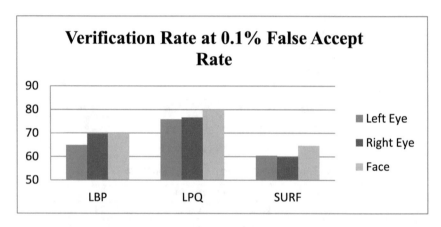

Fig. 7. Verification rate at 0.1% false accept rate (VR @ 0.1% FAR) for the LBP, LPQ and SURF methods

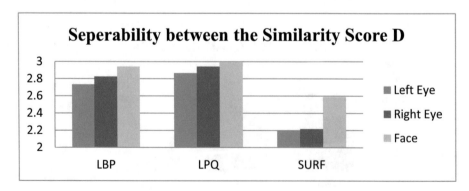

Fig. 8. D shows the separability between the similarity score distributions of the set of true matches and false matches for the LBP, LPQ and SURF methods

5   Conclusion

In this paper, multiple descriptors are used for deriving the discriminative features of the periocular region and city block distance is used to compute the similarity between the feature vectors. Feature extraction techniques employed on the periocular region are Local Binary Patterns (LBP), Local Phase Quantization (LPQ) and Speeded Up Robust Feature (SURF). Results of Periocular modality are then compared with the results of face modalities. Periocular region showed significant accuracies compared to face with only using 25% of the full face information. The experimentations were carried on FRGC database, and accuracies of both the periocular and face regions are compared.

References

1. Jain, A., Ross, A., Prabhakar, S.: An introduction to biometric recognition. IEEE Trans. Circ. Syst. Video Technol. **14**(1), 4–20 (2004)
2. Park, U., Jillela, R., Ross, A., Jain, A.: Periocular biometrics in the visible spectrum. IEEE Trans. Inf. Forensics Secur. **6**(1), 96–106 (2011)
3. Miller, P., Lyle, J., Pundlik, S., Woodard, D.: Performance evaluation of local appearance based periocular recognition. In: Proceedings of the IEEE International Conference on Biometrics: Theory, Applications, and Systems, pp. 1–6 (2010)
4. Merkow, J., Jou, B., Savvides, M.: An exploration of gender identification using only the periocular region. In: Proceedings of the IEEE International Conference on Biometrics: Theory, Applications, and Systems, pp. 1–5 (2010)
5. Juefei-Xu, F., Savvides, M.: Unconstrained periocular biometric acquisition and recognition using COTS PTZ camera for uncooperative and non-cooperative subjects. In: Proceedings of the IEEE Work-Shop on ACV, pp. 201–208 (2012)
6. Juefei-Xu. F., Luu, K., Savvides M., Bui, T., Suen, C.: Investigating age invariant face recognition based on periocular biometrics. In: Proceedings of the International Joint Conference on Biometrics, pp. 1–7 (2011)
7. Joshi, A., Gangwar, A., Sharma, R., Saquib, Z.: Periocular feature extraction based on LBP and DLDA. In: Advances in Computer Science, Engineering and Applications. Advances in Intelligent and Soft Computing, vol. 166, pp. 1023–1033. Springer, Heidelberg (2012)
8. Hollingsworth, K., Bowyer, K., Flynn, P.: Identifying useful features for recognition in near-infrared periocular images. In: Proceedings of the IEEE International Conference on Biometrics: Theory Applications and Systems, pp. 1–8 (2010)
9. Dreuw, P., Steingrube, P., Hanselmann H., Ney, H.: SURF-face: face recognition under viewpoint consistency constraints. In: Proceedings of the British Machine Vision Conference, pp. 1–11 (2009)
10. Givens, G.H., Beveridge, J.R., Draper, B.A., Grother, P., Phillips, P.J.: How features of the human face affect recognition: a statistical comparison of three face recognition algorithms. In: Proceedings of the International Conference on Pattern Recognition, 2004, vol. 2, pp. 381–388
11. Bharadwaj, S., Bhatt, H., Vatsa, M., Singh, R.: Periocular biometrics: when iris recognition fails. In: Proceedings of the IEEE International Conference on Biometrics: Theory, Applications, and Systems, pp. 1–6 (2010)
12. Bay, H., Ess, A., Tuytelaars, T., Van Gool, L.: SURF: speeded up robust features. Comput. Vis. Image Underst. **110**(3), 346–359 (2008)

13. Ojala, T., Pietikainen, M., Maenpaa, T.: A generalized local binary pattern operator for multiresolution gray-scale and rotation invariant texture classification. In: Second International Conference on Advances in Pattern Recognition, pp. 397–406 (2001)
14. Mahalingam, G., Kambhamettu, C.: Face verification with aging using AdaBoost and local binary patterns. In: Proceedings of the Seventh Indian Conference on Computer Vision Graphics and Image Processing, ICVGIP 2010 (2010)
15. Ling, H., Soatto, S., Ramanathan, N., Jacobs, D.W.: Face verification across age progression using discriminative methods. IEEE Trans. Inf. Forensics Secur. **5**, 82 (2010)
16. Ramanathan, N.: Face verification across age progression. IEEE Trans. Image Process. **15**, 3349 (2006)
17. Ahonen, T., Rahtu, E., Ojansivu, V., Heikkilä, J.: Recognition of blurred faces using local phase quantization. In: Proceedings of the International Conference on Pattern Recognition, pp. 1–4 (2008)
18. Ramanathan, N.: Computational methods for modeling facial aging: a survey. J. Vis. Lang. Comput. **20**, 131 (2009)
19. Ahonen, T., Hadid, A., Pietikäinen, M.: Face description with local binary patterns: application to face recognition. IEEE Trans. Pattern Anal. Mach. Intell. **28**(12), 2037–2041 (2006)
20. Ahonen, T., Hadid, A., Pietikäinen, M.: Face recognition with local binary pattern. In: Proceedings of the European Conference on Computer Vision, pp. 469–481 (2004)
21. Kumar, K.K., Rao, P.T.: Face verification across ages using the discriminative methods and see 5 classifier. In: Proceedings of First International Conference on ICTIS: Volume 2, Springer Smart Innovation, Systems and Technologies, vol. 51, pp. 439–448
22. Sato, K., Shah, S., Aggarwal, J.: Partial face recognition using radial basis function networks. In: Proceedings of the IEEE International Conference on Automatic Face and Gesture Recognition, pp. 288–293 (1998)
23. Savvides, M., Abiantun, R., Heo, J., Park, S., Xie, C., Vijayakumar, B.: Partial holistic face recognition on FRGC-II data using support vector machine. In: Proceedings of the IEEE Conference on Computer Vision and Pattern Recognition, p. 48 (2006)
24. Teo, C., Neo, H., Teoh, A.: A study on partial face recognition of eye region. In: Proceedings of the International Conference on MV, pp. 46–49 (2007)
25. Miller, P., Rawls, A., Pundlik, S., Woodard, D.: Personal identification using periocular skin texture. In: Proceedings of the ACM Symposium on Applied Computing, pp. 1496–1500 (2010)
26. Santos, G., Hoyle, E.: A fusion approach to unconstrained iris recognition. Pattern Recogn. Lett. **33**(8), 984–990 (2012)

Identifying Student for Customized Tutoring

Rishi Kumar Dubey and Umesh Kumar Pandey[✉]

MSIT MATS Unievrsity, Raipur, CG, India
rishi.rpr@gmail.com, umesh6326@gmail.com

Abstract. Improving quality of student is related with increasing their learning and skills in the concerned field. The benchmark used to measure the learning and skill is obtained score of student and participation in various activities. During a session, this session may be a semester in short term and in longer term it may be program duration, student generates huge size of data. If this data is studied properly then learning and skill easily grouped. After knowing a person's weakest zone a customized solution will provide to nurture him/her in required area of development. In this paper data will be studied to identify weaker student so that customized tutoring offered to him/her.

Keywords: Data mining · Customized tutoring · k-means · Clustering

1 Introduction

Customized tutoring offers a special attention and teaching methodology to a learner for increasing learning and enhancing skill. Learning involves two basic activity i.e. listening and understanding. A student may have problem in both or any one of them. But when it comes to measure students learning level a third factor also play role, and this is expression. This expression may be a verbal, non verbal, textual etc. Institutions organize tests and activities for enhancing student skill in the concerned field of study.

Data mining studies data and provide those hidden and unseen information. Data mining with its predictive and descriptive model tools has facility to classify, cluster, prediction, summarization, association. Use of data mining methodology will help in understanding information hidden inside the data generated by student during activities and tests. Most of the time these data are collected by institutions but due to lack of professional data analyst complex data analysis do not take place; lack of time destroys the value of that information; spent time and effort do not provide return on investment.

Core objective of education is to provide intellectual and skilled human resource to the various service offering sectors. Students have different ability of response on given problem. To find this ability an instructor led test will be organized. After collecting this data student are clustered in instructor decided number of groups. These groups will help instructor to provide customized aid for the student.

In this paper k-means clustering data mining method is proposed to cluster student by studying activity generated data. So, Institution will offer a customized tutoring to him/her.

© Springer International Publishing AG 2018

S.C. Satapathy and A. Joshi (eds.), *Information and Communication Technology for Intelligent Systems (ICTIS 2017) - Volume 2,* Smart Innovation, Systems and Technologies 84, DOI 10.1007/978-3-319-63645-0_70

2 Literature Review

Dongkuan Xu and Yingjie Tian [1] delineate about various modern clustering algorithm. They categorize several clustering algorithm on the basis of kernel, ensemble, swarm intelligence, quantum theory, graph theory, affinity propagation, density and distance, spatial data, data stream and large data scale. The objective of this review paper is to describe specific source of each algorithm and study advantage and disadvantage of each algorithm.

Qian Wei-ning, Zhou Ao-ying [2] analyzed clustering with different viewpoints. In this paper clustering algorithm are analyzed on three points i.e. criteria, representation and algorithm framework. They studied distance based, density based, Linkage based, representative points, dense area, cells based, probability based, optimization method and agglomerate clustering techniques. On the basis of algorithm framework they concluded that k-medoids are more robust than k-means but cost of k-medoids is also expensive.

Osama A A [3] in his paper did comparison between four algorithm i.e. k-means, hierarchical clustering algorithm, Self-organization Map (SOM) algorithm and Expectation Maximization. Reason behind choosing these algorithm were popularity, flexibility, applicability and handling high dimensionality. Factor which were consider for this comparison were: size of data set, number of cluster, types of dataset and type of software. For software he used LMKnet (Unix Environment) and cluster and TreeView (Windows Environment). He concluded that k-means and Expectation Maximization are good for huge dataset compare to other. But on quality front k-means and EM give poor quality. All of them give almost same result when run on any software because most software method adopts same procedure.

3 Methodology

Through observation, surveying literatures related to defining the performance factor of student and discussion with faculties of various discipline, it has been inferred that number of factors responsible for student performance. But expression is one of the common factors for poor performance among the student. Data, scores obtained under various characteristic, is studied using Rapidminer tool. To make cluster of student following steps will be followed (Fig. 1).

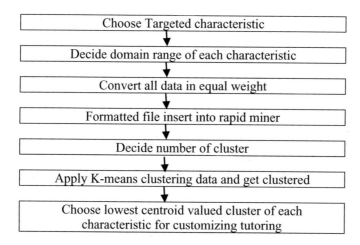

Fig. 1. Flowchart of methodology

3.1 Choose Targeted Characteristic

Targeted characteristic is the variable on which students are grouped into more than one cluster. If more than one characteristic will be chosen then at a time only one characteristic is used for the clustering. Characteristic selection depends upon evaluator. For example for this paper we have chosen five characteristic as per following table (Table 1).

Table 1. Characteristic and method of assessment to be evaluated

Method of assessment	Characteristic
Key point extempore speaking on topic	Spoken ability
Key point extempore writing on topic	Writing ability
Question given	Focused
Out of question given	Memory
Question not given but discussed	Studious

Data is stored in structure shown as Table 2:

Unique Id will be used to identify student. Spoken ability checked (focus given on fluency, pronunciation, relevancy etc.) from the topic. Writing ability (spelling, sentence structure, relevancy etc.) checked from the topic. In both characteristic either student will chose topic or evaluator will assign random topic. To evaluate focused characteristic student has been given fix number of question and the same question will be asked in classroom examination. To check memory characteristic m number of question will be given to student for preparation and n (less than m, nearly one third of m) number of same question asked in classroom examination. To evaluate studious characteristic of student random question asked in the classroom relevant to the broader subject area. Scores obtained in various characteristic stored in a table as shown in Table 2.

Table 2. Data structure to store scores

Unique Id	Spoken ability	Writing ability	Focused	Memory	Studious
–	–	–	–	–	–

3.2 Decide Domain Range of Each Characteristic

All scores must be in number either in integer or floating point. Decide the range of score for each characteristic. Preferably keep same range of score for each score. If two different ranges have been chosen for evaluation then convert them in same range by using formula:

$$\text{new score} = \frac{\text{old score} * \text{common maximum value}}{\text{old maximum score}}$$

It will not effect in study of data but while comparing one characteristic to another it will play an important role.

3.3 Decide Number of Cluster

Data, students score, used in this paper is collected from 100 students. Big question is "How many clusters are needed". Answer of this question is decided by evaluator. While deciding the number of cluster; student number and difference between lowest and highest must be considered. For suggestion minimum 4 clusters must be prepared.

Insert data file into rapid miner: Rapidminer can read data directly from MS-Excel sheet and user are much familiar with MS-Excel. Thus data set will be collected and

Row No.	Unique Id	Speaking
1	1	5
2	2	6
3	3	1
4	4	4
5	5	8
6	6	5
7	7	7
8	8	4
9	9	2
10	10	3

Fig. 2. Rapidminer data set for speaking data clustering

organized in MS-excel. Data is imported for each characteristic separately. This as case mentioned above, five different data sets are created in the Rapidminer. Following Fig. 2 shows one data set.

4   Model in Rapidminer

Above model is used to cluster the retrieved data. Speaking database Tab contain student score of spoken ability in the format of Fig. 2. If evaluator wants to cluster student on other characteristic, then replace speaking database with new characteristic database. This retrieved data goes "Generate ID" tab to assign an identity number for each record which later used for join purpose. After this data is sort in increasing order on the field of score. If data is not sorted then scores may be clustered in different order. Thus sorted data is clustered fast and accurately placed in the right cluster. This sorted data goes into two tabs i.e. join and select attribute. Clustering algorithm directly cannot be used. Before applying clustering algorithm, attribute on which clustering will be done, must be selected in select attribute tab and number of cluster in cluster tab. For this paper number of cluster is 4. In join tab sorted data and clustered data joined by applying inner join type. The output of this model goes for display as well as stored in MS excel file for later use.

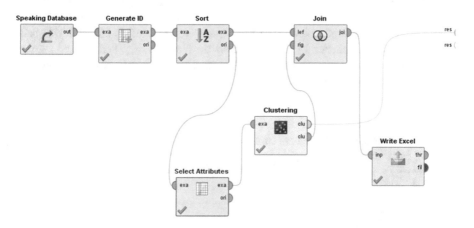

Fig. 3.   Clustering model in rapid miner

5   Result

Rapidminer randomly assign cluster number to each cluster on the basis of centroid value. Following Table 3 shows centroid value for each cluster in ascending order along with its Rapidminer assign cluster number.

Table 3. Clustered data in increasing centroid value and Rapidminer assigned cluster

Characteristic		Group 1	Group 2	Group 3	Group 4
Speaking	Rapidminer assigned cluster	Cluster 1	Cluster 0	cluster 2	Cluster 3
	Centroid value	2.059	4.5	6.533	8.22
Writing	Rapidminer assigned cluster	Cluster 3	Cluster 0	Cluster 2	Cluster 1
	Centroid value	1.5556	3.833	5.362	7.75
Focused	Rapidminer assigned cluster	Cluster 3	Cluster 0	Cluster 1	–
	Centroid value	2	4.9	7.645	–
Memory	Rapidminer assigned cluster	Cluster 2	Cluster 3	Cluster 0	Cluster 1
	Centroid value	2.154	4	5.37	7.857
Studious	Rapidminer assigned cluster	Cluster 0	Cluster 1	Cluster 3	–
	Centroid value	1.889	4.96	7.625	–

All centroid value arranged in ascending order and group number is assigned in first row. All characteristic has been studied on equal number of student as per score obtained in the evaluation process. But centroid value is different for each cluster in different characteristic study. Rapidminer clustering model represented in Fig. 3 forms same number (4 cluster) of cluster for each characteristic, but Speaking, Writing and memory has four different clusters whereas focused and studious has only three clusters, because of meanness value around centroid value.

Table 4. Clustered Number of student in respect of Rapidminer assigned cluster

Characteristic		Group 1	Group 2	Group 3	Group 4
Speaking	Rapidminer assigned cluster	Cluster 1	Cluster 0	cluster 2	Cluster 3
	Number of student	17	50	24	9
Writing	Rapidminer assigned cluster	Cluster 3	Cluster 0	Cluster 2	Cluster 1
	Number of student	9	24	47	20
Focused	Rapidminer assigned cluster	Cluster 3	Cluster 0	Cluster 1	–
	Number of student	13	70	17	–
Memory	Rapidminer assigned cluster	Cluster 2	Cluster 3	Cluster 0	Cluster 1
	Number of student	13	20	46	21
Studious	Rapidminer assigned cluster	Cluster 0	Cluster 1	Cluster 3	–
	Number of student	9	75	16	–

Table 4 shows number of student falling in the respective cluster. Group 1 shows number of student falling around lowest centroid value in each analyzed characteristic. Group 2, Group 3 and Group 4 shows number of student around next higher centroid value in increasing order.

6   Analysis

Group 1 belongs to lowest centroid value which has those students list who are not scoring good scores compare to other student. Group 4 list adept student for speaking, writing and memory whereas Group 3 list adept student for focused and studious, because for focused and studious no student fall in the group 4. Thus student near to lowest centroid value, who are assigned to Group 1, considered as weaker student whereas student near to highest centroid value, who are assigned to last group in the respective characteristic are considered as adept. Other groups falling between lowest and highest are considered to be average student. They also needed customized tutoring but priority must be given to those students who fall under group 1 of different characteristic.

7   Advantage

Most of the time evaluator applies manual method, guessing method, simple statistical method, or fixed length class. These methods may be either biased or evaluator dependent or simple in nature or static. K-means clustering groups data on centroid value which is calculated as per received data. Thus, there is no space for biasness, guessing of value. Clustering technique do not have any predefined length of class, thus it cluster all data in specified number of cluster, except those cluster in which no value will fall.

Second advantage is that K-means work with floating values as well as integer value. Most of the time evaluator try to award scores in integer value so that calculation can be made easy. But using this model evaluator can use both integer and floating values to differentiate scores among student.

8   Conclusion

Performance of student in a class varies because of variation in expression and study intensity. Above explained model using K-means clustering groups student into evaluator specified number of cluster (preferably 4). GUI feature of Rapidminer makes easy to use this model. Student falling in lowest mean cluster is considered for customizing tutoring. Now the ball is in the side of educationist to find and develop of new methodology to eradicate weakness of students.

References

1. Xu, D., Yingjie, T.: A comprehensive survey of clustering algorithms. Ann. Data Sci. 2(2), 165–193 (2015). doi:10.1007/s40745-015-0040-1
2. Qian, W., Zhou, A.: Analyzing popular clustering algorithms from different viewpioints. J. Softw. 13(8), 1382–1394 (2002). 1000-9825
3. Osama, A.A.: Comparison between data clustering algorithms. Int. Arab J. Inf. Technol. 5(3), 320–325 (2008)

A Clustering Techniques to Detect E-mail Spammer and Their Domains

Kavita Patel[1(✉)], Sanjay Kumar Dubey[1], and Ajay Shanker Singh[2]

[1] CSE Department, Amity University, Sec-125, Noida, Uttar Pradesh, India
kavita117@gmail.com, skdubey1@amity.edu
[2] School of CSE, Galgotias University, Greater Noida, Uttar Pradesh, India
ajay.shankersingh@galgotiasuniversity.edu

Abstract. The latest internet has become a collaboration and communications platform, in that e-mail system is one of the most reliable internet services. Sending a spam e-mail is an economically useful commerce for intruders, with the very good earning of millions of dollars. The spam e-mail has become a critical issue to web and society, to stop/reduce the spam e-mails filtering techniques is not sufficient. This paper proposes to recognize spam domain by reading spam e-mails. These spam domains are nothing but Uniform Resource Locator (URL) of the website that intruder is promoting. The approach is based on extracting mail content; links from URL injected e-mail and subject of spam e-mails. These extracted parameters are grouped together through clustering algorithms and evaluated. This proposed work can be help as additional accessory to already available anti-spam tool to recognize intruders.

Keywords: Spam · Spam e-mail · Data mining · Clustering algorithms · Spam URL

1 Introduction

E-mail has appeared as a necessary communication link for billions of people around world by its inexpensive and convenience. Spam email is unrelated or anonymous texts forwarded on Internet, usually to huge amount of peoples, with intention for phishing, advertising, propagate malware, etc. [1]. Spammer sends different types of hazardous and malicious activity using spam mails. Email can include URLs which exhibits to unwanted thinks directly or to open kits which first analyses the specifications of the visiting web browser and then selects an attack technique most likely to weaken that system [2]. Following two conditions is satisfied by spam email: the first is whenever email contents are found irrelevant and forwarded to many recipients and, second is, if someone is sending email to unknown recipient. At major level this e-mails are defined in 6 types: - E-mail, I-Messenger, Irrelevant content SMS, Comment, Waste Spam, or Social-Network [3]. As internet consumer is increasing now a day, a number of users are getting email communication a cheaper way to deliver their data and interact with their peers. With benefits also come some losses. Approximately, all website take email

© Springer International Publishing AG 2018
S.C. Satapathy and A. Joshi (eds.), *Information and Communication Technology for Intelligent Systems (ICTIS 2017) - Volume 2*, Smart Innovation, Systems and Technologies 84, DOI 10.1007/978-3-319-63645-0_71

id so as to finish their registration process, thus making users progressively vulnerable to get affected by the spam e-mails.

Earlier, the efficient way of limiting spam emails is filtering. Spam filtering is unsuccessful to stop spammers, because of no real penalty [4]. Now a day billions of spam emails are separated out by filter, and many of them are immediately deleted or saved to reach some limitation available on storage is crossed, and then discarded. Moreover, spam filters can only make difference between spam e-mails and non-spam but no idea of the roots of it. The best method is to decrease the spam is domain blacklisting. The researchers and anti-spam community is enhancing the effective model of detecting spam emails, the spammers are also rigorously making new methods to defeat the models [5]. This paper presents the analysis of spam emails. The system takes specific features from spam e-mails and then with help of different clustering techniques it group all regarding to their matching. By doing this spam domains are caught. The spam links of e-mails were found and told to do blacklist. After that, coming e-mails coming from black listed domains will be restricted. Strict action can be taken to close that spams. The remaining paper is designed in more four section namely: Sect. 2 as background and an overview of related work, Sect. 3, as proposed system, Sect. 4 shows analytical evaluations. And conclusions and future work in Sect. 5.

2   Overview and Related Work

2.1   Spam E-mail Writing Tactic

The instructions lie into three regions: [6] First that used to the text written in the e-mails, second is the email address DBMS, and last which used for tech-settings. Further, below these instructions are given in brief.

Email Text Instructions. Main motive of an intruder are to write the e-mail text as attractive as possible. Attractive text for an e-mail that seems alike to private talk between two persons. On basis of the type of service/product being advertised, it suggests several combinations of HTML and plain text.

Email Address Database Instructions. To successful spam campaign, there should be a lot of valid email address. Although, many of the e-mail ids are gathered from internet, so there is no assurance that it will really be reachable and active. Thus, it is necessary to verify unavailable e-mails databases. If number of e-mail ids on individual domain is less, things are not useful. In opposite to that, if the number of e-mail ids on individual domain is large, then possibility of caught and black-listing is increases simultaneously.

Technical Instructions. How to install their spam-bots on victims system is the most important issue for the spammer. They are ready to install bots by own or, take it from party. If 1000 bots are online in same instance of time then it generates good throughput of e-mails.

2.2 Anti-spam Techniques

In this portion, it give overview on the spam filtering technique is used segregate spam e-mail [7].

Signature Matching. Signature matching was the most efficient technique, in previous days of spam. This solution is depending on the database of the entire spam signature that is calculated. In contrast to that, spammers/intruders have enhanced their range of sophistication; so the use of anti-spam software (i.e. signature matching) has drastically decreased.

Heuristics. If filter is applied properly, then one of the most accurate anti-spam filtering methodologies is heuristics filter. In this approach rules have some value associated with it. The rules creation and set up is very easy, has constant rate of perfectness, and very hard for intruders to hide, when protocols will modified by a regular basis.

Bayesian Filtering. This filter is error free filtering technology for e-mail accounts where legitimate e-mail has important discriminate text than other spam. Huge memory consumption, CPU, disk uses may build it inappropriate for many web-sites, although largely helps another filter methods which have extreme level of perfectness.

Domain Name Server Blacklisting. With this 40% of e-mail spams are detected and destroyed from spam a mail box. It is alive until it has authorized mail, Domain Name Server black-listing is fruitful technique if it uses with some other anti-spam techniques.

Challenge/Response. It is a good answer theoretically, although in actual, this disturbs e-mail compare to spam does. Challenge/Response play a very tiny role in contrast to other filtering techniques.

2.3 Background

A profit of research has been depended on financial generated by e-mail spam. Currently, researchers are more interested in blocking spam e-mails rather than just detecting it. The same idea is supported and preceded by this paper. The goal is to use clustering method of to gather the similar spam e-mails and identify the spammer origin by their spam domains. The reviewed paper has anti-spam technology and some clustering techniques on data streams. Identification of spammer's domain can be increased by taking more parameters for clustering.

Li et al. [8], studied where spam e-mails are normally forwarded in bundle with uniqueness amongst them, according to URLs in email, to their corresponding domain or prototype which were used. Chun et al. [9], analyzed the spammer's unpredictable behavior and clustering same spam mails on basis of subject of it. Further, the IP address can be helpful to track spammers. In their research, they used fuzzy similarity. It is used in recognizing spam emails without having prior knowledge of spam emails. Negative selection works on antigen feedback mechanism as the detection engine [10]. By Halder et al. [4], spam e-mails had analyzed on semantics and similarity of mail content.

Criminal can be detected through grouping similar spam e-mails based on semantic, style and by combining both the features together. The resultant clusters are relating to the internet protocol and the 'whois' information provides the detail of source of spam with help of IP addresses. Web spam or Link spam is technique to increase rank of pages by making artificial popularity. To identify web spam by finding boosting pages created by spammer [11]. To identify variety of spam mails merely looking into the content won't suffice, the header of mail also needs to look upon. Several effective features are available in Yahoo Mail, Gmail and Hotmail's email header to filter spam email [12]. To identify text as well as image based spam emails, three algorithms used as: K- Nearest Neighbor or KNN algorithm, Naive Bayes algorithm and reverse DB SCAN algorithm. For comparison the evaluation factors are namely: precision, sensitivity, Specificity and accuracy [13]. Malicious mails are discriminated on the basis of supervised learning; unsupervised learning (this is faster, but level of accuracy has a low); or a hybrid learning. The random forest classifier is not required any prior learning. Hence, it is easy to use for detecting spammer [14].

The proposed paper has to cluster spam e-mails by taking three parameters so that it create certain clusters of spam e-mails. Identified clusters are later map with their own domains and after that informed to authority for legal actions, this way source of spam can be mitigate.

3   Methodology

The proposed technique to recognize also give detail of the intruders includes in 6 main sequences those are elaborated through following Fig. 1.

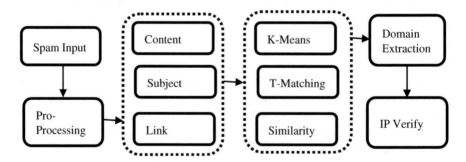

Fig. 1. Proposed technique to recognize spam domain

3.1   Spam Input

This is first step which accept the input as purely spam e-mail with extension .eml file. The spam e-mail input data are zipped through WinZip or any other compression software because, separate .eml file require more time to transfer and also system load increases.

3.2   Pre-processing

In this step, it pre-processes the spam e-mails which accepted as input in first step. In pre-process, the e-mail which contains attachments, images and also English (UK) lang's was removed. In similar step, this e-mails are scanned for below mentioning parameters like. UniqueId, FName, MsgId, MailDate, DateofDelivery, MailerFrom, ReplyTo, PathReturn, Ref, Subjects, and last is Content. Basic set of data had nearly 3800 non-ham e-mails of every languages with categories. In this process, approximately 64% of dataset are cleaned from the actual data used. This Filtered datasets are grouped in 3 other sizes. Data-sets are as described, first set has 600 spam e-mails, and then 1200 and 2400 spam e-mails are considered as the second and third dataset respectively.

3.3   Parameter Extraction

In this step, the parameters which require clustering are extracted. Main there parameters are considered viz. content of spam e-mails, subject of spam e-mails and Links placed in the spam e-mails.

Content of Spam E-mails. The goal of writing spam e-mails is to attract population to infect their system with virus/malwares and buy the products. Due to this purpose, spam e-mails are written by inserting links, Images and attachments with content. By considering, some of features which comprises: counting of words in the e-mail, counting of lines in mail, count of punctuations in content, count of e-mail-Ids, no of links in the content. From the content of e-mails bigram are also extracted.

Subject Extraction. A sequence of continuous characters is a subject, which specify as 'tokens', these are distinguishes by blank spaces. A count of tokens is known as 'length of subject'. Here, subject of spam emails are extracted from the prepared dataset. Later on stages, by considering string matching algorithm cluster will form.

Links Extraction. Usually, a links of website are inserted in spam e-mail. This websites have strict action which gives good profit to intruders. Generally, spam e-mails contain more than one hyperlinks compare to legitimate e-mails. Here, URLs that is links embedded in spam e-mail texts are extracted and later cluster are made according to these.

3.4   Clustering Algorithm

In this part, three different algorithms are used to cluster all the features individually. The three algorithms are namely; first is k-means algorithm for content of spam e-mails, second is string matching algorithm for subject-wise clustering and last is link similarity to cluster according to link present in content.

K-means Algorithm. In this section k-means algorithm is used to cluster the spam e-mails on basis of similar content. Choose any K points and places in spaces where this points known as starting group centroids.

1. Allocate all other object to their respective nearest centroid.
2. If all objects are assigned then again recalculate K centroids.
3. Reassigned Steps 2 and steps 3 till the centroids not moved. From which all the objects are grouped to their relevant/similar group.

String Matching Algorithm. In this algorithm, to detect similar subject e-mails string matching score is applied on spam e-mail dataset. Two different approaches are used to get subject similarity. First method to find subject similar rank is depend on half token ranking, in this similar of subject p and q evaluated by Kul's(p, q), p and q were 2 strings, individual token in p and q is reacted as a separate string of character.

$$Kul's(p, q) = \frac{ILD(p, q)/|p| + ILD(p, q)/|q|}{2} \tag{1}$$

In which, matching number between two different subject like subject's p and q is known as Inverse Levenshtein Distance (ILD). For example,

Token p1: T R I E D _
Token q1: T _ I E D

In above example total 4 similar letters, therefore ILD(p1, q1) = 4.

Many of subject in our datasets are different in size. Some of them subjects are smaller than others subject and contain fewer to-kens. Co-efficient reduces the value given to smaller subjects.

$$MatchRank(p, q) = Co * Kul's(p, q) \tag{2}$$

In which, $Co = \dfrac{\sqrt{\min(|a| + |b|)}}{2 * maxlen}, 1$

To perform the clustering based on subject match score above equation is required. For example,

Subject 1: Your Job Confirmed
Subject 2: Find Confirmed Job

From above example, it is to understand that when two subjects has similar length of words. In this example, if both the tokens p and q have the exact count of letters, suppose x letters: length (p) = length (q) = x, it calculates MatchRank(p, q) = y/x where, y means count of similar letters. If letters (p1, p2, ..., pn) in a and (q1, q2, ..., qn) in q are equal then MatchRank(p, q) = 1. So, the subject match score subject1 and subject2 in the above example is 0.78, because 'get' is not similar to 'is'.

Link Similarity. In this paper, only the spam e-mails that contains link in the text format are considered. The comparison of links between both different domain names becomes a set operation. To measure the similarity between two link sets, it uses the Kul's coefficient. The Kul's coefficient on sets X and Y is derived as follows:

$$Kul's(X, Y) = \frac{(|X \cap Y|/|Y|) + (|X \cap Y|/|X|)}{2} \tag{3}$$

In which, |X| and |Y| are the sizes of set X and Y. It brings the value between 0 and 1.

If two IP addresses of link belongs to the same subnet than it can be partially matched, which is identified by comparing the first three octets. The link similarity score will be calculated as follows:

$$S(X, Y) = C * Kul(X, Y) \tag{4}$$

where, $C = \sqrt{\min\left(\frac{|X| + |Y|}{2 * maxsize}, 1\right)}$

For example: if domain X has IP set $\{1.2.3.4, 4.5.6.8, 3.5.6.1\}$ and domain Y has IP set $\{1.2.3.4, 3.5.6.2\}$

$$S(X, Y) = 0.79 * (1.5/3 + 1.5/2)/2 = 0.49$$

3.5 Domain Extraction

The domain extraction idea is motivated by the fact that spammers/intruders use a lot of domain names to reduce the damage occurred by domain name blacklisting and maximizes site availability. In this paper, three other groups of clustering are created with help of already extracted features. At the same time, domain-names are extracted from those e-mails. To build groups using content of e-mail by considering as parameter, K-means algorithm is used [4]. For subject-wise clustering, use string matching algorithm. The last parameter considered as links injected in the content of spam e-mail uses the link similarity algorithm.

3.6 IP Address Verification

This segment, recognizes the domain-name from Sect. 3.5 which is transferred to WHOIS. The WHOIS is the information of the domain registrar which is sent to the legal office and any other related communities to take applicable punishment against intruders.

4 Result Analysis

The spam e-mails were taken from personal account e-mail account of known people. There are lot of open source tool available to collect spam mails, proposed system have "MailStore Home" tool to gather e-mails. To see, the final evaluations of all separate parameters are assumed on basis of purity %.

4.1 Purity (%)

It is a very easy and obvious calculation criteria. To evaluate individual cluster standard purity is required. The purity % is calculated by below written equation [3]:

$$Purity\,(\%) = \frac{\sum (accurately\ clustered\ mails)}{all\ no.\ of\ mails}$$

There are two different ways to calculate purity. First, in this it calculate high purity cluster from total of remaining cluster for each individual parameters so, it named as highest purity. The other is full pure which evaluates accurately clustered inputs from remaining number of evaluated cluster.

From Fig. 2, it is shown very clearly that the 2nd dataset of size 1200 gives good result; because, the spam e-mail in this set has not good text. From this it can be depicted that the length of spam e-mails also affects creation of clustering. Form Fig. 3, if consider subject as clustering parameter it gave us better results in all datasets in comparison to pervious parameters. Performed experiment shows that high purity gives 100% result in all three datasets.

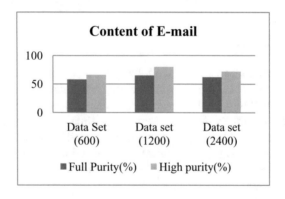

Fig. 2. Result from content of e-mail

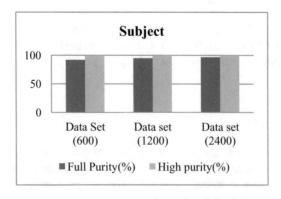

Fig. 3. Result of Subject-wise

By looking at Fig. 4 it is easy to depict that, if link of e-mail content consider as parameter than both purity is 100%. But in created dataset there are very less number of spam e-mails which contains link in body of message.

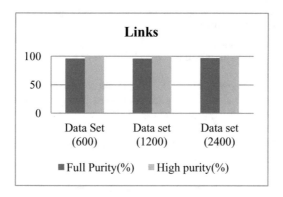

Fig. 4.   Result from Link's in mail

5   Conclusion and Future Work

This research mainly focuses on to assist the termination of spam emails by using clustering algorithm. Initially, the proposed system takes three features of spam e-mails. In which content of spam e-mail as parameter used to identify common writing pattern of spam campaigns and also give deep insight about the meaning of spam e-mails, subject as parameter of e-mails to detect similar templates originated by malware and last parameter links of e-mail content which spammer/intruder is trying to promote.

In further work, initial email scanning of spam and receiving feed based on input from receivers would be done simultaneously. In addition to that, the utilization of many parameters like, images presents in the e-mails, attachments and may more other parameter to detect spam domain. Identify that which parameter gives better results. Also, it could help the computer forensics experts to get hold of the primary spammer.

References

1. Harisinghaney, A., Dixit, A., Gupta, S., Arora, A.: Text and image based spam email classification using KNN, Naïve Bayes and reverse DBSCAN algorithm. In: 2014 International Conference on Reliability, Optimization and Information Technology, Noida, India, pp. 153–155. IEEE (2014)
2. Sirisanyalak, B., Sornil, O.: An AI based spam detection system. IEEE, Thailand (2007)
3. Likitkhajorn, C., Surarerks, A., Rungsawang, A.: An approach of two-way spam detection based on boosting pages analysis. In: 2012 9th International Conference on Electrical Engineering/Electronics, Computer, Telecommunications and Information Technology, Bangkok 10330, Thailand, pp. 1–4. IEEE (2012)

4. Li, F., Hsieh, M.-H.: An empirical study of clustering behavior of spammers and group-based anti-spam strategies. In: Third Conference on Email and Anti-spam, CEAS 2006, Mountain View, California, USA, July 2006
5. Iedemska, J., Stringhini, G., Kemmerer, R., Kruegel, C., Vigna, G.: The tricks of the trade: what makes spam campaigns successful? In: Security and Privacy Workshops, University of California, Santa Barbara. IEEE (2014)
6. Dong, J., Yuan, Z., Zhang, Q., Zheng, Y.: A novel anti-spam scheme for image-based email. In: First International Symposium on Data, Privacy and E-commerce, Lanzhou, China, pp. 520–522. IEEE (2007)
7. Chaitanya, T.K., Ponnapalli, H.G., Herts, D., Pablo, J.: Analysis and detection of modern spam techniques on social networking sites. In: Third International Conference on Services in Emerging Markets, Mexico, pp. 147–152. IEEE (2012)
8. Pingdom.com (n.d.). http://royal.pingdom.com/2013/01/16/internet-2012-in-numbers/
9. Deshmukh, P., Shelar, M., Kulkarni, N.: Detecting of targeted malicious email. In: 2014 IEEE Global Conference on Wireless Computing and Networking (GCWCN), Nashik, Maharashtra, pp. 199–202. IEEE (2014)
10. Process Software: Explanation of common spam filtering techniques, pp. 1–7 (2011)
11. Razak, S.B.A., Mohamad, A.F.B.: Identification of spam email based on information from email header. In: 2013 13th International Conference on Intelligent Systems Design and Applications, pp. 347–353. IEEE Conference Publications, Malaysia (2013)
12. Hadler, S., Tiwari, R., Sparague, A.: Information extraction from spam emails using stylistic and semantic features to identify spammers, pp. 365–372. IEEE, USA (2011)
13. Spamhaus DBL (2010). http://www.spamhaus.org/dbl/
14. Spamhaus PBL (2010). http://www.spamhaus.org/pbl/

SMABE (Smart Waist Belt) Using Ultrasonic Sensor

Hozefa Ali Bohra, Stuti Vyas, Garima Shukla, and Mayank Yadav[✉]

M.L.V. Textile and Engineering College, Bhilwara, Rajasthan, India
hozefaalibohra.786@gmail.com, friend.stutivyas@gmail.com,
garimashukla253@gmail.com, m.yadav12328@gmail.com

Abstract. This paper shows an electronic route framework for outwardly weakened and daze individuals (subject). This framework comprehends impediments around the subject up to 500 cm in front, left and right course utilizing a system of ultrasonic sensors. It successfully figures separation of the identified protest from the subject and plans route way likewise keeping away from obstructions. It utilizes discourse input to mindful the subject about the recognized deterrent and its separation. This proposed framework utilizes ATmega16 microcontroller based implanted framework to process continuous information gathered utilizing ultrasonic sensor organize. In view of bearing and separation of identified hindrance, important pre-recorded discourse message put away in APR9600 streak memory is conjured. Such discourse messages are passed on to the subject utilizing headphone.

Keywords: Visual impairment · Ultrasonic sensors · Vibrators · IR sensors for color detection

1 Introduction

India is by and by home to the world's greatest number of outwardly disabled people. Of the 37 million people over the globe who are outwardly debilitated, more than 15 million are from India. Visual inadequacy or visual impairment is a condition that effects various individuals far and wide. This condition prompts to the loss of the basic assumption vision. There are different course structures for clearly debilitated explorers to examine rapidly and securely against tangles and particular threats went up against. Generally, an outwardly disabled customer passes on a white stick or a bearing puppy as their convenience offer assistance. There are different standard and progressed navigational helpers are open for obviously incapacitated and apparently crippled individuals. Though many advanced electronic course aides are available these days for apparently crippled and outwardly weakened people, not a lot of them are being utilized. Thusly customer appropriateness assessment of such systems is basic. The most influencing parameters in such way are gauge, transportability, enduring quality, profitable functionalities, direct UI, get ready time, structure power and sensibility with respect to cost. Considering all these customer yearnings and essentials, a precisely fit straightforwardness and trustworthy course system is proposed in this paper for ostensibly obstructed and stupor people.

© Springer International Publishing AG 2018
S.C. Satapathy and A. Joshi (eds.), *Information and Communication*
Technology for Intelligent Systems (ICTIS 2017) - Volume 2, Smart Innovation,
Systems and Technologies 84, DOI 10.1007/978-3-319-63645-0_72

Objective

   The paper major objective to produce a guiding system to the blind people. We are going to develop a waist belt that helps the blind people for their navigation. We are using six ultrasonic sensors in the waist belt. Ultrasonic sensor detects the obstacle in the given direction at a particular distance. We have introduced small vibrator sensor here for deaf people, who can't hear but they can feel and with the help of vibrator sensor they can detect the obstacle. The vibrator will vibrate whenever there is an obstacle. Blind people use IR sensor for the color detection.

2   Description of the SMABE System

An implanted framework incorporating six ultrasonic sensors sets, VRTB V01 and IC APR9600 V02 IC 0514 sound recording and playback streak memory, headphone along microcontroller ATmega16. Figure 1 demonstrates the considered framework as outwardly hindered and dazzle route. In this wearable framework, two Ultrasonic sensors are utilized for snag identification and figuring of its versatile separation from the outwardly weakened individual. Ultrasonic sensors are utilized as a part of combine as handsets. One gadget which discharges waves is called as transmitter a well as other that one gets reverberate is known as beneficiary. Six ultrasonic sensor sets are utilized as a part of this framework. Midriff belt. These three ultrasonic sensor sets are put 12 cm separated confronting towards front left, focus and right bearing. Utilizing this situation and arrangement of ultrasonic sensor sets, subject can recognize hindrances from abdomen level stature to head level tallness in the scope of 500 cm in any course. These five sets of ultrasonic sensors gather ongoing information after each 20 ms. what's more, send it to ATmega16 microcontroller. In the wake of preparing this information, micro-controller summons important discourse message put away in blaze memory. APR9600 sound recording and playback streak memory is utilized for putting away previously

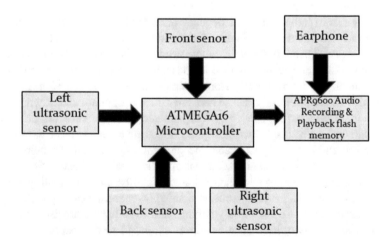

Fig. 1. Ultrasonic displays and midriff belt framework for outwardly weakened and daze individual

recorded discourse messages. Changeable term of discourse information up to 12 min. term can be put away in this memory. Genuine photos of the frameworks and its segments are appeared in Fig. 1.

3   Literature Review

While dazzle and outwardly hindered individuals had added to the collection of regular writing for hundreds of years, one eminent case being the creator of Heaven Lost, John Milton, the creation of personal materials, or materials particular to visual impairment, is moderately new. The vast majority know about Helen Keller, who was both visually impaired and hard of hearing, however there has been extensive advance since the production of her work. Daze and outwardly impeded individuals are off guard when they travel since they don't get enough data about their area and introduction as for movement and deterrents in transit and things that can effectively be seen by individuals without visual inabilities. The traditional methods for guide canine and long stick just help to maintain a strategic distance from obstructions not to recognize what they are. Route framework as a rule comprise of three sections to help individuals go with a more noteworthy level of mental solace and freedom detecting the prompt condition for snags and dangers, giving data about the area and introduction amid travel. Voice worked open air route framework for outwardly debilitated people done by Koley and Mishra [1]. Utilizes a stick furnished with ultra-sonic sensors, GPS and sound yield framework. The stick contains GPS which will have SD memory card which used to store diverse areas. The client can set the area by voice and the GPS will direct the individual to his/her widening. This framework will likewise give the speed and the rest of the separation to achieve the widening.

At the point when the ultra-sonic sensors recognize any deterrent straightforwardly the voice framework will initiate the alert voice. Another review in a similar field to help dazzle individuals utilizes the beat resound strategy keeping in mind the end goal to give a notice sound when distinguishing the hindrances. This system is utilized by the Assembled States military for finding the submarines. They utilized beat of ultrasound range from 21 kHz to 50 kHz which hit the hard surface to produce resound beats. By ascertaining the contrast between signs transmit time and flag getting time we can foresee the separation between the client and the hindrances. This framework is extremely touchy regarding recognizing the obstacles [2].

4   Obstacle Detection and Distance Calculation

Obstacle Detection

Ultrasonic sensors are used for tangle recognizable proof and figuring of its flexible partition from the apparently weaken person. Ultrasonic sensors are used as a piece of join as handsets. One contraption which releases sound waves is called transmitter or other who gets resonate is known as recipient. These sensors take a shot at a rule like radar and sonar which distinguishes to challenge with the help of echoes from sound waves. An estimation is completed in C-vernacular on ATmega16 microcontroller.

The time break between sending the banner and getting the resound is processed to choose the partition to a question. As those sensors utilize this waves as opposed to light for protest recognition, so can be serenely utilized as a part of encompassing open air application. Five ultrasonic sensor sets are utilized as a part of this framework. Before finishing up the hindrance separate from the subject, rehashed data inspecting and averaging is performed. As surrounding light conditions don't influence ultrasonic sensors, question location and separation estimation can be performed precisely.

5   Correspondence Between System and Subject

This framework can comprehend 500 m far off protest/obstruction in any utilizing discourse messages. To make remove seeing all the more engaging the subject, discourse messages can be put away in an all inclusive dialect.

Message Passing on Distinguished Condition to Object

Numerous scientists utilized vibrator cluster, bell depends sound recurrence clasps or content to discourse transformation for declaring any distinguished condition to the subject. This framework utilizes pre-recorded messages for passing on any identified location. This utilizes APR9600 recorder IC or streak memory. This recorder IC can store variable term discourse messages up to 12 min length. Number of information can be extended by reducing the traverse of each message. ATmega16 forms ongoing information gathered by ultrasonic sensor cluster and takes the right choice. In light of handled information, amend choice is taken and applicable message is conjured from the blaze memory and passed on to the subject through headphone.

Adaptability Towards Utilize Each Dialect as Discourse Cautioning Information

For discourse helped route, numerous scientists are utilizing content to discourse transformation. In such cases specialists are changing over content into English dialect as it were. As this framework utilizes APR9600 streak memory store the pre-recorder IC discourse information, there is no obstacle for utilization of any dialect. Each engaging all inclusive dialect can be utilized for recording discourse cautioning information. This framework action to a basic instrument and recorded to a certain discourse cautioning information.

6   Algorithm

A. Firstly we will collect all the requirement equipments and place them at one place. Now we first simulate the circuit on 'EXPRESS PCB'. It is simulation software from which we can simulate any circuit.
B. Now we do the coding for require task, then we import the coding in simulation circuit and check whether it is working or not.
C. If the coding performs right in simulation then we implement the circuit on PCB.
D. After making it, we import the program in microcontroller.

E. Now we transmit pulse ultrasonic signal and check whether the circuit working properly or not.
F. The front sensor detect the signal then he will play the sound, if the back one detect then it will produce the sound and so left and right too.

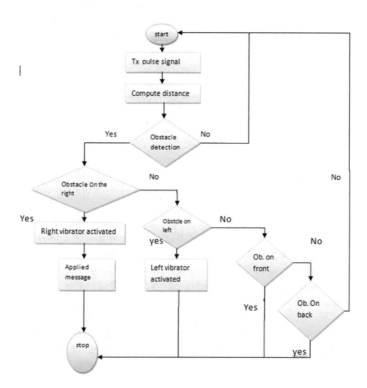

Innovation

A. We have introduced small vibrator sensor here for deaf people, who can't hear but they can feel and with the help of vibrator sensor they can detect the obstacle. The vibrator will vibrate whenever there is an obstacle.
B. We have use IR sensor for detecting the colors, with the help of this the blind people who want to paint can make their wish fulfilled. The IR sensor gives different ranges for different colors and for particular color the pre installed voice recorder informs them about the color and they can paint easily.

7   Results and Analysis

Test Methodology
Ultrasonic sensors, ATmega16 and APR9600 are tried separately and in addition an incorporated framework. As ultrasonic sensors chip away at standard of reverberate,

investigation of its appearance properties divider, static human, vehicle and iron. Surface assumes enter part new hindrance discovery. Smooth surface challenge can be recognized from most outrageous acknowledgment extent of ultrasonic sensors. Metal surface gives most noteworthy reflections and after that solid divider, iron and human. The above-mentioned four surfaces are considered for examination of object can run over whatever of the system amid route. Every one of the above- mentioned analysis are completed in lab condition and their study are recorded. Points of interest of test conveyed and their separation go results are given in Table 1.

Table 1. Ultrasonic sensor's response of unlike object

	Detection range in cm			
	Test 1	Test 2	Test 3	Test 4
Iron	439	423	412	380
Vehicle	415	426	414	350
Wood	300	302	398	402
Human	292	370	400	395

Results

We design SMABE (waist belt) using six ultrasonic sensors. This device is helpful in navigation for blind people. Two sensors are act as a front sensor which detect the obstacle in front of the people. There is left sensor which detect the obstacle in left. And one sensor is toward the right detect the right obstacle. Two more sensors are at back which detect the obstacle at the back of the blind people. There is a particular range we decided according to the ultrasonic sensor. And a previously recorded system also used which record the voice. IR sensor also attached to the device for the painting purpose. Vibrators are helpful for the deaf people because of their sensing ability.

Conclusion

This wearable electronic route framework is effectively tried on visually impaired collapsed objects in indoor and outdoor environment. Less time is to require utilize this framework. With the help of this project blind person detect the obstacle and easily walked like a normal human being. IT is used for the painting purpose. Blind people for different reasons have lost their visual sense so this device is used also helping for that type of persons. Considering the yearnings and necessities of the apparently hindered and blind people, this structure offers a straightforwardness, strong, reduced, low power and generous response for smooth course. In spite of the fact that the framework is light weight, yet hard set up with sensors and different segments. Facilitate wearable part of this framework can be enhanced utilizing remote availability between the framework segments. This framework is produced considering outwardly weakened and daze individuals in creating nations.

Advantage

- Low cost
- Path finder

- Easy to implement
- Security
- Less weight.

References

1. Koley, S., Mishra, R.: Voice operated outdoor navigation system for visually impaired persons. Glob. J. Eng. Trends Technol. (2012)
2. Liang, R.H., Ouhyoung, M.: A constant ceaseless motion acknowledgment system for communication via gestures. Face Gesture Recognit., 558–565 (2011)
3. Merletti, R., Parker, P.A.: Electromyography—Physiology, Building and Non-intrusive Applications
4. Aimone, C., Fung, R., Khisti, A., Varia, M.: Head Mounted Control System (in press)

Measurement (Data Mining) of Real Mobile Signals Data in Weka Tools for Interference Detection

Ravirajsinh S. Vaghela[✉], Atul Gonsai, and Paresh Gami

BCA Department, R. P. Bhalodia College, Manhar, Ploat-7, Rajkot, Gujarat, India
sunstate9999@gmail.com

Abstract. In this paper we have collected the data from selected population of Rajkot city by the way of android and iPhone application after collecting all radio signals data like Wi-Fi signal power, GPS signal power, 4g signal power, 3g signal power, and Signal to noise ratio in different mobile device in different geographical location we can apply datamining technique by which can measure the different type of the scenario. After applying different method we can find hidden pattern and many insight to deal with interference situation.

Keywords: SNR · WIFI · GPS · LTE · WEKA

1 Introduction

In today's world we are surrounding with many buzzing technologies like GPS, Wi-Fi, Bluetooth, GSM, 3GPP, CDMA, 4g LTE and all technologies based on Radio signals at different frequency and different type of modulation techniques and concepts. We are using all the technique simultaneously some time or many a time. Some research also try to create bridge between two more technologies to boost internet speed.

More often time simulation will not give exact idea of interference problem. But the real world problem may differ than simulation predefine situation. Many GPS receiver company find interfering by other radio signal in real world [1]. For the real world collection of data we here consider mobile devices are the best fit for our real world data collection tools in which android and IPhone mobile application are better, faster way to collect data from real world GPS, Wi-Fi, Bluetooth, GSM, CDMA, 4g LTE signal data and SNR data because all radio signal receiver are built in smart phone. By collecting all the device data from different cell of mobile towers, with WI-FI signal strength, GPS signal strength as row data of our measurement. The data collection process is voluntary how like to give feedback to our collection then simple install and wait for 5 min maximum in between our application collecting. After collecting the all the data from selected population we can measure in Datamining tools like WEKA by which we can measure some sort of classification and clusters.

The entire data collection period is significantly shortened, most of data collect within 3–4 days. But we can make best insight and classification and clustering by taking daily basis input from various mobile in which already installed.

© Springer International Publishing AG 2018
S.C. Satapathy and A. Joshi (eds.), *Information and Communication Technology for Intelligent Systems (ICTIS 2017) - Volume 2*, Smart Innovation, Systems and Technologies 84, DOI 10.1007/978-3-319-63645-0_73

2 Row Data Collection Procedures

For row data collection we have develop one android and iphone application based on angular js by which we can collect the signal data by just installing and make it open for just 3–5 s it will collect and submit to our scrver. So it is very easy to collect, not need to give more detail instruction to user for data collection. So data collection without user inference could give better data then user manual entry.

You could download from following links in google play store.

https://play.google.com/store/apps/details?id=com.ud.signalmeasurement

By Downloading installing following main screen will appear and after permission of user all GPS, WI-FI, mobile GSM, 4g sent to the webserver and server will fill the database [2, 3] (Fig. 1).

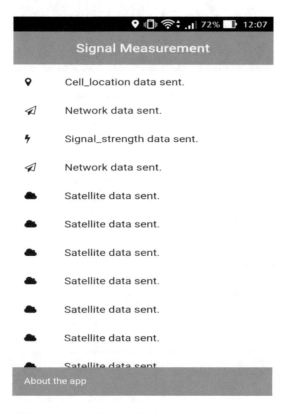

Fig. 1. Application transferrin the data to the server.

After filling the database we can see the area from data is coming by latitude and longitude got from user mobile. In following figure we can see google map balloon from we were got the data (Fig. 2).

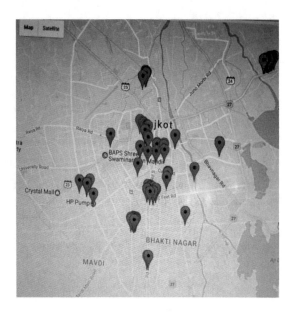

Fig. 2. In admin panel we can find the balloons of samples who share the data of signals.

After collecting the data admin can visualize the each n every balloon with detail information by just clicking on balloon (Fig. 3).

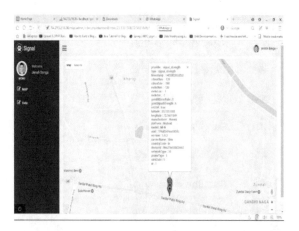

Fig. 3. In admin panel we can also find the balloon wise detailing.

Admin can also visualize data real-time when anyone who installed the app and start at current time. So by just refreshing grid (Fig. 4).

Fig. 4. we can find all the data on Dashboard data in grid.

3 Datamining Tools Setup and Configuration

Today's generation datamining is the activity by which one can easily estimate, find pattern, classify data, and find clustered from Dataset. Here we set our data into MySQL database then we export the data into the excel sheet (CSV) comma separated file from where we have to convert the file in arff file extension. After conversion we can apply the data into the Waikato Environment for Knowledge Analysis (WEKA) "arff" file format [4, 5].

4 Measurement of Data and Results

From Fig. 5 we just classified how many different manufacturer include in datasheet right now, we have highest data of xiaomi, and least data of coolpad.

Figure 6 chart we just indicate the no of records of different carrier provider in datasheet right now, we have jio 4g and idea, Vodafone with sufficient data to analysis.

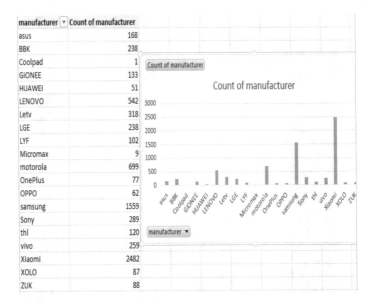

manufacturer ▾	Count of manufacturer
asus	168
BBK	238
Coolpad	1
GiONEE	133
HUAWEI	51
LENOVO	542
Letv	318
LGE	238
LYF	102
Micromax	9
motorola	699
OnePlus	77
OPPO	62
samsung	1559
Sony	289
thl	120
vivo	259
Xiaomi	2482
XOLO	87
ZUK	88

Fig. 5. Total android mobile from which get the data manufacturer wise.

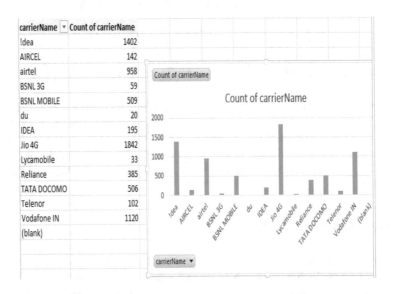

carrierName ▾	Count of carrierName
!dea	1402
AIRCEL	142
airtel	958
BSNL 3G	59
BSNL MOBILE	509
du	20
IDEA	195
Jio 4G	1842
Lycamobile	33
Reliance	385
TATA DOCOMO	506
Telenor	102
Vodafone IN	1120
(blank)	

Fig. 6. Total mobile operator company from which we gather data company name wise.

Figure 7 provide comparison of 4g jio carrier robustness with different mobile's data with different feature's value by can deduce that how evdoECIO effect the GSM bit error rate.

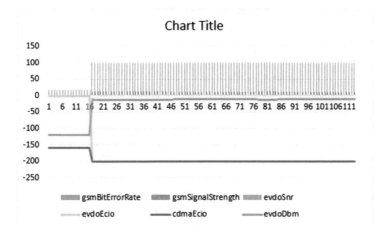

Fig. 7. Signal comparison combo chart of 4g jio mobile operator's gsmBiterrorrate, gsmsignal strength, evdoSnr, evdoECIO, cdmaECIO, ebdDbm.

Figure 8 provide comparison of Lenovo mobile data with different feature's value by can deduce that how evdoECIO effect the GSM bit error rate.

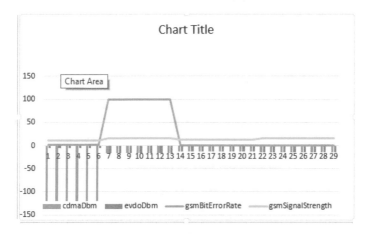

Fig. 8. Signal comparison combo chart of Lenovo mobile gsmBiterrorrate, gsmsignal strength, evdoSnr, evdoECIO, cdmaECIO, evdoDbm

5   Weka Tool Base Datamining

Weka tools Classify 59 rows with 7 attribute evdoDbm, evdoEcio, evdoSnr, gsmbitErorrate, gsmsignalStrength, Carriername(BSNL3G), SNR

```
=== Run information ===

Scheme:     weka.classifiers.rules.M5Rules -M 4.0

Relation:   whatever

Instances:  59

Attributes: 7

            evdoDbm

            evdoEcio

            evdoSnr

            gsmBitErrorRate

            gsmSignalStrength

            carrierName

            SNR

Test mode:  split 66.0% train, remainder test

=== Classifier model (full training set) ===

M5 pruned model rules

(using smoothed linear models) :

Number of Rules : 2

Rule: 1

IF
        gsmSignalStrength > 5

THEN

gsmBitErrorRate =

        0.9821 * gsmSignalStrength

        + 46.7508 [38/38.373%]

Rule: 2

gsmBitErrorRate =

        + 1 [21]

Time taken to build model: 0.02 seconds

=== Evaluation on test split ===

Time taken to test model on test split: 0 seconds

=== Summary ===

Correlation coefficient          0.9943

Mean absolute error              3.2421

Root mean squared error          5.0781

Relative absolute error          10.4391 %

Root relative squared error      15.9249 %

Total Number of Instances        20
```

Likewise we can create "4g Jio" file to calculate regression and M5Rules to get idea out this data. We can also generate the file of "Lenovo mobile" Model data and check same classifier on it and deduce some pattern on real time data [6].

6 Conclusion

By this research we can identify mobile devices manufacturer problem or Mobile operator to improve the overall performance. On above data mining we can give how the bit error rate and signal strength makes conditional change. This also help in future to develop intelligent mobile communication system.

References

1. Boulton, P., et al.: GPS interference testing: lab, live, and LightSquared. Inside GNSS **5**(4), 32–45 (2011)
2. Loibl, W., Peters-Anders, J.: Mobile phone data as source to discover spatial activity and motion patterns. G1_Forum 524–533 (2012)
3. Kanchana, N., Abinaya, N.: Mobile data mining-using mobile device management system (MDM). In: Proceedings of the UGC Sponsored National Conference on Advanced Networking and Applications (2015)
4. Siła-Nowicka, K., et al.: Analysis of human mobility patterns from GPS trajectories and contextual information. Int. J. Geogr. Inf. Sci. **30**(5), 881–906 (2016)
5. Blunck, H., Kjærgaard, M.B., Toftegaard, T.S.: Sensing and classifying impairments of GPS reception on mobile devices. In: International Conference on Pervasive Computing. Springer, Heidelberg (2011)
6. Kumar, R.P., Rao, M., Kaladhar, D.: Data categorization and noise analysis in mobile communication using machine learning algorithms. Wirel. Sens. Netw. **4**(4), 113 (2012)

Author Index

© Springer International Publishing AG 2018
S.C. Satapathy and A. Joshi (eds.), *Information and Communication Technology for Intelligent Systems (ICTIS 2017) - Volume 2*, Smart Innovation, Systems and Technologies 84, DOI 10.1007/978-3-319-63645-0